TESTING AND MEASUREMENT: TECHNIQUES AND APPLICATIONS

PROCEEDINGS OF THE 2015 INTERNATIONAL CONFERENCE ON TESTING AND MEASUREMENT:
TECHNIQUES AND APPLICATIONS (TMTA2015), 16–17 JANUARY 2015, PHUKET ISLAND, THAILAND

Testing and Measurement: Techniques and Applications

Editor

Kennis Chan

Advanced Science and Industry Research Center, Hong Kong

CRC Press
Taylor & Francis Group
Boca Raton   London   New York   Leiden

CRC Press is an imprint of the
Taylor & Francis Group, an **informa** business

A BALKEMA BOOK

CRC Press/Balkema is an imprint of the Taylor & Francis Group, an informa business

© 2015 Taylor & Francis Group, London, UK

Typeset by diacriTech, Chennai, India
Printed and bound in the UK and the US

Published by: CRC Press/Balkema
P.O. Box 11320, 2301 EH Leiden, The Netherlands
e-mail: Pub.NL@taylorandfrancis.com
www.crcpress.com – www.taylorandfrancis.com

ISBN: 978-1-138-02812-8 (Hardback)
ISBN: 978-1-315-68493-2 (eBook PDF)

Testing and Measurement: Techniques and Applications – Chan (Ed.)
© 2015 Taylor & Francis Group, London, ISBN: 978-1-138-02812-8

Table of contents

ix

Preface

The human world is basically built only on tens of common materials, but on the basis of these common materials, we have created hundreds of or even thousands of new things. In our creation of these things, various theories and technologies have been used for their testing and assessing. Because of the importance of understanding the properties of the things that we have created, like their reliability, durability and expansibility etc., it is safe to say that our test and measurement technology is just as important as the technology that has created them.

The 2015 International Conference on Testing and Measurement: Techniques and Applications (TMTA2015) is an international conference which precisely focuses on this aspect of our technology.

TMTA2015, with the participation of passionate researchers from the world, has been successfully held on January 16 and 17, 2015 in Phuket Island, Thailand. As someone who was present, I was especially glad to see the discussions between our participants and the presentations from them. I believe that these researches, which are about technologies and techniques in different fields or in the same field but are from different aspects, are of great pragmatic value.

TMTA2015, even it is only a short time after the opening, is already showing its value in its specialized field-many research achievements introduced at this convention have already been applied in small scale; I am positively certain that more accomplishments will come in no time.

With the help from CRC Press, we are able to present you in this book all the accepted papers selected from hundreds of contributions. To make it easy for reading and convenient for consulting, we have had them categorized into 6 sub-categories, namely:

– Microwave, Ultrasonic and Acoustic Measurement and Application
– Material Performance and Measuring and Testing Technique
– Laser, Optics Fiber and Sensor
– Industrial Autoimmunization and Measurement
– Artificial Intelligence and Application
– Image, Signal and Information Processing

Finally, I would like to thank all the people who have helped and supported TMTA2015, including all our contributors, the conference chairs and committees, the organizers and CRC Press. I hope that this support will continue until the next TMTA.

Organizing committee

General Chair

G. Fernando, *Huelva University, Spain*

Program Chair

K.S. Rajesh, *Defence University College, India*

Publication Chair

Kennis Chan, *Advanced Science and Industry Research Center, Hong Kong*

International Scientific Committee

S. Shakya, *Tribhuvan University, Nepal*
D. Pelusi, *University of Teramo, Italy*
I. Karabegović, *University of Bihac, Bosnia and Herzegovina*
Y.J. Mon, *Taoyuan Innovation Institute of Technology, Taiwan*
M. Mohammed Eissa, *Helwan University, Egypt*
R.B. Nazir, *Universiti Teknologi, Malaysia*
G. Singh, *Sharda University, India*
M. Longo, *Politecnico di Milano, Italy*
M. Subramanyam, *Anna University, India*
H. Yaghoubi, *Iran Maglev Technology (IMT), Iran*
C.F. Li, *Shanghai University, China*
P. Ravindran, *Anna University, Chennai, India*
O.P. Rishi, *University of Kota, Kota*
R.N. José, *University of Beira Interior, Portugal*
M.S. Chen, *Da-Yeh University, Taiwan*
R.R. Jorge, *Technological University of Ciudad Juarez, Mexico*
E.P. León, *Instituto Politécnico Nacional, Mexico*
E. Zalnezhad, *University of Malaysia, Malaysia*
B.H. Chen, *Fu Jen University, Taiwan*
V. Saetchnikov, *Belarusian State University, Belarus*
M.A. Saeed, *Universiti Teknologi Malaysia, Malaysia*
E. Pardo, *Institute of Electrical Engineering, Slovak*
E.P. Ng, *Universiti Sains Malaysia, Malaysia*

Microwave, ultrasonic and acoustic measurement and application

Testing and Measurement: Techniques and Applications – Chan (Ed.)
© 2015 Taylor & Francis Group, London, ISBN: 978-1-138-02812-8

Overlapped stepped frequency train of flam pulses for microwave imaging

K.V. Nikitin, A. Dewantari, S.Y. Jeon, T.Y. Lee, S. Kim, J. Yu & M.H. Ka
School of Integrated Technology, Yonsei Institute of Convergence Technology, Yonsei University, Korea

ABSTRACT: One way of improving the range resolution of microwave imaging systems is to synthesize the ultra wideband signal from a regular set of narrowband modulated pulses. A train of linear frequency modulated pulses (chirps) with equal carrier frequency steps between pulses is studied here. The frequency step is chosen to be relatively small in comparison with the chirp bandwidth to enable deep side lobes suppression by using bell-shaped window functions. As shown in the paper, the weighted chirp signal is effectively being applied to high-resolution imaging.

1 INTRODUCTION

Presently, there is a significant growth of interest in high-resolution microwave imaging systems for non-destructive studying of composed objects. The most successful experiments were made using ultra wideband pulsed signals [1, 2] and continuous wave stepped frequency (CWSF) signals [3, 4, 5, 6]. Recently, implementations of a synthetic bandwidth technique have been reported in [7, 8].

One of the important components of a microwave imaging system is a high-resolution short-range radio ranging unit, which includes the signal forming and processing part, transceiver, and antenna subsystem. For practical applications, the ranging unit must handle the signals which satisfy the following requirements:

1 High range resolution (1 cm and better);
2 Short minimum detection range (below 1 m);
3 Unambiguous range measurement;
4 Safety to biological objects and living tissues;
5 Robustness to unintentional jamming.

For purposes of hardware implementation of the ranging unit, it is desirable to keep the instantaneous bandwidth of the signal below 1 GHz.

The presented paper describes the stepped frequency train of a linear frequency modulated (LFM) pulses (chirps). Unlike [9, 10] and some references cited therein, the steps are overlapped by frequency for approximately 70% of the chirp bandwidth. This approach allows us to dramatically simplify the signal processing, to increase the variability of the signal parameters, and to use the amplitude modulation (windowing) to suppress the side lobes of the very short pulse signal.

2 THE TRANSMITTED PULSE AND PULSE PROCESSING

Consider the linear frequency modulated pulse

$$s^0(t) = w(t)\exp\left(i\omega t + i\beta t^2\right) \qquad (1)$$

where $w(t)$ is a window function, $\omega = 2\pi(f_c - \Delta f)$, $\beta = 4\pi\Delta f/T$, Δf is a frequency deviation, f_c is the center frequency, T is the pulse length. The example of transmitter signal with a Hamming window shape and the receiver response for this signal is shown in Figure 1.

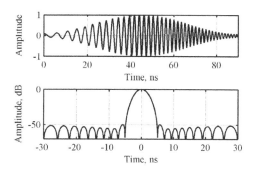

Figure 1. The transmitted pulse (top) and the envelope of the compressed pulse (bottom).

The minimum frequency $f_{min} = f_c - \Delta f$ occurs at the beginning of the pulse ($t \approx 0$) and the maximum frequency $f_{max} = f_c + \Delta f$ is at the end of the pulse ($t \approx T$). The output of the matched filter at the receiver is calculated as the convolution between the down-converted received signal and the pulse response h(t), which is

$$y^0(\tau) = \exp(-2i\pi f_c)\int_{t_1}^{t_2} s^0(t)\,h(\tau - t)\,dt \qquad (2)$$

If the sequence of K linear frequency modulated pulses is transmitted at stepped carrier frequencies $f_k = f_1 + (k-1)\cdot f_\Delta$, $k = 1..K$, the signal at the transmitter antenna is

$$s_k^{TX}(t) = A_k\,w(t)\exp\!\left(i\omega_k t + i\beta t^2\right) \qquad (3)$$

where A_k is the amplitude of the transmitted signal.

Assuming each echo signal is characterized by the range to the reflecting partial element r_n and radar cross section σ_n of the element, the received signal is represented as a superposition of echoes from each element of the studied object and echoes from other considerable objects

$$s_k^{RX}(t) = A_{1,k}\sum_{n=1}^{N}\left\{\frac{\sqrt{\sigma_n}}{r_n^2}\,s_k^{TX}\!\left(t - \frac{2r_n}{c}\right)\right\} \qquad (4)$$

where $A_{1,k}$ is the amplitude factor determined by the transmitter antenna gain G_{TX}, effective cross section of the receiver antenna A_{RX}, and losses factor L_k at the carrier frequency f_k

$$A_{1,k} = A_k\sqrt{\frac{G_{TX}(f_k)A_{RX}(f_k)}{(4\pi)^2\,L(f_k)}} \qquad (5)$$

The received signal is filtered out from the out-of-band components and down-converted to the intermediate frequency

$$x_k(t) = \sqrt{G_{RX}(f_k)}\,s_k^{RX}\exp\!\left(-it(\omega_k - \omega)\right) \qquad (6)$$

where G_{RX} is the receiver gain at the carrier frequency f_k. After pulse compression, the resulting data is

$$y_{p,k} = A_{2,k}\sum_{n=1}^{N}y^0\!\left(\frac{p}{f_s} - \frac{2r_n}{c}\right)\exp\!\left(-2i\pi f_k\frac{2r_n}{c}\right) \qquad (7)$$

where p is the index of the range sample, and $A_{2,k}$ is amplitude factor determined by (5), (6) and the calibration coefficient a_k as specified in (8).

If the calibration procedure is performed before the imaging experiment, it is assumed that

$$A_{2,k} = a_k A_{1,k}\sqrt{G_{RX}(f_k)} = 1 \qquad (8)$$

and the equation (7) is simplified to

$$y_{p,k} = \sum_{n=1}^{N}\left\{y^0\!\left(\frac{p}{f_s} - \frac{2r_n}{c}\right)\exp\!\left(-2i\pi f_k\frac{2r_n}{c}\right)\right\} \qquad (9)$$

Some important details of the analog-to-digital conversion and pulse compression are discussed in [11].For simplicity, assume that these effects are represented by the pulse response of the receiver.

The pulse compression output of the receiver is shown in Figure 2 as an amplitude of the range profile ($r = \tfrac{1}{2}\cdot c\cdot\tau$) of simulated objects.

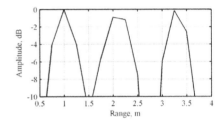

Figure 2. The range profile of a single chirp.

Although the present research aims to design the microwave ranger capable of operating at a frequency band of 2 to 18 GHz and higher, the modeling results provided here are performed for a simplified example for better visualization. Unless otherwise specified, the modeling was performed for the following signal parameters:

Chirp frequency deviation (Δf) .... 150 MHz;
DAC/ADC sampling frequency (f_s) ... 900 MHz;
First intermediate frequency (fc) .... 675 MHz;
Pulse length (T) .... 90 ns;
Lowest carrier frequency (f_1) .... 1 GHz;
Highest carrier frequency (f_k) .... 3 GHz;
Carrier frequency step (f_Δ) .... 100 MHz;
Number of steps of the carrier (K) ..... 21.

The simulated objects are placed at the distances of 1, 2, 2.15, 2.35, and 3.3 meters.

3 THE PULSE TRAIN PROCESSING

The example of a pulse train processing (PTP) input matrix is shown in Figure 3. The results of the pulse compression corresponding to each frequency are collected in the PTP input matrix and arranged so that each column represents the range profile formed at a certain carrier frequency while each row represents the responses received from the same range.

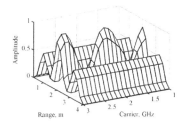

Figure 3. The amplitudes of PTP input elements.

Next, the windowed inverse Fourier transform is performed over each row of the matrix

$$Y_{p,m} = \frac{1}{K}\sum_{k=1}^{K}\left\{y_{p,k}W_k \exp\left(-i\frac{2\pi km}{K}\right)\right\},$$

$$m = 1...K .$$

After substitution of (9) into (10), the resulting matrix

$$Y_{p,m} = \frac{1}{K}\sum_{k=1}^{K}\left\{W_k\sum_{n=1}^{N}\left\{\frac{\sqrt{\sigma_n}}{r_n^2}y^0\left(\frac{p}{f_s}-\frac{2r_n}{c}\right)\times\right.\right.$$
$$\left.\left.\times\exp\left(2\pi i\left(\frac{km}{K}-\frac{2r_n}{c}f_k\right)\right)\right\}\right\}$$

(11)

possess the peaks if

$$m = \frac{2r_n}{c}\frac{Kf_k}{k} \text{ and } p = \frac{2r_n}{c}f_s$$

(12)

The amplitudes of the resulting matrix elements are shown in Figure 4. It should be noted that although all

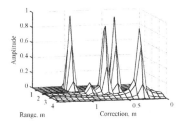

Figure 4. The amplitudes of the PTP output elements.

simulated partial elements of the studied object are quite visible, it is still necessary to convert the pulse-to-pulse processing output matrix to conventional range profile as a complex vector of the range sample index.

Using an auxiliary index j, the range profile could be calculated as

$$z[j] = Y_{p[j],m[j]}$$

(13)

$$r[j] = j\frac{c}{2K\Delta f}$$

(14)

Where indexes p[j] and m[j] are calculated as

$$p[j] = \text{floor}\left(\frac{jf_s}{2K\Delta f}\right)$$

(15)

Where the operator 'floor' denotes the largest integer number not greater than its argument, and

$$m[j] = j \bmod (K)$$

(16)

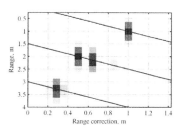

Figure 5. The reading order from PTP output.

where $j \bmod(K)$ denotes the remainder of integer division of j by K.

Figure 6 demonstrates the resulting range profile for the studied pulse train. The central fragment of this profile is shown in Figure 7 to specify the resolution of two partial elements with similar cross section as well as the discrimination of a smaller element.

A better accuracy of the amplitude estimation is achievable if the resulting value z[j] is interpolated by several points taken above and below the straight lines shown in Figure 5.

For the two-point linear interpolation, the equation (13) is corrected as follows:

$$z[j] = (1-\alpha[j])Y_{p[j],m[j]} + \alpha[j]Y_{(p[j]+1),m[j]}$$

(17)

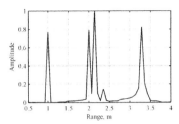

Figure 6.   The range profile of the pulse train.

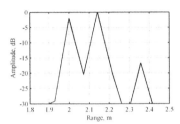

Figure 7.   The central part of the range profile.

where α is determined by

$$\alpha[j] = \frac{j f_s}{2 K \Delta f} - p[j] \qquad (18)$$

The range profile calculated using (17) is represented in Figure 8. All partial elements have the same cross section except for the fourth object which is five times smaller.

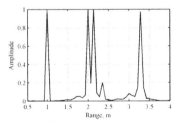

Figure 8.   The range profile plotted using the interpolated data.

4   CONCLUSIONS

One of effective ways of the high-resolution signal synthesis using narrow-band components has been studied.

The main advantage of the considered technique is its capability of independent optimization of the pulse and train parameters.

Chirp signal compression requires relatively high analog-to-digital conversion rate as well as enough computational resources.

ACKNOWLEDGMENT

This research was supported by the MSIP (Ministry of Science, ICT and Future Planning), Korea, under the "IT Consilience Creative Program" (NIPA-2014-H0201-14-1002) supervised by the NIPA (National IT Industry Promotion Agency).

REFERENCES

[1] Schacht, M., Rothwell, E.J., and Coleman, C.M. 2002. Time-Domain Imaging of Objects Within Enclosures. *IEEE Transactions on Antennas and Propagation*, vol. 50, no 6: 895–898.
[2] Jofre, L., Broquetas, A., Romeu, J., Blanch, S., Toda, A.P., Fabregas, X. & Cardama, A. 2009. UWB Tomographic Radar Imaging of Penetrable and Impenetrable Objects. Proceedings of the IEEE, Vol. 97, No. 2: 451–464.
[3] Weedon, W.H. Chew, W.C. & Mayes, P.E. 2000. A Step-Frequency Radar Imaging System for Microwave Nondestructive Evaluation. Progress In Electromagnetics Research, PIER 28: 121–146.
[4] Haotian, Y., Shuliang, W. & Zhen, C. 2009. Simutaneous Radar Imaging and Velocity Measuring with Step Frequency Waveforms. Journal of Systems Engineering and Electronics, vol. 20, no. 4: 741–747.
[5] Gurbuz, A.C., McClellan, J.H. & Scott, W. R. 2009. A Compressive Sensing Data Acquisition and Imaging Method for Stepped Frequency GPRs. IEEE Transactions on Signal Processing, vol. 57, no 7: 2640–2650.
[6] Zhang, L., Qiao, Z., Xing, M., Li, Y. & Bao, Z. 2011. High-Resolution ISAR Imaging With Sparse Stepped-Frequency Waveforms. IEEE Transactions on Geoscience and Remote Sensing, vol. 49, no 11: 4630–4651.
[7] Berens, P. 1999. SAR with Ultra-High Range Resolution Using Synthetic Bandwidth. Geoscience and Remote Sensing Symposium, 1999. IGARSS '99 Proceedings. IEEE: 1752–1754.
[8] Wang, Y., Abbosh, A.M., Henin, B. & Nguyen, P.T. 2014. Synthetic Bandwidth Radar for Ultra-Wideband Microwave Imaging Systems. IEEE Transactions on Antennas and Propagation, vol. 62, no 2: 698–705.
[9] Gladkova, I. 2009. Analysis of Stepped-Frequency Pulse Train Design. IEEE Transactions on Aerospace and Electronic Systems, vol. 45, no. 4: 1251–1261.
[10] Luo, Y., Zhang, Q., Qiu, C., Liang, X. & Li, K. 2010. Micro-Doppler Effect Analysis and Feature Extraction in ISAR Imaging With Stepped-Frequency Chirp Signals. IEEE Transactions on Geoscience and Remote Sensing, vol. 48, no 4: 2087–2098.
[11] Blankenship, P.E. & Hofstetter, E.M. 1975. Digital Pulse Compression via Fast Convolution. IEEE Transactions on Acoustics, Speech & Signal Processing, vol. ASSP-23, no. 2: 189–201.

Testing and Measurement: Techniques and Applications – Chan (Ed.)
© 2015 Taylor & Francis Group, London, ISBN: 978-1-138-02812-8

The automation of a resonant perturbation method to research electrodynamic characteristics of microwave devices

A.N. Savin, I.A. Nakrap, C.P. Vakhlaeva & V.V. Kornyakov
Saratov State University, Saratov, Russia

ABSTRACT: This paper presents the results of development of automated computer system intended to research electrodynamic characteristic of microwave devices, which uses vector analyzer R4-36 and IBM PC based on resonant perturbation method. Usage of phase locking to resonance frequencies of an object being investigated made possible higher precision and automation of field distribution measure process.

1 INTRODUCTION

Increasing accuracy and automation of measurements, microwave devices, combined with a rapid processing of receiving results using a computer is one of actual modern measuring engineering problems.

To calculate interaction efficiency of the waveguide system (WS) field with a bundle, the development of the WS of vacuum microwave devices requires accurate knowledge of field distribution.

The field distribution by experiment can be obtained using a resonant perturbation (RP) method, which consists in the pulling of the perturbing body along the desired direction of the WS resonant dummy and fixing the resonant frequency proportional to the sum of squares of field components (Ginzton 1957). In this case, the measurement process and the amount of information obtained for further processing involve the use of automated systems.

Earlier Work (Amato & Herrmann 1985) describes the improved method for measuring the electric field of the microwave structure by the RP - method through the use of a phase-locked-loop frequency control (PLLFC), supporting the frequency of the voltage controlled oscillator equal to the resonance frequency of the device with the introduction of the perturbing body. The device being researched is adjusted to the resonance frequencies with the help of consistency being turned to him microwave phase shifter.

The main realization difficulty for such circuits is a lack of stable, precise electrically controlled phase shifters; that results in instability of frequency and accordingly leads to the decrease of measurement resolution.

In this work, it is suggested to solve the above problems by applying standard measuring instruments of complex transfer factors which transfer the information about the amplitude and phase of a reference and measured microwave signals to the stabilized low frequency. At the same time, as a phase shifter that is included in the low-frequency channel reference voltage, a delay time digital unit (timer) can be used, which makes it easy to automate the tuning process at the resonant frequencies of researched device.

2 MAIN PART

To implement this idea, it is suggested to use a measuring instrument of complex transfer factors R4-36 gauge in the range of 4-12 GHz.

The microwave generator of R4-36 gauge is an analog voltage controlled, which makes it easy to organize a PLLFC.

The frequency of signals of reference and measured channels is constant (stabilized by a crystal oscillator) and equals to 100 kHz. Signals at this frequency can easily be digitized by fast comparators and further processed by suitable digital functional devices.

To research electrodynamic characteristic (EDC) of microwave devices by a resonant perturbation method, the automated computing-measuring system (CMS) has been developed based on R4-36 gauge controlled by IBM PC (Fig. 1).

R4-36 gauge is enabled to output digital information about measuring object, but gauge is not to be externally controlled. To solve this problem the interface device was developed, allowing total control of the gauge by PC.

In the PLLFC mode, the installation functions are as follows. Controlled from PC (1) through the interface device (2) R4-36 converter (3) forms reference (U_{on}) and measured (U_{in}) signals - voltages

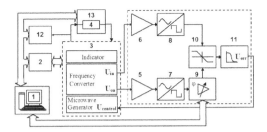

Figure 1. Block-diagram of developed CMS for measurement of EDC by RP-method. 1 – IBM PC, 2 – interface device, 3 – R4-36 gauge, 4 – measured object, 5 – amplifier-filter of reference voltage channel, 6 – amplifier-filter of measured voltage channel, 7 – comparator of reference voltage channel, 8 – comparator of measured voltage channel, 9 – digital control phase shifter, 10 – digital phase detector, 11 – low-frequency filter, 12 – frequency counter Ch3-54, 13 – the perturbing body transportation device.

with a constant frequency 100 kHz, containing the information about the amplitude and phase of an electromagnetic field in a sample object (4). From R4-36 converter U_{in}, U_{on} signals are directed to amplifier-filters (5, 6) to improve the signal/noise ratio and then transformed by comparators (7, 8) into a digital form. The output of the comparator (8) is directly connected to the input of the digital phase detector (10), and the comparator's (7) – through the digital phase shifter (9). The low-frequency filter (11) connected to the output of the phase detector (10), forms error voltage U_{err} proportional to a difference of phases U_{in}, U_{on}, added to the control voltage ($U_{control}$) of the R4-36 microwave generator. Thus, stabilization and frequency trim are performed when the perturbing body is moved into the sample structure. The R4-36 microwave generator frequency change, containing the information about the field magnitude, is detected by the Ch3-54 frequency counter (12) and is directed to the PC.

Control of the phase shifter, frequency counter and the perturbing body transportation device (13) are also performed by personal computer.

In PLLFC contour uses the digital phase detector with three output states, explicitly described in (Williams 1987). The phase change range is 4p, and sensitivity is 0.3 V/Rad.

The phase shifter (9) is a digital pulse edge delay device, which is able to record the delay time. Characteristics of the phase shifter are: pitch of phase realignment is 2.7^0, realignment range 720^0.

The software was developed in a graphical programming environment LabVIEW to operate a CMS, which is allowed to perform:

- Control and receive the data from R4-36 gauge;
- Control and receive the data from the Ch3-54 frequency counter;
- Control the perturbing body transportation device;
- Control the digital phase shifter.

The algorithm performing the automatic determination of resonance frequencies and tuning of PLLFC contour is developed.

LabVIEW was used to perform gathering, initial handling and accumulation of the received data about the measured device (amplitude-frequency characteristic, phase characteristic, complex input resistance, and electromagnetic field distribution of resonance frequencies).

First, we use experimental data about appropriate EDC of microwave devices and the embedded functions of the LabVIEW software package. After that, the following values are calculated and rendered graphically: for resonators – field structure and bypass resistance; for periodic structures – dispersive characteristics, field structure and its spectral structure, amplitude and spatial harmonic interaction impedance.

The LabVIEW application provides the system with the ability of operating in remote access mode.

The developed CMS applying PLLFC at resonance frequencies for researching microwave devices EDC by a small resonant perturbation method has the following performance:

- Frequency range – 4-12 GHz (it is defined by R4-36 gauge);
- Relative short-term stability of R4-36 microwave generator frequency ($\Delta f_{osc}/f_{osc}$) has increased from $2\cdot10^{-5}$ up to $5\cdot10^{-7}$;
- Long-term relative stability has increased from $2\cdot10^{-4}$ up to $2.5\cdot10^{-6}$;
- The relative accuracy of tuning to resonance frequencies ($\Delta f_{res}/f_{res}$) is not lower then $5\cdot10^{-6}$;
- The relative accuracy of resonance frequency measurement is $5\cdot10^{-6}$;
- The relative error of resonance frequencies drifts measurement ($\delta f_{osc}/\Delta f_{fc}$) does not exceed $1\cdot10^{-3}$.

The measurement accuracy of *S*-parameters is determined by R4-36 gauge characteristics.

3 CONCLUSIONS

Thus, the developed measuring system based on R4-36 gauge along with the PC and utilizing PLLFC during the measurements by resonant perturbation method, as well as the software developed basing on LabVIEW, allowed us to automate complex research of dynamic characteristics of the microwave devices with the increased accuracy and enabled with the remote access.

Work is performed with financial support from Ministry of Education and Science of Russian

Federation within the project of state assignment in the field of scientific activity #3.1155.2014/K.

REFERENCES

[1] Amato, J.C. & Herrmann H. 1985. Improved method for measuring the electric fields in microwave cavity resonators. *Review of Scientific Instruments* 56(5): 696–699.

[2] Ginzton, E.L. 1957. *Microwave Measurements*. New York: McGraw-Hill.

[3] Williams, A.B. 1984. *Designer's handbook of integrated circuits*. New York: McGraw-Hill.

Testing and Measurement: Techniques and Applications – Chan (Ed.)
© 2015 Taylor & Francis Group, London, ISBN: 978-1-138-02812-8

Resource assignment using quantized information in uplink based on OFDMA

J.D. Chimeh & N. Mokari
Iran Telecommunication Research Center

A. Asgharian & M.A. Esfahani
Communications Regulatory Authority (CRA) of Iran

ABSTRACT: Today, OFDMA is a dominant technique for advanced wireless telecommunication systems. Using complete timing channel information in the transmitter, we can dynamically assign network resources among users which improve the system performance. However, we cannot access the timing channel information easily and this information may be sent from the receiver which occupies the channel bandwidth in the downlink. Thus we use a quantized timing information procedure in this paper. We study a resource assignment technique in the uplink WiMax using the quantized timing information in this paper. We pay attention to maximizing the total ergodic transmitted data in the uplink regarding the bounded transmitted power of each user. Regarding the maximization of the transmitted data rate in the uplink, we find the channel timing information using BCDA algorithm. We found that through using the quantized channel timing information and number of bits for quantization we may reach to the total ergodic channel data rate resulted from the complete channel timing information.

1 INTRODUCTION

OFDMA technique is a solution with small complexity, but with high performance which diminishes Inter symbol interference in the band limited multipath frequency selective channels (Wang, 2000), (WiMax, 2008). Using the complete channel timing information in the transmitter we may enhance the system performance highly. In line with that, there are many works in the area of the complete channel timing information which maximizes the capacity and minimizes the transmitted power of the system. To reduce the power consumption and interference, wireless networks such as WiMax use power control. Power control procedures are performed using the training sequences. A mobile station and base station are considered as transmitter and receiver, respectively, in this paper. Ordinarily, after measuring the power of the training sequences in the receiver, the measured values are fed back to the transmitter and consequently the MS sets the optimum transmitted data rate and power (WiMax, 2008). The disadvantages of the above steps are as the following:

The complete channel timing information in the downlink makes some of the bandwidth wasteful. This information is used for power adjustment on the transmitter that is sent to the transmitter from the receiver.

- Periodic power control wastes the system time and reduces the processing gain.

A procedure to delete the above disadvantageous includes the following steps:
- Signal to noise quantization in some digital intervals.
- Calculating the transmission bit rate and power in each interval using the optimization methods.
- Saving them in a lookup table in the receiver.
- Regarding the received power in the receiver only an index is transmitted to the transmitter and receiver selects one bit rate and power from the predetermined lookup table.

The receiver needs to use channel timing information on the wireless mobile networks and since the channel timing information varies in the mobile channels, updating them is very difficult (Lapidoth, 2002; Medard, 2000). Providing complete channel timing information is very difficult since it needs high processing gain and bandwidth (Lapidoth, 2002; Medard, 2000). Therefore, to save the bandwidth in the downlink and processing time, the receiver had better to use quantized channel timing information. Here, the receiver sends only an index to the transmitter via a small number of bits (Mukkavilli, 2003), (Zhou, 2006).

The receiver first estimates the channel timing information in the introduced method, then determines the number of the interval of a few bits (by the name of channel codeword) and sends them to the transmitter. Then, the transmitter sets its optimum power and data rates from a lookup table based on the received codeword. Practically, using the quantized channel timing information is very useful for mobile wireless networks because this wastes only a small bounded bandwidth for the above information.

In this paper, we present a procedure to assign an optimum resources in Wimax OFDMA networks using the quantized timing information. Using the BCDA algorithm we maximize ergodic data rate for each user. The simulation showed that the more quantized regions, the total data increases and finally we may reach to the system data rate in the ideal case using this feedback channel information.

2 SYSTEM MODEL

In addition to considering a scenario of base station and K users and assuming the total bandwidth is B(MHz), a WiMaxsystem based on OFDMA in the uplink should be considered. Thus, each subscriber has bandwidth of $B_n = B/N$ (MHz) and the data rate of the user k on the n-th subcarrier is

$$R_{kn} = \log_2\left(1 + p_{kn}\gamma_{kn}\right) \tag{1}$$

In which P_{kn} is the transmitted power of user k on nth subcarrier. $\gamma_{kn} = \dfrac{h_{kn}}{N_0}$ indicates the ration of channel gain between user k and base station on the nth subcarrier over the received noise at the base station N_0.

3 DEFINING AND SOLVING THE OPTIMIZATION PROBLEM

3.1 Defining the optimization problem

We should define the intervals based on the optimization problem and save the results in a lookup table. The optimization problem is to maximize the total ergodic data rate regarding the bounded transmitted power of each user as:

$$\max_P \sum_{k=1}^{K}\sum_{n=1}^{N}\rho_{kn}\mathbb{E}_{h_{kn}}\left\{R_{kn}\right\} \tag{2}$$

$$S.t. \ \sum_{n=1}^{N}\rho_{kn}\mathbb{E}_{h_{kn}}\left\{p_{kn}\right\} \le P_k^T, \ \forall \ k \tag{3}$$

$$\sum_{k=1}^{K}\rho_{kn} \le 1, \quad \forall n \tag{4}$$

$$p_{kn} \ge 0, \ \rho_{kn} \in \{0,1\} \quad \forall \ k, \ n \tag{5}$$

in which (3) indicates the bounded transmitted power of each user that should be less than threshold power of each user $P_k^T, \ \forall \ k$. $\mathbb{E}(.)$ indicates the statistical mean function. ρ_{kn} is a binary variable which is used in the subcarrier assignment for each user somehow that $\rho_{kn} = 1$ means sabcarrier n is assigned to user k, otherwise $\rho_{kn} = 0$.

3.2 Quantized information based on BCDA algorithm

We divide the channel gain by noise γ_{kn} between user k and the base station to $J = 2^q$ intervals in which q is the required bits in quantization.

Therefore, we define the above intervals for channel gain to noise between base station and users as $\Omega^j = \left(\gamma_{knj}, \gamma_{knj+1}\right), \ j = 1,\dots,J$. Using the quantized intervals we rewrite the optimization problem as:

$$\max_P \sum_{k=1}^{K}\sum_{n=1}^{N}\rho_{kn}R_{knj}\left[F_{\gamma_{kn}}\left(\gamma_{knj+1}\right) - F_{\gamma_{kn}}\left(\gamma_{knj}\right)\right] \tag{6}$$

$$S.t. \ \sum_{n=1}^{N}\rho_{kn}p_{knj}\left[F_{\gamma_{kn}}\left(\gamma_{knj+1}\right) - F_{\gamma_{kn}}\left(\gamma_{knj}\right)\right] \le P_k^T, \ \forall \ k \tag{7}$$

$$\sum_{k=1}^{K}\rho_{kn} \le 1, \quad \forall n \tag{8}$$

$$p_{kn} \ge 0, \ \rho_{kn} \in \{0,1\} \quad \forall \ k, \ n \tag{9}$$

In which $F_{\gamma_{kn}}(.)$ Is the CDF of γ_{kn} and R_{knj} is as

$$R_{knj} = \log_2\left(1 + p_{knj}\gamma_{knj}\right) \tag{10}$$

4 SOLVING THE OPTIMIZATION PROBLEM

To solve the optimization problem, we use the Lagrange dual function. Lagrange function for optimization problem is as

$$\mathcal{L}(p,\rho,\mu,\beta) = \sum_{j=1}^{J}\sum_{k=1}^{K}\sum_{n=1}^{N}\rho_{kn}R_{knj}\left[F_{\gamma_{kn}}\left(\gamma_{knj+1}\right) - F_{\gamma_{kn}}\left(\gamma_{knj}\right)\right]$$
$$- \sum_{k=1}^{K}\mu_k\left(P_k^T - \sum_{j=1}^{J}\sum_{n=1}^{N}\rho_{kn}p_{knj}\left[F_{\gamma_{kn}}\left(\gamma_{knj+1}\right) - F_{\gamma_{kn}}\left(\gamma_{knj}\right)\right]\right) \tag{11}$$
$$- \sum_{n=1}^{N}\beta_n\left(1 - \sum_{k=1}^{K}\rho_{kn}\right)$$

In which μ_k and β_n are Lagrange coefficient. Thus we write the dual function as

$$\min_{\mu>0,\ \beta>0} \max_{p,\rho} \mathcal{L}(p,\rho,\mu,\beta) \qquad (12)$$

To find the threshold of the intervals we use BCDA algorithm in which all variables are assumed to be constant except one and we choose that variable as optimum. That algorithm will approaches to the optimum value after some repeat.

We firstly find the optimum relative power using KKT condition as

$$\frac{\partial \mathcal{L}(p,\rho,\mu,\beta)}{\partial p_{knj}}\Big|_{p_{knj}=p_{knj}^*} = \begin{cases} <0, & p_{knj}^* = 0 \\ =0, & p_{knj}^* > 0 \end{cases} \qquad (13)$$

Thus, the assigned power to the user k on the subcarrier n is

$$p_{knj}^* = \left[\frac{1}{\mu_k \ln(2)} - \frac{1}{\gamma_{knj}}\right]^+ \qquad (14)$$

Secondly, Lagrange coefficient values and p_{knj}^* are assumed to be constant and threshold values for the intervals will be found as

$$F_{\gamma_{kn}}\left(\gamma_{knj+1}\right) = F_{\gamma_{kn}}\left(\gamma_{knj}\right) + \frac{f_{\gamma_{kn}}\left(\gamma_{knj+1}\right)\ln(2)\left(1 + p_{knj}\gamma_{knj}\right)}{p_{knj}}$$
$$\left[log_2\left(\frac{1 + p_{knj}\gamma_{knj}}{1 + p_{knj-1}\gamma_{knj-1}}\right) - \mu_k\left(p_{knj} - p_{knj-1}\right)\right] \qquad (15)$$

To show that this algorithm approaches to an optimum solution, the second derivative should be non-positive. Thus

$$-\frac{p_{knj}^2}{\ln(2)\left(1 + p_{knj}\gamma_{knj}\right)^2}\left[F_{\gamma_{kn}}\left(\gamma_{knj+1}\right) - F_{\gamma_{kn}}\left(\gamma_{knj}\right)\right]$$
$$-\frac{\partial f_{\gamma_{kn}}\left(\gamma_{knj}\right)}{\partial \gamma_{knj}}\left[log_2\left(\frac{1 + p_{knj}\gamma_{knj}}{1 + p_{knj-1}\gamma_{knj-1}}\right) - \mu_k\left(p_{knj} - p_{knj-1}\right)\right] < 0 \qquad (16)$$

5 SUBCARRIER ASSIGNMENT

To assign subcarriers to a user, we use the following equation

$$k^* = \arg\max \sum_{j=1}^{J}\left(R_{knj} - \mu_k p_{knj}\right)\left[F_{\gamma_{kn}}\left(\gamma_{knj+1}\right) - F_{\gamma_{kn}}\left(\gamma_{knj}\right)\right]^K \qquad (17)$$

Updating the Lagrange coefficient
To update the Lagrange coefficient we use sub-gradient methods

$$\nabla\mu_k = P_k^T - \sum_{n=1}^{N}\rho_{kn}p_{knj}\left[F_{\gamma_{kn}}\left(\gamma_{knj+1}\right) - F_{\gamma_{kn}}\left(\gamma_{knj}\right)\right] \qquad (18)$$

$$\nabla\beta_n = 1 - \sum_{k=1}^{K}\rho_{fkn} \qquad (19)$$

6 SIMULATION RESULTS

The considered scenario includes a base station and K users. The number of subcarriers is N=64 and the bandwidth is B=1 MHz. We used MATLAB tool for simulation. Fig. 1 depicts the total ergodic transmitted data in the system versus the number of users, regarding the different quantized intervals. As it is seen the aggregated system data rate increases and approaches to the system data rate for the complete channel timing information while the number of intervals increases.

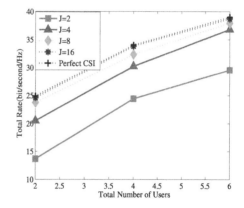

Figure 1. Aggregated transmitted data rate versus the users regarding the number of different quantized intervals.

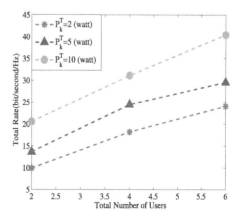

Figure 2. Aggregated transmitted data rate versus the users regarding the transmitted power of each user.

Fig. 2 depicts the total ergodic transmitted data versus the number of users for different amounts of power of each user. As is seen, increasing the transmitted power for each user increases the total ergodic data rate in the system.

7 CONCLUSION

We considered a base station and K users in the uplink WiMaxscenario based on OFDMA in this paper. Using BCDA algorithm we found channel timing quantized information, then using them, we determined the transmitted power and subcarrier assignment for each user. In the end, using the simulation results, we showed that by increasing the number of bits for quantized channel timing information, the total ergodic data rate approaches to the complete channel timing information.

REFERENCES

[1] Lapidoth and S. Shamai. 2002. Fading channels: how perfect need 'perfect side information' be *IEEE Trans. Inf. Theory*, vol. 48, no. 5, pp. 1118–1134.
[2] Medard M.2000. The effect upon channel capacity in wireless communications of perfect and imperfect knowledge of the channel.IEEE Trans. Inf. Theory, vol. 46, no. 3, pp. 933–946.
[3] Mukkavilli K. Sabharwal A.Erkip E.&Aazhang B. 2003.On beamforming with finite-rate feedback in multiple-antenna systems.IEEE Trans. Inf. Theory, vol. 49, no. 10, pp. 2562–2579.
[4] Wang Z. &Giannakis G. B. 2000. 1980. Wireless multicarrier communications: where Fourier meets Shannon. IEEE Signal Process. Mag., vol. 17, no. 3, pp. 29–48, May 2000.
[5] WiMax system evaluation methodology. 2008. WiMax forum, Ver .1.
[6] Zhou S.& Li B. 2006.BER criterion and codebook construction for finiterateprecoded spatial multiplexing with linear receivers.IEEE Trans. Signal Process., vol. 54, no. 5, pp. 1653–1665.

Impact of internal residual stresses to dissemination, shape and size of the ultrasound signal

M. Hatala, J. Zajac, D. Mital, Z. Hutyrova & S. Radchenko
Faculty of Manufacturing Technologies with the seat in Presov, Technical University of Kosice, Presov, Slovakia

J. Zivcak
Faculty of mechanical engineering, Technical University of Kosice, Kosice, Slovakia

ABSTRACT: Nowadays, engineering industry is full of modern and progressive technologies. Fast identification of residual stress is specially needed. The most common operation in component manufacturing is done by milling. The deformation are formed in internal layers during machining, which influences crystal lattice and do not disappear even after the discontinuation of external force effect on machining material. This paper is focused on using classic ultrasound defectoscope for identification of internal residual stress after machining material steel C45, which is a less economically demanding method of measuring residual stress. In the experimental part the results were verified by X-ray diffraction.

1 INTRODUCTION

Each technological operation performed on the material surface causes the rebuilt of atoms, molecules and grains in the internal layers of material. Technologies used to machine are directly influenced by energy, type, size, mechanical energy, thermal and chemical energy surface and internal layers. Residual stress is formed by elastic plastic deformation in the cutting zone (area of removing material). Main causes of residual stresses can be defined by four basic factors:

- Irregularity of plastic deformation distribution in material,
- Irregularity of thermal heat in the material during machining (which is reflected in dilatation or changing the length of material),
- Uneven changes in the structure caused by heat cutting forces formed during machining technology.

Residual stress or internal stresses are stresses in the internal layer of material, which remains in material even after the ceasing of force or load and they are closely related to plastic deformation. Formation of residual stress is caused by thermal heat of material in the area of cutting edges, when it is formed and is called phase stress.

Residual stresses are divided into two main groups: one is represented by tensile stress (referred to by symbol +) and the other is comprehensive stress (referred to by the symbol −). There are three factors to the forming residual stress:

- Plastic deformation,
- Phase transition,
- Local thermal heating.

Emerging of residual stress occurring by thermal heating formed during machining are shown in the next figure (Fig. 1).

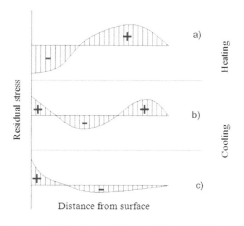

Figure 1. Residual stress formed during machining and cooling of material a) thermal heating b) c) cooling of the material + tensile − comprehensive.

Residual stress measuring methods can be divided into three main groups according to their influence on material as follows:

- Non-destructive,
- Destructive,
- Semi - destructive.

Measuring of residual stress by RTG diffraction is based on deformation of crystal lattices and its corresponding elasticity constant, where is assumed linearity in the elastic deformation in crystallographic planes of the material. The procedure is based on fundamental interactions between X-ray photon and crystal lattice of the material.

Geometrical conditions of diffraction in the case of 3D periodic structure are set in Lauhen diffraction conditions. Alternatives can be determined by Bragg law (Fig. 2), which is based on a conception of planes in the crystal lattice. Bragg law is implied from optic base on interference – track differences of two beams is integer multiplication of wavelength λ. Bragg angle is the angle between an incident x-ray beam and a set of crystal planes for which the secondary radiation displays maximum intensity as a result of constructive interference.

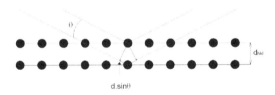

Figure 2.   Bragg angle.

$$\lambda = 2.d.sin\theta; \qquad (1)$$

where:

λ - wavelength of primary radiation,
d –planar distance between atoms,
θ - Bragg dispersing angle.

Pluralities of mutually parallel lattice planes are described by a trinity of Miller indices (hkl) and are characterized by planar distances between atoms d_{hkl}. In each crystal, it is possible to implement an infinite number of different lattice planes.

Basic method of RTG tensiometry is called method $sin^2\psi$ (Fig. 3). It utilizes the fact, that relative changes in planar distances between atoms are linear function $sin^2\psi$, where ψ is the angle between diffraction planes with surface normal of material surfaces. Coefficient $sin^2\psi$ contains surface components of the stress tensor.

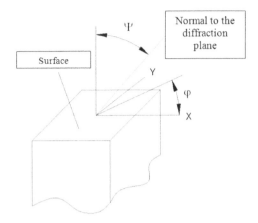

Figure 3.   Principle of the method $sin^2\Psi^2$.

$$\varepsilon_{\varphi\psi} = (d_{\varphi\psi} - d_0)/d_0 \qquad (2)$$

$$\varepsilon_{\varphi\psi} = f(\sigma_{i.j}, \varphi, \psi, v, E) \qquad (3)$$

where:

$\varepsilon_{\varphi\psi}$– deformation in direction,
$d_{\varphi\psi}$ – relative changes base on changes of angles φ,ψ,
σ_{ij} – stress tensor components in the measuring area,
E – Young's module,
v – Poisson ratio.

Relative change of $d_{\varphi\psi}$ in direction of measuring is characterized by angles φ a ψ and is identical with deformation in direction $\varepsilon_{\varphi\psi}$ and linear with function of $sin^2\psi$. Coefficient at $sin^2\psi$ contains Young´s model E and Poisson ratio in the sample and also a component of the stress tensor. Measuring is determined for various angles φ a ψ , where σ_{ij} is evaluated.

Cutting method is based on cross-section in the area of interest, where the cut is done perpendicular to the plane of the surface. Resulting gap (Fig. 4) is monitored, where on the surface of the cut is deformed by releasing residual stress. Tensile sensor detects result deformation as a dependence of depth

Consequently, the tension can be calculated by following equations:

$$\sigma x(y) = \sum_{i=0}^{n} A_i.P_i(y) \qquad (4)$$

where:

σx – tensile perpendicular to cut plane,
A_i , P_i – coefficients from polynomial, Fourier equations, and so on.

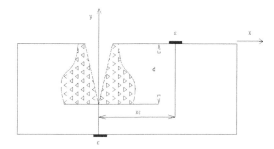

Figure 4.   Principle of cutting method.

Tensometric or drilling method is based on principle of tensile relaxation after drilling a hole with diameter R_0. Drilling is the process of how to release residual stress in the zone of cutting. Drilling method is derived from analytical solving on thin plate with hole. Based on this theory, it is possible to derive released radial deformation in point P and also their mathematical description as follows.

$$\varepsilon\_r = -(\sigma\_x\,(1+\mu))/2E\,[1/r^2 - 3\,r^4\;cos2\alpha + 4/(r^2\,(1+\mu))\;cos2\alpha] \qquad (5)$$

where:
 σ - tensile,
 E – Young´s module,
 r – ratio $r = R/R\_0$, where $R^3 R_0$,
 R – diameter of hole,
 R_0 – distance from the hole center.

For an assessing suitable alternative, it is necessary to have knowledge of parameters of the method. For choosing the method, it is necessary to know which of the methods is destructive or non destructive and their specifications (Table. 1).

Ultrasonic measuring of residual stresses was used for welded luminum plates. A 3D thermo-mechanical finite element analysis is used to evaluate residual stresses caused by friction stir welding of 5086 aluminum plates. The finite element (FE) model has been validated by the hole drilling method. Residual stresses obtained from the FE analysis are then compared with those obtained by ultrasonic stress measurement. The ultrasonic measurement technique is based on acoustoelasticity law which describes the relation between the acoustic waves and the stress of the material. The ultrasonic stress measurement is accomplished by using longitudinal critically refracted (LCR) waves which are propagated parallel to the surface within an effective depth. Through-thickness distribution of longitudinal residual stresses is evaluated by employing the LCR waves produced by four different frequencies (1 MHz, 2 MHz, 4 MHz and 5 MHz) of ultrasonic transducers. By utilizing the

Table 1.   Comparing the methods.

Method	Resolving power	Accuracy	Additional info
Drilling method	50 μm	±50MPa	type I,
Rtg. diffraction	20 μm	±20 MPa limited of linearity sin Ψ	type I and II
Hard rtg. rays	1mm	± 0,05 deformation + 10% real tensile	type II a III
Neutron diffraction	500μm	± 0,05 deformation + 10% real tensile	type II a III
Ultrasound	5mm	10% miss	Type I, II, III sensitive to microstructure
Magnetic	1mm	50MPa	Type I, II, III sensitive to microstructure

FE analysis along with the LCR method (known as FELCR method), the through-thickness distribution of longitudinal residual stress could be achieved. The comparison between ultrasonic and FE results shows an acceptable agreement, hence the FELCR method could be successfully applied to the FSW plates. It has been concluded that the longitudinal residual stresses of aluminum plates joined by friction stir welding can be evaluated by using the FELCR method. The good potential of FELCR as a nondestructive method is also confirmed in through-thickness stress measurement of aluminum plates.

Ultrasound can be defined as a type of waves, whose frequency is above the audible frequency in the elastic zone. The wave is induced by oscillation of elements in the environment. The wave is transmitted between elements in all directions and is characterized by propagation velocity.

$$a(t) = A_0 . sin. \omega . \left(t - \frac{x}{c}\right) \qquad (6)$$

where:
 a(t) - deviation of elements from equilibrium position in time,
 A_0 - maximal amplitude,
 ω - angular frequency ; w=2.p.f,   [Hz]
 f - frequency,   [Hz]

x - distance of oscillating points from the beginning,

c - propagation velocity.

Acoustic pressure is caused by the waves' propagation in the environment. Acoustic pressure depends on velocity of oscillating elements. Acoustic impedance is resistance of environment to ultrasound wave propagation.

$$p = z.v \qquad (7)$$

$$z = c.\rho \qquad (8)$$

where:

p - acoustic pressure,

v - velocity of elements oscillating around equilibrium position,

z - acoustic impedance of environment,

r - specific gravity of environment.

Principle of the ultrasound measuring device is based on a generator on impulse, which transmits to probe. Signal in probe is transformed by a piezoelectric transducer to ultrasounds waves.

Echoes which are received have too small amplitudes and have to be amplified and converted through A/D convertor to numerical code for further operation in the operating system.

The sampling frequency is dependent on the type of probe, but in general is not lower than 100MHz.

2 EXPERIMENTAL PART

Delivered semi finished steel C45 is 100x100 with a length of 1500. The first step of experimental preparation was to cut samples with the thickness of 20 mm, and after that they were machined by milling on CNC center Pinacle VMS 650 S.

Samples were cut on band saw Bomar type Ergonomic 275.230 DG (Fig. 4). Conditions of the

Figure 4.   Cutting the samples.

experiment are shown in the table below. Three variables (spindle speed, feed rate and depth of cut), were set in experimental parts whereby one of the parameters is variable and the remaining two are constant.

For milling of surface end mill FMPCM3063S with 6 teeth and the diameter D=63 mm was used (Fig.5).

Figure 5.   End mill FMPCM 3063S.

Table 2.   Technological conditions of experiment.

Sample number	Spindle speed n [min⁻¹]	Feed rate vf [mm.min⁻¹]	Depth of cut ap [mm]
1.	1000	500	1
2.	1500		
3.	2000		
4.	2500		
5.	3000		
6.	2000	600	1
7.		700	
8.		800	
9.		900	
10.		1000	
11.	2500	500	0,75
12.			1,3
13.			1,5
14.			1,7
15.			2

Measuring device used in the experiment was Olympus MX. It is a professional device used in companies by certified manufacturers for detection of welds defects, coat thickness, grooves, and in the aerospace industry to control outer shells, homogeneity of materials, and detections of luminary cranks.

The first step of the experiment was to set measuring system for material steel C45 and choose available probe. Configuration of measuring device consists of a loading measuring probe, time base and angle beam. Dead zone can be defined as length of transmitting impulse, where ultrasound is not able to detect defects. There are several reasons:

- Power of transmitting impulse (the greater the power, the greater the dead zone),
- Damping
- Frequency of the probe (low frequency – wide dead zone),
- Amplification of the signal – increasing of dead zone,

Probe with the frequency of 10MHz was chosen to experiment to secure sufficient short pulse and the suppression of the dead zone. The probe used in the industry was of type Parametrics – NDT 10x0,25 with frequency 10MHz. The probe used in experiments is direct, where a signal is transmitted from the probe perpendicular to the axis of the feed pressure of the probe to the measured surface and have to be used angle beam with angle 90° (we obtain LCR wave). LCR wave is the aim of this experiment, because we monitored its size, amplitude and amplification. Angle beam used in the experiment was Parametrics ABWL – 4T 90° - steel.

RTG diffraction was used as the verification method. RTG diffraction with type designation Proto XRD is measuring device provides measuring residual stress and also residual austenite. Anode of RTG lamp is made of Cr and Ka with cooling. RTG diffraction was used as a method to detect real values of internal method and was compared with changes of reflected echoes of ultrasound signal, which are dependent on residual stress.

Monitoring of residual stress by ultrasound was built for testing the hypothesis of the reflected echo, where monitor shape, size, amplification and amplitude of reflected echo were monitored. Identification of reflected signal was based on monitoring following parameters:

- width of reflected impulse (time), X axe
- Amplitude of reflected signal, (%), axe y,
- required amplification to display reflected signal.

Viewed echo on display was saved in internal memory, which was subsequently analyzed. Curves were saved in format JPEG and their processing was done by manual overwriting to the text editor. Analyses were done based on the width of impulse, reflected a time in microseconds and subsequently point of start and end echo signal, maximal and minimal amplitude of reflected signals (Fig. 6).

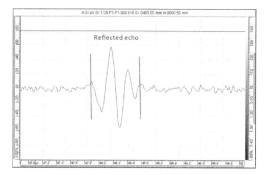

Figure 6.   Detail of reflected signal and marked area of the interest.

Calculation of carrier frequency has only informative character, as only values from the beginning and the end of the reflected signal were used. To improve quality, it would be necessary to use Fourier transform to exactly express carrier frequency.

$$f = \frac{1}{T} \tag{9}$$

where:

f - calculated carrier frequency of reflected echo [Hz],

T - time (mean pulse width) [s]

Samples tested in this study were the same as the samples verified on RTG difractometer. From graphical dependencies, it can be seen that amplitude of the reflected signal oscillating around the curve of real internal residual stress. Confirmation of initial hypothesis is impact of internal stress to wave propagation (Fig. 7).

The following step was to monitor the influence of technological cutting parameters to value of internal residual stress. From graphical dependences, it can be stated that curve of residual stress in response to spindle speed caused, that the shape of carrier frequency is similar to curve of residual stress in initial position 25 mm. Monitoring curves in the middle and at the end shows differences between carrier frequency and residual stress, where the curve of carrier frequency may be distorted by arithmetic averaging of signal

Table 3. Table of measured and calculated values of measuring residual stress by ultrasound.

Point	Time [µs]				Amplitude [-]			Amplification	Amplitude [-]		
	Start	End	Average time	Size	Start	End	Size		Start	End	Size
1	17,15	17,54	8,6725	0,39	-58	66	124	75	-65,5	73,5	139
2	16,88	17,28	8,54	0,4	-78	98	176	71	-95,5	115,5	211
3	16,6	17,68	8,57	1,08	-82	84	166	71	-99,5	101,5	201
4	30,1	30,4	15,125	0,3	-40	60	100	77	-42,5	62,5	105
5	30,2	30,78	15,245	0,58	-58	48	106	75	-65,5	55,5	121
6	29,68	30,1	14,945	0,42	-80	98	178	73	-92,5	110,5	203
7	48	48,44	24,11	0,44	-58	62	120	83	-45,5	49,5	95
8	47,88	48,2	24,02	0,32	-62	90	152	80	-57	85	142
9	47,9	48,5	24,1	0,6	-48	68	116	83	-35,5	55,5	91

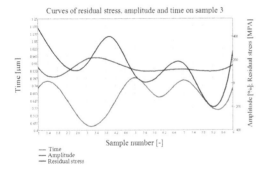

Figure 7. Graphical dependence of amplitude, width of impulse and real residual stress on sample 12.

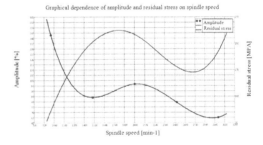

Figure 8. Graphical dependence of the amplitude and residual stress depend on spindle speed.

and carrier frequency. Curves were described by regression analysis and conversion coefficients were calculated to evaluate the values of residual stress by measuring parameters.

By monitoring the amplitude of the reflected echo signal, it can be again stated that the shape of amplitude has oscillating character around the curve of residual stress. Impact of residual stress on amplitude is proved in all measured points, what confirm an initial hypothesis. This phenomenon is also observed when analyzing middle and end lengths of the samples (Fig. 8).

Monitoring course of carrier frequency and amplitude depending on spindle speed, feed rate and depth of cut are oscillating around curves of real measured residual stress. Initial distances are characterized by the dependence of carrier frequency similar to curve of real residual stress, but the differences are more pronounced in the middle and end lengths. Graphical dependence of amplitude confirms the hypothesis that course of amplitude oscillating around residual stress in selected scale.

Table of conversion coefficients was created depend on geometrical position. In next step were done arithmetic averages to create general conversion coefficients to fast identify residual stress by monitoring parameters for material steel C45.

$$x = \frac{\sigma}{A(f)} \qquad (10)$$

where:

x - conversion coefficient,

σ - measured residual stress RTG [MPa],

A - amplitude of reflected signal,

f - value of carrier frequency.

Value of residual stress can be evaluated using following mathematical equations as a multiplying of conversion coefficients and measured parameter (carrier frequency or amplitude).

$$\sigma = x.f \qquad (11)$$

$$\sigma = x.A \qquad (12)$$

where:

x - conversion coefficient,

σ - calculated residual stress [MPa],

A - amplitude of reflected signal,

f - value of carrier frequency.

Range of conversion coefficients is from 1,9 to 2,1 depending on spindle speed for evaluation through amplitude and interval of values from 6,9 to 28 for the conversion of carrier frequency. Analyzing of conversion coefficients for different parameter of feed rate is the range from 2,56 to 2,76 for amplitude and interval of coefficients values from 8,4 to 30,66 for evaluation residual stress using carrier frequency. Interval of coefficients values for evaluating residual stress depend on the depth of cut is in the range from 1,52 to 2,25 for amplitude and from 12,87 to 16,87 for conversion based on a carrier frequency of the reflected echo signal.

Values of conversion coefficients depend on the geometrical position (position of the tool). Conversion coefficients are evaluated depending on the geometrical position of measured point. Conversion coefficients are different in depend monitored parameter (A - amplitude and f - frequency):

- Line 1 conversion coefficient for amplitude A=2,5 and carrier frequency f=8,5
- Line2conversion coefficient for amplitude A=2,5 and carrier frequency f=13,2
- Line3conversion coefficient for amplitude A=2,5 and carrier frequency f=1,95

From graphical dependences (Fig. 9), it is apparent that influence of residual stress is more significant on amplitude than to frequency respectively carrier frequency. Calculated residual stress based

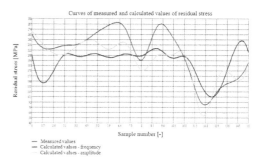

Figure 9. Graphical dependences of calculated and measured values of residual stress.

on conversion coefficients for amplitude provides sufficient information value, where the maximal difference between calculated residual stress and the measured value is 52MPa.

3 CONCLUSION

Monitoring of internal residual stress using ultrasound confirms the influence of residual stress on size, amplitude of the reflected signal. This article was also presented the influence of internal residual stress on ultrasound signal velocity. The probe used in the experiment was with frequency 10 MHz and acoustic medium was glycerin.

Evaluation of residual stress is to use conversion coefficients calculated for the amplitude of a reflected echo signal. Conversion coefficients for carrier frequency and calculated residual stress have informational character.

Correlation index of calculating curve of residual stress through conversion coefficient for amplitude isw 87–89%.

REFERENCES

[1] HRIVŇÁK, Ivan: Fractography (textbooks), Slovenská technická univerzita, Bratislava (2004).

[2] STEJSKAL, Vladimír – BŘEZINA, Jiří – KŇEZU, Jiří. Mechanic, České vysoké učení technické.

[3] G. Totten. Handbook on residual stress. SEM, Bethel 2005, ISBN: 978-0871707291; 1: 417.

[4] HRIVŇÁK, Ivan. At al.Experimental method of material study I, II; residual stress, their measuring and ways of elimination, Trnava, 2010, 71 strán, ISBN 978-80-8096-120-6.

[5] SEYEDALI, S. at al.Using ultrasonic waves and finite element method to evaluate through-thickness residual stresses distribution in the friction stir welding of aluminum plates In: Materials and design 2014, Iran, pages 27–34, ISSN 0261-3069.

[6] KOPEC, B. at al. Non destructive testing of material and constructions. Brno 2008, 384 pages, ISBN 978-80-7204-591-4.

[7] PEKLENIK, J., KISIN, M. An Investigative of Material Structure Transformation in Cutting Process. Annals of CIRP, 1998. Vol. 47/1, s.3–68.

[8] DUPLÁK J. et al.: Comprehensive identification of durability for Selected Cutting Tool Applied on the base of Taylor Dependence 2013. In: Advanced Materials Research. Vol. 716 (2013), p. 254–260. - ISSN 1022-6680.

[9] RIMÁR, M. et al.: Experimental investigation of choice technological material and qualitative parameters in technology of laser cutting 2005. In: Nonconventional Technologies Review. No. 2 (2005), p. 33–38. - ISSN 1454-3087.

[10] HATALA, M. et al.:Balance equation - an essential element of the definition of the drying process 2014. In: Advanced Materials Research. Vol. 849 (2014), p. 310–315. - ISSN 1022-6680 Spôsob prístupu: www.scientific.net/AMR.849.310].

Testing and Measurement: Techniques and Applications – Chan (Ed.)
© *2015 Taylor & Francis Group, London, ISBN: 978-1-138-02812-8*

Study on the attenuation laws of ultrasonic propagation in the inhomogeneous media

L. Yao, Q.X. Zhang & K. Wu
School of Information Science & Engineering, Shenyang University of Technology, Shenyang, Liaoning, P. R. China

ABSTRACT: In the exploration of industry, medium parameters detection is of great significance. Ultrasonic technique as the media parameter analysis is given more and more attentions for its advantages. In view of the medium parameters detection, the dynamic characteristics were researched on ultrasonic propagation in the inhomogeneous media. Based on the attenuation expression of ultrasonic propagation, the ultrasonic attenuation characteristics were analyzed under the influence of media parameters. MATLAB was used to simulate the ultrasonic attenuation of the inhomogeneous media. According to the simulation curve of attenuation coefficient, it was concluded that the greater the medium density is, the smaller attenuation becomes; the larger particle size is, the greater attenuation becomes. And the experimental platform has been built up, and the experimental results and the numerical simulation have the same trend. The results of the study of ultrasonic in inhomogeneous media parameter measurement provide a theoretical foundation for the further application.

1 INTRODUCTION

Ultrasonic methods are widely used in the industrial test because the outstanding advantages over penetrating ability, instruments, operation and testing cost. The inhomogeneous media parameters have an important effect on the attenuation characteristics. MATLAB has powerful function, and it can undertake numerical symbols and efficient calculation. Also, it can undertake numerical simulation. Therefore, MATLAB software was used for the numerical simulation of ultrasonic attenuation characteristics in the inhomogeneous media to get the trend which parameters it affects.

2 THE FLUCTUATION OF ULTRASONIC IN INHOMOGENEOUS MEDIA

Ultrasonic wave is a kind of mechanical vibration. So the study of the attenuation regularity and characteristics of ultrasonic should be instituted from the ultrasonic equation. Wave equation of ideal media is:

$$\frac{\partial^2 p}{\partial x^2} = \frac{1}{c^2}\frac{\partial^2 p}{\partial t^2} \tag{1}$$

Where c= $(k_s/\rho)^{(1/2)}$, ρ is the medium density. k_s=-vdP/dV, the adiabatic bulk modulus of elasticity. V is the volume. X is the transmission distance. T is the transmission time. Pressure is represented by P, so P=dP=-k_sdV/V, dV/V=dρ/ρ=dξ/dx, and we can get p=-k_s dξ/dx.

When the velocity of adjacent particles is not the same, the mud's certain viscosity will produce relative motion. According to the Newton law of internal friction in fluid mechanics, it's known that the viscous force per unit area on a one-dimensional plane wave is p'=T'=-ηdv/dx, η is the coefficient of viscosity. It's substituted into the acoustic equation ρdv/dt=-dp/dx. As a result, the acoustic equation can now be written as follows:

$$\rho\frac{\partial^2 \xi}{\partial t^2} = K_s\frac{\partial^2 \xi}{\partial x^2} + \eta\frac{\partial^3 \xi}{\partial x^2 \partial t} \tag{2}$$

3 THE ULTRASONIC ATTENUATION

When ultrasound spread through the inhomogeneous media, the sound pressure or sound energy will gradually increase or decrease with the propagation distance, and this is called ultrasonic attenuation. The sound attenuation can be divided into diffusion, viscous, thermal conductivity and scattering attenuation.

3.1 *Diffusion attenuation*

Diffusion attenuation is caused by the sound intensity weakening owing to the expansion of the wave front. It only depends on the waveform and the beam of the sound radiation, regardless of the media properties, so the attenuation in this article shall not be considered.

3.2 Viscous attenuation

Viscous attenuation is caused by the sonic velocity gradient owing to the effect of viscosity of the media, resulting in loss of internal friction between the media. If $H=\eta/k_s$, due to $K=k_s+j\omega\eta$, so:

$$K = K_S(1 + j\omega H) \qquad (3)$$

Due to the wave number $k'=\omega\,(\rho/k_s)^{(1/2)}=\omega/c\text{-}j\alpha\eta$, we substitute them into (3):

$$\alpha_\eta = \omega\sqrt{\frac{\rho}{K_S}}\sqrt{\frac{(\sqrt{1+\omega^2 H^2}-1)}{2(1+\omega^2 H^2)}} \qquad (4)$$

Since the viscosity force is much smaller than the elastic force of the media, that is, $\omega H \ll 1$, so the viscous attenuation coefficient is:

$$\alpha_\eta = \frac{\omega^2 \eta}{2\rho c^3} \qquad (5)$$

Where η = viscosity of media; ω = angular frequency; and c = propagation velocity.

3.3 Thermal conductivity attenuation

Media particle will be compressed or inflated when sound waves going through the media. The temperature of compression zone will rise, and the temperature of expansion zone will decrease. The temperature difference of the adjacent compression and expansion area will lead to a part of heat transferring from high to low temperature area. The thermal conductivity attenuation coefficient is:

$$\alpha_\xi = \frac{\omega^2 \chi}{2\rho c^3}\left(\frac{1}{c_v}-\frac{1}{c_p}\right) \qquad (6)$$

Where χ = coefficient of heat conduction; c_v = specific heat at constant volume; and c_p = specific heat at constant pressure.

3.4 Scattering attenuation

Scattering attenuation is due to the unevenness of media, resulting in slurry interface having much small different acoustic impedance which scatters in different directions. These particles are used as a completely rigid pellet. If the quantity of scattering particles is n of unit volume, then the scattering attenuation coefficient is:

$$\alpha_s = \frac{2}{9}nk^4\pi a^6 \qquad (7)$$

Where k = the number of ultrasonic wave; a = radius; and n = the number of particles per unit volume.

The total attenuation coefficient of ultrasonic spread through the media is:

$$\alpha = \alpha_\eta + \alpha_\xi + \alpha_s$$
$$= \frac{\omega^2}{2\rho c^3}[\eta + \chi(\frac{1}{c_v}-\frac{1}{c_p})] + \frac{2}{9}(\pi a)^2 (ka)^4 n \qquad (8)$$

From (8) it can be seen that, in the process of ultrasonic signal transmission, the attenuation coefficient is related to media density, viscosity, particle size and the acoustic frequency.

4 NUMERICAL SIMULATION

4.1 Numerical simulation of dust attenuation

We use dust in the air as the research object, and dust viscosity is 0.000018Pa.s, velocity of sound waves is 350m/s, acoustic frequency is 200 kHz, coefficient of heat conduction is 0.025W/m.°C, the specific heat at constant volume is 1004J/kg.°C, and Specific heat at constant pressure is 717.4 J/kg.°C.

4.1.1 The influence of dust density on the acoustic attenuation

Assuming the dust particle size is 50um, according to (8), the curve of dust density and attenuation coefficient is shown in Figure 1:

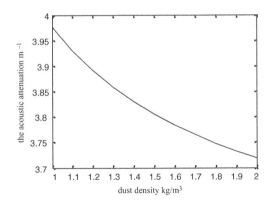

Figure 1. The curve of dust density and attenuation coefficient.

The attenuation coefficient of ultrasonic signal is inversely proportional to the dust density.

4.1.2 The influence of dust particle size on the acoustic attenuation

Assuming the dust density is 1.6Kg/m³, according to (8), the curve of dust particle size and attenuation coefficient is shown in Figure 2:

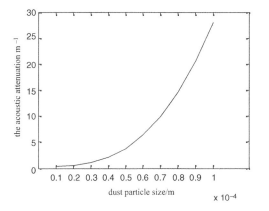

Figure 2. The curve of dust size and acoustic attenuation.

The attenuation coefficient of ultrasonic signal is proportional to the dust size.

4.2 Numerical simulation of mud attenuation

The mud is made by a certain percentage of water and clay. Mud viscosity is 1Pa.s, velocity of sound waves is 1850m/s, coefficient of heat conduction is 0.65W/m.°C, the specific heat at constant volume is 10392J/kg.°C, Specific heat at constant pressure is 7873 J/kg.°C, and acoustic frequency is 200 kHz.

4.2.1 The influence of mud density on the acoustic attenuation

Assuming the particle size is 20um, according to (8), the curve of mud density and attenuation coefficient is shown in Figure 3:

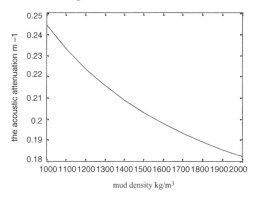

Figure 3. The curve of mud density and attenuation coefficient.

The attenuation coefficient of ultrasonic signal is inversely proportional to the mud density.

4.2.2 The influence of mud particle size on the acoustic attenuation

Assuming the mud density is 1.25g/cm³, according to (8), the curve of mud particle size and attenuation coefficient is shown in Figure 4:

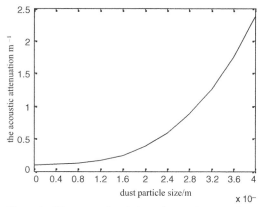

Figure 4. The curve of mud size and acoustic attenuation.

The attenuation coefficient of ultrasonic signal is proportional to the particle size of mud.

Characteristics of ultrasonic attenuation in crude oil were simulated, and the curve of acoustic attenuation is similar to the mud.

5 EXPERIMENTAL VERIFICATION

In this paper, mud is used as inhomogeneous media in the experiment. The platform of experimental device is mainly composed of the 555 trigger circuit, the ultrasonic ASIC chip LM1812, a 200 kHz ultrasonic transducer, a 12V power supply and an oscilloscope. In this experiment, with the same four stainless steel containers holding the different mud, the reflecting surface function is realized by the bottom of the stainless steel container, and the ultrasonic transducer distance from the reflecting surface is about 30mm.

5.1 Mud density and attenuation

Assuming the particle size is 150um. Matched the mud density is about 1.21g/cm³, 1.35g/cm³, 1.42g/cm³ and 1.51g/cm³. Mud viscosity is measured about 1.16Pa.s. Attenuation coefficients are calculated for 42.28m⁻¹, 37.15 m⁻¹, 32.69m⁻¹ and 30.03m⁻¹. The curve of the mud density and attenuation coefficient is shown in Figure 5:

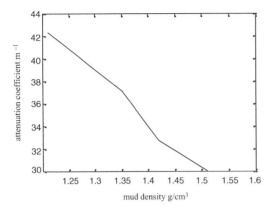

Figure 5. The curve of mud density and attenuation coefficient.

As indicated by the experimental results, the greater the mud density is, the larger the ultrasonic echo amplitude becomes, and then the smaller the attenuation is.

5.2 *Mud particle size and attenuation*

We use the soil sieves whose equivalent pore diameters are 80mm, 150mm, 200mm and 300mm to sieve soil. And we configure the four samples of mud of density, which is about 1.21g/cm³. Attenuation coefficients are calculated for 30.03m⁻¹, 32.69 m⁻¹, 38.77m⁻¹ and 48.36m⁻¹. The curve of the mud particle size and the attenuation coefficient is shown in Figure 6:

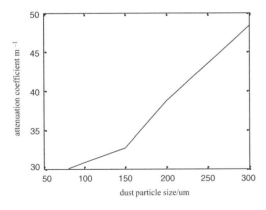

Figure 6. The curve of mud size and acoustic attenuation.

As indicated by the experimental results, the greater the mud particle size is, the smaller the ultrasonic echo amplitude becomes, and then the larger the attenuation is.

6 SUMMARY

Based on theoretical analysis and experimental research, we discussed the law of ultrasonic propagation and attenuation in inhomogeneous media and the factor causing the ultrasonic signal attenuation is determined by us. Through research and analysis, the experimental results and the theoretical simulation of conclusion are of the same tendency.

1 In the fixed ultrasonic frequency, the same material and substantially the same particle size of the mud, if the mud density becomes greater, the amplitude of ultrasonic echoes becomes greater, and the attenuation coefficient gets smaller.
2 In different mud particle sizes, same material and the same density of the mud, if mud particle size becomes larger, the amplitude of ultrasonic echoes becomes smaller, and ultrasonic attenuation gets greater.

REFERENCES

[1] Chen, X.Z. 2012. Experimental Study of Transmission and Water Treatment Performance of Ultrasonic. Beijing, China: D. Beijing University of Technology.
[2] Feng, Y. 2013. Research on The Detection of Casing in Mixed Media Based on Ultrasonic. Harbin, China: D. Harbin Institute of Technology.
[3] Liu, F. Fu, J.H. Zhang, Z & Xu, L.B. 2012. Research of Ultrasonic Attenuation Theory in Drilling Fluid, J. Oil Drilling & Production Technology 34(1): 57–60.
[4] Meng, Y.Q. 2012. The Research of Pulp Concentration Detection System Based on Ultrasonic Technology. Jinan, China: D. Shandong Polytechnic University.

Testing and Measurement: Techniques and Applications – Chan (Ed.)
© 2015 Taylor & Francis Group, London, ISBN: 978-1-138-02812-8

Ultrasonic defect detection of a thick-walled filament-wound composite tubular structure

H.L. Bu, J.L. Chen, G.Y. Gao & Z. Li

Department of Mechanics and Engineering Science, State Key Laboratory of Turbulence and Complex Systems and College of Engineering, Peking University, Beijing, P.R. China

ABSTRACT: In this paper we propose an ultrasonic defect detection method for thick-walled filament-wound composite tubular structures based on the Continuous Wavelet Transform (CWT). Defect locations and sizes in the composite layers and composite-steel interface as well as the steel thickness are evaluated. Moreover, a Data Acquisition System (DAQ) is built and the collection, storage, processing and visualization of the C-Scan testing results are all automatically achieved by the system. The results show that the Gabor wavelet performs well for defect detection in composite layers and interface. The detection results of defect locations and sizes are in consistence with the real information of embedded defects.

1 INTRODUCTION

Composite tubular structures are increasingly used in the particular corrosive, mechanical, or thermal loading conditions [1]. Among these structures, thick-walled filament-wound composite tubular structures are widely used in the petrochemical, medical, space and aeronautical areas because of their unique properties such as high specific stiffness and good corrosion resistance [2]. However, the manufacturing process of composite tubular structures may result in the presence or introduction of unwanted defects (voids, resin, rich areas, and inclusion) [3]. The defects will seriously affect the overall structural integrity and make the inspection become a critical task.

In the last decade, ultrasonic techniques have been shown to be very promising for non-destructive testing (NDT) of composite materials [4]. Aymerich et al. [5] has proved that normal and oblique incidence ultrasonic techniques with full waveform acquisition are very sensitive to matrix damages and delamination. Roman et al. [6] has developed a one-side access air-coupled ultrasonic measurement technique for detection and visualization of inhomogeneities in composite materials. Orazio et al. [7] used a neural network-based analysis of ultrasonic data for the automatic detection of internal defects in composite materials and recognized the defect locations successfully. Kersemans et al. [8] investigated the feasibility of both amplitude and time-of-flight based pulsed ultrasonic polar scan (P-UPS) as a sophisticated non-destructive damage sensor for fiber reinforced composites, the results showed that P-UPS can detect and quantify local deviations in the prescribed stacking sequence.

Although ultrasonic NDT techniques are effective in defect detection of composite materials or structures, proper measurement and signal processing methods are necessary to obtain precise results. Pagodinas et al. [9] described the ultrasonic signal processing methods used in different areas of nondestructive testing of materials, including the transform-domain ultrasonic signal processing, the complex cepstrum domain analysis, the autoregressive cepstrum model, the split-spectrum processing. The wavelet transform seems to be more effective for ultrasonic data processing, especially for detection of defects in grainy materials. Legendre et al. [11] applied two wavelet-transform-based methods in the processing of Lamb-wave signals, computed the signal wavelet transforms, and found the best wavelet coefficients in terms of discriminating ability. Furthermore, He [10] proposed an interpretation procedure of the wavelet coefficients based on the selection of preponderant wavelet coefficients and a windowing process. The results showed that the NDE image of the composite structure can be obtained directly from the wavelet coefficients [11], without requiring further signal reconstruction [12]. Jeong et al. [13] found that the wavelet transform using the Gabor wavelet was an effective tool for the experimental analysis of the dispersive waves in the composite laminates, and has proved that the arrival times of each frequency component needed in the velocity calculation can be determined from the peak of the magnitude of wavelet transform data on the time-frequency domain. Yu et al. [14] proposed a crack detection method for fiber reinforced composite beam based on the CWT, and chose the Gabor and Morlet wavelets for wave signal processing owing to their excellent time-frequency characteristics. The results showed that the Gabor wavelet can locate the cracks more accurately by virtue of its higher time resolution. Su et al. [15] developed an identification approach for delamination, locating in laminated composites, based on the Lamb wave propagation; it achieved good diagnostic results employing the wavelet transform-based signal processing technique in the identification scheme. Wang et al. [16] proposed a defect imaging method which is based on spatial filter and complex Shannon wavelet transform, and applied it in the composite structural without structural parameters.

All above results are mainly limited to ultrasonic A-Scan or B-Scan testing of fiber reinforced composite laminates. Ultrasonic C-Scan testing techniques of filament-wound composite tubular structures for defect evaluation without baseline have rarely been proposed, due to the complexity of the structure, the diversity of defects, the sophisticated of C-Scan techniques.

In this study, an ultrasonic defect detection method for thick-walled filament-wound composite tubular structures based on the CWT was presented. A DAQ system was developed based on LabVIEW. Then, the ultrasonic C-Scan testing of the composite tubular structure was conducted.

2 THE COMPOSITE TUBULAR STRUCTURE

The thick-walled filament-wound composite tubular structure shown in Figure 1 was cut from a three-meter-long filament wound cylinder. The structure is 421mm of inside diameter and 261mm of total length, consisting of steel liner, carbon/epoxy composite layers and coating from inside to outside. The ply sequence is $[90°_3/30°_3/90°_3]_s$ and the thickness of each ply is 0.5mm or 0.8mm alternatively. The carbon fiber reinforcement is T700SC-12K and the epoxy resin is TDE-86. In order to verify the experimental results, defects had been embedded in Coating 1 region while locations and sizes is listed below.

Table 1. Defects in the composite layers and on the composite-steel interface.

Defects	location	size		location (x)
	cm	mm		cm
Composite	(36, 3.5, 9)	15	Composite-steel	
Layers	(50, 4.5, 7)	20	Interface	
	(66, 5, 5)	30		20-40
	(79, 2.5, 3)	25		60-80

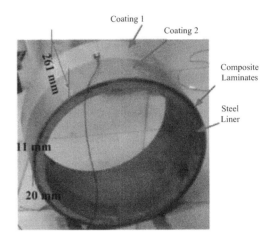

Figure 1.   The composite tubular structure.

3 ULTRASONIC TESTING METHOD

3.1 *Ultrasonic method of the structure*

In the homogeneous materials, the presence of defects will cause discontinuation in materials, resulting in inconsistencies of the acoustic impedance and generation of interfaces. When the pulse is transmitted across these interfaces, reflections will occur [17]. As shown in Figure 2, when a defect exist in the specimen, the probe will receive a reflected signal between the front-wall and backwall reflections, the depth of the defect can be determined based on the time of flight (ToF).

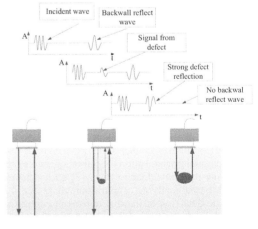

Figure 2.   The basic principle of ultrasonic testing.

Based on this theory, the ultrasonic pulse-echo ultrasonic testing method for the composite tubular structures (Fig.3) was developed. A probe, placed

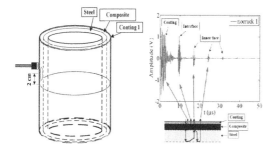

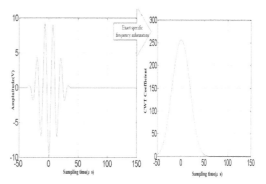

Figure 3.   The ultrasonic pulse-echo testing.

vertically on the measuring point, transmitted the incident wave and subsequently received the reflected wave. The interface reflection signals are all displayed based on time series.

3.2   Continuous wavelet transform method

Assuming that $f(t)$ and $\Psi(t)$ are square-integrable functions, satisfying the admissibility condition

$$\int_R |\psi(\omega)|^2 \Big/ \omega d\omega < \infty, \tag{1}$$

the CWT definition of $f(t)$ in time and frequency domains is described as [18]

$$W_f(a,b) = \langle f(t), \psi_{a,b}(t) \rangle = \frac{1}{\sqrt{a}} \int_R f(t)\bar{\psi}(\frac{t-b}{a})dt, \tag{2}$$

$$W_f(a,b) = \langle f, \psi_{a,b} \rangle = \frac{1}{2\pi} \langle \hat{f}, \hat{\psi}_{a,b} \rangle = \frac{\sqrt{a}}{2\pi} \int_R \hat{f}(\omega) \exp(i\omega b)\overline{\hat{\psi}(a\omega)}dt. \tag{3}$$

Where i is the imaginary unit, ω is the circular frequency, a and b are the scale and translation parameters, respectively.

A crucial problem in CWT application is the mother wavelet selection. In this study, the Gabor wavelet function was chosen for continuous signals, because of their good abilities in feature extraction [14]. Considering that the waveform is

$$u(x,t) = \frac{1}{2\pi} \int_R A(\omega) \exp[i(\omega t - kx)]d\omega, \tag{4}$$

the absolute wavelet coefficient based on the Gabor wavelet can be gained as

$$|W_f(a,b)| = \frac{1}{\sqrt[4]{\pi}\sqrt{a}} \sqrt{\frac{\omega_0}{\gamma}} \left| A(\frac{\omega_0}{a}) \right| \exp\left[-\frac{(\omega_0/\gamma)^2}{2a^2}(b - \frac{x}{c_g})^2 \right]. \tag{5}$$

By adjusting the scaling factor a, we can extract the arrival time and amplitude of each frequency component of the wave. As shown in Figure 4, specific frequency information was exacted. Thus, peak time was obtained more accurately.

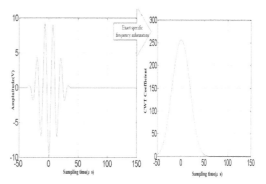

Figure 4.   CWT of the ultrasonic signal.

4   THE C-SCAN EXPERIMENT SYSTEM

4.1   Ultrasonic method of the structure

In this study, the ultrasonic C-Scan testing method was used to obtain defect distribution of the thick-walled filament-wound composite tubular structure (Coating 1 region). During the C-Scan process (Fig.5), the surface of the structure was scanned by the manual operated ultrasonic probe (parameters are shown in Table 2). Measured data were transferred via the AL8200 DAQ card to the DAQ software, which is programmed based on LabVIEW (Fig.6). The data collection, data storage, data processing based on CWT and visualization were all automatically achieved through this DAQ software. In the software, MATLAB subroutines were also used to analyze data by embedding in LabVIEW Mathscripts.

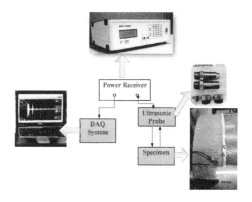

Figure 5.   Schematic representation of the hardware system.

Table 2. Parameters of the hardware system.

Hardware	Parameters
A109S-RM Probe	Frequency:5 MHz; transmitter :13 mm
APR-S300T Power Receiver	Pulse: Negative Spike Pulse; Rise Time 10% to 90%: 40 ns; Energy: 100 μjoules; Mode: Pulse-Echo, Through Transmission; External PRF:0 ~ 100 KHz; Bandwidth: 1 KHz ~ 35 MHz; Gain: -20 dB ~ 60 dB; HPF: 300 K,1 M,5 MHz; LPF: 10 M,15 M,25 MHz
AL8200 DAQ card (DAQ system)	Sampling Rates ◆ Sampling rates: 200 MHz Memory ◆ 64 M Samples on-board acquisition memory A/D Converter ◆ 8 bits resolution, single channel Analog Section ◆ 1 analog channel, two inputs; AC/DC coupling Input ranges: 50 m Vp-p - 5Vp-p Bandwidth: DC to 80 MHz -3dB Trigger Sources ◆ Software trigger; Encoder Trigger; Internal trigger connector, TTL Bus Interface ◆ PCI interface, 32 bits, 33 MHz ◆ PCI burst transfer rates up to 133 M Bytes/sec

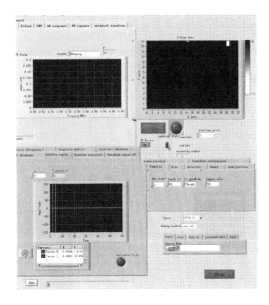

Figure 6.   The front panel of the DAQ software.

5   RESULTS AND DISCUSSION

In order to recognize the defect locations, the longitudinal wave velocities (Table 3) of the Coating 1, the carbon/epoxy composite and the steel liner were obtained based on the equation:

$$C = \frac{d}{\Delta t}, \qquad (6)$$

where C is wave velocity, d is the thickness, Δt is the peak interval.

Table 3.   Wave velocities.

	Longitudinal wave
Materials	(m/s)
Coating 1	3500
Carbon/epoxy composite	1724
Steel liner	5890

5.1   *Defect detection of the composite layers*

Firstly, original signals of defects in different depth of the composite layers (Fig.7) and their CWT at a specific frequency were obtained through single point detection. By comparison, defect signals can be obviously seen between the incident signals and the signals reflected by the composite-steel interface. With depths of defects increase, peak arriving times delay. When defects are large enough, the signals reflected by the composite-steel interface will disappear, and the second reflections of defects will be captured.

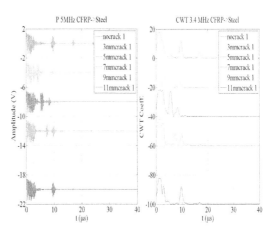

Figure 7.   Defect signals and their CWT of different depth in composite layers.

Moreover, it can be seen obviously that the peak arriving time of defects are recognized more accurately after CWT.

2D contours of defects in the composite layers were drawn after the C-Scan testing of the whole structure. Coordinates of the typical defects are (36 cm, 3.5 cm), (49 cm, 4.5 cm), (66 cm, 5.5 cm) and (80 cm, 3 cm), the diameters are 12 mm, 22 mm, 27 mm and 25 mm, respectively. These results are in consistence with the locations of pre-embedded defects shown in the Table 1. In addition, small random defects mainly caused by manufacturing processing, exist at (94 cm, 5 cm), (123 cm, 1.7 cm).

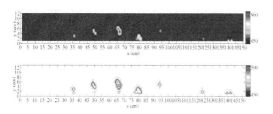

Figure 8.   2D contours of defects in the composite layers.

5.2   *Defect detection of the composite-steel interface*

When a defect exists in the composite-steel interface, signals reflected by the composite-steel interface will be stronger, and the signal transmitting into the steel will weaken or even disappear. By comparing the intact and defect signals, the interface debonding can be determined.

As is shown in Figure 9, compared with the intact case, the peak arriving time of the composite-steel interface in the defect signal are the same (about 9.6 MS), but the signal transmitting into the steel (about 17 MS) weakened significantly.

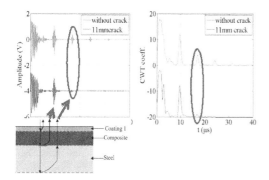

Figure 9.   Signals and CWT of defect in composite-steel interface.

2D contours of defects in the composite-steel interface were drawn by calculating the energy ratio of the interface reflection with the reflection of inner surface of the steel. Interface debonding areas, located at 20cm -35 cm and 63cm- 84 cm along x axis, are in consistence with the locations of pre-embedded defects shown in the Table 1.

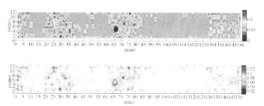

Figure 10.   2D contours of defects in the composite-steel interface.

5.3   *Evaluation of the steel thickness*

By the test above, Longitudinal wave velocity in the steel liner was about 5892 m / s. According to the Equation 6 and the peak intervals, 2D contours of the steel thickness were drawn (Fig.11). Steel liner is thicker from 35cm to 100cm. The thickness ranges from 18mm to 21.6 mm while the average thickness is 20mm.

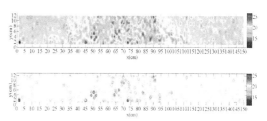

Figure 11.   2D contours and XY plot of steel thickness.

6   CONCLUSIONS

In this paper, an ultrasonic defect detection method for thick-walled filament-wound composite tubular structure based on the CWT was presented. A DAQ was built based on LabVIEW and the C-Scan testing of the composite tubular structure was conducted. Then, we can draw the following conclusions:

1 The Gabor wavelet performed well for the defect detections of the composite layers and the composite-steel interface.

2 The C-Scan testing of the composite tubular based on the DAQ system was completed successfully. Both defect locations and sizes in composite layers and composite-steel interface as well as the steel thickness were evaluated. The results are in consistence with the real information of embedded defects.

ACKNOWLEDGMENTS

This research was supported by the National Natural Science Foundation of China (Grant No.11472012). The authors would like to thank Dr. Kezhuang Gong in the Navy Academy of Armament of China for his help with modifying the equipment.

REFERENCES

[1] Martins, L. A. L. & Bastian, F. L. & Netto, T. A. 2012. Structural and functional failure pressure of filament wound composite tubes. *Materials & Design* 36: 779–787.

[2] Xia, M. & Takayanagi, H. & Kemmochi, K. 2001. Analysis of multi-layered filament-wound composite pipes under internal pressure. *Composite Structures* 53(4): 483–491.

[3] Cantwell, W. J. & Morton, J. 1992. The significance of damage and defects and their detection in composite materials: a review. *The Journal of Strain Analysis for Engineering Design* 27(1): 29–42.

[4] D'orazio, T. & Leo, M. & Distante, A. 2008. Automatic ultrasonic inspection for internal defect detection in composite materials. *NDT & E International* 41(2): 145–154.

[5] Aymerich, F. & Meili, S. 2000. Ultrasonic evaluation of matrix damage in impacted composite laminates. *Composites Part B: Engineering* 31(1): 1–6.

[6] Kažys, R. & Demčenko, A. & Žukauskas, E. 2006. Air-coupled ultrasonic investigation of multi-layered composite materials. *Ultrasonics* 44: e819–e822.

[7] D'orazio, T. & Leo, M. & Distante, A. 2008. Automatic ultrasonic inspection for internal defect detection in composite materials. *NDT & E International* 41(2): 145–154.

[8] Kersemans, M. & Baere, I. D. & Degrieck, J. 2014. Nondestructive damage assessment in fiber reinforced composites with the pulsed ultrasonic polar scan. *Polymer Testing* 34:85–96.

[9] Pagodinas, D. 2002. Ultrasonic signal processing methods for detection of defects in composite materials. *Ultragarsas* 45(4): 47–54.

[10] Legendre, S. & Goyette, J. & Massicotte, D. 2001. Ultrasonic NDE of composite material structures using wavelet coefficients. *Ndt & E International* 34(1): 31–37.

[11] Legendre, S. & Massicotte, D. & Goyette, J. 2000. Wavelet-transform-based method of analysis for Lamb-wave ultrasonic NDE signals. *Instrumentation and Measurement, IEEE Transactions on* 49(3): 524–530.

[12] Staszewski, W. J. & Pierce, S. G. & Worden, K. 1997. Wavelet signal processing for enhanced Lamb-wave defect detection in composite plates using optical fiber detection. *Optical Engineering* 36(7): 1877–1888.

[13] Jeong, H. & Jang, Y. S. 2000. Wavelet analysis of plate wave propagation in composite laminates. *Composite Structures* 49(4): 443–450.

[14] Yu. L & Zheng, L. & Zhang, W. 2010. Crack detection of fibre reinforced composite beams based on continuous wavelet transform. *Nondestructive Testing and Evaluation* 25(1): 25–44.

[15] Su, Z. &Ye, L. & Bu, X. 2002. A damage identification technique for CF/EP composite laminates using distributed piezoelectric transducers. *Composite structures* 57(1): 465–471.

[16] Wang,Y. & Yuan, S. & Qiu, L. 2011. Improved Wavelet-based Spatial Filter of Damage Imaging Method on Composite Structures. *Chinese Journal of Aeronautics* 24(5): 665–672.

[17] Charlesworth, J. P. & Temple, J. A. G. 1989. *Engineering applications of ultrasonic time-of-flight diffraction*. Somerset: Research Studies Press.

[18] Mallat, S. 1999. *A wavelet tour of signal processing*. Academic Press.

Testing and Measurement: Techniques and Applications – Chan (Ed.)
© 2015 Taylor & Francis Group, London, ISBN: 978-1-138-02812-8

Monitoring of the radio-frequency atmospheric pressure plasma jet by hairpin probe

W. Yan, J.Z. Xu & J.F. Wang
Department of Applied Physics, College of Science, Donghua University, Shanghai, P.R. China

ABSTRACT: A hairpin resonance probe was used firstly to monitor the radio-frequency-driven He or He/O2 atmospheric pressure plasma jet. The reflected microwave spectrum from the probe was recorded by a vector network analyzer. It is observed that the microwave resonant frequency is not affected by the applied power, and displayed the same resonant frequency in air or no applied power. However, the full width at half maximum or the resonance width of the resonance spectrum varies with the applied power or oxygen admixture. The electron density at the plasma jet exit is estimated to be about 10^{15} m^{-3} by a method combing hairpin probe and optical emission spectrum.

KEYWORDS: Atmospheric pressure plasma jet, Diagnostics, Hairpin probe, Microwave resonant spectrum.

1 INTRODUCTION

Nonthermal atmospheric pressure plasma (APP) and atmospheric pressure plasma jet (APPJ) have gained much more focus recently due to its potential applications in material processing [1]. In order to have a good understanding of the property of APPJ, we need to know the electron temperature and electron density. Although many diagnostic techniques such as the Langmuir probe, microwave resonant probe and light absorption have been widely applied in low pressure plasmas, these techniques are rarely used for atmospheric pressure plasma especially for plasma jet as most of them become useless at higher pressure. So far, optical emission spectroscopy (OES) [2, 3] is the main tool to monitor the APP and APPJ.

In this paper, we present a diagnostic method based on the hairpin probe for the radio-frequency-driven APPJ (rf-APPJ). Section 2 describes the experimental setup and diagnostic method. Section 3 gives the results and discussions.

2 EXPERIMENT SETUP AND DIAGNOSTIC METHOD

The rf-APPJ source is generated by AtomfloTM 250 plasma applicator (13.56MHz). The working gas is helium (5~30 L/min) or a helium-oxygen mixture. The discharge chamber is shown schematically in figure 1 [4].

The hairpin probe is a quarter wave transmission line with one short-circuited end and opened at other [5], which in this paper its length, width and the wire diameter is 22mm, 3.5mm and 0.85mm, respectively.

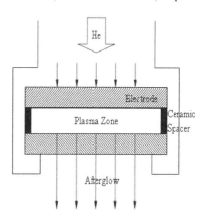

Figure 1. Schematic of the rf-APPJ source: AtomfloTM 250.

A low-power microwave signals from a vector network analyzer (VNA) drives an amplitude time-varying current to excite the floating probe through a small two-turn loop antenna [6], and the current is swept over a frequency range about 0.5GHz. To perform the monitoring, the hairpin probe is fixed underneath the plasma exit of less than 3mm, the reflected signal or the power reflection coefficient defined as $S_{11} = 10\log (Pr /Pi)$ is recorded by VNA (R&S ZVL6, frequency up to 6GHz). The diagram of the experimental setup to monitor the rf-APPJ is shown in figure 2.

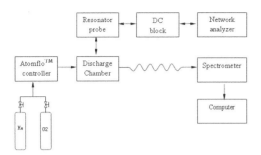

Figure 2.   A schematic diagram of the experimental setup.

The length of the probe and the dielectric properties of its surrounding medium affect the characteristics of the probe resonant spectroscopy such as the resonant frequency f_r and the resonance width Δf or the quality factor Q due to the collision-plasma absorption of the microwave power [5]. When the probe length timing the wave-numbers satisfies $\beta l = (2k + 1)\pi/2$, the probe works in a resonant state. The fundamental resonant frequency of the probe is

$$f_r = \frac{1}{4l\sqrt{LC}} = \frac{c}{4l\sqrt{\varepsilon_{pr}}} \qquad (1)$$

Where c = the speed of light; l = the length of the resonator probe; $\varepsilon_{pr} = 1 - \omega_p^2/(\omega^2 + v_m^2)$ is the permittivity of plasma and ω = the plasma frequency. For the case of low pressure plasma, when hairpin probe is immersed into a collision-less plasma (v_m is negligible), the resonant frequency fr will shifted to a higher compared with the resonant frequency in vacuum f_0, and satisfy the relation of $f_r^2 = f_0^2 + (\omega_p/2\pi)^2$. which is used to calculate directly the spatially averaged electron density.

For high-pressure plasma (more than 40Torr), the electron-neutral collision frequency v_m is in the order of 100GHz. The resonant frequency of the hairpin probe remains the same when it is immersed into the high pressure plasma because of $\varepsilon_p \sim 1$ [5,7]. However, the resonance width normalized by resonant frequency is proportional to plasma conductivity as [7].

$$\left(\frac{\Delta f}{f_r}\right)_p \propto \frac{240l\sigma_p}{\sqrt{\varepsilon_{pr}}} \approx 240l\sigma_p \qquad (2)$$

For high-pressure plasma ($v_m \ll \omega$), the plasma conductivity σ_p is

$$\sigma_p = \frac{\varepsilon_0 \omega_p^2 v_m}{\omega^2 + v_m^2} \approx \frac{\varepsilon_0 \omega_p^2}{v_m} = \frac{e^2 n_e}{m_e v_m} \qquad (3)$$

The electron number density is then determined by the normalized width of the resonant spectroscopy

when the probe is immersed in open air and plasma respectively [7].

$$n_e = 1.48 \times 10^{17} \frac{v_m}{l}\left[\left(\frac{\Delta f}{f_r}\right)_p - \left(\frac{\Delta f}{f_r}\right)_0\right] \qquad (4)$$

Where the electron-neutral momentum-transfer collision frequency v_m in plasma is as

$$v_m = \frac{p}{kT_g}\pi r^2 \sqrt{8kT_e/\pi m_e} = 1.809 T_e^{1/2}\, p/T_g \qquad (5)$$

The electron temperature T_e is in unit eV, and estimated by OES as described flowing. The atmospheric pressure $p = 1.0133 \times 10^5$Pa, and the gas temperature T_g in Kelvin.

The optical emission spectrum of rf-APPJ is collected by an optical fiber probe connected to a spectrometer (Ocean Optics, USB 2000, USA). The recorded spectrum is shown in figure 3. The electron temperature of the rf-APPJ was estimated to be $T_e = 0.161$ eV according to the reference [3].

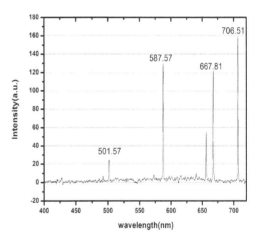

Figure 3.   Optical emission spectra of the helium rf-APPJ.

3   RESULTS AND DISCUSSION

Figure 4 shows the reflection coefficient versus frequency for helium rf-APPJ under the applied powers of 0, 50, 70, 90, 110W, respectively. The flow rate is 30 SLM. The results show that the resonant frequency does not change with the applied power, and are the same as the resonance frequency in the air, in other words the plasma parameters do not affect the resonance frequency 3.38GHz, which coincides with formula (1) with $\varepsilon_{pr} \sim 1$ for atmospheric pressure plasmas. However, the effect of the applied power on FWHM was observed clearly. From Figure 4, it is shown that the depth of ripple zero is at –35.2 dB for no applied power (0W), and decreases with input power to a value of –23.01dB at 110W. It is also found that the FWHM becomes wider (from 0.049GHz at 0W to 0.058GHz at 110W) with increasing applied power. This is consistent with the formula (3), and demonstrated by electromagnetic field finite difference time domain (FDTD) simulation [7] and the newly report [8].

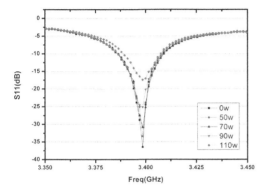

Figure 4. Reflection coefficient versus frequency for helium rf-APPJ at different applied powers.

From the resonant frequency and FWHM at the power of 0W and 110W in figure 4, the electron density is estimated by formula (4) to be 2.832×10^{15} m^{-3} at the RF power of 110W.

Figure 5 shows that the reflection coefficient versus frequency for rf-APPJ in feed gas of helium and oxygen mixture at the rf power of 65W. The result shows that when molecular oxygen is added from 0 to 40 sccm, the magnitude of the depth of ripple increases. As oxygen is electronegative gas, it will absorb the electrons in plasma.

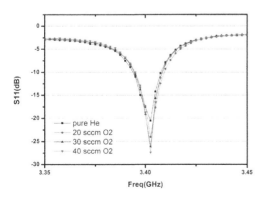

Figure 5. Frequency response to the helium rf-APPJ when admixture molecular oxygen.

4 CONCLUSION

Through experiment and analysis, we can find the resonator frequency for APPJ does not change and the full width at half maximum to be affected by applying power and the amount of oxygen in the feed gas. With the applied power increasing, the FWHM becomes wider due to the collisional absorption of microwave energy in higher electron density, which is consistent with the theory. As molecular oxygen admixture to the helium plasma, the loss of electrons reduces the microwave energy absorption. This work demonstrated that hairpin probe could be applied for monitoring atmospheric pressure plasmas. By combing with OES, we can also obtain approximately the plasma density.

ACKNOWLEDGMENTS

This work was supported by the National Nature Science Foundation of China (No.11075033).

REFERENCES

[1] Park, G.Y., Park, S.J & Choi, M.Y. et al. 2012. Atmospheric-pressure plasma sources for biomedical applications, *Plasma Sources Science and Technology*, 21:043001.
[2] Kohel, J.M., Su, L.K. & Clemens, N.T. et al. 1999. Emission spectroscopic measurements and analysis of a pulsed plasma jet, IEEE Transactions on Magnetics, 35:201–206.
[3] Xiong, Q., Nikiforov, A.Y. & González, M.Á. et al. 2013. Characterization of an atmospheric helium plasma jet by relative and absolute optical emission spectroscopy, Plasma Sources Science and Technology, 22:015011.
[4] Nowling, G.R., Babayan, S.E. & Jankovic, V. et al. 2002. Remote plasma-enhanced chemical vapour deposition of silicon nitride at atmospheric pressure, Plasma Sources Science and Technology, 11:97–103.
[5] Sands, B.L., Siefert, N.S. & Ganguly, B.N. 2007. Design and measurement considerations of hairpin resonator probes for determining electron number density in collisional plasmas, Plasma Sources Science and Technology, 16:716–725.
[6] Piejak, R.B., Al-Kuzee, J. & Braithwaite, N.S.J. 2005. Hairpin resonator probe measurements in RF plasmas, Plasma Sources Science and Technology, 14:734–743.
[7] Xu, J.Z., Nakamura, K. & Zhang, Q. et al. 2009. Simulation of resistive microwave resonator probe for high-pressure plasma diagnostics, Plasma Sources Science and Technology, 18:045009.
[8] Law, V.J., Daniels, S. & Walsh, J.L. et al. 2010. Non-invasive VHF monitoring of low-temperature atmospheric pressure plasma, Plasma Sources Science and Technology, 19:034008.

Testing and Measurement: Techniques and Applications – Chan (Ed.)
© 2015 Taylor & Francis Group, London, ISBN: 978-1-138-02812-8

The influence of the oblique shot to localization precision and the method to revise

S.L. Wu, J.X. Liu & L. Li
Faculty of Mechanical Engineering, Xi Jing University, Xi' an, China

ABSTRACT: The traditional acoustic target of the shot position request that the ballistic trajectory should plumb the target surface. In some occasion the ballistic trajectory does not always plumb the target surface, so the influence of the oblique shot should be considered. Based on aerodynamic theory and the characteristics of the Shockwave around a supersonic projectile, the influence of the oblique shot on localization precision is analyzed, then a circular array is designed to revise the error brought by the oblique shot, and the error correction expression is given according to the geometrical relation of the circularity. Finally, the error correction expression is proved to be feasible through numerical calculation, and in order to get close to the real angle of incidence, the numerical calculation result is fitted. The results show that the method is feasible.

KEYWORDS: Oblique Shot, Circular Array, Shockwave.

1 INTRODUCTION

The reached times difference between ballistic waves detected by arranging transducers is generally used in precision target positioning system, combining the geometric relationships of the transducers to determine the impact point coordinates of the target surface. It has been the critical equipment for scientific research, production and firing range acceptance check. Up to now, there are two kinds of acoustic target positioning system, the bar-type and the dot matrix. The former is of much more precision and the other is easily used to be suitable for a large area target surface. Geometric relationships of transducers' positions in space are usually used to drive the location functions in positioning with traditional planar arrays. The functions are valid when the projectile is vertical incidence to the target surface [2]. In some shooting situations, however, the direction of projectile incidence is not always definitely vertical to the target surface. Therefore, the influence of the oblique shot to localization precision is necessary to be considered in practice. Particularly, the effect brought by oblique incident should be taken into account when the ballistic trajectory is used in positioning the impact point coordinates. In the condition of oblique shot, the produced error in positioning accuracy should be corrected when the traditional planar transducer array is utilized. In this paper, a circular transducer array is designed and the correcting methods for the errors produced by oblique incident is presented.

2 POSITIONING ERROR ANALYSIS FOR OBLIQUE INCIDENCE

Because of the kinetic energy transfer of a high-speed flight, bullet in the air, a detached shock wave will be produced around the bullet head, which is a trajectory shock wave H. In a case, where the bullet speed exceeds the speed of sound in the current medium, the molecules become highly compressed and bend around the bullet. As a consequence, a very sharp shock wave front is formed. The phenomenon is illustrated in Figure 1, where a cone-shaped shock wave pattern forms behind the bullet. The angle of the shock wave front with respect to the bullet trajectory is called a Mach-angle, μ. Generally, the shock wave shapes its pattern as a straight line alone its outside profile, while bending in the head nearby.

When the bullet vertically access to the target surface (the incident angle is 90 degrees), the area detected by transducers array, the incident angle has no effect on positioning accuracy. While it access to the target surface at an angle α, the incident angle will proine a certain error for positioning [5]. As is shown in Figure 2, the incident angle α is the angle between the coordinate axis X and trajectory direction. As shown

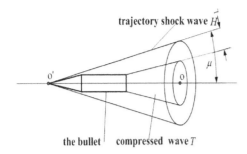

Figure 1.   The pill aerodynamics model.

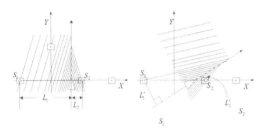

Figure 2.   The projection drawing of the shock wave sweeps the sensor array (left: shoot plumb, right: shoot at α).

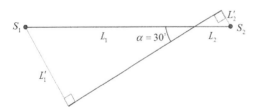

Figure 3.   The projection drawing of the shock wave.

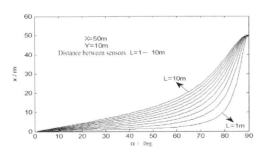

Figure 4.   The incidence angle's influence to the coordinates (3 sensors array).

in figures 2 and 3, the projection drawings when the shock wave sweeps the sensor array in angle α.

The distance from sound source to transducer S_1 could be calculated as function (1).

$$L_1' = L_1 \sin \alpha \qquad (1)$$

Where, L_1 is the distance from sound to transducer S_1 when the bullet vertically access. α is the angle between the incident direction and the target surface.

As a result, transmission route of trajectory shock wave is different between vertical incidence and oblique incidence. The effect of oblique incidence on positioning accuracy based on time delay should be considered.

As is shown in figure 4, the three sensor position array is used to calculate the effects produced by the incident angle α to the coordinate x. The family curves show the results as change of distance between transducers L=1-10m. The coordinates of transducers S_1, S_2, S_3 are $(-L,0)$, $(L,0)$, $(0,2L)$, respectively.

It can be derived from figure 4 that increasing the distance L can decrease the error produced by oblique incidence. The effectiveness is not obvious yet. It is necessary to design a method for correcting the error produced by oblique incidence.

3   ERROR CORRECTING METHOD FOR OBLIQUE INCIDENCE

A new sensor array is designed for correcting positioning error produced by oblique incidence, in which a circular and homogeneously arranged array is used. Its principle is illustrated in figure 5, in which the vertical situation is shown.

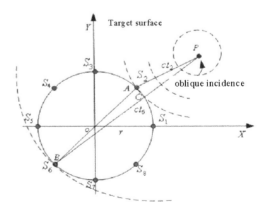

Figure 5.   The circular array to revise the influence of the oblique shot.

As shown in figure 5, a rectangular coordinate system is set and the original point is located at the center of the circle. Transducer S_2 is set on point A, S_6 on B, and C is the intersection point of PB and the circle. The radius of the circle is r and N transducers $S_1, S_2, \cdots, S_N$ (8 transducers in figure 5) are arranged homogeneously alone the circle. The coordinates of transducer position can be obtained as followings:

$$x_i = r\cos((i-1)\times\theta)$$
$$y_i = r\sin((i-1)\times\theta) \tag{2}$$

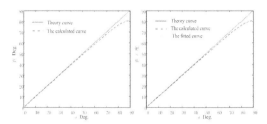

Figure 6. The contrast of the fitted curve (left) to the count (right) and the true curve.

Where, $i = 1, \cdots, N$, $i = 1, \cdots, N$

Suppose that the shock wave from impact point P arrives firstly at transducer S_2 in the situation of vertical shot, which means the arriving time to S_2 is the shortest. Then $t_{min} = t_2$, and the time shock wave arrives transducer S_6 from impact point P is the maximum, $t_{max} = t_6$. It can be derived that PA and PC, AB and BC are nearly equal to each other. So an approximate condition can be introduced as a function (3).

$$\begin{cases} PA \approx PC \\ AB \approx BC = 2r \end{cases} \tag{3}$$

Then $2r = AB \approx BC = PB - PC \approx PB - PA = c(t_6 - t_2) = c(t_{max} - t_{min}) = c\tau$ and $2r \approx c(t_{max} - t_{min}) = c\tau$.

As the bullet is put in an angle of α into target surface, the time from impact point P to the transducer S_2 is $t_2' = t_2 \times \sin\alpha = t_{min} \times \sin\alpha$ while the arriving time to S_6 is $t_6' = t_6 \times \sin\alpha = t_{max} \times \sin\alpha$. The time t_2' and t_6' are absolute values received by transducers S_2 and S_6 respectively. Because $\tau' = t_6' - t_2'$ is known value through the measuring system, it can be derived out that $c\tau' = c\tau \times \sin\alpha = 2r \times \sin\alpha$ as analyzed from figure 3.

Therefore,

$$\alpha = \arcsin(\frac{c\tau'}{2r}) \tag{4}$$

And the approximate formula $\sin\alpha = \dfrac{c\tau'}{2r}$

The approximate value of α calculated by the formula has error to some extent to the actual one. As is shown in figure 6, the diagrams are the relationship between the approximate angle calculated by the formula (4) and the actual one, in which incidence point is fixed and suppose that the angle between bullet and target surface changes from $1° \sim 90$, $r = 1m$ and sound velocity $c = 340 m/s$. The solid line represents the theory value and the dashed curve is calculated by the formula (5). From figure 6, it can be derived that the calculated incident angle is almost equal to the actual value as the incidence angle less than, while the actual angle is a little larger than the calculated as the incidence angle is larger than 70. The results are almost the same if the different incidence point is used.

The incident angle calculated through the approximate formula should be correct in order to reduce the error. The relationships shown as figure 6 are fitted, in which polynomial function is used to fit the curves. The relationships between incident angles are calculated from a circle transducer array and the corrected angles.

$$\alpha' = 1.0282 \times \alpha - 0.5984 \tag{5}$$

Where, α' represents the corrected angle and α is the calculated one. As is shown in figure 6 (right) is the contrast of the fitted curve to the calculated curve. It is clear that the fitted curves are much more approaching to the theoretical curves, which means the error is smaller and more acceptable.

4 CONCLUSIONS

In this paper, the effects of oblique shot on positioning accuracy have been analyzed in detail.

The travel path of ballistic trajectory is quite different from the situation of vertical incidence. The effects of oblique shot should be considered when the time delay is used to locate the target. A new circular transducer array is designed to correct the error produced by the oblique shot. The correcting method can be summarized as follows, three steps.

1 Take the maximum time t_{max} and minimum time t_{min} among the times received by transducers set on the circle. Then calculate the time delay $\tau' = t_{max} - t_{min}$.
2 Calculate the approximate incident angle α according to the function (4).
3 Correct the incident angle α calculated as step (2) into α' by fitted function (5). The α' acts as the final bullet incident angle, which is used to correct the error produced by oblique shot.

ACKNOWLEDGMENTS

This paper is financially supported by the Major Project for Edu. And Teaching Reform (ShannXi Prov.) "Reform of the Teaching Content and Course System for Training Skill & Application-Oriented Talent of Mechanical Manufacture Specialty" No.11BY99.

REFERENCE

[1] Xiao F, Li H C. Sound, Weapons and Measurements [M]. BeiJing: National Defence Industry Press, 2002.
[2] E. Danicki. The shock wave-based acoustic sniper localization[J]. Nonlinear Analysis. 2006, 65(5): 956–962.
[3] Satoshi Kagami, Hiroshi Mizoguchi. Microphone array for 2D sound localization and capture[J]. Springer Tracts in Advanced Robotics, 2006, 21(1): 45–54.
[4] Zhang F M, Ma C M. Accuracy Analysis for the Measuring System of Target Deviation for Projectiles Shooting an Acoustic Target[J]. ACTA ARMAMENTARII, 2000, 21(1): 23–26.
[5] Mumolo, Enzo. Algorithms for acoustic localization based on microphone array in service robotics [J]. Robotics and Autonomous Systems, 2003, 42(2): 69–88.
[6] ZHANG W P, WANG W C. The Analyze of Passive Acoustic Location in Plane by Random Shape Array of Three Sensor[J]. Journal of Detection & Control, 2003.9, 25(3):54–57.
[7] SUN S X, GU X H, SUN X Study on Location of Acoustic Target with a Rectangular Pyramid Array[J]. Applied Acoustics, 2006.3, 25(2): 102–107.

Testing and Measurement: Techniques and Applications – Chan (Ed.)
© 2015 Taylor & Francis Group, London, ISBN: 978-1-138-02812-8

Three novel scenarios for coverage drive tests in WIMAX

J.D. Chimeh
Iran Telecommunication Research Center

A. Asgharian & M.A. Isfahani
Communications Regulatory Authority (CRA) of Iran

ABSTRACT: To verify the mobile operators' performance, regulations should measure network's coverage and Quality of Service (QoS) parameters like RSSI, CINR, Jitter. Thus, measuring these parameters is a main disturbance for regulatory. These disturbances are due to the test accuracies and both their time and cost effectiveness. Totally, drive test includes a vehicle, a laptop, measuring software and a modem (of indoor/outdoor/ USB type). In this paper, we have gathered new important equations for drive test and introduced three new field test methods and compared them with the existing ones. It showed our methods are time and cost effective in all cellular mobile networks and are applicable in WIMAX networks.

1 INTRODUCTION

Field test is accomplished in two manners as drive tests and fixed point tests. Tools establishes in a moving vehicle in the drive test while measuring is going on. In the fixed point test we locate tools in a still point before accomplishing the test. We may divide the tests into coverage and QoS tests branches. RSSI defines regional coverage and CINR, Jitter, throughput, etc. define QoS.

We surveyed field tests in the eleven countries including Italy, Germany, Portugal, USA, Norway, Finland, Taiwan, Malaysia, Tanzania, Bahrain, and Saudi Arabia. Although they use different tools to test, but all of them accomplish tests in the fixed and drive test methods. In these countries drive tests and fixed tests have been accomplished irregularly without any specific strategy in them (Betzaya, 1011 to Teletopix, 1978)). In Portugal we found only a drive test that has been accomplished circularly around the BTS (Eliamani, 2010). On the whole none of them have any systematic method or strategy with the aim of saving time and cost that is our aim in this paper.

XCAL, XCAP, Epitiro, etc. are some of the measurement tools which may be used for coverage and QoS measurement depending on the regulatory demand. To accomplish definite and precise coverage measurement, we showed test intervals should be defined properly in a big city. Besides, the number of test points and their distances ought to be defined. Thus, in section II we study our test procedures and the number of windows or bins. Then, using a suitable statistical distribution, we compute the number of test points in each bin. In the end, we introduce three practical methods in the field measurement.

2 DEFINING SPECIFIC MEASUREMENT METHODS

Measuring method includes all practical and theoretical methods to evaluate coverage in the drive test. Here, we consider some measuring method and their practical processes. Drive test has been done for two reasons: firstly in order to test operator's coverage in a city which may confirm its coverage undertaking and secondly in order to test special or blind points. The latter may occur as a result of the complaint of customers. To accomplish drive test in a city we must first divide its region into smaller zones.

Figure 1. Dummy windows and regional coverage in a city.

Practically, we divide the region into windows or bins and accomplish fixed tests or drive tests in them (Fig. 1).

Thus, to accomplish the test, we should divide the whole region into distinct windows and make driving or fixed tests in them. Since coordination of the points are at hand through a digital map, we can get the coverage map from the operator and make tests in the green areas which are the areas which have suitable coverage. Here we continue to define the number of windows, the length of each window, the number of test points and their distances in each window as follow:

2.1 Windowing

Taking into account the channel fading and shadowing to evaluate the window size x we find the relation: in which λ is the carrier wavelength (Teletopix, 1978).

$$40\lambda < x < 1500 \tag{1}$$

Assuming s is the coverage area, we may find the number of windows in the area N as:

$$N = \frac{S}{x^2} \tag{2}$$

Assuming x = 1500m and also s = 727km^2 for Tehran we may find the number of windows for Tehran as:

$$N = \frac{R^2}{x^2} = \left(\frac{27}{1.5}\right)^2 = 324 \tag{3}$$

3 DRIVE TEST SCENARIOS FOR COVERAGE EVALUATION

The driving test is easier and faster than fixed point test, thus we prefer to accomplish drive test. In addition, we may accomplish it more efficient than fixed point test. We accomplish our test due to the operator's coverage map and verify its honesty. We explain four the capturing method here, which all are applicable in all wireless networks. First of them is applicable in the fixed point method. It is a customary method which has been used before as well. Three other methods are our new introduced methods. We describe each one as follow:

3.1 First method: fixed point method

This is a customary method which is used by the regulators. After windowing the coverage area and determining the coordinates of the center of them, regulator asks the contractor to measure the coverage

parameters only in those points, i.e., in this method contractor only captures one sample in the middle of each window (Fig. 2).

Although only one measurement is accomplished in each window in this method, but the test takes too long time in the big cities like Tehran because of the too many windows. This phenomenon results in wasted time and investment. It should be noted that the precision of this method grows as the number of samples in each window increases.

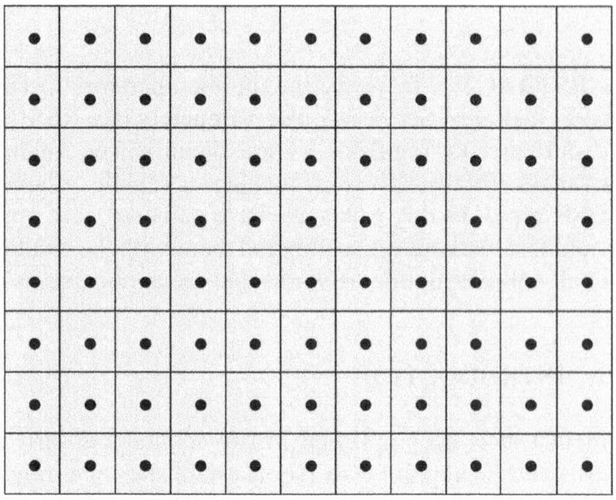

Figure 2.   Current test method.

3.2 New introduced drive tests

Since cities should have full coverage, urban and suburban wireless cells should constitute integrated and connected networks. This is feasible through directional / omnidirectional antennas which mostly haven't narrow beams and their beams may also overlap. Thus, at a first glance, we may assume that the whole of the network has enough coverage except for a special points.

Therefore, it is not necessary to test all the windows one by one and if the borders of the network have full coverage, we may conclude that the interior points have also suitable coverage. Using this idea, we propose three time and cost effective methods to test the coverage of a mobile wireless network (Chimeh, 2013).

3.2.1 Second method: linear method

If border paths of the cities are linearly surveyed and have suitable coverage over an integrated and connected region, we can conclude that the areas between the paths have also suitable coverage. This property is because of the wide and connected network coverage in the cities. However, we can test some points between paths if necessary. In this method a test vehicle with 5–30 km speed moves straightly on the paths on the border of a region while testing. Regulations may provide these paths and dictate them to the contractor (Fig. 3).

As is seen when the vehicle moves linearly, it traverses 11+11+11+10=43 windows. Assuming that the total number of windows is 88, the contractor may save 88−43=45 windows while it is accomplishing drive test and it is the benefit of this method. Thus we may save time and cost in this method. The percent of saving is 28/88=%51. We may accomplish additional fixed tests on the important windows that include important cites (refer to two points in Fig. 3).

A region

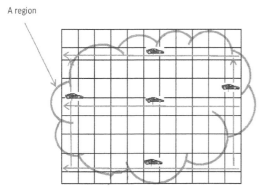

Figure 3. Straight paths are surveyed by the test vehicle.

3.2.2 Third method: circular-linear method
This method is also based on this fact that if the borders of a region have enough coverage its interior points also have suitable coverage. Thus, we move around the region circularly while capturing the samples in this method. Besides, the vehicle moves in two linear directions as cross diagonal while capturing the samples (Fig. 4).

A region/
Path

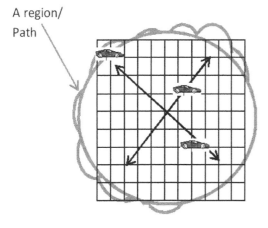

Figure 4. Circular-linear method.

As is seen when the vehicle moves circularly, it traverse 28 windows and when it moves linearly, it traverse 18 windows, totally vehicle traverses 45 windows. Thus, the contractor may save 88−46=42 windows while it is accomplishing drive test. The percent of saving is 28/88=%36.

3.2.3 Fourth methods: concentric-circles method
In line with the previous idea we use concentric circles in this method. Figure 5 depicts 3 consecutive circles that cover whole of the region.

If the outer circle has a radius of R the interior circles may have a radius of R/2 and R/4 respectively. As is seen in the case of 3 circles, the total numbers of windows are 28+20+12=60. Thus 88–60=28 windows have been saved in this method. The percent of saving is 28/88=%32. We should note that we may use two concentric circles in this method.

A region/
Path

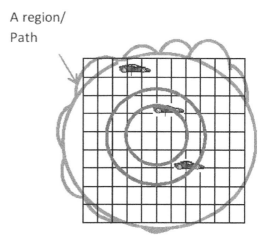

Figure 5. Concentric circles method.

Note 1: we may accomplish additional fixed point test on some of the windows that drive test is not done before in them. Besides, we may make additional tests in some important and critical windows e.g. governorship.

Note 2: Drive test is more precise than the fixed point test since in the former the software captures more than one sample and makes average over all of them automatically.

4 DOFTWARE REQUIREMENT IN DRIVE TEST

Provided that the test tool has enough capability (capturing rate) we may accomplish drive test properly. We should note that sampling rate may vary respective to the vehicle speed. Maximum speed is limited by the tool's sampling rate and the length of the window. In an actual scenario for 1500m window size and four test points and 30Km/h speed of the vehicle (8.3m/s) we get sampling rate n2 as:

$$\frac{sample}{sec} = \frac{sample}{m} * \frac{m}{sec}$$

and

$$n1 = \frac{4}{1500}\left[\frac{m}{sample}\right] = 0.0027 \ m \,/\, sample \qquad (4)$$

Thus, the least sampling rate is:

$$n2 = n1.v = 8.3*0.0027 = 0.02 \ \frac{sample}{sec} \qquad (5)$$

i.e., it is only necessary for the tools to capture a sample during 50 seconds. For example, since TEMS tool is capable to capture a sample in each second we can use this tool in this scenario very well.

5 ANALYSIS

Since the first method has the least efficiency, we abandon it and only compare the next three methods from the saving and precision points of view. The time and cost saving parameters are compared in each method and depicted in figure 6. It shows that the second method has the highest saving and the fourth method has the least saving. Besides, since in the fourth method the vehicle turns circularly in the internal regions and covers all the cell's directions it has the most precision. The second method has the least saving and precision.

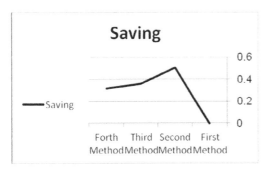

Figure 6.   Comparison of the time and cost saving.

6 CONCLUSION

We showed that the drive test method saves test time and cost relative to the fixed test since we make use of a vehicle and a test tool in a drive test that captures samples automatically by a suitable rate. Drive test speed is also high enough relative to the fixed test method. We showed our methods are time and cost effective in all cellular mobile networks. We could reach to 51% time and cost saving in site surveying. We derived new equations for the drive test and introduced three new fixed test methods and compared them with the existing ones.

REFERENCES

[1] BatzayaGanbat 2011. *Path Loss Measurement for WiMAX in Campus Environment*. In Master Thesis.
[2] Chimeh. J. D. 2013. Three novel time and cost effective evaluation coverage drive test of mobile networks by Regulatory. Wireless Personal Communication.
[3] EliamaniSedoyekaZiadHunaiti, Daniel Tairo. 2010. Evaluation ofWiMAXQoS in a Developing Country's Environment. Computer Systems and Applications (AICCSA), 2010 IEEE/ACS International Conference on.
[4] FuraihAlshaalan, SalehAlshebeili, AbdulkareemAdinoyi. 2010. On the Performance of Mobile WiMAX System: Measurement and Propagation Studies, Int. J. Communications, Network and System Sciences, Vol 3, PP. 863–869.
[5] RohaniBakar, 2012. Muhammad Ibrahim & D.M. Ali, Performance Measurement of VoIP over WiMAX 4G Network. 2012 IEEE 8th International Colloquium on Signal Processing and its Applications.
[6] http://www.tra.org.bh/en/pdf/Broadband_Report_Q1_2011_F.pdf.http://www.teletopix.org/gsm/what-and-why-drive-test-in-rf-networkistics, Edward Arnold Publisher, London, 1978.

Testing and Measurement: Techniques and Applications – Chan (Ed.)
© 2015 Taylor & Francis Group, London, ISBN: 978-1-138-02812-8

For-EMD regularized deconvolution in acoustic measurement

Y. Dai, J.H. Yang & X.H. Cao
School of Automation, Northwestern Polytechnical University, Xi'an, China

H. Hou
School of Marine Engineering, Northwestern Polytechnical University, Xi'an, China

ABSTRACT: This paper presents a hybrid Fourier-EMD Regularized Deconvolution (For-EMDRD) algorithm. When the impulse response is estimated, the Fourier shrinkage and EMD shrinkage follow the ideal deconvolution result. Simulated and experimental results show that For-EMDRD not only avoidsill-posed problem in deconvolution, but also improves signal edge character and Signal-to-Noise Ratio (SNR). For-EMDRD can be applied in signal estimation of acoustic wideband pulse measurement, removing the effect of transient response of a loudspeaker, eliminating the noise effect, which makes a better separation of the incident wave and reflected wave and improving the measurement accuracy of absorption coefficient.

1 INTRODUCTION

Echo-pulse method is an accurate acoustic measurement method, whose principle is to generate a wideband frequency sound pulse, to separate the incident wave and reflected wave, and to measure all the acoustic parameters in frequency range. Inverse filter was used to generate a wideband sound pulse and eliminate the transient effect, measuring the absorption coefficient in full range (Sun et al. 2009, Sun et al. 2010). That method was proved to be simple and reliable.

When we use the echo-pulse method, the signal received by the microphone is the convolution of echo-pulse signal, the impulse response of the loudspeaker and the transfer function of pipe. If we want to acquire the real echo-pulse signal, deconvolution is needed. Deconvolution, is also called inverse filter, which can be defined as a solution to the inverse problem of convolving an input signal $x(n)$ and impulse response $h(n)$. The model can be interpreted as (Norcross et al. 2004):

$$y(n) = x(n) * h(n) + \eta(n) \qquad (1)$$

Where $y(n)$ is the measured signal, $*$ denotes the convolution operation and $\eta(n)$ is zero-mean additive white Gaussian noise (AWGN). This approach is based on the concept that one can "undo" the filtering caused by the measured impulse response (IR) with its deconvolution. This approach strives to correct both the magnitude and the phase of the system. In the frequency domain, Eq. (1) can be expressed as:

$$Y(f) = X(f) * H(f) + N(f) \qquad (2)$$

If the frequency response $H(f)$ has no zeros, an estimation of $x(n)$ can be obtained by using an ideal deconvolution. However, when $H(f)$ has values closer to zero, the noise is highly amplified, thus leads to incorrect estimates. In such cases, it is necessary to include a regularization parameter in the deconvolution which reduces the variance of the estimated signal. The most used regularized deconvolution for stationary signal is the Wiener deconvolution.

Wiener deconvolution can be deemed as a type of shrinkage of ideal deconvolution in Fourier domain. The problem is that the Fourier domain does not economically represent signals with singularities. Neelamani et al. proposed a wavelet-based regularized deconvolution (For-WaRD) technique to solve the deconvolution process (Neelamani et al 2004). For-WaRD solves the problem of signals with singularities. However, the wavelet transforming depends on the wavelet base function which is hard to choose. Different wavelet base functions may lead to different results.

EMD is a new type of time-frequency processing method which was addressed by Huang (Huang et al. 1988). The EMD method breaks signals down into a number of amplitude and frequency modulated (AM/FM) zero-mean signals, termed intrinsic mode functions (IMFs). Recently, a lot of denoising methods based on EMD were introduced (Tsolis et al. 2011,Khaldi et al. 2008,Kopsinis et al. 2009). Inspired by the For-WaRD and EMD denoising, we come up with the Fourier-EMD, which proceeds with shrinkage of ideal deconvolution in EMD domain. This method improves the details of signals after Wiener deconvolution, increasing SNR and resolution of estimated signals. The method is self-adaptive and robust because it doesn't require any regularization parameters setting.

2 FOR-EMDRD ALGORITHM

2.1 *Wiener deconvolution*

Once the impulse response $H(f)$ is estimated, Wiener deconvolution can be expressed as follows

$$\hat{X}(f) = \frac{Y(f)H^*(f)}{\left|H(f)\right|^2 + S_n(f)/S_s(f)} \tag{3}$$

Where $S_n(f)$ denotes the power spectrum of the noise, and $S_s(f)$ denotes the power spectrum of the estimated signals. When the SNR is very large, that means $S_n(f)/S_s(f) \approx 0$, Eq.(3) becomes $\hat{X}(f) = \frac{Y(f)}{H(f)}$, which is called ideal deconvolution. However, lack of prior information of signals and noise makes it impossible to get the value of $S_n(f)/S_s(f)$. Farhang Honarvar sets it as $max(\left|H(f)\right|^2)/100$ (Honarvar et al. 2004). Instead of this approach, we use a median absolution deviation (MAD) estimator, which estimates the noise variance on the intrinsic mode functions (IMF).

2.2 *Fourier-EMD regularization deconvolution*

In prior to Fourier-EMDRD process, we should know the information about the impulse response (IR) of the system. Here, MLS (Maximum Length Sequence) is used to measure and estimate the IR of the system. EMD procedure is beyond the scope of this paper, so we will not discuss here. For-EMDRD can be summarized as:

1 Noise variance estimation
   The observed signal is decomposed into IMF_s. Noise variance for the first IMF_1 is estimated using the estimator as (Khaldi et al. 2008) :

$$\sigma_1 = 1.4826 \times Median\left\{\left|IMF_1(t) - median\{IMF_1(t)\}\right|\right\} \tag{4}$$

According to the relationship among the IMF_s of Gaussian white noise, the rest levels of IMF_s can be calculated as (Kopsinis et al. 2009):

$$\sigma_k = \frac{\sigma_1}{\sqrt{2}^{k-1}}\ k \ge 2 \tag{5}$$

2 The regularization parameter τ for Fourier shrinkage was calculated in the range

$[0.01 \sim 10] N\sigma_n^2 / \left|y - \mu_y\right|^2$, where N is the length of y, and μ_y is its mean. The regularization parameter τ is chosen to achieve the best results in the mean square error (MSE) sense (Henry et al. 2006).

3 Calculating the Tikhonov shrinkage in Fourier domain.

$$X_{\lambda_1}(f) = \frac{Y(f)}{H(f)}\left[\frac{\left|H(f)\right|^2}{\left|H(f)\right|^2 + \tau}\right] \tag{6}$$

4 Obtaining the first estimation of the input signal through an inverse Fourier transformation (IFT).

$$x_{\lambda_1}(t) = IFFT\left\{X_{\lambda_1}(f)\right\} \tag{7}$$

5 Doing EMD with the first estimated signal for two times. The first one is used to estimate the IMF shrinkage parameter, while the second is used for denoising IMF shrinkage.
   The hard threshold of shrinkage parameter can be expressed:

$$IMF_j = \begin{cases} 1, & if\ \left|IMF_j\right| \ge \rho_j\sigma_j \\ 0, & f\ \left|IMF_j\right| < \rho_j\sigma_j \end{cases} \tag{8}$$

where $\rho_j = \sqrt{2\ln N}$, σ_j is the noise standard deviation of IMF_j.

6 Calculating the parameter for EMD shrinkage.

$$\lambda_j = \frac{|IMF|_h}{|IMF|_h + \sigma_j^2} \tag{9}$$

7 Making a EMD shrinkage.

$$IMF_{sh} = IMF_2\lambda_j \tag{10}$$

8 Adding all the IMF_s with shrinkage to obtain the final estimation of the input signal.

3 SIMULATION RESULTS

In order to identify the algorithm, we used the simulated data to compare different methods. The input signal is $x(n) = 2.1e^{-0.5((n-70)/5)^2} - 0.5e^{-0.5((n-70)/5)^2}(130-n)$. The impulse response is $h(n) = n0.9^n \sin(0.2\pi n)$. Accordingly, the output is $y(n) = x(n) * h(n) + \eta(n)$, and $*$ denotes convolution operation.

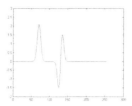

Figure 1.   Input signal.

Figure 2.   Observed signal SNR=44.62.

Ideal deconvolution, Wiener deconvolution and Fourier-EMDRD were adopted to estimate the observed signal $y(n)$. Estimation results are shown as:

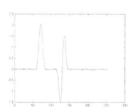

Figure 3.   Fourier-EMDRD.

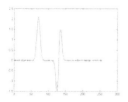

Figure 4.   Wiener deconvolution (regularization parameter is 0.01).

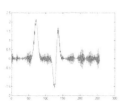

Figure 5.   Ideal deconvolution (no regularization).

As we can see from these figures, ideal deconvolution is almost failed. Wiener deconvolution is closer to Fourier-EMDRD. However, when we set different parameters for Wiener regularization, results differ a lot. If improper parameters are set, the deconvolution may be worse. In contrast to Wiener regularization, Fourier-EMDR doesn't need any setting of parameters, and the algorithm will adjust the regularization parameter according to the noise variance automatically. Results of different regularization parameters and different deconvolution methods are shown in Table 1, where the signal-to-noise ratio (SNR) of the estimation is used to quantify the quality of the estimation.

Table 1.   Results of different methods.

Deconvolution methods	Regularization parameter	SNR (dB)
Fourier-EMDRD	self-adaptive	26.60
Ideal deconvolution	0	7.12
Wiener deconvolution	0.01	25.85
Wiener deconvolution	0.1	23.18
Wiener deconvolution	1	8.72

From the above results, if the regularization is not used, then the Gaussian white noise may cause failure for deconvolution. Regularization for deconvolution can avoid the effect of noise to zero point of impulse response. However, deconvolution results are sensitive to regularization parameter. When the parameter is set to 0.01, Wiener deconvolution can yield the best result, nevertheless, it is still worse than Fourier-EMDRD.

4   EXPERIMENTAL RESULTS

4.1   *Experimental setup*

As shown in Fig.6, the experimental setup includes a computer, a NI-DAQ6062E data acquisition board, a BNC2029 adaptor, a power amplifier, a signal conditioner, a loudspeaker-pipe system and a 1/4in MPA416 condenser microphone. The loudspeaker is a 5in Hi-Vi mid-frequency woofer which is positioned in a column steel enclosure whose length is 15 cm. The sound wave from the loudspeaker is guided through a 20 cm conical reduction to a straight pipe 27mm in diameter. The microphone, which is flush mounted into the pipe wall and placed at 24cm from the open end of the tube, is used to measure the sound pressure of incident and reflected pulse. The sample is placed vertically at the end of the straight pipe, backed by a rigid steel cover. The AI and AO channels of the NI DAQ 6062E board are programmed to be synchronized and have the scanning and sampling rate of 100k Hz.

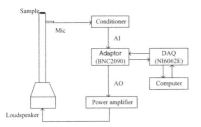

Figure 6. The sound source configuration and the test equipment.

4.2 *Fourier-EMDRD results*

Fourier-EMDRD can be applied in acoustic measurement to obtain the original signal that is blurred with the loudspeaker and white noise. Sound wave signals before and after Fourier-EMDRD are shown if Fig.8 and Fig.9. The incident wave and reflected wave can be separated from the estimated signal, and the reflection coefficient and absorption coefficient can be calculated(Sun et al. 2010).

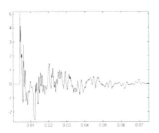

Figure 8. Wave signal observed by microphone.

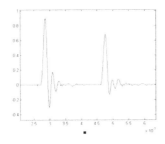

Figure 9. Incident wave and reflected wave estimated by For-EMDRD.

5 CONCLUSIONS

In this paper, a novel Fourier-EMD algorithm is presented whichexhibitsan enhanced performance compared to Wiener deconvolution in the case where the signal is blurred with Gaussian white noise and the system response is ill-posed. Some practical experiments are also carried out to identify the new technique. These preliminary results suggest further efforts for improvement of Fourier-EMD when SNR of signal is lower.

ACKNOWLEDGEMENT

This work was financially supported by the National Natural Science Foundation of China with GrantNo.11474230 andNo.11204242.

REFERENCES

[1] FarhangHonarvar, Hamid Sheikhzadeh, Michael Moles, Anthony N. Sinclair.Improving the time-resolution and signal-to-noise ratio ofultrasonic NDE signals. *Ultrasonics*, 2004, 41(9):755–763.

[2] George Tsolis and Thomas D. Xenos. Signal Denoising Using Empirical Mode Decomposition and Higher Order Statistics[J].*International Journal of Signal Processing, Image Processing and Pattern Recognition*, 2011, 4(2):91–106.

[3] HERRERA Roberto Henry, OROZCO Rubén, RODRIGUEZ Manuel. Wavelet-based deconvolution of ultrasonic signals innondestructive evaluation [J]. *Journal of Zhejiang University SCIENCE A*,2006, 7(10):1748–1756.

[4] Kais Khaldi, Abdel-OuahabBoudraa, Abdelkhalek Bouchikhi, andMoniaTurki-Hadj Alouane1.Speech Enhancement via EMD [J].*EURASIP Journal on Advances in Signal Processing*, 2008,2008 :1–8.

[5] N.E.Huang, Shen Z, S.R. Long, et al. The empirical mode decomposition and the Hilbert spectrum for nonlinear and non-stationary time series analysis. *ProcRsocLond*, 1988, 454:56–78

[6] Ramesh Neelamani, Hyeokho Choi, and Richard Baraniuk. Fourier-Wavelet Regularized Deconvolution for Ill-Conditioned Systems [J].*IEEE TRANSACTIONS ON SIGNAL PROCESSING*, 2004,52(2) 418–433.

[7] S. G. Norcross, G. A. Soulodre and M. C. Lavoie, Subjective Investigations of Inverse Filtering [J].*Journalofthe Audio Engineering Society*, 2004,52: 1003–1028.

[8] Sun liang, Houhong, Dong liying,Wanfangrong. Sound Absorption Measurement of Acoustical Material and Structure Using the EchoPulseMethod[C].*2009 International Conference on Computer Engineering and Technology*, 2009:281–285.

[9] Sun liang, Houhong, Dong liying,Wanfangrong. Sound Absorption Measurement in a CircularTube Using the Echo-Pulse Method [J].*ACTA ACUSTICA UNITED WITH ACUSTICA*, 2010, 96:1–4.

[10] YannisKopsinis, Stephen McLaughlin.Development of EMD-Based DenoisingMethodsInspired by Wavelet Thresholding [J].*IEEE TRANSACTIONS ON SIGNAL PROCESSING*, 2009,57(4):1351–1362.

Testing and Measurement: Techniques and Applications – Chan (Ed.)
© 2015 Taylor & Francis Group, London, ISBN: 978-1-138-02812-8

Acoustical analysis of a *Jing* based on beat phenomena

S.J. Cho
Automobile/Ship Electronics Convergence Centre, University of Ulsan, Nam-gu, Ulsan, Republic of Korea

K.H. Lee & Y.J. Seo
School of Electrical Engineering, University of Ulsan, Nam-gu, Ulsan, Republic of Korea

ABSTRACT: This paper presents acoustical characteristics of the *Jing*, Korean traditional percussion instrument. The analysis focuses on the beat phenomena according to various blow strength and drive point. The blow strength is classified into 'very strong', 'strong', and 'weak', and the drive points are specified into 'centre', 'up', and 'right'. The spectrogram is utilized to analyze beat phenomenon and the beat of the *Jing* can be categorized into early beat and late beat. Furthermore it is possible to figure out blowing at the side triggers diverse beats in broader frequency band compare to one at the centre. In addition, there are unique components that are around harmonics and affect early beat for the case of blowing at the centre.

1 INTRODUCTION

Recently, *Samulnori* is a world-famous musical performance played with four different percussion instrument. The *Jing* is a Korean traditional percussion instrument that generates a very unique sound, which is very soft, low tone gliding up, and is playing important roles in *Samulnori* (Sung 1994).

There are many kinds of metallic percussion instrument in the world: Chinese *Gong*, Balinese gamelan *gong*, and so on. Among them, the representative percussion, similar to the *Jing* is a Chinese *Gong* and many researches of that one have been reported (Rossing & Fletcher 1983, Fletcher 1985, McLachlan et al. 2012). The researches on Balinese gamelan *gong* have been also reported until recently (Krueger et al. 2010, Rerrin et al. 2014). However, it is rarely found literatures about *Jing*. Up to now, the representative research that covers acoustic, physical and material characteristics of the *Jing*; the metal percussion is done by the National Centre for Korean Traditional Performing Arts in 1999 (National Gugak Center 1999). This book includes the Jing and Korean small gong researches on vibration and sound analysis and materials used over manufacturing and its process (Lee et al. 1999). Apart from The National Centre for Korean Traditional Performing Arts, study seeks for elements determine Jing's sound characteristics had carried out years before. The study presented vibration mode and dominant frequency using planar acoustic holography and accelerometer. In result, rotation energy phenomena occur in certain frequency (Kwon et al. 1997) Later on, tone research about sound analysis depends on deepness change of rim (Sohn & Bae 2004). An EEMD (ensemble empirical mode decomposition) algorithm has introduced for detecting main frequency element (Kwon & Cho 2014). Therefore, this paper presents acoustical characteristics of the *Jing* is about beat phenomena caused by various blow strength and drive point.

2 JING, KOREAN PERCUSSION INSTRUMENT

Jing is a Korean traditional metallic percussion instrument which has a shallow spherical shell with a rim bent inward as shown in Figure 1. Intangible cultural asset brassware maker generally makes the Jing. The Jing is made of alloy with a ratio of copper and tin by 72:28. Entire forging process demands hundreds of hammering (Sung 1994). Table 1 shows size comparison of Jing used in references with one used in this paper. Manufacture of *Jing* is totally upon homemade, therefore it does not have an exact standard, but Table 1 could offer rough size.

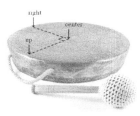

Figure 1. Shape of a *Jing*, Korean metallic percussion instrument.

Table 1. Measurement of the *Jing*.

	This	Reference*
Outer diameter (mm)	389	385
Inner diameter (mm)	359	342
Length of rim (mm)	93	87.4
Angle of rim (°)	83	76

*(National Gugak Centre, 2008).

3 ACOUSTICAL ANALYSIS

All sound data used in this paper are recorded in an anechoic chamber. The distance between *Jing* and microphone was set in 1 m. The height of the microphone and the middle point of the *Jing* were synchronized. The microphone and digital recorder used are c1000s from AKG and DA-P1 (sampling 44.1 kHz, noise level –60dB) from TASCAM. The blow strength was classified into 'very strong', 'strong', and 'weak'. The drive point was specified into 'centre', 'up', 'right' and all these information are given in Table 2.

Table 2. Targets and symbols.

		Symbol
Blow strength*	Very strong	VSTR
	Strong	CNTSTR
	Weak	WEAK
Drive point**	Right	RTS
	Centre	CNTSTR
	Up	UPS

* centre-drived, ** strong-striked.

To analyze *Jing* sound, spectrogram function from Matlab was used, which is able to freely adjust the resolution in time and frequency. The maximum frequency range is set to 1.5 kHz. The result shows over this range, after blow, components had a tendency of rapid decrease. Window sizes of the spectrogram set to 0.05 s, and overlap set to 90 %.

The fundamental frequency of the *Jing* used in this paper is 114 Hz and harmonics are precisely maintained integer multiple frequency. The beat of *Jing* appears to blow and exists till sound fades away. In this paper, beat show up with a blow and disappears rapidly will be named with 'early beat', beat from the middle of the sound till fade away call 'late beat'.

In Figure 2(a), early beat occurred over full band and mostly disappeared before 3 s. The late beat showed up around 3rd and 4th harmonic, 6 Hz for former and 3 Hz – 4 Hz for the latter. The beat in the higher frequency band is belong to early harmonic and 6 Hz – 12 Hz of beat has detected.

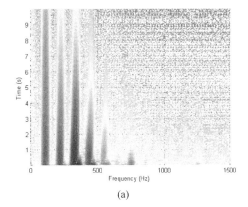

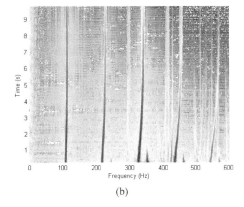

(a)

(b)

Figure 2. Spectrogram for (a) bear and (b) softening of the VSTR sound.

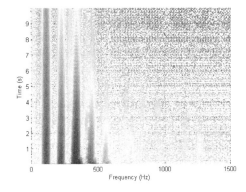

Figure 3. Spectrogram of the CNTSTR sound.

The beat in CNTSTR (Fig. 3) is alike one in VSTR. The spectrogram from VSTR in Fig. 1(b) shows the frequency components showed up for short with strong power, this also shown in CNTSTR. The components are 240 Hz, 351 Hz, 460 Hz and 572 Hz;

these mixed with from 2nd to 5th harmonic showed early beat. For those after blow, range 19 Hz – 20 Hz shows beat but with softening, till all components fade away, beat frequency has decreased.

The late beat of CNTSTR has tendency just like VSTR in area around 3rd and 4th harmonic. 4Hz from 338 Hz and 342 Hz has maintained most strong, 8 Hz from 408 Hz and 416 Hz maintained well till sound fades away. The beat in high band is early beat with range of 5 Hz – 11 Hz but mostly extinct within 3 s.

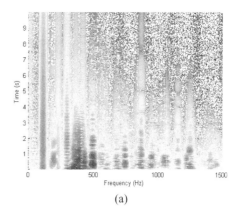

(a)

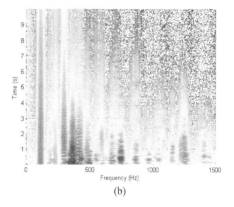

(b)

Figure 5. Spectrogram for (a) RTS sound and (b) UPS sound.

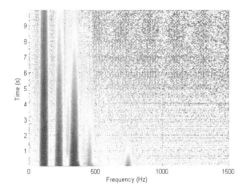

Figure 4. Spectrogram of the WEAK sound.

WEAK in Figure 4 noticeably decreased compared to VSTR and CNTSTR, however maintained beat with frequency range from 300Hz to 400Hz. Frequency has recorded 9Hz that is greater than VSTR and CNTSTR. The Beat after 700 Hz is barely found and ones detected were even lasted less than 1 s, this classified as early beat.

For RTS and UPS as shown in Figure 5(a) and (b), more beat has found when side had blown compare to middle. In general, beats mentioned above are shown over the whole frequency band after blow till 2 s and extinct rapidly but had tendency of remains some major components. These components include 351 Hz frequencies only lasted for 1 s when blown in the middle. However, when side had blown, components lasted about 7 s and also affected late beat. In high frequency component higher than 700 Hz showed components those affects late beat. This result is in contrast from center blow and which component higher than 700 Hz is classified as early beat and faded quickly. Furthermore, more than two beats were founded in the same component. For instance, 8 Hz appeared from 1155 Hz and 1163 Hz, 1.3Hz showed up from 1155 Hz and 1156.3 Hz at the same time. This phenomenon has much more commonly founded in higher frequency then in low frequency.

4 CONCLUSIONS

In this paper, the acoustical characteristics of the *Jing* has described by perspective of various blow strengths and drive points. The beat differentiated in early and late beat than analyzed. Especially looked up for characteristic of early beat. It was possible to find out blow at the side has more beat in the wider frequency band comparing to blow in the middle. Also, blow in middle triggers components around harmonic that affects early beat. However, more problems needed to be solved, thus continuous research is needed for scientific investigation.

ACKNOWLEDGMENTS

This research was supported by the Basic Science Research Program through the National Research Foundation of Korea (NRF) funded by the Ministry of Education, Science and Technology (No. 2012R1A1B6002600).

REFERENCES

[1] Sung, K. 1994. Characteristics of Korean metallic percussion instruments, *Proc 5th Western Pac. Reg. Acoust. Conf., Seoul, Korea.*

[2] Rossing, T. D. & Fletcher, N. H. 1983. Nonlinear vibrations in plates and gongs, *Journal of the Acoustical Society of America* 73: 345–351.

[3] Fletcher, N. H. 1985. Nonlinear frequency shifts in quasispherical-cap shells: Pitch glide in Chinese gongs, *Journal of the Acoustical Society of America* 78: 2069–2073.

[4] Legge, K. A. & Fletcher, N. H. 1989. Nonlinearity, chaos, and the sound of shallow gongs, *Journal of the Acoustical Society of America* 86: 2439–2443.

[5] McLachlan, N., Adams, R. & Burvill, C. 2012. Tuning natural modes of vibration by prestress in the design of a harmonic gong, *Journal of the Acoustical Society of America* 131: 926–934.

[6] Krueger, D. W., Gee, K. L. & Grimshaw, J. 2010. Acoustical and vibrometry analysis of a large Balinese gamelan gong, *Journal of the Acoustical Society of America* 128:EL8–EL13.

[7] Rerrin, R. Elford, D. P., Chalmers, L., Swallowe, G. M., Moore, T. R., Hamdan, S. & Halkon, B. J. 2014. Normal mode of a small gamelean gong, *Journal of the Acoustical Society of America* 136:1942–1950.

[8] National Gugak Center. 1999. *A study on metal percussion instrument: Jing and Kkwaenggwari*(1991–1998), Seoul, National Gugak Center.

[9] Lee, H. J., Park, H. & Lee, D. N. 1999. processes and acoustics of Jing and Kkwaenggwari, *Bul. Kor. Inst. Met. & Mater.* 12:10–18.

[10] Kwon, H. S., Kim, Y. H. & Rim, M. 1997. Acoustical characteristics of the Jing; An experimental observation using planar acoustic holography, *Journal of the Acoustical Society of Korea* 16:3–13.

[11] Sohn, J. H. & Bae, M. J. 2004. A study on the sound amplitude and decaying time of the Jing depending on the depth of rim, *Journal of Broadcasting Engineering* 9:424–433.

[12] Kwon, S. & Cho, S. 2014. Analysis of acoustic signal based on modified ensemble empirical mode decomposition, *Transactions on Engineering Technologies* 2014:377–386.

[13] National Gugak Center. 2008. *Korean Traditional Musical Instruments Measurement Series 1*, Seoul, National Gugak Center.

Testing and Measurement: Techniques and Applications – Chan (Ed.)
© 2015 Taylor & Francis Group, London, ISBN: 978-1-138-02812-8

Application of multi-geophysical approach for detection of cavern

M. Bačić, D. Jurić Kaćunić & M.S. Kovačević
Department for Geotechnics, Faculty of Civil Engineering, University of Zagreb, Croatia

ABSTRACT: This paper presents the application of multi-geophysical approach for detection of a cavern between two lanes of the Rijeka - Zagreb highway in Croatia. Applied geophysical methods were Ground Penetrating Radar (GPR) and seismic refraction method. While the ground penetrating radar method gave information about the degree of karstification of rock mass under the highway, seismic refraction gave an estimation of cavern dimensions. Basic principles of both methods are given in paper as well as the results of investigation works which consisted of 21 GPR profiles and one seismic refraction profile. Also, cavern remediation works, which are based on results on investigation works, are presented in the paper.

1 INTRODUCTION

Republic of Croatia is situated in southeast Europe and, according to Kovačevič et al. (2011), more than half of the its area of (or over 70% if taking into account the Croatian Adriatic seabed) is situated on karstic terrain (Figure 1).

Figure 1. Karst region in Croatia (Kovačević et al., 2011).

As a consequence of karstification, a large number of karstic features, such as caverns, are linked with engineering activities and directly influence existing infrastructure, especially national highways and roads.

In February 2014, on the 9th kilometer of the Rijeka-Zagreb highway, in the direction of Zagreb, an opening appeared between the two lanes, as is shown in Figure 2.

Figure 2. Cavern on 9th km of Rijeka – Zagreb highway.

An opening had a depth of approximately 3 m and aperture area of 10 m². The appearance of such features is common for karst terrains where the surface material collapses in the caverns which are located at greater depths.

In order to determine the size and dispersion of the cavernous system, geological and geophysical investigations were carried out. Applied methods of geophysical investigations and results are presented in this paper.

2 INVESTIGATION METHODS

Geophysical methods in this particular situation were chosen according to the geological structure of the ground and geotechnical nature of the problem related to the project requirements. Two investigation geophysical methods were chosen – ground penetrating radar (GPR) and seismic refraction.

2.1 *Ground Penetrating Radar (GPR)*

GPR method was firstly developed for military purposes, but in a short period it has found application in many other sectors of human activities such as civil engineering, geology, etc. The method is based on emission of high frequency electromagnetic waves in subsurface by using, according to Marčič et al. (2013), a configuration which consists of suitable antenna, a control unit (which is 'responsible' for generation of the radar signal and for detection of received signals as a function of time), and the processing system (a laptop). Taking this into consideration, it is clear that the system for GPR survey is very easy to operate with.

In the subsurface, an emitted wave can be attenuated, reflected or refracted. After reflection, a wave returns to surface where it is received by second antenna (so called bistatic system), as shown in Figure 3, or it is received by same antenna which was used for emission (so called monostatic system).

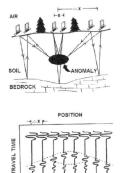

Figure 3. Principle of GPR investigation (Annan, 2003).

GPR configuration can be also classified as air-coupled or ground-coupled depending on the position of the antenna during investigation. In this investigation, a monostatic, ground coupled system was used. According to Kovačević et al. (2013) ground penetrating radar (GPR) imaging provides a high - image of dielectric characteristics of rock mass from a depth of a few centimeters to tens of meters.

When talking about GPR investigation, it is necessary to mention the terms of horizontal resolution and vertical resolution (Alvarez Cabrera, 2013). Vertical resolution is presented by smallest distance in vertical direction at which two phenomena can be apart in order to see and distinguish them as separate phenomena. Horizontal resolution is the minimum horizontal distance between two phenomena at the same depth before the radar merges them out into one single event.

Resolution, as well as depth of investigation, is determined by the frequency of the antenna. When higher frequencies are used, lower investigation depth can be achieved, but with higher resolution. Using lower frequencies, subsurface investigation will result in lower resolution, but larger depths can be investigated.

Most important parameter for the conduct of GPR investigation is the dielectric constant of material which influences velocities of the electromagnetic waves. The high dielectric contrast between two materials will result in stronger reflection in GPR image.

After the data are collected, a processing phase follows, in which the filtration is carried out in order to remove noise, interference and unwanted effects. This is done in order to emphasize the phenomena which is the subject of research. In many cases it is possible to interpret the results of investigation with very little post-processing.

2.2 *Seismic refraction method*

The main objective of the seismic methods is to define the profile of the elastic wave velocities in depth which are in direct correlation with the elastic stiffness properties of the material through which they travel. Such elastic waves spread through the soil or rock after they were generated by an impulse or controlled vibration on the surface (Kovačević et al., 2013). When they arrive at the border of layers of different seismic velocity, waves are reflected or refracted, and after they travel back to the surface. Refraction and reflection of waves are conducted according to Snell's law of dispersion of rays of light in a stratified medium (Figure 4).

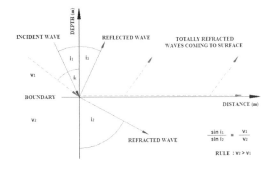

Figure 4. Snell's law of optics.

Consider the wave that passes through the top layer with the velocity of v_1. If the wave arrives on the boundary between two layers at angle i_1, it is partly reflected (with i_1 angle) and it returns to the surface with v_1 velocity. Also, the wave is partly refracted (with i_2 angle) and it goes through the second layer with velocity v_2. In the event of $v_2 > v_1$, an equation $i_2 > i_1$ is valid, while in the event of $v_2 < v_1$ an equation $i_2 > i_1$ is valid. The ratio of angle sinuses of the incident and refracted wave is equal to the ratio of the velocities at which the wave travels in the lower and upper layer. However, if the wave arrives at the boundary at angle i_c, it will spread on the boundary plane and return to surface after some time with same i_c angle.

The arrival of the wave on the surface is detected by measuring sensors - geophones, which are placed on pre-defined positions. They measure the travel time of the wave. In order to collect data on the field, following equipment is necessary:

• generator of waves
• geophones (digital; in case of analog geophones it is necessary to have the A / D converter)
• laptop

The main task of the generator is to generate an impulse by means of mechanical impact with a large frequency band and the high wave energy. Different materials have different stiffness properties and therefore, in order to cover a large range of frequencies, different types of generators are recommended. For induction of the deeper layers, or in order to generate lower frequencies, a very large wave energy is necessary (usage of dynamites or larger loads falling from a crane).

After recording the times of the first arrivals, knowing the distance from the geophones to the generator, pairs of distance-time are formed. Using the method of least squares, an x–t diagram (distance - time graph) is formed which is then used for interpretation of results and the formation of the velocity profile.

To determine the wave velocity profile, that is to interpret collected data, a series of methods were developed and the following ones are mostly used: Generalized Reciprocal Method (Palmer, 1981), Delta-tV Method (Gebrandl and Miller, 1985) and Diving-wave Tomography (Sheriff et al., 1982).

3 RESULTS OF INVESTIGATION

At the location of the cavern, geophysical investigation consisted of seismic refraction investigation (1 profile, RP 1, with a total length of 120 m') and ground penetrating radar investigations (21 profile, with a total length of 1455 m'). This 3-dimensional type of GPR data acquisition gave better insight in potential anomaly zones. The situation of investigation profiles is shown in Figure 5 where seismic refraction profile is marked in blue, while GPR profiles are marked in red.

Since GPR investigations are fast, a larger number of these profiles were investigated. The main task of this investigation was to provide an information on

whether there are karstic related anomalies below the lanes which could endanger functionality and safety of the highway. All profiles were investigated in only a few hours and the results were interpreted in only one day.

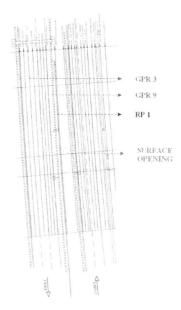

Figure 5.   Situation of investigation works.

Two distinctive GPR profiles are shown on Figure 6. First one is profile GPR 3 which is located in the middle of the highway. A relatively 'clean' image can be seen which suggests that there are no larger anomalies under the highway. By contrast, a GPR 9 profile, which is located in the vicinity of surface opening, suggests that cost-linked features are presented. Therefore, GPR investigation has fulfilled its main task and also provided a useful information about where it is optimal to conduct seismic refraction investigations. It can be said that GPR tests have 'focused' refraction investigations to the area near the surface opening.

A seismic refraction investigation was conducted using 60 geophones on 2 m separation. This provided investigation depth up to 26 meters. A shot was generated on every 4 m along the line. After interpretation by using delta–tV tomography, a velocity profile was obtained and it is shown on Figure 7. A feature which can be easily seen is an area of reduced velocity of seismic waves between 30th and 45th meter of profile and on 6th to 14th meter in depth. This area is assigned to cavern which caused the collapse of material on the surface. In addition, on the 10th meter of profile, another area of lower velocity was noticed which pointed out to another potential cavern which could cause problems

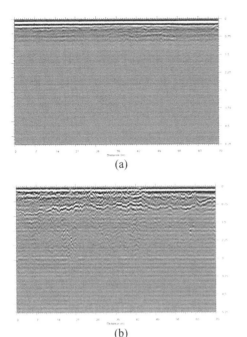

(a)

(b)

Figure 6. Results of GPR investigation: GPR 3 (a) and GPR 9 (b).

in future. Seismic refraction therefore gave information on approximate size of cavern which was estimated to be 160 m³. It should be noted that precise dimensions are practically impossible to obtain, but seismic refraction can give satisfactory preliminary results necessary for the conduction of remediation works.

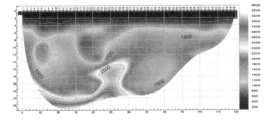

Figure 7. Seismic refraction profile RP 1.

4 REMEDIATION OF CAVERN

The remediation of cavern consisted of two phases (Figure 8):

- Phase 1 - Filling of the cavern with fine-grained concrete.
- Phase 2 - Stabilization of the cavern surrounding area with 'contact grouting' method.

After completing these phases, a surface hole was filled with gravel and reinforced slab was constructed on surface with dimensions of 4.5 × 5.5 × 0.15 m.

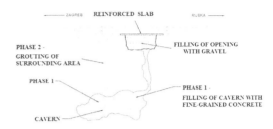

Figure 8. Phases of cavern remediation.

As it was already mentioned, by conducting seismic refraction investigations, a volume of cavern was estimated to be 160 m³. During construction of cavern a total amount of fine-grained concrete which was used for filling of cavern was 130 m³. This can be considered as good estimation and therefore it verifies the application of multi-geophysical approach for detection of caverns in karstic terrain.

5 CONCLUSION

As a result of being constructed in karstic terrain, subjected to a process of karstification, an opening of an area of approximately 10 m² and 3 m deep appeared between two lanes of Rijeka – Zagreb highway in Croatia. The collapse of surface material is common for karst and it is caused by the caverns that are located at greater depths.

In order to determine the position and size of the cavern which caused the collapse of material, two methods were used - ground penetrating radar (GPR) method and seismic refraction method. By using the GPR method, it was shown that the potential location of karstic features is in the vicinity of the surface opening and such features were not noticed under the highway. On the other hand, seismic refraction method was used to estimate the dimensions of the cavern. By conducting seismic refraction investigations, a volume of the cavern was estimated to be 160 m³ and during remediation works, total consumption of fine-grained concrete was 130 m³. This can be considered as good estimation and therefore it verifies the application of multi-geophysical methodology for detection of caverns in karstic terrain. The main advantage of this methodology is to obtain relatively precise information about presence, distribution and size of cavernous features in a fast and simple manner.

REFERENCES

[1] Alvarez Cabrera, R. 2014. *GPR Antenna Resolution*, Application notes, www.geoscanners.com (28.10.2014.).
[2] Annan, A.P. 2003. *Ground Penetrating Radar - Principles, Procedures & Applications*. Sensors & Software Inc., Mississauga, Canada.
[3] Gebrande, H. & Miller, H. 1985. Refraktionsseismik (in German). *Angewandte Geowissenschaften II. Ferdinand Enke*, ed. F. Bender, Stuttgart, 226–260.
[4] Kovačević, M.S., Jurić-Kaćunić D. & Simović R. 2011. Determination of strain modulus for carbonate rocks in Croatian karst (In Croatian). *Gradevinar* 33 (1), 35–41.

[5] Kovačević, M.S., Marčić, D. & Gazdek, M.. 2013. Application of geophysical investigations in underground engineering. *Technical Gazette* 20 (6), 1111–1117.
[6] Marčić, D., Bačić, M. & Librić., L. 2013. Using GPR for detecting anomalies in embankments, *13th International Symposium on Water Management and Hydraulic Engineering*, Bratislava, 695–710.
[7] Palmer, D. 1981. An Introduction to the generalized reciprocal method of seismic refraction interpretation. *Geophysics* 46, 1508–1518.
[8] Sheriff, R.E. & Geldart, L.P. 1982. *Exploration Seismology: History, Theory, and Data Acquisition*. Cambridge University Press, Cambridge.

Testing and Measurement: Techniques and Applications – Chan (Ed.)
© 2015 Taylor & Francis Group, London, ISBN: 978-1-138-02812-8

Noise monitoring system for construction workers using smart-phone

S.H. An
Department of Architectural Engineering, Daegu University, Korea

C.B. Son
Department of Architectural Engineering, Semyung University, Korea

B.S. Kim
Department of Civil Engineering, Kyungpook National University, Korea

D.E. Lee
School of Architecture and Civil Engineering, Kyungpook National University, Korea

ABSTRACT: The aim of this research is to develop a noise monitoring system for the health of construction workers using smartphones. The current regulation might fit in the manufacturing industry; however, it is difficult to apply the same regulation to the construction industry. Construction workers tend to move contiguously into a construction field, and work within confined work spaces. As a result, it is not avoidable that a crew shares same work space with a noise maker. It disturbs the communication among crew members and causes accidents. This study, therefore, presents an approach for identifying and monitoring noise level and developing a safety management system to avoid accidents caused by the noise.

KEYWORDS: Construction noise, safety, monitoring system, smart-phone.

1   INTRODUCTION

According to the accident analysis report from the Korea Occupational Safety & Health Agency (KOSHA), construction industry has the highest portion of accidental death as 43.3% among overall industries. Fall, collision and narrowness, are the major reasons which cause fatal accidents (KOSHA 2012). Apart from other industries, various crews share identical workspaces in the construction site. Therefore, when one crew makes a loud noise, another crew is subject to interfere with the communication among members. This can cause fatal accidents in the construction site. Since current noise regulation focuses on the manufacturing industry, it is hard to apply the regulation to the construction industry. Most of existing studies relative to the noise management focus on the hearing loss of workers. Therefore, this research proposes noise management method and noise, standard, taking account of the characteristics of the construction site.

2   NOISE MANAGEMENT METHOD

2.1   *Proper criterion for noise management*

According to the noise regulation (KOSHA 2012), employers should establish hearing safety program where the noise level exceeds 90dB. This regulation might be fit into manufacturing work sites. However, this is not appropriate for the construction industry because the construction site shares work space among crews. For instance, when the two crew share a work space and one work task generates loud noise and other work task requires elaborate communication among laborers, even the noise under 90dB could cause fatal accidents. To manage these accidents caused by the noise, real-time measuring and assessment of the noise environment is necessary which enables a safety manager to warn workers. To solve this issue, a new standard of noise level, which reflects the degree of conversation interference, is necessary. Preferred Speech Interference Level

(PSIL) and Articulation Index (AI) are used for measuring the degree of the conversation interference. In this study, the degree of the conversation interference is acquired by measuring the PSIL. Webster (1969) showed the relationship between possible distances to talk between the speech interference levels as figure 1. By using this relationship and when we know the distance between workers, it is possible to know PSIL. To acquire real-time noise environment, two assumptions were needed as follows. First, it is possible to know the distance information between crew members and is measured in advance. Second, the condition that is possible to talk is assumed as "Raised" voice at figure 1. For example, when working distance between crew members is 2.44m, PSIL is 57dB at "Raised". In this study, Allowable Ambient Noise Level (AANL) is used for the noise management standard.

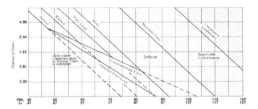

Figure 1. Relationship between PSIL and distance.

2.2 Preventing noise accidents & noise monitoring system

To apply AANL at the site, real-time measurement of the PSIL must be preceded. For this reason, it is necessary that real-time noise management system to acquire the noise level data exposed to site-worker. Also, it is necessary to determine whether PSIL of each worker is over the AANL or not to transfer the information of the worker (e.g., name, affiliation, cell phone number, etc.) to the safety manager. In this study, smart-phone is used as a device for measuring the noise environment of the construction workers. Noise data that are measured via a worker's smart-phone in real-time (the sound pressure level), and is converted to the PSIL using Eq. 1.

$$PSIL = \frac{LP_{500} + LP_{1000} + LP_{2000}}{3} dB \qquad (1)$$

where, LP_{500}, LP_{1000}, LP_{2000} is the values of sound pressure level representing 500 Hz, 1000 Hz, and 2000 Hz, respectively.

Based on the number of times, which exceeds the AANL measured by real-time measurement of the

site-worker, safety manager is able to judge the safety level of workers. When a situation is determined to be urgent, safety manager can force the site-worker move to the workplace without noise by calling directly. On the other hand, if it is determined not to be an emergency, a safety manager can visit the site and perform safety measurement directly based on the position data of the worker. Thus, it is expected that noise, providing the noise environment of workers to the safety manager in real-time could be extremely helpful to prevent potential accidents caused by noise in advance.

3 CASE STUDY

In order to verify the performance of the developed system, the parallel working condition was assumed. Table 1 shows the noise environment data when frame installation and cutting operations are performed simultaneously in the same workplace. Frame installation team moves from A-205 to A-305. At this time, a wall cut crews work in the A-305. The frame installation team has been experiencing a problem with communication due to the loud noise about 90dB occurred by the wall cutting. If a safety manager can acquire data of worker's noise environment using smart-phone, the safety manager can know 'who' needs safety action, 'where' is the location, and then, he/she can visit the dangerous site and take safety action. On top of that, when a safety manager makes full use of noise management and analysis system for preventing potential risk, he/she can catch who needs safety action rapidly. Therefore, this system can contribute to prevent accident caused by noise.

Table 1. Frame installation workers' noise environment data.

No.	Workplace	Work	Time	PSIL	AANL	Exceed times
1	A-205	Frame	15:03	43dB	65dB	1
2	A-205	Frame	15:11	44dB	65dB	1
3	A-205	Frame	15:13	46dB	65dB	2
4	A-205	Frame	15:22	47dB	65dB	1
5	A-205	Frame	15:25	57dB	65dB	2
6	A-205	Frame	15:47	56dB	65dB	1
7	A-305	Frame	15:50	84dB	65dB	4
8	A-305	Frame	16:08	85dB	65dB	5
9	A-305	Frame	16:13	89dB	65dB	6
10	A-305	Frame	16:18	88dB	65dB	7
11	A-305	Frame	16:27	85dB	65dB	8
12	A-305	Frame	16:28	87dB	65dB	9

4 CONCLUSION

In this study, we found that it is hard to apply current noise regulation in the construction workplace directly. Since the construction industry has various work tasks, which share same work place simultaneously. A new method is necessary to ensure the safety of construction workers. To solve this issue, microphone of smart-phone was used to measure the noise environment of the site-worker, and the noise monitoring system was developed to prevent potential accidents. However, the average distances between crew members were used to measure and store to the system database. Therefore, it is commendable to consider different distances in future work.

ACKNOWLEDGMENT

This research was supported by the Basic Science Research Program through the National Research Foundation of Korea (NRF) funded by the Ministry of Science, ICT & Future Planning (No. 2012R1A1A2042752).

REFERENCES

[1] Korea Occupational Safety & Health Agency (2012), *Analysis of fatal accident.*

[2] Han, M-H., Oh, Y-K., Lee, T-G., & Kim, S-W. (1990), A Study on the Physical Parameters of Assessing the Spatial Speech Transmission Quality Under Variable Noise Conditions, *Architectural Institute of Korea*, 6(6), 203–210.

[3] Kim, J-S. (1997), A Study on the Regulation and Noise Indication System of Construction Noise, *Architectural Institute of Korea*, 41(9), 76–79.

[4] Cho, G-S., Chin, S-Y., & Kim, Y-S. (2003), A Study on the Cause of Construction Space Conflicts in Apartment Housing Projects, *Architectural Institute of Korea*, 19(6), 161–168.

[5] Yun, J-M., Shin, Y-M., Kim, D., Lee, G-H., & Park, P. (2010), Sound management of the worker-oriented guidelines For the USN-based monitoring system, *Ergonomics Society of Korea*, 281–284.

[6] Kang, S-H., & Lee, Y-H. (2010), A Development of an Environmental Noise Management Program focus to the Employee's Auditory Integrity in Nuclear Power Plants, *Ergonomics Society of Korea*, 162–166.

[7] Kim, Y-J., Yum, S-K., Lee, S-H., Song, D-S., & Kim, Y-S. (2010), Improvement of Speech Privacy in Open Plan Offices Using Articulation Index, *Architectural Institute of Korea*, 26(8), 309–316.

[8] Kim, J-S., & Ju, D-H. (2012), Analysis on the Noise Damage Reduction Effect by Working Environment Noise Regulation Standard Proposal, *Society of Air-conditioning and Refrigerating Engineers of Korea*, 519–522.

[9] Lee, Y-S., & Lee, D-H. (2013), A study on awareness rate of the fire alarm sound using Background noise in the factory, *Korea Safety Management & Science*, 409–420.

Material performance and measuring and testing technique

Testing and Measurement: Techniques and Applications – Chan (Ed.)
© 2015 Taylor & Francis Group, London, ISBN: 978-1-138-02812-8

Acoustic microscopy characterization of nanostructured carbon-ceramic composites

V.M. Prokhorov & D. Ovsyannikov
Technological Institute for Superhard and Novel Carbon Materials, Moscow, Troitsk, Russia

V.M. Levin & E. Morokov
Emmanuel's Institute of Biochemical Physics, Russian Academy of Science, Moscow, Russia

ABSTRACT: Acoustic microscopy is used to visualize the bulk microstructure and internal defects and to measure the local values of ultrasonic velocities of solids based on which the elastic moduli can be calculated. Two types of impulse acoustic microscopes fabricated in the AM-laboratory of Emmanuel's Institute of Biochemical Physics, Russian Academy of Science, were used to measure the local values of ultrasonic velocities (microacoustic technique) and to visualize the bulk microstructure of a sample (scanning acoustic microscopy). One of them is a WFAM wide-field pulse scanning acoustic microscope ($25 \div 100$ MHz) designed in 1996. The other one is a SIAM scanning impulse acoustic microscope ($50 \div 200$ MHz) designed in 2010. As an example of the acoustic microscopy measurements, we present the results obtained on the samples of nanostructured fullerene-ceramic composites, such as B_4C/C_{60}, $c\text{-}BN/C_{60}$, and Bi_2Te_3/C_{60}.

1 PRINCIPLE OF ACOUSTIC MICROSCOPY

The principles and methods of high-frequency ultrasound imaging have developed for many past decades. The basic ideas of acoustic microscopy have been formulated by the Russian scientist, professor Sokolov (1934a, b, c). Experiments on a prototype acoustic microscope that implemented the principles to obtain acoustic images began to develop rapidly since the mid-1970s.

The first scanning acoustic microscope was created by Lemons & Quate (1974) at Stanford University forty years ago.

In Russia, the first laboratory prototype of a scanning acoustic microscope was set up at the Physics Department of the Moscow State University (Berezina et al. 1976). In the Acoustic Microscopy Laboratory, Russian Academy of Science, works on the creation and development of methods for acoustic microscopy have been conducted since 1978 (Maev 1988). By this time, several generations of pulsed acoustic microscopes was created with operating frequencies from 25 MHz up to 3 GHz (resolution up to 300 nm).

The principle of acoustic microscopy is illustrated by figure1. Acoustic lens with ultrashort pulse (1.5-2 oscillations in the pulse) of high-frequency focused ultrasonic probe the specimen. The pulse is reflected from the front surface, elements of internal structure and the specimen bottom. Echoes reflected from the sample, radiate from a narrow focal spot of the probe beam whose diameter is given lateral resolution.

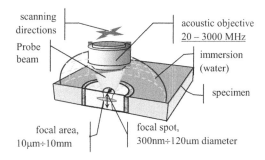

Figure 1.   Impulse acoustic microscopy - raster imaging by focused ultrasonic pulses.

Oscillogram of the reflected signal is an ultrasonic A-scan; short pulse allows to select the signals coming from the desired depth and find delay times between signals. 1D- or 2D- scanning of the probe beam over the specimen surface results in raster formation acoustic images (B-and C-scans, respectively). C-scans are formed as gray-scale images displaying the amplitude of the reflected signal coming from a definite depth. Electronic gate is the tool of acoustic image formation. Its position and width on A-scan are corresponding depth and thickness of imagining layer inside the specimen bulk.

In our experiments, we used the scanning impulse acoustic microscopes fabricated in the AM-laboratory of Emmanuel's Institute of Biochemical Physics, Russian Academy of Science, to measure the local values of ultrasonic velocities (microacoustic technique) and to visualize the bulk microstructure of a sample (scanning acoustic microscopy). Ultrashort probing ultrasonic 30-40 ns pulses were used for measurements. An acoustic microscope provides the following options: a $30 \div 60$ µm lateral resolution; a $30 \div 40$ µm transverse resolution; an $8^0 \div 37^0$ angle aperture of acoustic objectives. The data on the sound velocities of longitudinal V_L and transverse V_T waves were obtained with an accuracy of ~1%; elastic moduli ~2-3%.

2 EXAMPLES OF AM-INVESTIGATIONS

2.1 *Boron carbide/C_{60} (B4C/C_{60}) composites*

Nanostructured boron carbide/C_{60} (B_4C/C_{60}) carbon-ceramic composites were prepared by the synthesis and sintering methods:

Boron carbide B_4C powder (cleared of excess carbon fraction with an average particle size of 100 nm) is mixed with powdered molecular C_{60} (average grain size of 1 µm) in a weight ratios in the range from 80/10 to 50/50 wt.% in a vibratory mill. The mixture was then subjected to grinding in a ball mill (700 rpm for 1 minute).

The nanostructured powders were charged into a pressure chamber, fixed to the load pressure between 1.0 and 5.0 GPa, and heated to 1000 °C with a holding time of 60-100 seconds (that is the synthesis and sintering hard form of C_{60}).

For AM measurements, we used the samples in the shape of disks with a 15-17 mm diameter, a 2.5-6.0 mm height and nonflatness of the opposite faces of ±1 µm/cm.

The bulk acoustic wave velocities were measured with an acoustic microscopy method using the A-scan data. The results of the measurements are shown in figure 2. A-scan shows that the time interval of the echo-signals reflected from the front and back surfaces (double specimen thickness) is 690 nanoseconds. The sample thickness is 2.9 mm. Hence the velocity of longitudinal waves is equal to 8.4 km/s.

L-signal from bottom.
$\Delta t_L = 690$ ns, $V_L = 8.4$ km/s.

LT-signal from bottom.
$\Delta t_{LT} = 690$ ns, $V_T = 5.2$ km/s

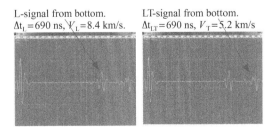

Figure 2. SIAM measurement the bulk acoustic wave velocities.

Using the date of sound velocities (longitudinal and shear) and densities of investigated materials the elastic modulus were calculated. For some samples, the data are presented in Table 1. Note that the sound velocities of the sample No2H (50 wt.%C60) is measured in the directions that are both parallel and perpendicular to the pressing direction of the bar sample that was cut from the prepared wafer. So we set a significant anisotropy of the elastic properties.

Table 1. Elastic moduli of B4C/C60 carbon-ceramic composites.

Sample	No1	No2H1	No2H2	No3
C_{60} wt.%	50	50	50	10
ρ g/cm^3	2.30	2.23	2.23	2.34
V_L km/s	8.95	8.30	8.05	8.16
V_T km/s	5.06	5.03	4.66	4.33
E GPa	149	136	121	113
K GPa	105	79	80	99
G GPa	59	56	48	43
σ	0.265	0.214	0.245	0.309

Figure 3 shows A-scans with electronic gate and corresponded acoustic images (C-scans) of the sample No1. Different position of the gate allows visualized internal defects on separate depth from the front surface. For example $h_1 = 400$ µm, and $h_2 = 1.3$ mm. The large electronic gate between signals reflected from front and back surfaces shows integral bulk structure of the sample. Defects of internal structure look like bright dotes and ports on dark background.

As a result, on C-scans, we see bright spots with different sizes. These spots represent voids or pores in the sample. Their size reaches 500 microns.

Under surface	Middle	Integral

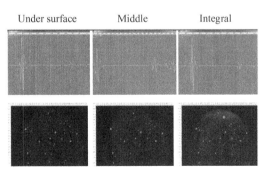

Figure 3. SIAM A-scans and acoustic images (C-scans) of the sample No1.

2.2 *Boron nitride/C_{60} (c-BN/C_{60}) composites*

Nanostructured boron nitride/C_{60} (c-BN+C_{60}) carbon-ceramic composites were prepared by the same synthesis and sintering methods as the nanostructured B_4C/C_{60}.

Elastic moduli of the samples are as follows: Young's modulus E = 55-80 GPa, bulk modulus K = 37-58 GPa, shear modulus G = 22-33 GPa. Low elastic moduli indicate the absence of the chemical bond between the c-BN particles, and the resulting carbon material prepared by the synthesis. The hardness of the material obtained by us is within 10-30 GPa, and the material is sufficiently highly rigid.

Figure 4 shows the acoustic images of carbon nitride/C_{60} composite microstructure. The data illustrates the possibilities of the acoustic microscopic inspection.

A-scan   B-scan   C-scan

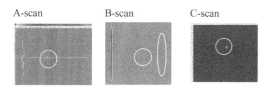

Figure 4. SIAM scans of c-BN/C_{60}. Such kind of reflection from the bottom of the sample on B-scan indicates its great heterogeneity.

Circle marked an ultrasonic signal reflected from the defect and its image on B-and C-scans. Oval area on the B-scan allocated position of echo reflected from the specimen bottom. Such kind of reflection from the bottom indicates strong specimen heterogeneity.

High amplitude of the echo reflected from internal defect allows us, with sufficient accuracy, to determine the pore position in the sample bulk.

2.3 Bismus telluride/C_{60} (Bi2Te3/C_{60}) composites

The samples of Bi_2Te_3-based nanostructured thermoelectric alloys were synthesized by mechanical alloying of the mixture of the nano- and micro-sized powders of parent materials followed by hot pressing sintering (HPS). By using the acoustic microscopy, we investigate the exfoliations and pores distributions in the bulk of the lapped wafers after sintering and in the bars which were cut out of these wafers.

The obtained acoustic microscopy images for the high-pressure sintering samples showed the pores and the exfoliations the planes of which are oriented perpendicularly to the loading axis during the sintering.

Figures 5 and 6 show a comparison of acoustic microscopy images of the same sample obtained using a WFAM and a SIAM acoustic microscopes.

Considerable anisotropies in the elastic properties and porosity of the sintered materials were established.

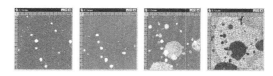

Figure 5. WFAM-images (C-scans) at different depths and at the bottom of the HPS (250 MPa, 450 °C, 5 min) sample (disk Ø30 × 2.8 mm).

Figure 6. SIAM-images (C-scans) at different depths of the same sample.

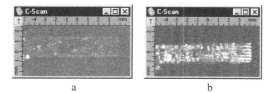

a   b

Figure 7. WFAM-images (C-scans) of the Bi2Te3/C60 composite sample in the form of a bar 1.89×1.98×10 mm3. Internal layer structure a) in plane parallel and b) perpendicular to the pressing direction. They show pore orientation anisotropy.

Last acoustic microscopy images (Fig.7) shows pores orientation anisotropy in parallel and perpendicular directions to the pressing direction for bar sample that was cut from prepared wafer. So we set a significant anisotropy of sintered materials. Anisotropy of pores distribution and orientation in samples are corresponded with anisotropy of thermal conductivity that was confirmed in study by Xie et al. (2011) and our research (Prokhorov & Pivovarov 2011, Prokhorov et al. 2012).

Anisotropy of elasticity in the directions parallel ($\parallel$) and perpendicular ($\perp$) to the axis of the applied load during a high pressure/high temperature treatment, is manifested in the difference between the sound velocities in these directions. For the specified sample in Figure 7, these differences are as follows: $V_L(\parallel)$=3.13 km/s, $V_L(\perp)$=2.86 km/s, and $V_T(\parallel)$=1.74 km/s, $V_T(\perp)$=1.70 km/s.

This anisotropy must be taken into account at thermoelectric properties measurements when making thermoelectric devices.

Elastic characteristics of the Bi_2Te_3/C_{60} composites prepared with widely differing parameters are presented in Table 3.

Table 3. Elastic moduli of the Bi2Te3/C_{60} composites. HPS parameters: A − P = 250 MPa, T = 300 − 450 °C, t = 10 min. B − P = 7.7 GPa, T = 600 − 700 °C, t = 2.0 min.

Parameters	A		B	
C_{60} wt.%	0.5	1.5	0.5	1.5
ρ g/cm^3	5.80	5.80	7.71	7.64
V_L km/s	2.10	2.10	2.5	2.4
V_T km/s	1.25	1.25	1.3	1.2
E GPa	22.0	22.0	36.5	30
K GPa	12.6	12.6	31	30
G GPa	9.0	9.0	16	11
σ	0.210	0.210	0.30	0.33

In this case, we observe a strong dependence of the elastic characteristics of the ceramic on the preparation conditions (parameters of thermobaric treatment).

3 CONCLUSIONS

Acoustic microscopy was used to visualize the bulk microstructure and internal pores and cracks and to measure the local values of ultrasonic velocities of the nanostructured carbon-ceramic composites prepared by a high-energy pre-treatment of the powder parent materials followed by a high-pressure/high-temperature treatment.

The measuring of sound velocity of longitudinal and shear bulk acoustic waves were made with an acoustic microscope as well. Elastic moduli (E - Young's modulus, K - bulk modulus, G - shear modulus, and σ - Poisson's ratio) have been calculated based on the experimentally measured density and values of sound velocity in the samples.

Anisotropy of the sound velocities and of the elastic properties of the nanostructured carbon-ceramic composites was established in the directions that are both parallel and perpendicular to the pressing direction of samples that were cut from a prepared wafer.

The data has been used to optimize the parameters of the nanostructured composites synthesis.

ACKNOWLEDGMENTS

This work was supported by the Ministry of Education and Science of the Russian Federation (grant 14.577.21.0090); the work was done using the Shared-Use Equipment Center "Research of Nanostructured, Carbon and Superhard Materials" FSBI TISNCM.

The authors acknowledge Lomakin R., Perfilov S., Popov M. for many assistance in samples preparation and useful discussions.

REFERENCES

[1] Berezina, S.I., Lyamov, V.E., Solodov I.Yu. 1977. Acoustic microscopy. *Vestnik Mosk. Univ. Fiz. Astr.* 18: 3–6.
[2] Lemons, R. A. & Quate, C. F. 1974. Acoustic microscope – scanning version. *Appl. Phys. Lett.* 24(2): 163–165.
[3] Maev, R.G. 1988. Scanning acoustic microscopy of polymeric materials and biological substances. Review. *Tutorial Archives of Acoustics* 13(1–2): 13–43.
[4] Prokhorov, V.M. & Pivovarov, G.I. 2011. Detection of internal cracks and ultrasound characterization of nanostructured Bi₂Te₃-based thermoelectrics via acoustic microscopy. *Ultrasonics* 51: 715–718.
[5] Prokhorov, V.M. et al. 2012. Internal cracks distribution in hot-pressed and spark plasma sintered nanostructured Bi₂Te₃-based thermoelectrics: acoustic microscopy versus optical microscopy. *IOP Conf. Series: Materials Science and Engineering* 42: 012003–4.
[6] Sokolov, S. Ya. 1936. Description of the device to determine irregularities in the solid, liquid and gaseous media with ultrasonic vibrations. Patent USSR No 49426.
[7] Sokolov, S. Ya. 1937a. Improvements in and relating to the detection of faults in solid, liquid or gaseous bodies. Patent GB No 477,139.
[8] Sokolov, S. Ya. 1937b. Means for indicating flaws in materials. Patent USA No 2.164.125.
[9] Xie, W.J. et al. 2011. Investigation of the sintering pressure and thermal conductivity anisotropy of melt-spun spark-plasma-sintering (Bi,Sb)₂Te₃ thermoelectric materials. *J. Mater. Res.* 26(15): 1791–1799.

Testing and Measurement: Techniques and Applications – Chan (Ed.)
© 2015 Taylor & Francis Group, London, ISBN: 978-1-138-02812-8

Method of calculation volume of the color gamut body

L. Varepo, A. Golunov, A. Golunova & O. Trapeznicova
Omsk State Technical University, Omsk, Russia

I. Nagornova
Moscow State University of Printing Arts, Moscow, Russia

ABSTRACT: The paper presents a new approach for the assessment of the color gamut of printing system using the indicator of the volume of the color gamut body. Accuracy of the calculation allows the use of this index as the primary when color reproduction is evaluated. Experiment demonstrated high correlation between volume of the color gamut body and surface roughness of materials for printing.

1 INTRODUCTION

Color gamut of printing is one of the most important characteristics of the conclusions drawn about the quality of the print impression. Color gamut is assessed using scales as a test object containing the control elements applied to the printed material paint [1]. The colors are not included in the color gamut of the press, and can not be reproduced within the printing system in use.

The most accurate color gamut can be assessed using body coverage flowers in space CIELab. Color system CIELab is possible to assess color gamut and their comparison for different chromatogenic systems [2]. Color gamut body of the printing system is given by the set of points on the surface in color space CIELab. Coordinates of points in space were removed from the field scale color gamut, and printed from the print system on the test material. Then the neighboring points on the surface of the body are connected by lines that form a closed surface.

2 METHOD

The proposed method for calculating the volume of the color gamut body (V) is based on the approximation of the body surface by triangulation. Surface of color gamut body represent the union of triangles forming a closed surface that is limited by the range of colors reproducible printing system on the test material (Figure 1).

The surface of each triangle is given by equation of a plane passing through three points $A_i\ (x_i; y_i; z_i)$; $B_j\ (x_j; y_j; z_j)$; $C_k\ (x_k; y_k; z_k)$:

$$\begin{vmatrix} x - x_i & y - y_i & z - z_i \\ x_j - x_i & y_j - y_i & z_j - z_i \\ x_k - x_i & y_k - y_i & z_k - z_i \end{vmatrix} = 0 \tag{1}$$

or in expanded form:

$$
\begin{aligned}
Z =\ & \frac{(y_3 - y_1) \times (z_2 - z_1) - (y_2 - y_1) \times (z_3 - z_1)}{(x_2 - x_1) \times (y_3 - y_1) - (x_3 - x_1) \times (y_2 - y_1)} \times x + \\
& + \frac{(x_3 - x_1) \times (z_3 - z_1) - (x_3 - x_1) \times (z_2 - z_1)}{(x_2 - x_1) \times (y_3 - y_1) - (x_3 - x_1) \times (y_2 - y_1)} \times y + \\
& + \frac{x_1 \times ((y_2 - y_1) \times (z_3 - z_1) - (y_3 - y_1) \times (z_2 - z_1))}{(x_2 - x_1) \times (y_3 - y_1) - (x_3 - x_1) \times (y_2 - y_1)} - \\
& - \frac{y_1 \times ((x_2 - x_1) \times (z_3 - z_1) - (x_3 - x_1) \times (z_2 - z_1))}{(x_2 - x_1) \times (y_3 - y_1) - (x_3 - x_1) \times (y_2 - y_1)} + z_1
\end{aligned}
$$

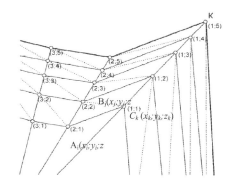

Figure 1. Surface of color coverage body in color space CIELab.

Volume of the color gamut body is defined as the difference between the amount of cylindrical bodies, the first of which is limited to the upper part of the color gamut body and its projection on the plane XOY, and the second is limited to the lower part of the body color gamut and its projection on the plane XOY (Fig. 2).

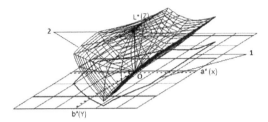

Figure 2.   The color gamut body (2) and its projection (1) on the plane XOY.

Volume of the color gamut body bounded by the upper part, is calculated as the sum of the volumes of the elementary cylindrical bodies, the upper portion of each of which is a triangle plane given by the equation (1), the lower part of the elementary cylindrical body is a projection of the triangle Δ ABC to the plane XOY. Generating elemental cylindrical body is parallel to the axis OZ. Similarly, a limited amount of body lower part of the body surface and the plane of the color gamut XOY are expected.

Elementary volume of the cylindrical body (V_1) is calculated using equation (2) of the double integral of the function $Z = z(x, y)$ given by equation (1) and the triangle is the projection Δ $A_iB_jC_k$ onto the plane XOY. Each of the integrals is generally broken down into two integrals as top (or bottom) of the projection of the triangle defined by two straight lines.

$$V_1 = \iint (a_i x + b_i y + c)dxdy = I_{i1} + I_{i2} =$$

$$= (\int_{xA}^{xB} dx \int_{yAB}^{yAC} (a_i x + b_i y + c_i)dy) + \newline + (\int_{xB}^{xC} dx \int_{yBC}^{yAC} (a_i x + b_i y + c_i)dy \quad (2)$$

As a result of computing the values of any of the elementary volume of the cylindrical body of limited arbitrary triangle and the projection of the triangle on the plane XOY.

$$V_1 =
\begin{pmatrix}
\left(\left(\dfrac{y_j - y_i}{x_j - x_i}\right) - \left(\dfrac{y_k - y_i}{x_k - x_i}\right)\right) \times \left(\dfrac{a_i - x_j^3}{3} - a_i x_i \dfrac{x_j^2}{2}\right) + \\[2mm]
+\left(c_i \dfrac{(x_j - x_i)^2}{2} - a_i \dfrac{x_i^3}{3} - \dfrac{a_i x_i^3}{2}\right) + \\[2mm]
+b_i y_i \left(\dfrac{y_j - y_i}{x_j - x_i} - \dfrac{y_k - y_i}{x_k - x_i}\right) \times \left(\dfrac{(x_j - x_i)^2}{2}\right) + \\[2mm]
+\dfrac{b_i}{2}\left(\dfrac{(y_j - y_i)^2}{x_j - x_i} - \dfrac{(y_k - y_i)^2}{x_k - x_i}\right) \times \dfrac{(x_j - x_i)^3}{3}
\end{pmatrix} +
$$

$$+a_i \frac{x_k^2 - x_j^2}{2} \times (y_j - y_i) +$$

$$+a_i \times \frac{y_k - y_j}{x_k - x_j}\left(\frac{x_k^3}{3} - x_j \frac{x_k^2}{2} + \frac{1}{6}x_j^3\right) - a_i \times \frac{y_k - y_i}{x_k - x_i} \times$$

$$\times \left(\frac{x_k^3}{3} - \frac{x_k^2 x_i}{2} - \frac{x_j^2}{3} + \frac{x_j^2 x_i}{2}\right) +$$

$$+c_i(y_j - y_i)(x_k - x_j) + c_i \times \left(\frac{y_k - y_j}{x_k - x_j}\right) \times \qquad (3)$$

$$\times \frac{(x_k - x_j)^2}{2} - c_i \frac{y_k - y_i}{x_k - x_i} \times \left(\frac{(x_k - x_i)^2}{2} - \frac{(x_j - x_i)^2}{2}\right) +$$

$$+\frac{b_i}{2}\left(y_j^2(x_k - x_j) + 2y_k \frac{y_k - y_j}{x_k - x_j} \times \frac{(x_k - x_j)^2}{2}\right) +$$

$$+\left(\frac{y_k - y_j}{x_k - x_j}\right)^2 \times \frac{(x_k - x_j)^3}{3} -$$

$$-
\begin{pmatrix}
y_i^2(x_k - x_j) + 2y_k \dfrac{y_k - y_j}{x_k - x_j} \times \dfrac{(x_k - x_j)^3}{3} - \\[2mm]
\left(y_i^2(x_k - x_j) + 2y_i \dfrac{y_k - y_j}{x_k - x_i} \times \dfrac{(x_k - x_i)^2}{2}\right) - \\[2mm]
-\dfrac{(x_j - x_i)^2}{2}
\end{pmatrix} +
$$

$$+\left(\frac{y_k - y_i}{x_k - x_i}\right)^2 \times \left(\frac{(x_k - x_i)^3}{3} - \frac{(x_j - x_i)^3}{3}\right)$$

Volume of the solid color gamut of the printing system on the test material is calculated as the difference between the sum of volumes of elementary cylindrical bodies which make up the body, calculated using the expression (3), formed by the upper part of the body color gamut and the XOY plane ($V_{1,\ top}$) and a body formed by the lower part of the body color coverage and the xy-plane ($V_{1,}$ lower.).

$$V_{1,top} = \sum_{l=1}^{n} V_l, \text{ where}$$

70

n - number of triangles that form the surface of the upper part color gamut body;

$$V_{l,lower} = \sum_{l=1}^{m} V_l, \text{ where}$$

m — количество треугольников, образующих поверхность нижней части тела цветового охвата;

Then the total amount of body color gamut (V)

$$V = V_{l,\,top} - V_{l,\,lower}.$$

To otain an analytical expression of the elementary volume of the cylindrical color coverage body in color space CIELab, color a and b coordinates are denoted by (x) and (y), respectively, luminance component L in equation of the plane (2) introduced as (z). This substitution is introduced for convenience, designate derive analytical expressions for the volume element in connection with the fact that designation of color coordinates of an arbitrary triangle on the surface of the body color gamut in the color space CIELab, and the coefficients in the equation of the plane (2) are identical.

Thus, this method provides a more accurate calculation by application of methods of mathematical analysis. The method was tested in the measurement of the volume of the color gamut body when we assess the quality of the printed impression offset sheet-fed printing on the surface roughness of the printing material. Characteristics of the materials are presented in Table 1 as the main indicator of the quality of the print volume of the body proposed color gamut (V) printing system as a set of graphics, the tone and the color accuracy of the execution of the original. The color gamut is shown graphically (Fig. 4) as a projection into the plane ab (XOY).

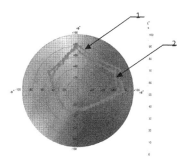

Figure 4. Projections of section of the color gamut body on a plane ab (XOY) 1 -material № 11 (130 g/m²); 2 -material № 3 (120 g/m²).

As a result of determining the volume of the color gamut body of the printing system, the proposed method and numerical values of the volume are expressed in units ΔE^3 (Table 1). Figure 4 shows the color gamut of printing systems with minimal (2) and maximum (1) gamut.

Analysis of the results in Table 1 shows that there is a correlation between the ratio of the surface roughness and the volume of the color gamut of the printing system. The correlation is described by a hyperbolic analytical model by a factor 0.9 approximation, graphical interpretation of this dependence is shown in Fig. 5.

3 CONCLUSIONS

Thus, we propose a method for calculating the amount of body color gamut of the printing system. The method is characterized by high accuracy, and tested in practice in the assessment of the color gamut of substrates with different surface roughness. Table 1

Table 1. Characterization of substrates.

№	Massm², g	Ra, µm	V,ΔE³
1	260	1,91	92046
2	260	2,16	95540
3	120	3,75	75678
4	120	3,79	88524
5	120	2,38	80798
6	230	0,529	92481
7	225	0,444	119199
8	220	0,588	122244
9	220	0,801	96877
10	220	0,522	94547
11	130	0,222	147633
12	55	2,8	83644
13	220	5,11	81579
14	250	2,78	84100

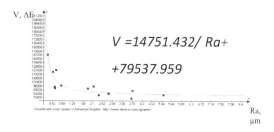

$$V = 14751.432/\, Ra + 79537.959$$

Figure 5. Schedule dependence of the surface roughness Ra and volume of the color gamut body.

REFERENCES

[1] Stefanov, S. 2005, Color,its name and scale like a standard of color, Journal of technology in printing arts and print advertising, No. 3:p. 8–12.
[2] Shashlov, B.A. 2013, Color and its reproduction Moscow: Kniga.

Testing and Measurement: Techniques and Applications – Chan (Ed.)
© 2015 Taylor & Francis Group, London, ISBN: 978-1-138-02812-8

Thermal cyclic tests of shrink polymeric products with the shape memory

A.P. Kondratov & G. Zachinjaev
The Moscow State University of Printing Arts named by Ivan Fedorov

ABSTRACT: Improved device of the automated laboratory bench for thermomechanical testing of solid materials with "shape memory". It is shown that during the heating of the air environment in the heat chamber samples of shrink polymer films release internal stress, which cause the effect of "shape memory" that reaches $1,5 \div 2$ MPa. During the thermostating process under isometric conditions the tensions are reduced due to the relaxation processes occurring in the thermoplastic polymer with a linear structure. This relaxation process corresponds to the common rheology laws of viscoelastic systems. However, if the samples are cooled in the clamps of a standard test device, tension rises above the maximum and it does not have any theoretical explanation, and may occur due to unaccounted interaction of samples with clamps of the testing machine. The article shows that due to the use of construction materials with known different linear coefficient of thermal expansion; the choice of configuration and installation of additional elements for the clamps with adjustable length; leveling can be provided for systematic errors in the tension-measurement during thermal cycle of shrink polymer films and also can be made a quantitative research of the unusual thermomechanical phenomena in polymeric materials.

1 INTRODUCTION

Thermal cycling tests of materials are widely used in industrial and research laboratory practice to determine the mechanical and relaxation characteristics of samples [1,2] as well as the service life of the finished product [3].

To improve the accuracy and reproducibility of the measurement results ceteris paribus, determining importance is shape of the sample [4] and the method of its attachment in the grips of the test stand. [5]

On the reliability of the results obtained under cyclic thermo mechanical studies of rigid materials that undergo structural changes when heated during thermal cycling is significantly affected by the interaction of the sample with the grips of the testing machine. For the first time the effect of the interaction of the sample with the testing machine grips detected and analyzed in, for example, single-crystal samples [6]. Influence of interaction of the samples heat-resistant nickel alloys with grips of the testing machine on the deformation characteristics of the material had been so great that they forced scientists to question the reliability of the results obtained without taking into account this effect and recommended to improve the technique for thermal cycling [6].

The purpose of this research is an elimination of systematic errors in the measurement of stress during thermal cycling of shrink polymer films through the usage of construction materials with various known linear coefficient of thermal expansion and configuration of additional elements of the mount samples.

2 THE EXPERIMENTAL PART

To study the relaxation processes occurring in the rigid polymer films,it designed and manufactured special stand [7], which allowed a very high accuracy to produce sequential programmable operation, tension, contraction, heating and cooling of the materials under isometric conditions and at different speeds. With the help of this experimental stand itdiscovered and investigated the effect of unusual option – "shape memory" on the shrink PVC films [8], for a quantitative description of which the common known physical and mathematical models of viscoelastic bodies are not suitable [9,10]. To avoid the systematic errors of the true stress measurement in the sample during thermal cycling, due to its unaccounted interaction with the grips of the testing machine, which was previously shown in this paper[6], it is necessary to provide thermomechanical compensation of the stress.

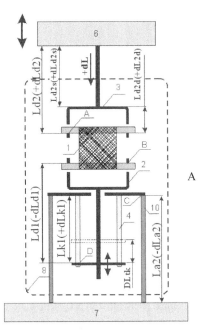
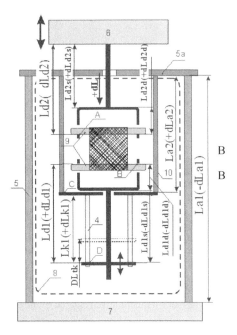

Figure 1. Variants of constructions of the experimental stand for testing solid materials with changing temperature.
A – direct mounting of temperature compensator racks (10) to the frame (7); B – mounting of temperature compensator racks (10) to the frame (7) through the side (external) racks (5) and a transverse girder (5a).

The essence of thermo-mechanical compensation is to compensate the thermal expansion and subsequent contraction of the top sample holder and the rod, connecting it to the load's sensor by a unidirectional movement of the lower sample holder with a cyclic temperature variation of the power plant elements made of different metallic materials.

The stand construction [7], to eliminate distortion of the true stress in the material, which is a consequence of interaction between the sample and grips of the testing machine, could be achieved by setting a consistent force elements with different coefficients of thermal expansion (thermal linear expansion factor). Installing the optional power components shall be done in accordance with their linear dimensions so that the multidirectional thermal increase in length of one element is equal to the reduction of others. There are two ways for solving the problem. The first – on the basis of the selection of structural materials to determine the geometric dimensions of the grip force components of the testing stand. Second – for a given geometrical dimensions of grip force components of the testing stand to determine materials with the desired

thermal properties. The combination of both these approaches in the design of the stand for the thermal cycling of samples under load is our vision task.

In this research, on an example functioning laboratory stand [7] has developed the constructive compensation scheme of mechanical stresses caused by thermal expansion of the materials; determined the optimum ratio of the longitudinal sizes of the main and complementary parts of grip force components; and found the condition to achieve a constant distance between the sample holders (points A and B in Figure 1) within a predetermined range of temperature in the test area.

For settlement and subsequent testing of the stand, completed with new grip force components, in order to identify and determine the magnitude of the potential error measurement during thermal cycling upper and lower sample holders were fastened by pin.

The results of calculations presented in Figure 2 show that for values of the length of the rack (10) La2 = 0.119 m and the increment of the total length of the elements 2 and 4 to 2.0 mm thermomechanical compensation of a relative change in the distance between points A and B occurs, which is caused by the thermal expansion of the power components of lab stand, placed in a heat chamber.

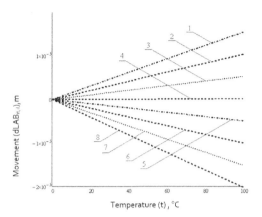

Figure 2. Relative movement of the points A and B in Figure 1 as a function of temperature at a controlled in-crease in the value of the total length of the elements of the lower sample holder (1 and 4 in Figure 1) is equal to 3.5 mm (1); 3 mm (2); 2.5 mm (3); 2.0 mm (4); 1.5 mm (5); 1.0 mm (6); 0.5 mm (7); 0.0 mm (8).

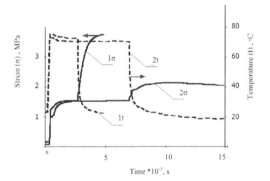

Figure 3. Changes in mechanical stress (1 σ, 2 σ) and temperature (1t, 2t) in a sample of polyvinylchloride (PVC) shrink film is heated and cooled. 1 - "direct" loading and measurement on the bench [7], 2 - measurement of voltage in the sample at the booth through the temperature compensator.

This is reflected in the value of zero change in the distance between points A and B (Figure 2)

It should be noted that the calculations do not take into account the temperature dependence of the coefficient of linear thermal expansion, therefore finish setting of thermomechanical compensator is made empirically: choosing the exact length of the rack 4 (Figure 1).

3 EXPERIMENTAL VERIFICATIONS

Figure 3 shows a stress change in the sample thermo shrink film when heated at a rate of 5 °C / minute to a temperature of 75 °C, followed by incubation for 10 minutes and cooled to room temperature of the laboratory. Thermal cycling processes were carried out under comparable conditions at the stand [7] without compensation of thermomechanical reaction of sample holders (curve 1σ) and using the offered compensation scheme (curve 2 σ). For clarity, the thermostatic time of samples heating under tension at 75°C using thermo compensation scheme doubled (curve 2 t).

It is shown that when the sample of PVC shrink film is cooled and an abnormal increase in the stress [7] occurs, it exceeds the stress shrink by 2.5 times and reachs 3.7 MPa at its "direct" measurement by a strain gauge without the use of the compensation scheme. If we use the proposed scheme of thermo compensation effect of increasing the voltage at cooling is significantly lower, at around 30% of the shrinkage stresses.

To determine the validity of a significant increase in stress in the sample PVC shrink film with the effect of "shape memory" during cooling pilot tests of power metal elements of the anchorage of the sample were carried out. Tests were carried out by thermo-cycling terminals without the polymer film under a load commensurate with the magnitude of shrinkage stresses in polymers (1,5 ÷ 2 MPa). To this sampler holder grips 3 and 2 (Figure 1) come close to each other and are connected to a horizontal steel pin. They are subjected to tensile reinforcement elements of power with an estimated length, made of materials with known thermomechanical characteristics.

In the absence of thermo-mechanical compensator tensile stress of the material in the power metal retention components of the sample made of steel with a high coefficient of linear thermal expansion (12 * 10^{-6} K^{-1}) by heating is reduced to zero or changes sign and cooling will not be returned to the previous level. Bias voltage measurement after one cycle of up to 25%. When used as a mother's power metal retention components of the sample invar alloy [11] with a low coefficient of linear thermal expansion (1.5 x 10^{-6} K^{-1}), the voltage at the heating is reduced by 75%, while cooling is restored, but it does not return to previous level, and decreases monotonically. At the same time there is a negative bias voltage, which reaches 20%. The use of thermo-mechanical compensator to avoid systematic errors in the measurement of stress in the process of heating and cooling of the samples and to obtain consistent results with a deviation of not more than 15%, with a possible deviation of the voltage measurement is directed toward reducing stress.

From these results it is clear that without the use of the proposed scheme thermomechanical voltage compensation from use of these stress relaxation with varying temperature is not correct, and quantitative estimates of the stresses in the polymer sample with the effect of "shape memory" needs to be clarified.

To those found in the works [7–9] abnormal increase by $20 \div 50\%$ of the voltage at cooling shrink films subjected to thermal cycling lawfully amended by 15–20% upwards.

4 CONCLUSIONS

The use of construction materials with various known linear coefficient of thermal expansion and the installation of additional elements of the mount models with adjustable length eliminate the systematic error in the measurement of stress during thermal cycling shrinkable plastic films and conduct a quantitative study of the thermo-mechanical phenomena in polymer materials.

ACKNOWLEDGMENT

The present work is supported by the Ministry of Education and Science of Russian Federation, Contract # 2014/87-1064 of the 30th of January 2014

REFERENCES

[1] Du, H., & Zhang, J. 2010. Shape memory polymer based on chemically cross-linked poly (vinyl alcohol) containing a small number of water molecule. Colloid Polymer Science, 288, pp. 15–24.

[2] Daniel R., Tinga T., Henderson M. B., and Ward T. 2002, J. Deformation Modelling of the single crystal superalloy CM186 LC/ Proc. Conf. "Materials for Advanced Power Engineering 2002", at Liege on October 2002, J. Lecompte-Beckers et. Al. (Eds.) Julich, Part 1. pp.293–302.

[3] Abdel-Naby, A., & Hefnawy, M. E. 2003. Improvement of the thermal and mechanical properties of poly (vinyl chloride) in presence of poly (ethylene succinate). Polymer Testing, 22(1), 25–29. DOI:10.1016/S0142-9418(02)00044-2.

[4] Bruno L., Felice G., Pagnotta L. et al. 2008. Elastic characterization of orthotropic plates of any shape via static testing / Solids and Structures. Vol. 45. pp. 908 –920.

[5] Doig M., Kraska M., Wyrwich Pet al. 2002. Inverse parameter identification by out-of-plane bending tests / Material of Forming Technology Forum, ETH Zurich, Switzerland. 20.

[6] Sidohin EF, Azizov TN, Tikhomirov EA. 2013 Thermal cycling tests of single-crystal samples / Factory Laboratory,v. 79. № 2. pp 59–62.

[7] Zachinyaev GM, Kondrashov AP. 2002. Physical modeling of the fixation process for shrink labels on cylindrical containers applicators / News of higher educational institutions. Problems of printing and publishing № 3. S. 31–39.

[8] Kondratov A.P., Konovalova M.V. et al. 2014. Relaxation processes in the interval shrinkable polyvinylchloride films with tactile marking of shrinkable labels, 46th Annual International Conference on Graphic Arts and Media Technology Management and Education 25–29 May, Athens and Corinthia, Greece, p. 26.

[9] Kondratov A.P., Tishchenko A. S. 2012. Interval-shrink material with a "shape memory" for the protection printed products and packaging against counterfeiting .Journal of International Scientific Publications: Materials, Methods & Technologies, v/ 6: pp.289–290.

[10] Courtney T. H. 1990. Mechanical behavior of materials. New York: McGraw-Hill. 399 p.

[11] Uvarov A.I., Sandovskii V.A., Kazantsev V.A. et al. 2006. Influence of preliminal plastic deformation on the structure and physico-mechanical properties of the alloy invar N30K10T3 / Physics of Metals and Metallography. Vol. 101. № 4. pp. 392–399.

Testing and Measurement: Techniques and Applications – Chan (Ed.)
© 2015 Taylor & Francis Group, London, ISBN: 978-1-138-02812-8

FRP hot-stick flashover testing under freezing conditions

M. Ghassemi & M. Farzaneh

NSERC/Hydro-Quebec/UQAC Industrial Chair on Atmospheric Icing of Power Network Equipment (CIGELE)
and Canada Research Chair on Engineering of Power Network Atmospheric Icing (INGIVRE), Université du
Québec à Chicoutimi (UQAC), Chicoutimi, QC, Canada

ABSTRACT: After four separate FRP hot-stick flashover incidents occurred in Canada under steady-state system conditions at the peak of the voltage negative half-cycle during cold and freezing conditions, many tests were carried out at Manitoba Hydro, Hydro-Quebec Research Institute (IREQ), Kinectrics and CIGELE to reproduce the flashovers. To the best of our knowledge, the most reliable reproduction has been achieved at CIGELE at a voltage stress of 105 kV/m at -1.04 °C, RH of 109 % with visible fog and 2.8 $\mu g/cm^2$ ESDD during a series of "true" cold fog tests reported in this paper.

1 INTRODUCTION

Manitoba Hydro's 500-kV ac interconnection to the US runs 532.7 km from the Dorsey Converter Station, located about 8 km north west of Winnipeg, Manitoba, to the Forbes substation near Duluth, Minnesota. The line is conducted with triplex 54/19 ACSR Pheasant with sub-conductor spacing of 45.7 cm. The center-phase of the line is suspended by V-string insulators with a tower window while the outer phases employ I-string (McDermid et al. 1999, Swatek 2013).

On 27 October 1997, a FRP hot stick flashover occurred under steady-state system condition at the peak of the negative half-cycle of voltage waveform during live-line working for the purpose of replacing V-string insulators. The ambient temperature, relative humidity, average wind speed and the peak wind speed (gusts) were recorded at Winnipeg Airport, and about 44 km WNW from the accident site at the time of the accident were 1 °C, 57%, 52 km/h and 76 km/h, respectively.

Following the flashover, a series of tests was carried out at Manitoba Hydro including wet high-potential and watts-loss tests based on IEEE Std. 978-1984 (IEEE Std. 978), and hydrophobicity measurements according to the IEC TS 62073 Method C (IEC Tech. Spec. TS 62073). The researchers attributed the flashover to poor surface condition (i.e. hydrophobicity) of the hot-stick surface. Thus, some corrective measures were introduced to improve FRP maintenance and test procedures at Manitoba Hydro (McDermid et al. 1999). Despite performing the above mentioned rigorous maintenance program, Manitoba Hydro experienced a second flashover during similar live line work in 2002 (McDermid et al. 2003). The ambient temperature, relative humidity, average

wind speed and the peak wind speed (gusts) measured at the work site at ground level were -13 °C, 46%, 22 km/h and 28 km/h, respectively. Equivalent salt deposit density (ESDD) measured on the 1997 and 2002 accident sticks were 1.9-2.4 $\mu g/cm^2$ and 3 $\mu g/cm^2$, respectively. Although these values of ESSD are considered a typical background level of outdoor work, the investigation led to the conclusion that 2-3 $\mu g/cm^2$ ESDD values could no longer be considered "clean" (McDermid et al. 2004).

In February and December 2012 and in common with Manitoba Hydro, SaskPower experienced two flashovers of "clean" FRP hot sticks during cold weather live-line work on a 230-kVac line under steady-state condition in a neighboring province, Saskatchewan. The ambient temperature and relative humidity recorded in the town of Antler, at 6 km of the accident site were -19 °C and 84%, respectively, for the December 2012 flashover, whereas those for the February 2012 flashover were -1 °C and 80%, respectively. The SaskPower and Manitoba Hydro flashover incidents occurred with a nominal stress of 71 kV/m and 90 kV/m, respectively.

These four incidents led to a series of tests at Manitoba Hydro, Hydro-Quebec Research Institute (IREQ), Kinectrics and CIGELE to investigate factors that may have contributed to the flashovers.

To the best of our knowledge, the most reliable reproduction of the incidents has been achieved at CIGELE at a voltage stress of 105 kV/m at -1.04 °C, RH of 109 % with visible fog and 2.8 $\mu g/cm^2$ ESDD during a series of "true" cold fog tests (Farzaneh & Chisholm 2013-2014). The paper reports test facilities, test method and some test results carried out at CIGELE which led to the most reliable reproduction of the occurred flashovers.

2 TEST ARRANGEMENT

2.1 *Cold fog flashover test method*

The cold-fog test (Chisholm et al. 1996) was developed to reproduce widespread station insulator and bushing flashovers that occurred in December 1989, after a full month of unseasonably cold weather and road salting in Ontario, Canada. The flashovers occurred right at the freezing point. The cold-fog test reproduces the insulator flashovers at line voltage by applying uniform pollution to all surfaces and then exposing the insulators to a controlled temperature rise from -2 to +1 °C under moderate fog density and wind speed. Figure 1 shows the details of the test plan.

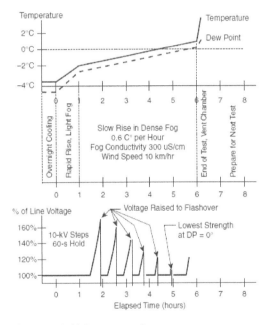

Figure 1. Cold-fog test procedure.

2.2 *Test facilities*

For the cold-fog tests carried out at CIGELE, new unwaxed 32-mm diameter universal sticks and a waxed 38-mm sister stick provided by SaskPower were used. ESDD in the range of 2 to 15 µg/cm² was applied to the tool surfaces using a leather pad, saturated with NaCl solution. No NSDD, such as kaolin or other inert material, was used in the pre-contamination solution.

CIGELE test facility is the sole source for combined capability of low-temperature icing and high voltage in North America. For example, the Kinectrics cold chamber can barely be chilled to -10 °C, and it is designed to work best in the range of -2 to +1 °C with high relative humidity.

There are actually four separate icing chambers at CIGELE, all sharing a common refrigeration system and a central core of high voltage supplies. One chamber, an icing wind tunnel, supports modest applied voltage to establish the icing rate and morphology on energized conductors. Two smaller chambers are equipped with polymer bushings rated at 300-kVac. One of these is best for testing vertical insulators of 1-m length, and the other is laid out horizontally, mainly to study ice accretion and melting on conductor loops carrying high current. The third full-scale chamber has an 800-kV rated bushing suitable for icing tests at a full rated 350- kVac on vertical insulators.

The cold-fog test phase was carried out using a 120-kVac, 240 kVA supply combined with cold-temperature capability to -24 °C. The central chamber at CIGELE offered the greatest advantage for the live-line tool testing. With a horizontal orientation of the tool, experience could be gained in the lift/drop sequence to apply and withdraw the tool from the energized conductor. Air pressure in the chamber is normalized to ambient by mechanical flaps, but with no pressure difference. The chamber is sealed, allowing for regulation of relative humidity.

The cold-fog conditions were provided by special instruments with excellent response characteristics including a pair of ultrasonic nozzles to produce water particle with a diameter in the range of 5 to and a Lord MicroStrain EH-Link to recode the ambient temperature and relative humidity.

A National Instruments digitizer card was used to sample the line voltage and leakage current. The leakage current was monitored across a 10-Ω high-power resistor in the ground return to the transformer. The voltage signal was derived from a potential divider, consisting of two 100 pF / 100 kV Messwandler-Bau capacitors in series with an 8.9 µF low voltage arm and matched by a 54 Ω resistor.

The signals were optically isolated for equipment protection using Analog Devices 5B41 devices. The voltage and current signals and computer were enclosed in a Faraday cage, powered by an isolation transformer.

Real-time display of the line voltage, the leakage current, the averaged leakage current over ten cycles and the Lissajous figure of the line voltage (horizontal axis) versus line current (vertical axis) were presented, along with the readouts of the line voltage and leakage current parameters.

The leakage current values for the Labview monitoring system were compared with reference rms values measured using a Keithley 5 ½ digit volt/ ammeter. The values agreed within 1% for a wide range of sine and pulse signals, and were generated using a bipolar Avtech instrument.

78

2.3 Test set-up

For the test, a 32-mm brand-new AB Chance link stick was cleaned, pre-contaminated with the glove wipe method and mounted on the test bus at a 60° angle to horizontal. It was grounded with a metal band at 80 cm, the same distance as the horizontal 38-mm link stick.

Several series of tests were carried out at CIGELE. In this paper, however, only the results from the most reliable reproduction of flashovers are reported. The latter occurred during a "true" cold fog test at a voltage stress of 105 kV/m at -1.04 °C, RH of 109 % with visible fog and 2.8 $\mu g/cm^2$ ESDD.

For this test, the chamber was pre-chilled overnight to a set point of -10 °C. The temperature was allowed to rise naturally, with fans at 40 rpm giving a wind speed of 4 m/s on the horizontal 38-mm link stick and an average of 3 m/s on the vertical 32-mm link stick behind it. When the chamber temperature reached -4 °C, the fog nozzles were turned on. The temperature rose again to near the freezing point. Partial frosting and continuous, weak partial discharge activity was noticed and photographed in darkness. This combination of changes gave the "true" cold fog conditions with wetting rather than frost, leading to a series of flashovers on both horizontal sister stick and vertical link stick.

3 EXPERIMENTAL RESULTS

In the conditions mentioned in Section 2.3, after the 32-mm link stick at a 60° angle withstood -84 kV across 80 cm (105 kV/m) for several minutes, the air pressure system was used to drop the 38-mm SaskPower stick onto the energized conductor as seen in Figure 2(a).

There were discharges from the head of the tool to the conductor even with the tool in its up position as seen in Figure 2(b).

The discharge activity was initiated at the band as ground electrode, which has the highest electric field as seen in Figure 2(c). It was dealt with using a numerical model based on the finite element method (FEM) to compute the voltage and electric field distribution around an FRP hot stick (Ghassemi et al. 2014a, Ghassemi et al. 2014b).

The video frame rate in Figure 2 is 30 FPS and, Figure 2(a) is considered as reference time. Figures 2(b)-(d) show the sequences for arc propagation at t=1/30, 19/30 and 20/30 s, respectively.

The partial discharge activity takes four video frames (0.13 s) to move about 40 cm, a speed extension of 3 m/s. The final jump to flashover is at least four times faster, as it occurs in a single video frame (Figure 2(c) and 2(d)).

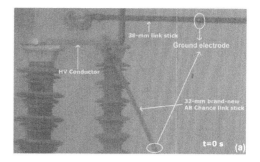

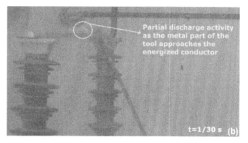

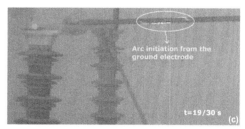

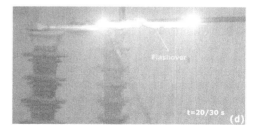

Figure 2.   FRP live-line tool flashover.

Figure 3(a) shows the peak of the applied voltage wave and the peak-to-peak of leakage current for the horizontal 38-mm link stick. The hypothesis from high leakage currents noted as the link stick head approached the conductor as seen from Figure 3(b) was that the flow of partial discharge current was sufficient to bring the iced pollution layer temperature up to just below freezing, where the cold-fog flashover mechanism prevails.

The coupled Computational Fluid Dynamics (CFD) and Heat Transfer model for an ice-hot stick was elaborated in (Ghassemi et al. 2014c, Ghassemi

79

& Farzaneh 2014)), could well explain why the flow of partial discharge current could be sufficient to raise the temperature of an iced pollution layer just below freezing, where the cold-fog flashover mechanism prevails.

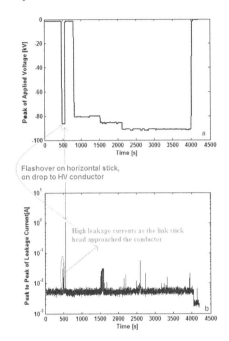

Figure 3. Peak of the applied voltage wave and peak-to-peak of leakage current of the horizontal 38-mm link stick.

4 CONCLUSIONS

This paper reported test facilities, test method and some test results carried out at CIGELE laboratories which resulted in the most reliable reproduction of four separate FRP hot-stick flashover incidents that occurred in Canada under steady-state system conditions during cold and freezing conditions. This reproduction occurred during a "true" cold fog test at a voltage stress of 105 kV/m at -1.04 °C, RH of 109% with visible fog and 2.8 μg/cm^2 ESDD.

ACKNOWLEDGMENT

This work was carried out within the framework of the NSERC/Hydro-Quebec/UQAC Industrial Chair on Atmospheric Icing of Power Network Equipment (CIGELE) and the Canada Research Chair on Engineering of Power Network Atmospheric Icing (INGIVRE) at Université du Québec à Chicoutimi (UQAC). The authors would like to thank the CIGELE partners (Hydro-Québec, Hydro One, Réseau de Transport d'Électricité (RTE), General Cable, K-Line Insulators, Dual-ADE, and FUQAC) whose financial support made this research possible. This project was also inspired and financially supported by SaskPower.

REFERENCES

[1] Chisholm W. A., Ringler K. G., Erven C. C., Green M. A., Melo O., Tam Y., Nigol O., Kuffel J., Boyer A., Pavasars I. K., Macedo F. X., Sabiston J. K. & Caputo R. B. 1996. The cold fog test. *IEEE Trans. on Power Delivery*, Vol. 11(4): 1874–1880.
[2] Farzaneh M. & Chisholm W. A. 2013-2014.Report: electrical flashover testing of live-line tools under freezing conditions, *NSERC/Hydro-Quebec/UQAC Industrial Chair on Atmospheric Icing of Power Network Equipment (CIGELE) and Canada Research Chair on Engineering of Power Network Atmospheric Icing (INGIVRE), Université du Québec à Chicoutimi (UQAC)*.
[3] Ghassemi M., Farzaneh M. & Chisholm W. A., 2014a. Three-dimensional FEM electrical field calculation for FRP hot-stick during EHV live-line work. *IEEE Trans. on Dielectric and Electrical Insulation, accepted.*
[4] Ghassemi M., Farzaneh M., ChisholmW. A. & Beattie J., 2014b. Potential and electric field calculation along a FRP live-line tool under cold and icing conditions. *Proc. of IEEE International Symposium on Electrical Insulation, Philadelphia, USA, 8-11 June, pp.218–222.*
[5] Ghassemi M., Farzaneh M. & Chisholm W. A., 2014c. A coupled computational fluid dynamics and heat transfer model for accurate estimation of temperature increase of an ice-covered FRP live-line tool. *IEEE Trans. on Dielectric and Electrical Insulation, accepted.*
[6] Ghassemi M. & Farzaneh M., 2014. A study of wind speed and direction effects on temperature increase of an ice-covered FRP live-line tool via a coupled computational fluid dynamics and heat transfer model. *IEEE Trans. on Power Delivery, submitted.*
[7] IEC Tech. Spec. TS 62073. Guidance on the wettability of insulator surface, *Method C.*
[8] IEEE Std. 978.1984. IEEE Guide for In-Service Maintenance and Electrical Testing of Live-Line Tools.
[9] McDermid W., Bromley J.C., Dodds D.J. & Swatek D.R. 1999. Investigation of the flashover of a FRP hot stick while in use for live line work at 500 kV. *IEEE Trans. Power Delivery*, Vol. 14(3): 1158–1166.
[10] McDermid W., Swatek D. R. & Bromley J. C. 2003. FRP hot stick flashovers during EHV live line work. *Proc. of Electrical Insulation Conf. and Electrical Manufacturing & Coil Winding Conf., Indianapolis, USA, 23-25 Sept.*
[11] McDermid W., Swatek D. R. & Bromley J. C. 2004. Progress in resolving flashovers of FRP hot sticks during EHV live line work," *Proc. of IEEE International Symp. on Electrical Insulation, Indianapolis, USA, 19-24 Sept.*
[12] Swatek D., McDermid W. M. & Laninga J. 2013. Experience with live-line insulator maintenance in freezing conditions. *INMR World Congress, Vancouver, Canada, 8-11 Sept.*

Testing and Measurement: Techniques and Applications – Chan (Ed.)
© *2015 Taylor & Francis Group, London, ISBN: 978-1-138-02812-8*

Compression after impact on titanium honeycomb sandwich structures

Z.H. Xie, W. Zhao & J.F. Sun
Northwestern Polytechnical University, Xi'an, China

X.S. Yue
AVIC Beijing Aeronautical Manufacturing Technology Research Institute, Beijing, China

ABSTRACT: The titanium honeycomb core structures are gradually used in several modern aircrafts recently in China. In the manufacturing process and service, low velocity impacts from foreign objects would quite likely happen and could not be avoided. In order to evaluate the influence of low-velocity impact damages on titanium honeycomb sandwich panels, Sandwich Compression After Impact tests on both damaged and undamaged panels were conducted. Numerical model was also generated. The results showed a 20% to 30% decrease in compressive strength of titanium honeycomb sandwich panel due to the low-velocity impact damage and the impact energy levels show little influence on the residual compressive strength of titanium sandwich panel.

1 INTRODUCTION

The honeycomb core sandwich structures are widely used in the aerospace applications due to their high specific stiffness and strength. The common honeycomb cores are mainly made by Nomex paper or aluminum foils. Recently, the titanium honeycomb core structures are gradually used in several modern aircrafts in China. Compared to the traditional aerospace sandwich panels, the titanium sandwich structures possess superior strength, stiffness and much better noise, thermal and corrosion resistance. Thus, they are used in the aircraft structures, such as fuselage, wing, engine cabin and air exhaust nozzle, etc (Zhang 2003, Yue 2009, Huang 2004).

In the manufacturing process and service, low velocity impacts from foreign objects would quite likely happen. They may lead to serious decrease in structural integrity and strength, especially the in-plane compressive strength of sandwich panels (Abrate 1998). Therefore, it is necessary to study the Compression After Impact properties of titanium sandwich panels to ensure they are safely used in the engineering applications.

The previous tests on composite sandwich panel showed that the compressive strength will sharply drop 30 to 40 percents or even more once the sandwich panel was hit by a low-velocity impact (Kassapoglou 1998). In the past twenty years, scientists and engineers had conducted a lot of Sandwich Compression After Impact (SCAI) tests on composite sandwich structure made by laminated face sheets and honeycomb or foam cores. Some of them also developed various finite element models or some analytical models to predict the residual strength and failure modes of composite sandwich structures with impact damage under in-plane compression (Aminanda 2009, Cvitkovich 1999, Gilioli 2014, Schubel 2007, Xie 2005, 2006, 2009, 2012). However, in the published literature, few work involves the compression after impact properties of titanium sandwich structures.

This paper introduces the experimental and numerical study on the performance of titanium honeycomb core structures with low-velocity impact damage subject to uniaxial compression. The test study involves the SCAI on titanium sandwich panels with and without impact damage.

2 SCAI TESTS

SCAI tests on titanium honeycomb core sandwich structures were carried out according to ASTM D7137-05(ASTM 2007). The specimens were made by TC1 titanium honeycomb core and two TC4 titanium face sheets. The dimensions of specimens were 150mm by 100mm by 16.6mm. The face sheets of sandwich structure were made by TC1 titanium plate with thickness of 0.8 mm. The honeycomb core was made by TC4 titanium foil with 0.05mm in thickness. The cell diameter was 4.8mm and its thickness is 15mm.

Low velocity impact damages were inflicted onto one side of the sandwich panel by a 12 mm diameter impactor with impact energy levels from 6 Joules to 12 Joules. The depth and radius of the corresponding indentation were measured with their average values shown in Table 1. In the SCAI tests, the strain gages locations were shown in Fig. 1. The load, displacement and the corresponding strain gage reading were all recorded during the test.

Table 1. Depth and radius of indentation.

Impact energy/J	depth of indentation δ_{av}/mm	radius of indentation R_{av}/mm
0	0	0
6	0.96	21.4
8	1.25	23.1
10	1.51	23.7
12	1.61	25.7

Figure 1. Strain gauge positions (mm).

A typical test curve of far field stress vs. local strains is shown in Figure 2. The far field stress is defined as the compressive load divided by the cross-section areas of titanium honeycomb core sandwich structures. The local strains stand for the strain gage reading of strain gages located along the damage propagation path. The transition points are corresponding to the moments

Figure 2. Typical test curve of far field stress vs. local strains.

when the edge of the damage propagation reaches the strain gage locations. It can be seen that the far field stress corresponding to the transition points would increase with the increase of the distance between the strain gage and the center of the impact damage, which exactly demonstrates the fact that the damage propagate along the width direction to the edge of the panel under increasing unidirectional compressive as observed in the SCAI test (Figure 3).

(a) on the impacted side

(b) on the other side

Figure 3. Typical failure modes in SCAI tests.

3 NUMERICAL ANALYSIS

A parametric finite element model for the full size titanium honeycomb sandwich panel with low velocity impact damage was generated automatically in ABAQUS by using a code developed in Python. By using this code, one can just input the geometric parameters for the titanium face sheets and honeycomb core, and the geometric model for the whole titanium sandwich panel will be automatically generated. The meshed finite element model with its boundary conditions is shown in Figure 4. Shell elements were used for the honeycomb cells and solid elements C3D8R were used for the face sheets. An indentation with refined elements was included in the model in the middle of one side of the panel with depth and radius measured from the real specimens, as shown in Figure 4 and Figure 5. The material properties used in

Table 2. Mechanical properties used in the modeling.

Materials	Tensile modulus (GPa)	Poison's ratio ν	Shear modulus (GPa)	Yield strength (MPa)	Ultimate Strength (MPa)
TC1	120.47	0.358	44.36	635.08	689.39
TC4	108.48	0.30	41.72	860.59	967.12

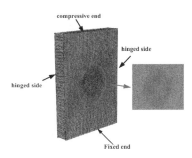

Figure 4. The finite element model for a titanium sandwich panel in SCAI test.

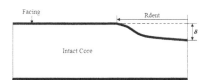

Figure 5. Indentation with its parameters.

Figure 6. The typical failure modes of titanium sandwich panels without damage.

(a) facesheet buckling

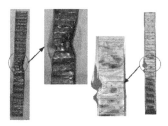

(b) Core crushing

Figure 7. The typical failure modes of low-velocity impacted titanium sandwich panels.

the finite element modeling is shown in Table 2. In the analysis, an uniformly distributed deformation was applied on the top end of the panel while the other end was fixed. The two sides of the panel are hinged supported similar to that in the SCAI test.

According to the finite element results, the titanium sandwich panel would follow the following failure process similar to what was observed in the SCAI tests.

For the panel without impact damage, under the increasing compressive load, the face sheet close to the end of the panel would buckle first, then lead to the collapse in the honeycomb core nearby and finally cause the catastrophic failure of the panel, as shown in Figure 6.

For the panel with an impact damage, under the increasing compressive load, the face sheet on the edge of indentation would buckle first, which leads to the damage propagation along the transverse direction and the final catastrophic failure in the impacted face sheet and the total buckling for the facesheet on the other side, as shown in Figure 7.

The typical compressive force vs. displacement curves from the finite element analysis is shown in Figure 8. One can see that, the low velocity impact damages introduced in this study show little influence

on the elastic performance of titanium sandwich panels, while they do lead to the change in failure mode and the corresponding ultimate strength by roughly 20%.

The comparison between the experimental and analytical results can be shown in Table 3. One can see that a good correlation between the test data and numerical prediction has been reached.

Table 3. The residual strength comparison between the experimental and analytical data.

Impact energy (J)	Depth of indentation (mm)	Experimental Results (MPa)	FEA Results (MPa)	Relative Error (%)
0	–	97.2	102.2	5%
6	0.96	88.61	81.25	−8%
8	1.25	86.49	80.37	−7%
10	1.51	84.38	79.29	−6%
12	1.61	82.83	79.14	−5%

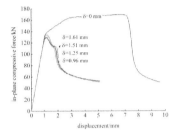

Figure 8. The typical compressive force vs. displacement curves (from FEA).

4 CONCLUSIONS

The experimental and numerical study on the Compression After Impact performance of titanium honeycomb sandwich panel with low-velocity impact damage show that:

1 The low velocity impact damages show little influence on the elastic performance of titanium sandwich panels.
2 For the sandwich panel with an impact damage, under the increasing compressive load, the indentation will propagate along the transverse direction with continuous core crushing on the path, and lead to the final catastrophic failure in the impacted face sheet and then total buckling in the other facesheet.
3 The low-velocity impact damage will cause the change in failure mode and the corresponding ultimate compressive strength by roughly 20%. The impact energy level shows little influence on the ultimate compressive strength.
4 Finite element model generated in this study can successfully predict the ultimate compressive strength of impact damaged titanium sandwich panel with relatively good accuracy.

REFERENCES

[1] Abrate, S. 1998. *Impact on composite structures.* Cambridge: Cambridge University Press.
[2] Aminanda, Y. & Castanié, B. & Barrau, J.J. & Thevenet, P. 2009. Experimental and numerical study of compression after impact of sandwich structures with metallic skins. *Compos Sci* 69: 50–59.
[3] ASTM D 7137/D 7137M-07. 2007. Test method for compressive residual strength properties of damaged polymer matrix composite plates. ASTM International, W. Conshohocken, Pa.
[4] Cvitkovich, M.K. & Jackson, W.C. 1999. Compressive failure mechanisms in composite sandwich structures. *Journal of American Helicopter Society* 44(4): 260–268.
[5] Gilioli, A. & Sbarufatti, C. & Manes, A. & Giglio, M. 2014. Compression after impact test (CAI) on NOMEX™ honeycomb sandwich panels with thin aluminum skins. *Composites: Part B* 67: 313–325.
[6] Huang, X. & Richards, N.L. 2004. Activated diffusion brazing technology for manufacture of titanium honeycomb structures-A Statistical Study. *Welding Research* 3: 73–81.
[7] Kassapoglou, C. & Jonas, P.J. & Abbott, R. 1998. Compressive strength of composite sandwich panels after impact damage: an experimental and analytical study. *Journal of Composite Technology & Research* 10: 65–73.
[8] Schubel, P.M. & Luo, J.J. & Daniel I.M. 2007. Impact and post impact behavior of composite sandwich panels. *Compos Part A: Appl Sci* 38: 1051–1057.
[9] Xie, Z. & Vizzini A.J. 2005. Damage propagation of a composite sandwich panel subject to increasing uniaxial compression after a low-velocity impact. *Journal of Sandwich Structures & Materials* 7(4): 269–288.
[10] Xie, Z. & Vizzini A.J. 2006. On residual compressive strength prediction of composite sandwich panel after low-velocity impact damage. *Acta Mechanica Solida Sinica* 19(1): 9–17.
[11] Xie, Z.H. & Su, Ni. 2009. Damage propagation behavior of composite honeycomb sandwich panels under low-velocity impact. *Journal of Nanjing University of Aeronautics and Astronautics* 41(1): 30–35.
[12] Xie, Z.H. & Tian, J. & Zhao, J. & Li, W. 2012. Study on the residual compressive strength of a composite sandwich panel with foam core after low velocity impact. *Advanced Materials Research* 430–432: 484–487.
[13] Yue, X.S. & OuYang, X.L. & Hou, J.B. & Ding, L.M. 2009. Brazing process of titanium alloy honeycomb sandwich panel structure. *Aeronautical Manufacturing Technology* 10: 96–98.
[14] Zhang, M. & Yu, J.M. 2003. Development of the mental sandwiched panels and their manufacturing methods. *Welding Technology* 32(6): 21–23.

On bearing strength of protruding and countersunk bolted joints of composite laminates

X. Li & Z.H. Xie
Northwestern Polytechnical University, Xi'an, China

J.P. Guo, X. Xiong & X.J. Dang
Chengdu Aircraft Design & Research Institute, Chengdu, China

ABSTRACT: Experiments on the bearing strength of protruding and countersunk composite-steel bolted joints were performed. The correction coefficients for countersunk composite bolted joints were obtained. Test results showed that the bearing strength of countersunk bolted joints was roughly 90% of that of the protruding bolted joints. Finite element models were developed for protruding and countersunk composite-steel bolted joints to predict the two bolted joints bearing strength. Hashin failure criteria and Tan-Camanho damage propagation rules were adopted in the analysis. The offset strength and ultimate strength were both obtained from the finite element analysis and compared with the test data. A good correlation between the numerical prediction and the test results was obtained with acceptable difference, less than 10% and 15% for the offset strength and the ultimate strength, respectively.

1 INTRODUCTION

Modern aircraft structures require extensive use of advanced materials such as composites in order to reduce the weight and to improve aircraft performance. Bolted joints especially countersunk bolted joints are of particular interest for use in skin joints where aerodynamic efficiency is important. The analysis of countersunk bolted joints has important implications. Few literatures(Egan 2012a, b, 2014, Chishti 2012, Ireman 1998) have been found to study the countersunk bolted joints in detail. Egan (Egan, 2012a) modeled a nonlinear finite element model using Abaqus and the radial stresses of clearance fit joint and neat fit joint were compared. Besides, the compressive through-thickness stresses are present at damageable region of the countersunk hole, but the strength of countersunk bolted joints was not involved. Chishti(Chishti 2012) conducted an experimental investigation into the damage progression and joint strength with countersunk fasteners and the effect of the countersunk geometry on the load-carrying capacity in comparison to the straight-shank case.

The differences in load-carrying capacity of countersunk head bolted joints and protruding head bolted joints are presented here and the relative correction factor of countersunk bolts was determined.

2 SPECIMEN CONFIGUARATION

The joint geometry was shown in Figure 1 with composite-steel single-lap joint type. Values of the specimen dimensions are as follows: $D = 5$mm, $e = 18$mm, $W = 36$mm, $t = 4.8$mm, $L = 135$mm and $L_f = 75$mm. The 100° countersunk head bolts were used to connect the plates with a torque of 2.5N•m. The laminate material was T700/QY9611, with the layer-ups consisting 20 plies in stacking sequence $[45/0/-45/0/45/90/-45/0/45/-45]_S$, $[45/0/-45/0/45/90/0/-45/0/-45]_S$ and $[45/0/-45/0/90/0/45/0/-45/0]_S$, designated as 30/60/10, 40/50/10 and 50/40/10, respectively. For each layer-ups, both protruding head and countersunk head bolted joints were tested.

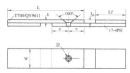

Figure 1. Countersunk joint geometry.

In the single-lap joints test, the specimens were gripped on the fixture and loaded in tension. An extensometer was placed between the two splices as shown in Figure 2. The specimens were loaded in displacement control at 1mm/min until ultimate failure.

After the test, the failure modes of each specimen were recorded and bearing strength was calculated according to recommendations in the ASTM standard, which involved determination of the offset strength and ultimate strength. The bearing stress was determined by dividing the applied load by the bearing area $(D \times t)$ and the bearing strain was determined by (δ/D). The applied stress versus the bearing strain was shown to describe the bearing behavior of the bolted joints.

Figure 2. Test fixture.

3 NUMERICAL ANALYSIS

3.1 Model description

Abaqus was used to model the single bolt and preload forces were ignored. The finite element model was shown in Figure 3. In order to improve the computational efficiency, one half of the model was built by using symmetry boundary condition. Surface-to-surface contact was defined in the contact region as shown in Figure 4. All the parts were meshed by using C3D8R elements. The material properties were shown in Table 1.

Table 1. Material properties of T700/QY9611.

E_{11} GPa	$E_{22} = E_{33}$ GPa	$v_{12} = v_{13}$	v_{23}	$G_{12} = G_{13}$ GPa	G_{23} GPa
128	10.3	0.289	0.3	5.98	3.96

Figure 3. Finite element model.

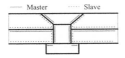

Figure 4. Definition of contact region.

3.2 Failure criteria and degradation rules

The stress distribution around bolt contact area has a three-dimensional effect obviously, especially for countersunk head bolted joint. This paper adopts three-dimensional Hashin failure criteria(Hashin 1980) to simulate the composite plate damage and to predict the strength of the joints.

Fiber tension-shear failure $(\bar{\sigma}_{11} > 0)$

$$r_1^2 = \left(\frac{\bar{\sigma}_{11}}{X_T}\right)^2 + \left(\frac{\bar{\sigma}_{12} + \bar{\sigma}_{13}}{S_{12}}\right)^2 \quad (1)$$

Fiber compression failure $(\bar{\sigma}_{11} < 0)$

$$r_2^2 = \left(\frac{\bar{\sigma}_{11}}{X_C}\right)^2 \quad (2)$$

Matrix tension failure $(\bar{\sigma}_{22} + \bar{\sigma}_{33} > 0)$

$$r_3^2 = \frac{1}{Y_T^2}\left(\bar{\sigma}_{22} + \bar{\sigma}_{33}\right)^2 + \frac{\bar{\sigma}_{23}^2 - \bar{\sigma}_{22}\bar{\sigma}_{33}}{S_{23}^2} + \frac{\bar{\sigma}_{12}^2 + \bar{\sigma}_{13}^2}{S_{13}^2} \quad (3)$$

Matrix compression failure $(\bar{\sigma}_{22} + \bar{\sigma}_{33} < 0)$

$$r_4^2 = \frac{1}{Y_C}\left[\left(\frac{Y_C}{2S_{23}}\right)^2 - 1\right](\bar{\sigma}_{22} + \bar{\sigma}_{33}) + \frac{1}{4S_{23}^2}(\bar{\sigma}_{22} + \bar{\sigma}_{33})^2$$
$$+ \frac{1}{S_{23}^2}(\bar{\sigma}_{23}^2 - \bar{\sigma}_{22}\bar{\sigma}_{33}) + \frac{1}{S_{12}^2}(\bar{\sigma}_{12}^2 + \bar{\sigma}_{13}^2)^2 \quad (4)$$

where σ_{ij} are the components of stress, X_T, X_C, Y_T, Y_C are the axial fiber tension strength, fiber compression strength, matrix tension strength and matrix compression strength, respectively and S_{ij} are the shear strength. After one or several of the damage criteria described above have been reached in a PDM analysis, the elastic properties of the material must be modified by using a set of degradation rules. This modification is accomplished by reducing stiffness variables in the material. Several damage response theories have been proposed by authors who have used the PDM (Chang 1987, Tserpes 2002, Tan 1991, Camanho 1999, Matzenmiller 2005).

The first one was Chang-Chang (Chang 1987) degradation rules and was extended to three dimensions by Tserpes et al. (Tserpes 2002) by assuming a complete and brittle fracture at a material point. For example, a fiber fracture was assumed to indicate complete failure of the material's tensile and shear stiffness and matrix failure leaded to a full reduction of shear and transverse tensile stiffness. The second one was Tan-Camanho (Tan 1991, Camanho 1999) degradation rules which reduced material properties in a more conservative fashion. A structural material may still have load bearing capacity after damage initiation. For example, when fiber failed, the fiber tension module was modified as $E_1^d = D_1^L E_1$, where D_1^L is called stiffness reduction factor. The third one was the Matzenmiller-like degradation rules (Matzenmiller 2005). The stiffness was assumed to reduce in a gradual manner and $D_i = f_i(r_i)$

(i=1,2,3,4). The material could not bear load until the stiffness was reduced to zero. Previous studies have shown that the load displacement was better predicted by using Tan-Camanho degradation rules. Therefore, the second one are used in the present model.

To describe the damage initiation and propagation, field variables were defined by user defined field subroutine (USDFLD, Abaqus 2009). USDFLD defines material properties through its field variables values (FVi=0 or 1, i=1,2,3,4). 0 means that the material is undamaged, and 1 means the material is totally damaged. The USDFLD was written by using Fortran language and FVi values were determined in the subroutine according to Hashin failure criteria. The damaged material properties are modified as D_1=0.05, D_2=0.1, D_3=0.1 and D_4=0.2 when the corresponding damage mode was detected.

4 RESULTS

4.1 Test results

Figure 5 shows the bearing stress and bearing strain of the two joints. Strength results were shown in Table 2. It can be seen that the stiffness of countersunk head bolted joints is much lower than that of protruding head bolted joints. This is mainly due to the fact that the weak lateral support makes the bolt easy to tilt to ones side in the loading process and makes the stiffness lower.

After examination of the specimen failure modes, it was found that except 2 of 15 specimens, the failure mode is mainly bearing failure for countersunk bolted joints. The left 2 are bolts shear failure. However, for protruding bolted joints all the specimen failure mode is bearing failure.

From Table 2, it was found that when the proportion of 0 degrees increase from 30% to 50% and the proportion of ±45 degrees decrease from 60% to 40%, the ultimate strength almost keeps unchanged.

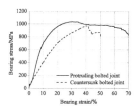

Figure 5. Typical bearing stress-strain history for protruding and countersunk bolted joints.

According to the test results, it can be seen that, the strength of countersunk is lower than that of protrude bolted joints. Referenced to protrude bolted joints, the correction factor of countersunk bolted joints is about 0.86~0.93.

Table 2. Test results.

Ply proportion (%)	Bolt head types	Ultimate strength (MPa)	Offset strength (MPa)
30/60/10	C	919	656
	P	1029	761
	C_f	0.89	0.86
40/50/10	C	871	633
	P	951	694
	C_f	0.92	0.91
50/40/10	C	956	639
	P	1023	742
	C_f	0.93	0.86

*C-countersunk; P-protruding; C_f-correction factor.

4.2 Analysis results

The bearing stress/strain curves of protruding bolted joints were shown in Figure 6. The bearing stress increased linearly with bearing stress till 650MPa, at which point the damage initiates around the hole edge. After the point, the plate internal damage propagates outwards continuously, and the joint experienced catastrophic failure at around 1000MPa.

Figure 7 shows the bearing stress/strain curves for countersunk bolted joints. It can be seen that the bearing behavior of the three laminates had little difference. Because of the convergence problem, the solver is terminated when the bearing stress is up to 800MPa. It is unable to obtain the joint ultimate bearing strength, but the offset strength is available.

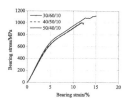

Figure 6. Bearing stress/strain history for protruding bolted joints.

The damage modes in the plate at ultimate strength are shown in Figure 8, including fiber tension and shear failure, fiber compression failure, matrix tension and shear failure and matrix compression. The damage area is found to be in the front of the hole edge and −90°~90° direction. The fiber tension and shear failure mainly occurs in the 90° direction. When fiber tension and shear damage occur, the bearing stress drops rapidly. The tensile-shear failure would bring serious numerical problem which makes the arithmetic easy divergence and unable to get the ultimate strength of the joints.

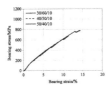

Figure 7. Bearing stress-strain history for countersunk bolted joints.

(a) Fiber tension-shear failure

(b) Fiber compression failure

(c) Matrix tension failure

(d) Matrix compression failure

Figure 8. Damage modes of protruding bolted joints.

Table 3. Analysis results.

Ply proportion (%)	Head types	Ultimate strength(MPa)		Offset strength(MPa)	
		Test	FEM	Test	FEM
30/60/10	C	919	/	656	674
	P	1029	1000	762	803
40/50/10	C	871	/	633	678
	P	951	994	694	798
50/40/10	C	956	/	639	655
	P	1023	1117	742	827

The predicted ultimate strength and offset strength, compared with test data, are shown in Table 3. It can be seen that the ultimate strength prediction error is less than 10% and offset strength prediction error is less than 15%. Besides, it is found that the prediction values falls between the maximum value and minimum value. Considering the discreteness of test data, the prediction model has high precision.

5 CONCLUSION

Experiment and finite element method were used to study the effect of the two ends of the bolts on the joint performance of the composite joints.

1 Countersunk head bolted joints strength is about 90% of that of protruding head bolted joints.
2 0 degree plies and ±45 degree plies have similar strength when used in countersunk head bolted joints.
3 Hashin failure criteria and Tan-Camanho damage propagation rules can be used to predict the strength of bolted joint. The ultimate strength error is within 10% and the offset strength error is within 15% compared with test data. The prediction results are in good agreement with experimental data.
4 Due to the computational convergence problem, using the present finite element model can only obtain ultimate strength when applied to countersunk bolted joints. Further improvements will be needed.

REFERENCES

[1] Abaqus, Inc. 2009. *Abaqus user manual, Version 6.10.* Pawtuckat, Dassault Systems Simulia Corp.
[2] Camanho, P.P. & Matthews, F.L. 1999. A progressive model for mechanically fastened joints in composite laminates. *Composite Materials* 33(24): 2248–2280.
[3] Chang, F.K. & Chang, K.Y. 1987. A progressive damage model for laminated composites containing stress concentrations. *Composite Materials* 21(9): 834–855.
[4] Chishti, M. & Wang, W.H. & Thomson, R.S. & Orifici, A.C. 2012. Experimental investigation of damage progression and strength of countersunk composite joints. *Composite Structures* 94: 865–873.
[5] Egan, B. & McCarthy, C.T. & McCarthy, M.A. & Frizzell, R.W. 2012(a). Stress analysis of single-bolt, single-lap, countersunk composite joints with variable bolt-hole clearance. *Composite Structures* 94: 1038–1051.
[6] Egan, B. & McCarthy, C.T. & McCarthy, M.A. & Gray, P.J. & Frizzell, R.M. 2012(b). Modelling a single-bolt countersunk composite joint using implicit and explicit finite element analysis. *Computational Materials Science* 64: 203–208.
[7] Egan, B. & McCarthy, M.A. & Frizzell, R.M. & Gray, P.J. & McCarthy, C.T. 2014. Modelling bearing failure in countersunk composite joints under quasi-static loading using 3D explicit finite element analysis. *Composite Structures* 108: 963–977.
[8] Hashin, Z. 1980. Failure criteria for unidirectional fiber composites. *Journal of Applied Mechanics* 47: 329–334.
[9] Ireman, T. 1998. Three-dimensional stress analysis of bolted single-lap composite joints. *Composite Structures* 43: 195–216.
[10] Matzenmiller, A. 1995. A constitutive model for anisotropic damage in fiber-composites. *Mechanics of Materials* 20: 125–152.
[11] Tserpes, K.I. & Labeas, G. & Papanikos, P. & Kermannidis, Th. 2002. Strength prediction of bolted joints in graphite/epoxy composite laminates. *Composite Part B* 33(7): 521–529.
[12] Tan, S.C. 1991. A progressive failure model for composite laminates containing openings. Composite Materials 25(5): 556–577.

Study on the shrinkage stress and temperature stress of cement stabilized macadam base

W. Wang & Z.S. Ren
School of Transportation of Southeast University, Nanjing, China

L. Zhang
Intelligent Transportation System Research Center of Southeast University, Nanjing, China

X. Lin & P. Peng
School of Transportation of Southeast University, Nanjing, China

ABSTRACT: Cement stabilized macadam base can easily suffer from shrinkage cracks under large temperature variation and heavy vehicle traffic. In order to improve the cracking resistance, different test schemes were proposed through changing the amount of expansive agent and shrinkage reducing agent, and both mechanistic test and numerical simulation were carried out to obtain the optimum mixture design. Through this research, the following conclusions were reached: the optimum contents of expansive agent and shrinkage reducing agent are 6% and 2% by weight. When both the drying shrinkage and thermal shrinkage are considered, the longitudinal stress and transverse stress of the optimum will decrease by 24% and 26% respectively compared with the specimen without additives. Furthermore, thermal stress significantly affects the shrinkage crack when temperature decrease suddenly. Therefore, low temperature during the cold season has a bad effect on the road.

1 INTRODUCTION

Cement stabilized macadam base provides excellent support for asphalt pavement. The stabilized base material is stronger, and more compressively resistant than the un-stabilized one. However, much attention was paid to that the cement stabilized macadam base actually shows more disadvantages with the increase in heavy vehicle traffic. For example, cement stabilized macadam base can be the source of shrinkage cracks in the base layer, which can reflect through the asphalt surface layer. The cracks tend to be even worse due to the effect of severe environmental temperature in winter, and gradually damage the pavement. This paper mainly focuses on improving the performance of cement stabilized macadam base, and analyzing the shrinkage stress and the temperature stress.

To improve the performance, Cho, Y. H developed a cement-treated base with a lower shrinkage and thereby prevented cracks. In this research, two mixture designs were selected: a mixture containing 25% fly ash and a mixture containing 25% fly ash with 10% expansive additive. The tests suggested that the mix design of 25% fly ash and 10% expansive additive was the optimal mixing alternative [1]. Collepardi, M showed the advantages of the combined use of shrinkage reducing agent (SRA) and Cao-based expansive agent to produce shrinkage- compensating concrete

even without an adequate wet curing [2-3]. Huang analyzed the relationship between tensile strength, shrinkage, self-desiccation shrinkage and flat plate restraint cracking, the experimental results showed that physical shrinkage-reducing of SRA and chemical expansion of expansion agent (EA) enhance the volumetric stability of concrete and reduce the probability of cracking [4]. Wang found the optimal application method of shrinkage reducing agent (SRA). The test results showed that the effect of SRA by surficial painting was the same as by mixing. However the latter can improve the workability of fresh concrete. Furthermore, using SRA and expansion agent (EA) together could improve shrinkage-reducing performance and decrease the risk of concrete crack which often occurred while only adding EA [5]. Cheng proposed the selection method of anti-crack materials, and the mechanical properties of cement stabilized macadam base was also studied, with additives of both the expansive agent and shrinkage reducing agent [6]. Button, J. M and Lytton, R. L studied on the guidelines for using geosynthetics, definitions of the various types of geosynthetics that were commercially available along with some of their advantages and disadvantages [7].

Despite these literary achievements, there is little research on the combined admixture of expansive agent and shrinkage reducing agent. This paper provided the different test schemes through changing amount of expansive agent and shrinkage reducing agent after the raw material and graduation were determined. Moreover, shrinkage stress and temperature stress of cement stabilized macadam base were analyzed based on the basic theory of heat transfer with the finite element method.

2 PERFORMANCE TEST OF HIGH QUALITY CEMENT STABILIZED MACADAM

2.1 Test scheme

Graduation of cement stabilized macadam in Jiangsu province, which had crack resistance, was adopted as the new graduation. The graduation was shown in Table 1.

Table 1. The graduation.

Sieve size/ mm	Passing ratio /%							
	315	26.5	19	9.5	4.75	2.36	0.6	0.075
Range/%	100	95~100	68~86	44~62	27~42	18~30	8~15	0~4.5

The maximum dry density and the optimum moisture content were obtained through changing the amount of expansive agent and shrinkage reducing agent after the raw material and graduation were determined. The cement dosage was 4.5%. The expansive agent and shrinkage reducing agent used UEA-H and BT-5001, respectively, the test scheme was shown in Table 2.

Table 2. The proportion of the different additives.

Numbering	Cement/%	Expansive agent /%	Shrinkage reducing agent /%
1	4.5	–	–
2	4.5	4	3
3	4.5	6	2
4	4.5	8	1

2.2 Test results

Fig. 1(a) showed the un-combined compression strength (UCS) values with 95% confidence, Fig. 1(b) showed the indirect tensile stress values with 95% confidence.

Fig. 1(a) shows that the values of USC at 7 day is group 2 > group 3 > group 4> group 1; the values of group 1~3 at 28 day have different increments, compared with 7 day. It suggests that USC increases

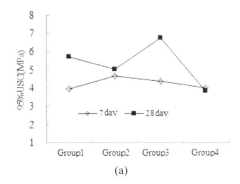

(a)

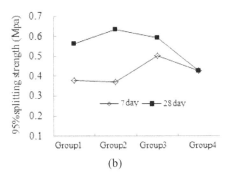

(b)

Figure 1. The test results.

firstly and then decreases along with the increasing of proportion of the additives. Also, group 3 is the optimum. In other words, the optimum contents of expansive agent and shrinkage reducing agent are 6% and 2%, respectively. Fig 1(b) shows that the values of indirect tensile stress at 7 day is group 3 > group 4 > group 1> group 2; the values of group 1~3 at 28 day have different increments compared with 7 day. It suggests that the development of indirect tensile stress is influenced along with the increasing of proportion of the expansive agent. Also, group 3 is the optimum. In other words, the optimum proportion of expansive agent and shrinkage reducing agent is 6% and 2%, respectively. Table. 3 shows total dry shrinkage coefficient.

Table. 3 shows that total dry shrinkage coefficient of group 3 is the smallest, and the following is group 4, group 2 and group1.

Table 3. Total dry shrinkage coefficient.

	Group 1	Group 2	Group 3	Group 4
Total dry shrinkage coefficient /%	0.7404	0.6656	0.5587	0.6034

3 THE MODELING AND SIMULATION OF THE CEMENT STABILIZED MACADAM BASE

3.1 The basic theory

The 3-D equation of heat conduction is as follows:

$$\frac{\partial u}{\partial t} + div(U_u) = k(\frac{\partial^2 u}{\partial x^2} + \frac{\partial^2 u}{\partial y^2} + \frac{\partial^2 u}{\partial z^2}) = k(u_{xx} + u_{yy} + u_{zz}) \quad (1)$$

Where $u = (t, x, y, z)$ is temperature, which is a function of time t and space (x, y, z); ∂_u / ∂_t is the rate of temperature vs time; k is determined with thermal conductivity, density and specific heat.

The equation of solar radiation q(t) is as follows.

$$q(t) = \begin{cases} 0 & 0 \le t \le 12 - \dfrac{c}{2} \\ q_0 \cos m\omega(t - 12) & 12 - \dfrac{c}{2} \le t \le 12 + \dfrac{c}{2} \\ 0 & 12 + \dfrac{c}{2} \le t \le 24 \end{cases} \quad (2)$$

Where q_0 is the biggest radiation, $q_0 = 0.131m$ Q, $m = 12/c$; Q is the total solar radiation of a day (J/m^2); c is the time of sunshine (h); ω is angular frequency, $\omega = 2\pi / 24$.

To make the function become smooth and continuous, equation (2) was expanded based on the Fourier series.

$$q(2t) = \frac{a_0}{2} + \sum_{k=1}^{\infty} a_k \cos \frac{k\pi(t-12)}{2} \quad (3)$$

$$a_0 = \frac{2q_0}{m\pi} \quad (4)$$

$$a_k = \begin{cases} \dfrac{q_0}{\pi} \left[\dfrac{1}{m+k} \sin(m+k) \dfrac{\pi}{2m} + \dfrac{\pi}{2m} \right] & k = m \\ \dfrac{q_0}{\pi} \left[\dfrac{1}{m+k} \sin(m+k) \dfrac{\pi}{2m} + \dfrac{1}{m-k} \sin(m-k) \dfrac{\pi}{2m} \right] & k \ne m \end{cases} \quad (5)$$

The equation of temperature is as follows considering the influence of solar radiation.

$$T_a = \overline{T_a} + T_m \left[0.96 \sin \omega(t - t_0) + 0.14 \sin 2\omega(t - t_0) \right] \quad (6)$$

Where $\overline{T_a}$ is average temperature of a day (°C), $\overline{T_a} = \frac{1}{2}(T_a^{max} + T_a^{min})$; T_m is the amplitudes of temperature (°C), $T_m = \frac{1}{2}(T_a^{max} - T_a^{min})$; T_a^{max} and T_a^{min} are the maximum and minimum temperature; t_0 is initial phase.

Figure 2. The finite element model.

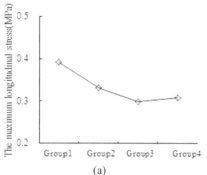

(a)

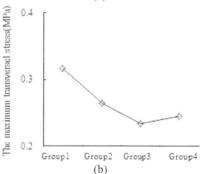

(b)

Figure 3. The simulation results.

The equation of heat transfer coefficient and winds is as follows:

$$h_c = 3.7v_w + 9.4 \quad (7)$$

Where h_c is the heat transfer coefficient $(W / (m^2 \, ^\circ C))$; v_w is winds (m/s).

3.2 Simulation results

The model consists of three parts: the surface course, cement stabilized base and under layers. The finite element model was shown in Fig. 2.

Considering both the drying shrinkage and thermal shrinkage, Fig. 3(a) shows the longitudinal stress, and Fig. 3(b) shows the transverse stress.

Fig. 3(a) and Fig. 3(b) show the value of four groups is: group 1 > group 2 > group 4 > group 3. It suggests that the emergence and development of longitudinal shrinkage crack and transverse shrinkage crack are prevented by adding the expansive agent and shrinkage reducing agent. group 3 is the minimum. In addition, the transverse stress is smaller than the longitudinal stress. The maximum longitudinal stress of group 1 increased by 107%, and group 2~4 increased by 83%, 101% and 94%, respectively. It suggests that shrinkage crack is influenced by temperature stress when the temperature is suddenly dropped. Therefore, low temperature during the cold season is one of the reasons for the road distresses. From the above, group 3 is the optimum, when both the drying shrinkage and temperature shrinkage are considered, the longitudinal stress and transverse stress will decrease by 24% and 26%, respectively.

4 CONCLUSIONS

This paper provided the different test schemes through changing the amount of expansive agent and shrinkage reducing agent, and both the dry shrinkage stress and the thermal stress of cement stabilized macadam base were simulated. It was found that the optimum contents of expansive agent and shrinkage reducing agent are 6% and 2% according to the compressive strength test and indirectly tensile strength test, and the same results were obtained by analysis of the stress field. The longitudinal stress and transverse stress considering both the drying shrinkage and thermal shrinkage decrease by 24% and 26% respectively. compared with the specimen without additives.

ACKNOWLEDGMENTS

This work was financially supported by the China Natural Science Foundation (51278518) and the Fundamental Research Funds for the Central Universities (2242013R30015).

REFERENCES

[1] Cho, Y.H. Lee, K.W. & Ryu, S.W. 2006. Development of Cement-Treated Base Material for Reducing Shrinkage Cracks. *Transportation Research Record*: 134–143.
[2] Collepardi, M. Borsoi, A. & Collepardi, S. 2005. Effects of shrinkage reducing admixture in shrinkage compensating concrete. *Cement & Concrete Composites* 27:704–708.
[3] Maltese, C. Pistolesi .C & Lolli, A. 2005. Combined effect of expansive and shrinkage reducing admixtures to obtain stable and durable mortars. *Cement & Concrete Composites* 35:2244–2251.
[4] Huang, Y. F. Wang, Y. F. Wang S, N. & Zhang, B. L. 2006. Co-effecton concrete anti-cracking by SRA and EA. *Port & Waterway Engineering* 4: 19–22.
[5] Wang, Y. F. Ma, B. G. & Huang, Y. F. 2006. Effective experimental study on performance of anti-cracking agent. *Journal of Wuhan University of technology* 28(12): 41–44.
[6] Cheng, J. Sun, Y. C. & Xu, Z. H. 2007. Study on new anti-crack materials in cement-stabilized macadam. *Shanghai Highways* 4:49–53.
[7] Button, J.W. & Lytton, R.L. 2004. Guidelines for using geosynthetics with hot-mix asphalt overlays to reduce reflective cracking. *Journal of the Transportation Research Board* (1): 111–119.

Testing and Measurement: Techniques and Applications – Chan (Ed.)
© *2015 Taylor & Francis Group, London, ISBN: 978-1-138-02812-8*

A performance comparison study of the Sorptive Building Materials with MEMS-based formaldehyde gas sensor from the aspect of reduction efficiency of indoor formaldehyde concentration

H. Huang
Department of Architecture, National Cheng Kung University, Tainan, Taiwan

C.C. Chen
Department of Imterior Design, Tung Fang Design Institute, Kaohsiung, Taiwan

C.T. Tzeng
Department of Architecture, National Cheng Kung University, Tainan, Taiwan

ABSTRACT: The problem of fugitive pollutants arising from the extensive use of indoor building materials in air-conditioned buildings in Taiwan directly affects the indoor air quality. Thus, how to balance the building indoor "healthy air quality" has become an important issue. It also probes into the application of the Sorptive Building Materials (SBMs) in indoor space to adsorb the pollutants in the air (formaldehyde) by calculating "sorption flux per time per area", "sorption efficiency" and "sorption equivalent per area". According to the standard method of ISO 16000, this study conducted the building material fugitive and absorption experiments with MEMS-Based Formaldehyde Gas Sensor by testing with the building materials (0.099m3) environmental chamber (25°C, 50% RH, Loading Factor 0.4 m2/m3). The research results showed that the sorption equivalent per area of SBMs is also the most effective. If used, it can maintain indoor air quality.

1 INTRODUCTION

According to the survey (P.C. Wu, et al.,2003), indoor formaldehyde and VOCs concentrations are relatively high in Taiwan due to the hot and humid climate, emission of chemical substances in interior building materials, and lack ventilation, thus causing high health risk hazards. Formaldehyde and volatile organic compounds (VOCs) are specifically the main pollutants in indoor, which negatively influence people's health, comfort and productivity (Kim et al., 2001; Godish,2001). It has been found that various building materials are sources of indoor formaldehyde and VOCs (Cox et al., 2002). Moreover, there is also evidence that building materials can act as a sink of formaldehyde and VOCs by sorption, and the re-emission of sorbed formaldehyde and VOCs can dramatically elevate indoor concentrations for months or even years (Tichenor et al., 1988; Sparks et al., 1994). Therefore, this study conducted a chamber experiment to evaluate the lowering of the indoor formaldehyde and VOCs concentrations and the removing efficiency by changing the "ventilation amount". The results were compared with the sorptive building materials (SBMs) in terms of the performance of absorbing

formaldehyde in order to find out the efficiency of its equivalent ventilation rate per area.

The concentrations of indoor formaldehyde and volatile organic compounds (VOCS) are relatively high in Taiwan (P.C Wu, et al., 2003) due to warm-humid climate, emission of chemical substances from interior building materials and poor ventilation. Formaldehyde and VOCs are the main pollutants in the indoor environment, which can negatively affect resident's health, comfort and productivity (Kim et al., 2001; Godish, 2001). It has been found that various building materials are the sources of indoor formaldehyde and VOCs (Cox et al., 2002). Moreover, certain building materials can absorb and re-emit the formaldehyde and VOCs, and the re-emission can dramatically elevate indoor concentrations for months or even years (Tichenor et al., 1998; Sparks et al., 1994). This study conducted a chamber experiment with different level of the ventilation amount to evaluate the reduction of formaldehyde and VOCs concentrations, as well as the removing efficiency. In order to find out the efficiency of its equivalent ventilation rate per area, the results were compared with the sorptive building materials (SBMs) in terms of the performance of absorbing formaldehyde.

2 EXPERIMENTAL

Based on ISO 16000 (part-9, 3,6,23,24) and ASTM-D5116 & D6670 standards, this study used the chamber experiment to conduct the building material effusion test and absorption test. In the 72-hour test under the conditions of temperature of 25°C, relative humidity of 50 % and loading factor is 0.4 (m2/m3), the building materials and SBMs were tested. The experiment with building materials analyzed the efficiency of reduction and removed formaldehyde and VOCs concentrations by increasing the ventilation rate (ACH 0.25 to 2.0). The experiment with SBMs analyzed the sorption flux per area and equivalent ventilation rate per area by changing the supply of formaldehyde of different concentrations (0.1–0.2 ppm) (ISO 16000-23).

The standard testing methods for aldehydes and ketone substances, and VOCs substances used in this study are referenced from the ISO 16000-9 regulaons by small scale chamber (0.099m3) test method, nd the ISO16000-3 (DNPH /HPLC) and ISO16000-6 Tenax-TA/GC-Mass) regulated aldehyde and ketone ubstances and VOC sampling and analysis methods. he Sorptive Building Materials (SBMs) test method s from the ISO 16000-23,24.

Figure 1.   The small scale test chamber (0.099m3).

Table 1.   Design of the experiment.

		board			
Sorptive Building Materials (SBMs) (Small scale chamber)	SBMs	latex paint	0.5 ACH (h⁻¹)	0.2 ppm	25°C,50%
		carbonation plywood	0.5 ACH (h⁻¹)	0.1 ppm 0.2 ppm	25°C,50% 30°C,50%
		Diatomite plate	0.5 ACH (h⁻¹)	0.1 ppm 0.2 ppm	25°C,50% 30°C,50%
		controlling humidity brick	0.5 ACH (h⁻¹)	0.2 ppm	25°C,50%

The methods of detection for formaldehyde gas may be divided into three main categories: GC/MS, optical detection devices, and MEMS based gas sensors. Gas chromatography–mass spectrometry (GC–MS) is a method that combines the features of gas–liquid chromatography and mass spectrometry to identify different substances within a test sample. Although it provides high sensitivity and selectivity, the drawbacks of high preventive costs and ponderous uses cannot be ignored. Numerous researchers have studied optical sensor with formaldehyde quantification applications. Even though the optical sensors are capable of simultaneous samplings and have instantaneous analyzing time, the associated optical arrangements tend to be rather bulky and elaborate. In the last decade, emerging MEMS and micro-machining techniques have led to the development of miniaturized sensing instrumentation that is capable of accessing information at a micro scale level. More importantly, the functionality and reliability of these micro sensors can be increased through their integration with mature, logic IC technology or other sensors (Lee et al. 2007).

There is a new process for the fabrication of a MEMS-based formaldehyde sensing device comprising of a micro heater and electrodes with Pt resistance heaters and a sputtered NiO/ Al2O3 layer. The experimental data indicated that by decreasing the thickness of the sensing layer in the sputtering process and increasing the area of the sensing layer contacting with the surrounding gas and co-sputtered NiO/ Al2O3 sensing layer, significantly increased the sensitivities of the gas sensor and improved its lowest detection limit capabilities (Lee et al. 2007).

Figure 2.   MEMS-based formaldehyde gas sensor.

3 RESULTS AND DISCUSSION

3.1 *The reduction efficiency of indoor formaldehyde concentration by using SBMs*

This study used five SBMs (wood fiber board, latex paint, carbonation plywood, Diatomite plate, and controlling humidity brick) for adsorption performance tests. The test was conducted in the small building environmental control box (0.099m3) under the conditions (25°C, 50% RH, Loading Factor 0.4 m2/m3) to examine the performance of SBMs by formaldehyde concentration and temperature. The performance benefits were calculated by adsorption rate, sorption flux per time per area.

Table 2. The sorptive building materials (SBMs).

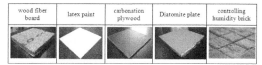

wood fiber board	latex paint	carbonation plywood	Diatomite plate	controlling humidity brick

3.2 Comparison of different SBMs adsorption performance

This study tested the adsorption performance of five SBMs (wood fiber board, latex paint, carbonation plywood, Diatomite plate, controlling humidity brick) under the conditions of (25°C, 50% RH, Loading Factor 0.4 m²/m³, formaldehyde 0.2ppm). The adsorption performance results showed that the mean absorption rates after 72-hour adsorption test are: wood fiber board is 66.26%, latex is 42.15%, humidity brick is 57.04%, carbonization plywood is 50.45% and Diatomite plate is 53.76%.

Table 3. The adsorption performance of Sorptive Building Materials (SBMs).

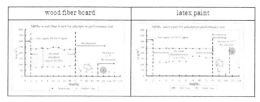

wood fiber board	latex paint

Table 4. The adsorption performance of Sorptive Building Materials (SBMs).

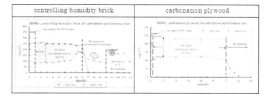

controlling humidity brick	carbonation plywood

3.3 Comparison of SBMs adsorption performance under different concentrations

This study tested two SBMs (carbonation plywood, Diatomite plate) in adsorption performance (comparing formaldehyde 0.2ppm and 0.1ppm, 50% RH, Loading Factor 0.4 m²/m³), and formaldehyde concentration decreased from 0.2ppm to 0.1ppm. According to the adsorption performance results, under the condition of 0.1ppm, the average absorption rate decreased compared to 0.2ppm. The decrease rates are: 30.31% for the carbonized plywood and 7.1% of the Diatomite plate. If converted into equivalent ventilation volume,

the formaldehyde concentration will decrease from 0.2ppm to 0.1ppm, and the equivalent ventilation volumes of SBMs will also decrease.

Table 5. The reduction efficiency of indoor formaldehyde concentration by using SBMs.

Experiment	materials	Supply Gas HCHO	Temp	RH	F_{in} μg (m²*h)	F_{50} m² (m²*h)	Sorption rate, %
Sorptive Building Materials (SBMs) (Small scale chamber)	wood fiber board	0.2 ppm	25°C	50%	192.95	2.43	66.26
	latex paint	0.2 ppm	25°C	50%	83.03	0.6	42.15
	controlling humidity brick	0.2 ppm	25°C	50%	175.03	1.643	57.04
	carbonation plywood	0.1 ppm	25°C	50%	40.3	0.42	25.14
		0.2 ppm	25°C	50%	148.59	1.26	50.45
		0.2 ppm	30°C	50%	144.05	1.12	47.52
	Diatomite plate	0.1 ppm	25°C	50%	77.86	1.08	46.66
		0.2 ppm	25°C	50%	168.04	1.44	53.76
		0.2 ppm	30°C	50%	148.85	1.25	50.23

3.4 Comparison of SBMs adsorption performance under different sensor

This study tested two sensors (BK, MEMS-based formaldehyde gas sensors) in adsorption performance (comparing formaldehyde 0.4ppm and 0.2ppm, 50% RH, Loading Factor 0.4 m²/m³) formaldehyde concentration decreased from 0.3ppm to 0.15ppm.

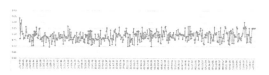

Figure 3. The adsorption performance of BK.

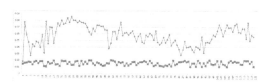

Figure 4. The adsorption performance of MEMS-based formaldehyde gas sensor.

The results showed that the performances of the MEMS-Based Formaldehyde Gas Sensor and BK are similar.

4 CONCLUSION

According to the comparison of ventilation removal and SBMs in terms of the reduction efficiency of indoor formaldehyde and VOCs concentrations,

High-EBMs of increased ventilation rate to remove the contaminants are poorer in removal efficiency as compared with Low-EBMs. However, in the environment of high concentration formaldehyde, SBMs have higher efficiency in terms of absorption efficiency and equivalent ventilation rate per area, indicating that ventilation can be used with SBMs for better efficiency under the condition of high concentrations of formaldehyde and VOCs.

In terms of reduced efficiency of indoor formaldehyde and VOCs concentrations, this comparison study concluded that with the increased ventilation rate, High- EBMs showed lower removal efficiency than Low-EBMs. However, in the environment with high formaldehyde concentration, SBMs are more efficient in terms of absorption efficiency and equivalent ventilation rate per area, which indicate that under the condition of high concentrations of formaldehyde and VOCs, ventilation can be used with SBMs for better performance.

REFERENCES

[1] Cox, S.S., Little, J.C., Hodgson, A.T., (2002). Predicting the emission rate of volatile organic compounds from vinyl flooring. Environmental Science & Technology 36, 709–714.

[2] Godish, T., (2001). Indoor Environmental Quality. Lewis Publishers.

[3] Kim, Y.M., Harrad, S., Harrison, R.M., (2001). Concentrations and sources of VOCs in urban domestic and public microenvironments. Environmental Science & Technology 35, 997–1004.

[4] Sparks, L.E., Tichenor, B.A., White, J.W., Jackson, M.D., (1994). Comparison of data from the IAQ test house with predictions of an IAQ computer model. Indoor Air , vol. 13, pp. 577–592.

[5] Tichenor, B.A., Sparks, L.E., White, J.W., Jackson, M.D., (1988). Evaluating sources of indoor air pollution. In: The 81st Annual Meeting of the Air Pollution Control Association, Dallas, TX.

[6] Wu PC, Li YY, Lee CC, Chiang CM, Su HJ. (2003) Risk assessment of formaldehyde in typical office buildings in Taiwan. Indoor Air, vol. 13, pp. 359–363.

[7] Yu-Hsiang Wang, Chia-Yen Lee, Che-Hsin Lin, Lung-Ming Fu., (2008). Enhanced sensing characteristics in MEMS-based formaldehyde gas sensors. Microsyst Technol (2008) 14:995–1000,pp.995–1000.

Testing and Measurement: Techniques and Applications – Chan (Ed.)
© 2015 Taylor & Francis Group, London, ISBN: 978-1-138-02812-8

Scarf repair of composite laminates

Z.H. Xie & X. Li
Northwestern Polytechnical University, Xi'an, China

Q. Yan
Shenyang Aircraft Design & Research Institute, Shenyang, China

ABSTRACT: The use of composite materials, such as Carbon-Fiber Reinforced Plastic (CFRP) composites, aero-structures has led to an increased need of advanced assembly joining and repair technologies. Adhesive bonded repairs as an alternative to recover full or part of initial strength were investigated. Tests were conducted with the objective of evaluating the effectiveness of techniques used for repairing damage fiber reinforced laminated composites. Failure loads and failure modes were generated and compared with the following parameters: scarf angles, roughness of grind tool and a number of external plies. Results showed that scarf angle was the critical parameter and the largest tensile strength was observed with the smallest scarf angle. Besides, the use of external plies at the outer surface could not increase the repair efficiency for large scarf angle. Preparing the repair surfaces by sanding them with a sander ranging from 60 to 100 grit numbers had a significant effect on the failure load. These results allowed the proposal of design principles for repairing CFRP structures.

1 INTRODUCTION

Adhesively bonded scarf repairs provide a significant recovery of residual strength in damaged composite structure and are compared with external patch repairs, and scarf repairs have relatively high levels to minimize aerodynamic disturbance (Whittingham 2009). For highly loaded with advanced composite structures, scarf angles ranging from 20:1 to 60:1 are often required to restore a damaged structure to its as-designed ultimate strength.

The restore efficiency are affected by many parameters, typically including surface preparing, scarf angles and external plies. Harman and Wang (2006) investigated the use of a low stiffness patch to repair carbon-fiber epoxy composites for scarf joint repairs. They found that a low stiffness patch improves the uniformity of adhesive stresses along the joint and consequently enhances the joint strength. Liu and Wang (2007) used a 3D progressive damage model to investigate the effects of several repair parameters on the failure initiation strength, ultimate strength and failure mechanism of repaired open-hole composite plates in tensile tests. Their research showed that overlay thick patches will deteriorate the strength of the repaired structures. Campilho (2009) investigated the tensile behavior of adhesive-bonded carbon/epoxy scarf repair using scarf angles ranging from 2° to 45°. Two distinct failure modes were observed experimentally: type A and type B failures. Type A was observed for the repairs with 15°, 25° and 45° scarf angles, consisting of an entirely cohesive failure of the adhesive layer, which occurred abruptly without visible crack initiation. Type B failure occurred for the repairs with 2°, 3°, 6° and 9° scarf angles, consisting of cohesive and interlaminar/intralaminar failure. Kumar (2006) investigated adhesively bonded aerospace composite scarf joints, with scarf angles ranging from 0° to 5°, in uniaxial tension both experimentally and numerically. Results showed that fiber fracture and pullout failure occurred for scarf angles less than about 2° and cohesive shear failure of the adhesive occurred for scarf angles more than 2°. The knockdown in tensile strength was most prominent for scarf angles less than about 1°. Charalambides (1998a, 1998b) and Pinto (Pinto 2010) investigated the use of over-laminating plies covering the repaired region at the outer or both repair surfaces as an attempt to increase the repair efficiency. Sung (1998) found that preparing the repair surface by sanding them with a diamond sander ranging from 60 to 400 grit number does not significantly affect failure load.

The current experiments for scarf repairs are designed to study these parameters on the repair performance. The tests were conducted at room temperature. The failure load and failure strength were used to evaluate the repair efficiency.

2 EXPERIMENTAL

Plain weave carbon fiber fabric prepreg plied (with T700 fiber and CYCOM970 resin system) were used in the construction of the parent laminate and the repair patch. The mechanical properties are shown in Table 1.

The plates are manually scarfed to the required scarf angle using a die grinder having a detachable abrasive pad of #60 grade and #100 grade. The scarf surfaces prior the repair are cleaned.

Two separate plates were joined together with the scarf techniques. Figure 1 shows the procedure sequence in the repair process of a damaged laminate. Initially a back plate was bonded at the back of the laminate in order to position the two parts. Then prepreg plies with tailored dimensions are placed on top of the plate. The patch materials are bonded to the laminate using adhesive film. One external ply was placed on the top and its boundaries are sealed with some adhesive. The electrically heated blanket is maintained until the reinforcement (patch) adherence to the parent component, awaiting solidification.

The manufacture process consisted of hand lay-up followed by curing in a hot-plates with 0.8 bar pressure. The heat cycle initiated warming up the plate at 2°C/min, followed by two hours curing at 180°C. Cooling was performed at 3°C/min up to room temperature (Figure 2).

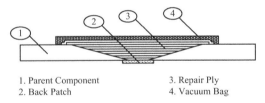

1. Parent Component   3. Repair Ply
2. Back Patch   4. Vacuum Bag

Figure 1.   Schematic of the vacuum patch repair process.

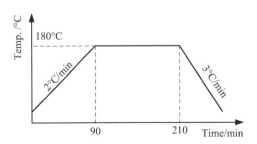

Figure 2.   The temperature cycle used in making the repair.

The coupons are prepared by cutting the panel into the standard coupon sizes (ASTM D3039). The coupons dimensions were shown in Figure 3. Three different scarf angles were conducted, and α=2°, 4° and 6°. In order to

investigate the external plies effect on the repair efficiency, two different laminates were designed with N=0 and 1. Besides, undamaged specimens, noted as scarf angle α=0°, were tested to verify the repair recovery efficiency. A total of five specimens was tested in tension for each case described above. The complete test matrix was shown in Table 2.

The experiments were performed in a displacement control mode. All the specimens were loaded in tension until failure. Based on the failure load, the failure strength is calculated using Eq. (1).

$$\sigma_{max} = F_{max}/Wt \qquad (1)$$

Strain gages were placed at several locations on the specimen as shown in Figure 4. The strain gage readings were used to analyze the damage initiation and damage growth path.

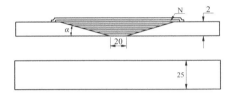

Figure 3.   Dimensions of the scarf repairs coupon.

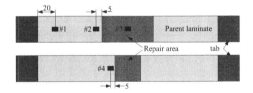

Figure 4.   Strain gauge position of the scarf joint coupon.

Table 1.   Mechanical properties of carbon-fiber/epoxy.

E_1(GPa)	E_2(GPa)	G_{12}(GPa)	v12	
54	54	13	0.044	
X_t(MPa)	X_c(MPa)	Y_t(MPa)	Y_c(MPa)	S(MPa)
537	549	604	724	261

3 TESTING RESULTS

The load-displacement curves for the repaired specimens and undamaged specimens are shown in Figure 5. The repair specimens stiffness reduces as the adhesive modulus is much lower than the parent laminate. Besides, stiffness reduction experienced by the repairs prior to failure was caused by the softening of the adhesive layer at the bond edges.

Table 2. Overview of the scarf joint specimens used in the tests.

Scarf angle	Grit number	External plies	σ_{max}/MPa
0°	–	–	455.1
2°	#60	1	232.6
	#100	1	281.8
4°	#60	1	239.2
	#100	1	192.4
6°	#60	1	168.0
	#100	1	133.6
	#100	0	155.8

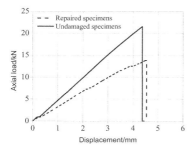

Figure 5. Typical load-displacement curves of the specimens.

The strains measured at critical locations were shown in Figure 6. For the 6° repair structure, the strain increase rate at location #4 decreases with the load whilst it increases at location #3 which indicates that the damage initiates from the bottom of the adhesive and grows along the bond line (see in Figure 7).

Figure 8 shows the typical damage modes observed in the tests conducted. For the case of 4°~6° repair structure, it was seen that the primary damage mode is adhesive debond. However, for the case of 2° repair structure, intra-laminar failure mode was also found.

Figure 9 and Table 2 show the failure strength obtained from the scarf repair approach with different repair parameters. The failure strength of the parent material was approximately 455MPa.

Testing of the scarf joints/repair recovered 29%~62% of the parent laminate strength. The failure strength increases as the scarf angle decreases. A slight decreasing trend in failure strength was observed to increase the scarf angle, even though with some variations.

Customarily, prior to repair the repair surface of the base laminate is sanded. To evaluate the influence of sanding on the failure load of repaired specimens, the repair surfaces were sanded with different grit diamond sanders. The failure loads of specimens repaired using different grit sanders are shown in Figure 9. The data indicate that the grit number of the sander has a significant effect on the failure load.

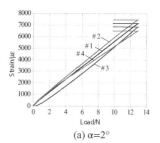

(a) α=2°

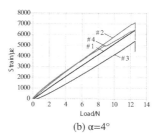

(b) α=4°

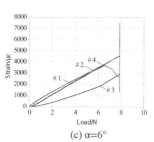

(c) α=6°

Figure 6. Strains measured at critical locations.

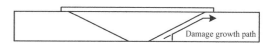

Figure 7. Damage initiation and growth path.

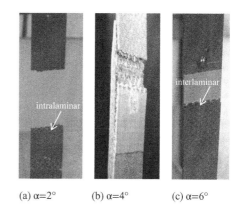

(a) α=2°   (b) α=4°   (c) α=6°

Figure 8. Failure modes in repaired specimens.

For the material and type of repair used in this study, a sander with greater grit number could increase the failure strength for lager scarf angle and inverse trend was observed for small scarf angle.

The addition of external plies on the surface of the specimen resulted in decreased loads. The repair patch applied on one side contributes to out-of-plane bending by affecting the symmetry which resulted in secondary bending effects that begin to add the adhesive peel stress. However, reverse effect was observed in a Sung's work (Sung 1998). This was because that in Sung's work, a smaller scarf angle was used. For the smaller values of α, σ_n normal stresses diminish in relative magnitude to τ_{avg}. Under these conditions, τ_{tn} shear stresses govern the repairs behavior (Campilho 2007).

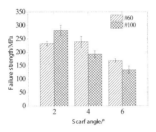

Figure 9. The failure strength of the specimens with different scarf angles and grit number of sanders.

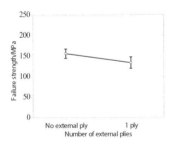

Figure 10. The effect of the number of external plies on the failure strength in scarf repair.

4 CONCLUSION

The strength of scarf joints in composite structures has been thoroughly investigated. A series of scarf repair tests were performed to obtain their strength. The mechanical response and failure behavior of the specimens with different scarf angles and surface treatment method were evaluated via the comparison of experimental data. The experimental data show that the optimal results are obtained from 2 degree using 1 layer for the external plies. The 2 degree repair results in an intralaminar failure mode, which is a higher strength failure mode. Repairs restored the strength to within 62% of the pristine panels. However, it should be noted that external plies would reduce failure load for large scarf angle repairs as a result of a secondary bending effect.

ACKNOWLEDGMENTS

Project supported by the State Key Program of National Natural Science of China(Grant U1233202).

REFERENCES

[1] Campilho, R.D.S.G. & de Moura, M.F.S.F & Pinto, A.M.G. & Morais, J.J.L. & Domingues, J.J.M.S. 2009. Modelling the tensile fracture behaviour of CFRP scarf repairs. *Composites: Part B* 40: 149–157.
[2] Campilho, R.D.S.G. & de Moura, M.F.S.F & Domingues. J.J.M.S. 2007. Stress and failure analyses of scarf repaired CFRP laminates using a cohesive damage model. *Journal of Adhesive Science and Technology* 21: 855–870.
[3] Charalambides, M.N. & Hardouin, R. & Kinloch, A.J. & Matthews, F.L. 1998a. Adhesively bonded repairs to fibre composite materials I: experimental. *Composites Part A: Applied Science and Manufacturing* 29A: 1371–1381.
[4] Charalambides, M.N. & Hardouin, R. & Kinloch, A.J. & Matthews, F.L. 1998b. Adhesively-bonded repairs to fibre composite materials II: finite element modelling[J]. *Composites Part A: Applied Science and Manufacturing* 29A: 1383–1396.
[5] Harman, A.B. & Wang, C.H. 2006. Improved design methods for scarf repairs to highly strained composite aircraft structure. *Composite Structures* 75(1–4): 132–144.
[6] Kumar, S.B. & Sridhar, I. & Sivashanker, S. & Osiyemi, S.O. & Bag A. 2006. Tensile failure of adhesive bonded CFRP composite scarf joints. *Materials Science and Engineering B*, 132: 113–120.
[7] Liu, X. & Wang, G. 2007. Progressive failure analysis of bonded composite repairs. *Composite Structures* 81(3): 331–340.
[8] Pinto, A.M.G. & Campilho, R.D.S.G. & de Moura, M.F.S.F. & Mendes, I.R. 2010. Numerical evaluation of three-dimensional scarf repairs in carbon-epoxy structures. *Adhesion and Adhesives* 30: 329–337.
[9] Sung, H.A. & Springer, G.S. 1998. Repair of composite laminates-I: test results. *Journal of Composite Materials* 32(11): 1036–1074.
[10] Whittingham, B. & Baker, A.A & Harman, A. & Bitton, D. 2009. Micrographic studies on adhesively bonded scarf bonded scarf repairs to thick composite aircraft structure. *Composite Part A: Applied Science and Manufacture* 40(9): 1419–1432.

Testing and Measurement: Techniques and Applications – Chan (Ed.)
© 2015 Taylor & Francis Group, London, ISBN: 978-1-138-02812-8

Pile material and soil damping effects on Pile Integrity Test (PIT)

R.B. Nazir
Associate Professor, Faculty of Civil Engineering, Universiti Teknologi Malaysia

O. El Hussien
PhD Student, Faculty of Engineering, Universiti Teknologi Malaysia

ABSTRACT: Pile-soil interaction effect on Pile Integrity Test (PIT) is not properly identified. However, it may lead to inconclusive results when piles are cast in hard soil strata. In integrity testing of piles, soil within which the pile is embedded causes attenuation of the propagated stress wave. Numerical solution of wave equation was utilized to arrive to a theoretical model, which accommodates pile material and soil damping effects on wave propagation in pile body. Results obtained from the model were compared with real in situ PIT results and were found to have good similarity. In addition, the model is used to assess the damping effects on the PIT results.

1 INTRODUCTION

1.1 Low strain integrity test of piles (PIT)

The low strain integrity test is used worldwide to assess piles integrity after pile construction. This test method is economic, fast, easy to apply and relatively reliable. The test is carried out by striking pile top with a small hammer. Stress wave will be generated as a result of the impact, which travels down pile body to pile bottom where it reflects back. The reflected wave generates a measurable pile top displacement, and the pile is considered to be free from major damages when the stress wave is received at the correct time at pile top and no earlier reflections are received. The impact force is measured by the instrumented hammer itself. Hence, the velocity record and to a lesser degree the force record will obtain information about pile non-uniformities. The increase in soil resistance force results in a decrease in the measured pile top velocity (Johnson & Rausche, 1996).

The velocity record can be used to assess pile quality assuming a proportional relation between force and velocity. This type of integrity tests is based on the theory of wave propagation and the name "low strain test" is based on the fact that when a light dynamic force is applied to a pile it generates a low strain compression wave that propagates down pile body at a constant speed. Measurable strains of around 10-5 micro strain will be produced as a result of the impact load (Hussein and Garlanger, 1992). Changes in pile cross sectional areas such as a reduction in size or material deficiencies such as a void and cracks in concrete produce wave reflections, which can be used for pile integrity assessment.

The impact force produces a one dimensional stress wave into the pile that propagates at speed "c", where "c" is a function of the pile density "ρ" and pile material modulus of elasticity "E". The data process is usually carried out under an assumed wave speed. Pile impedance "Z" is defined as the elastic modulus "E" times the cross sectional area "A" divided by material wave speed "c" (i.e. Z=EA/c).

Impedance changes are attributed to changes in the pile cross sectional area and material quality. A sample record for cast in place pile is shown in Figure 1.

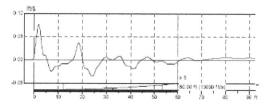

Figure 1. Typical Pile Integrity Test (PIT) record for cast in place pile, (Massoudi and Teferra, 2004).

1.2 Pile-soil interaction in PIT

Pile integrity testing method was introduced by the Center of Experimental du Batiment et des Travaux Publics (C.E.B.T.P.) in France through combining soil effects and the mobility calculated at the pile top (Paquet, 1992). In this method, the pile is divided into small segments where the characteristic impedance of pile segment and shear wave velocity of soil is considered to be the main characteristics of the analysis. This method is only used in specific conditions as a correcting technique.

Interpretation of pile integrity test was carried out by Rodriguez and Restrepo (2008) using an analytic solution of a modified wave equation, which includes soil friction and energy dissipation terms. The analytic solution was based on soil stiffness and impedance change effects on the propagated wave. The analytic simulation was found to produce comparable results with real integrity test of a defective pile. However, the deficiencies associated with pile integrity test when hard soil strata surrounds the pile, were not evaluated. In addition, the model is not correlated with the standard soil parameters, which are identified from laboratory and field tests.

2 MODELING PROCEDURE

2.1 Mathematical background

Assuming a bar having density (ρ) is subjected to dynamic force (F) resulted in linear displacement (u) in the (z) direction at time (t). The basic mathematical formulations will be:

$$-\sigma + \sigma + \frac{\partial \sigma}{\partial z} dz = \rho\, dz \frac{\partial^2 u}{\partial t^2} \tag{1}$$

$$v_i = \frac{u_i(t + \Delta t) - u_i(t)}{\Delta t} \tag{2}$$

$$\varepsilon = \frac{\partial u}{\partial z} \tag{3}$$

Substituting Equation 2 and Equation 3 in Equation 1, the partial differential equation, which is known as the wave equation, can be arrived:

$$\frac{\partial^2 u}{\partial t^2} = c^2 \frac{\partial^2 u}{\partial z^2} \tag{4}$$

Where,

$$c = \sqrt{\frac{E}{\rho}} \tag{5}$$

Equation 4 describes wave motion within the bar body. This equation is considered the basic equation used to analyze piles subjected to dynamic tests.

2.2 Model formulation

Based on the wave equation concept, pile material and soil damping effects were introduced into the equation.

During integrity test, the pile is subjected to an impact with a light hammer. Upon hammer strike, the impact will produce the following:

1 Downward displacement of pile body.
2 Mobilization of soil friction at the surface area of the pile.
3 Stress wave propagation through pile body, which will be:
  a. Attenuated by pile material damping effect.
  b. Attenuated by soil friction.
  c. Partially reflected at the pile toe after being attenuated by soil resistance at pile toe.

Assume the pile shown in Figure 2 having density (ρ), cross sectional area (A), circumference area (A_s) elastic modulus (E) and pile material damping coefficient (C) is subjected to a dynamic force (F_0) resulted in linear displacement (u) in the (z) direction at time (t).

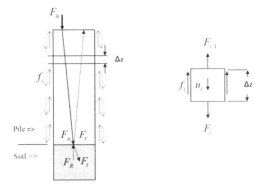

Figure 2.   Schematic diagram for a pile subjected to impact on top.

Wave motion within pile body can be described by the wave equation after incorporating soil and material damping effects as follows:

$$\rho A \frac{\partial^2 u}{\partial t^2} = E A \frac{\partial^2 u}{\partial z^2} - M u A_s - C \frac{\partial u}{\partial t} \tag{6}$$

where, (M) is the proportionality constant between soil frictional force and pile segment displacement, which is calculated as follows Kraft et al. (1981):

$$M = \frac{G_s}{r_0 \ln\left(\frac{r_m}{r_0}\right)} \tag{7}$$

where (r_0) is pile radius, (G_s) is soil shear modulus and (r_m) is the lateral distance from the pile center where soil vibration caused by the dynamic force applied on the pile diminishes.

The numerical solutions for the above equation are as follows:

$$F_i - f_i A_{si} - C v_i = \rho \, A \, \Delta z \, \frac{v_i(t+\Delta t) - v_i(t)}{\Delta t} \tag{8}$$

$$v_i = \frac{u_i(t+\Delta t) - u_i(t)}{\Delta t} \tag{9}$$

$$F_i - f_i A_{si} - C v_i = E \, A \, \frac{u_{i+1} - u_i}{\Delta z} \tag{10}$$

where (f_i) is the skin friction generated at pile segment, (v_i) is the pile segment velocity and (A_{si}) is the surface area of that segment.

Based on the analysis, the velocity received at pile top for the wave, which traveled a distance of $(2L/c)$ is pile body, can be calculated as follows:

$$v_{2L/c} = [F_0 - f_{downward} + \left(\frac{2 Z_s}{(Z_p + C_c) + Z_s}\right)(F_n - F_R)$$
$$+ f_{upward}] / (Z_0 + C_c) \tag{11}$$

where, $(v_{2L/c})$ is the velocity received at pile top, (F_0) is the applied force at pile top, $(f_{downward})$ is the total skin friction force for the stress wave propagating down wards, (Z_s) is impedance of soil below pile toe, (Z_p) is impedance of pile toe, (F_n) is the incident force at pile toe, (F_R) is the soil reaction at pile toe, (f_{upward}) is the total skin friction force for the stress wave propagating upwards, (Z_0) is the impedance at pile top and (C_c) is the damping coefficient of concrete.

3 MODEL VERIFICATION

3.1 In situ testing program

Pile integrity test (PIT) was carried on (30) newly constructed concrete piles in the selected project. The equipment used for testing is shown in Figure 3.

Figure 3.   Pile integrity tester and accessories.

Generally, a structurally sound shaft is indicated by a clear reflection form pile toe and the absence of reflections caused by impedance changes. Negative velocity reflections are often caused by soil resistance effect or by "bulges" in pile body, where softer soils or auger wobble cause enlargement of pile diameter. An impedance reduction due to decreases in the area or concrete quality is indicated by a positive velocity reflection before the toe.

3.2   Model – in situ results comparison

Based on piles characteristics and the information revealed from the geotechnical investigation, the parameters shown in Table 1 were inputted in the Model data sheet.

Table 1.   Pile and soil data.

Item	Value
Concrete Density, ρ_c (t/m³)	2.5
Pile elastic modulus, E (kN/m²)	4x10⁷
Pile Radius, r_0 (m)	0.45
Soil Density, ρ_s (t/m³)	1.80
Rock Density, ρ_s (t/m³)	2.30
Rock Poisson's Ratio, ρ_r	0.40
Wave velocity in concrete, m/s	4000
Shear wave velocity in soil, m/s	275
Shear wave velocity in rock, m/s	2000
Soil Reaction Coefficient	2.00
Attenuation coefficient(γ) for soil	2.00
Attenuation coefficient(α) for soil	0.13
Attenuation coefficient(γ) for rock	2.00
Attenuation coefficient(α) for rock	0.10
Concrete damping coefficient	0.015

Velocity amplitudes at pile toes, which were received at pile top, produced from in situ tests and the Model are shown in Figure 4 using similar impact loads, soil conditions and piles characteristics. As is shown in the graphical comparison, the Model predicted values are generally close to the in situ values.

The statistical analysis showed strong correlation between real in situ results and Model predicted results. In addition, it had been noticed that the highest impact force applied on pile top during field testing, which is the impact applied on pile P-3, was identified as outlier based on the statistical analysis. This is attributed to the unrealistic low response received for pile toe resulted from the high impact load.

The relationship between in situ integrity test results and Model predicted values was investigated using different regression models. The regression

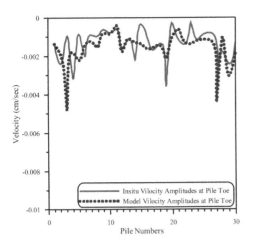

Figure 4. Velocity amplitudes produced from in situ PIT and the model.

model and the associated R-squared value are shown in Figure 5. The R-squared obtained from the relationship is equal to 0.86 after elimination of the outer.

The best fit based on the regression analysis for the 30 tested piles has the following formula:

$$V_{a(Model)} = 0.89 \, V_{a(Insitu)} \qquad (12)$$

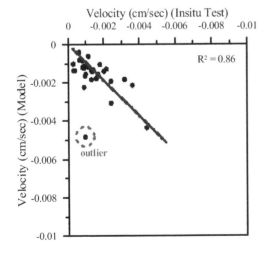

Figure 5. Velocity amplitudes at the pile toe produced from insitu tests versus model predicted results after elimination of outlier.

where, $V_{a(Model)}$ is the Model calculated velocity amplitude at pile toe and $V_{a(Insitu)}$ is the measured velocity amplitude in insitu test.

4 MATERIAL AND SOIL DAMPING

A computer program had been written using MATLAB to facilitate fast processing of Model calculations. The program is designed to process the computations automatically to provide outputs in both tabulated and graphical formats.

Using this program, the damping forces measured for a pile subjected to light impact force on top during pile integrity test, were quantified as shown in Table 2.

Table 2. Material and soil resistance forces values.

Force type	Amount (kN)
Force applied at pile top	12.000
Friction and material damping forces	9.140
Soil reaction at pile toe	2.748
Force received at pile top (t=2L/c)	0.122

5 CONCLUSION

Pile material and soil resistance effects in terms of skin friction and soil effect on pile toe were quantified using a numerical solution of wave equation after incorporating soil resistance terms.

The mathematical model is based on introducing pile material damping and soil resistance at pile surface and the pile toe in the basic wave equation.

Skin friction is assumed to be proportional to pile displacement. This assumption is considered to be more suitable for dynamic type of loading, while small and instantaneous mobilized frictional forces can be quantified.

The Model produced results were found to be comparable to real in situ test results.

This model can be used to evaluate the PIT test limitations based on the pile and subsurface soil conditions and the PIT tester sensitivity.

ACKNOWLEDGMENT

The authors would like to thank the Ministry of Education Malaysia (MOE) and the Research Management Centre of Universiti Teknologi Malaysia (UTM) for providing financial support through research vote: Q.J130000.2522.06H52 for this research for this research, and Pile Dynamic Incorporation for providing the technical support.

REFERENCES

[1] Hussein, M. H., Garlanger, J., June 1992. Damage Detection for Concrete Piles Using a Simple Nondestructive Method. First International Conference on Fracture Mechanics of Concrete Structures. Breckenridge – Colorado, USA.

[2] Johnson, M., Rausche, F., "Low Strain Testing of Piles Utilizing Two Acceleration Signals". September, 1996. Proceedings of the Fifth International Conference on the Application of Stress-wave Theory to Piles 1996: Orlando, USA, FL; pp. 859–869.

[3] Kraft, L. M., Ray, R. P., and Kagawa, T. (1981). Theoretical t-z curves, Journal Geotechnical Engineering Division, Proceedings Paper 16653, American Society of Civil Engineers, Vol 107(GT11).

[4] Massoudi, N., Teferra, W., April, 2004. Non-Destructive Testing of Piles Using the Low Strain Integrity Method. Proceedings of the Fifth International Conference on Case Histories in Geotechnical Engineering: New York, USA, NY. (CD-ROM).

[5] Paquet, J., 1992. Pile integrity Testing – The CEBTP Reflectogram, Piling, European practice and worldwide trends Proceedings of a conference organized by Institution of Civil Engineers, London, England.

[6] Rodriguez J. and Restrepo, V., September, 2008. Analytic Solution for One-Dimensional Propagation of Waves and its Adaptation to Results Interpretation of Pile Integrity Test (PIT). Proceedings of the Eighth International Conference on the Application of Stress Wave Theory to Piles 2008: Lisbon, Portugal.

Testing and Measurement: Techniques and Applications – Chan (Ed.)
© *2015 Taylor & Francis Group, London, ISBN: 978-1-138-02812-8*

Study of packing media of flotation columns

M. Zhang, J.H. Shen, Y.P. Wei, J.X. He & T.L. Xue
College of Resources and Metallurgy of Guangxi University, Nanning, Guangxi, China

ABSTRACT: Packing in a flotation column is one way to increase separation efficiency and effective way of descending flotation-column height. In this paper, the flaws of fillings packing in flotation columns were pointed out in industrial application, and a change of flotation process and its effect on flotation results are reported. Based on sieve-plate packing, a high-efficiency mixed packing mode consisting of screen-plate packing and honeycomb-tube packing has been proposed, with the packing mode optimized in the cyclonic static micro-bubble flotation column in the separation zone. In the mixed packing mode, the honeycomb tubes are used to reduce the vortex effect at the base of the column and weaken the radial dispersion of jetting bubbles, and the sieve plates are used to comminute bubbles in the separation zone, lengthen the bubble migration path, and to stabilize foam layer. The copper ore flotation was applied. Based on the test data, it was found that the sieve packing has the potential to yield a product of approximately 25.83% copper concentration with 91.78% recovery from a feed concentration of approximately 0.727%. If mixed packing was used, the product concentration could be further increased to 27.60% from a feed of approximately 0.739% concentration with 93.38% recovery.

KEYWORDS: Flotation column; Mineral separation.

1 INTRODUCTION

The idea of a packed flotation column came from the wide use of tower equipment in chemical processes. Its main features are: 1) the fluid phase in the flotation column can be varied and controlled, and plug flow can be achieved in the flotation process; 2) mechanisms used to encourage countercurrent collisions become effectively stabilized (Liu, 2000). 3) The foam layer is supported by filling, so that the selection processes of collision and mineralization can work efficiently; 4) the conversion of gas into bubbles is encouraged, and the bubbles merge with some difficulty into the flotation column.

In recent years, the filling of flotation columns has been widely studied. The flotation column is divided to create a narrow space in the axial direction. Slug flow is formed by slurry flowing from the top to the bottom and bubbles running from bottom to top. The bubble distribution in the radial direction is thus improved. In this way, merging flows strong back mixing and unstable operation can be effectively prevented.

However, filling a flotation column has many disadvantages. For example, filling is costly, installation and maintenance workload is heavy, and anti-blinding performance is poor. Furthermore, corrugated packing is ineffective at curbing large-scale eddies in the vertical direction (Yang, 1991; Zhou, 2001).

Aisaier (Aisaier, 2002) proved that sieves can prevent liquid from mingling in the radial direction and decrease bubble coalescence. The sieve plates can improve the radial diffusion of bubbles and the rapid outflow of slurry along the axial direction of the column, leading to the formation of slug flow as described earlier and an increase in the separation effect.

2 FLOTATION COLUMN PACKED SEPARATION ZONE

The particular gradient-optimization separation structure of cyclonic static micro-bubble column flotation, column flotation has been successfully applied in china-clay mineral processing. Gradient-optimization separation takes place in three stages: column flotation, cyclone separation and pipe-flow mineralization. The distinguishing characteristics of cyclone separation are the use of the suspending liquid to perform separation and the use of a cyclone as a separation environment. According to the principle of nozzle flow, pipe-flow mineralization depends on drawing air into solution and compressing it into air bubbles. This exemplifies the three recycle flow systems along the pipe and achieves turbulence and mineralization (Liu, 1998).

2.1 Sieve packing

As the cyclonic field rises and expands, lifting air bubbles and slurry in the column separation zone, the bubbles and slurry are unavoidably drawn into a circular motion, causing a turbulent current. The apparent contradiction between the static separation performed by column flotation and the high-turbulence current created by the liquid rotating at the bottom of the column has pose a major technical obstacle to the development of cyclonic static micro-bubble column flotation. To address these concerns, creation of a packed flotation-column separation zone was proposed. Figure 1 is the plane structure of the sieve packing.

Figure 2 shows the liquid phase in the column before and after sieve packing. Before sieve packing, in the large vortex, the bottom currents (containing large bubbles) are directly carried up to the column inlet, creating a great disturbance in the fluid due to the influx of large floating bubbles. Most bubbles and mineral particles will follow the fast water flow. There are no floating bubbles or descending mineral particles in the slurry, and so reverse collision and mineralization do not occur. No color change is observed in the separation process along the axial direction, and the tailings are of a homogeneous gray or black color. After sieve packing, the large-scale vortex and bubbles have disappeared. The flotation column is separated into several small narrow spaces by the sieve plate. Although small vortices still exist in each space, the turbulence is reduced significantly as the flow moves up the flotation column. The slurry color changes from dark to light in the course of the coal-separation process (Liu, 1998).

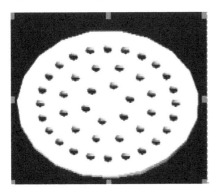

Figure 1. Plane structure of the sieve packing.

2.2 Mixed packing

Theoretically, the greater the open area in the sieve plates is, the greater the flotation flux will be. If the sieve-plate position is close to the cyclonic zone, recovery inside the column will be weak; on the other hand, the inhibition of radial cyclonic flow will

Figure 2. Fluid effects of sieve filling.(a: Large vortex; b:"static" separation).

also be weak. Therefore, the improved performance brought about by the sieve plate does not solve all operating problems.

Figure 3 shows the structure of the honeycomb-tube packing media for cyclonic static micro-bubble column flotation. The honeycomb-tube is the deduction of sieve plate by increasing thickness and enlarging open area, as many straight-pipes are bundled together, like honeycomb. The characteristics of this mixed-packing approach are that the honeycomb tubes are used to reduce vortex formation in the base of the column and weaken the radial dispersion of jetting bubbles, and the sieve plates are used to comminute bubbles in the separation zone, lengthen the migration path of the bubbles, and stabilize the foam layer (Zhang, 2007).

Figure 3. Structure of honeycomb-tube(a: overhead view of hocomb tube;b:lateral view of honeycomb tube).

Figure 4 is the comparison of packing effect of empty flotation column, sieve plate packing flotation column, and mixed packing flotation column. The mixed packing further reduces the turbulence and back-mixing phenomenon, forms more stable gradient grade, improves selectivity, and achieves higher degree concentration in shorter cylinder.

Figure 4. Comparison of the packing effects in three types of flotation columns (a: Empty flotation column; b: Sieve-plate flotation column; c: Mixed flotation column).

3 EXPERIMENTAL SECTION

The pilot-scale flotation tests were carried out in a copper mine in Yunnan, China. The copper ore sample was crushed and pulverized to a particle size less than 0.07 μm (d_{80}), and kept in plastic bags before the flotation tests. The proximate analysis of the copper is given in Table 1.

Table 1. Proximate analysis of the mineral matter.

Mineral	Chalcopyrite	Bornite	Covelline	Malachite	Magnetite
Percentage (%)	1.90	0.16	dram	dram	24.48
Mineral	Iron pyrite	Hematite	Limonite	Biotite	Plagioclase
Percentage (%)	0.14	0.24	dram	25.17	14.15
Mineral	Dolomite	Quartz	Chlorite	Calcite	other
Percentage (%)	16.15	11.29	3.02	1.18	2.03

3.1 *Experimental parameters*

1 Capacity: 20–23 t/d;
2 Feed density: 45–49%;
3 Pharmaceutical: air potato 10–12 g/t, oil of pine 45–55 g/t;
4 liquid level (the reading showed by control cupboard): rougher 0.750–0.820, concentrate 0.400–0.600;
5 Circulating pump working pressure: roughening 0.18–0.21 Mpa, concentrating 0.17–0.20 MPa.

3.2 *Experimental equipment and flow chart*

The experimental equipment and flow chart are shown in Figure 5. "One-to-one" refers to a rougher flotation column with a concentrated flotation column with a ratio volume of 1:1.

3.3 *Packing medium parameters*

Sieve-plate parameters: circular; diameter of aperture, 18 mm; proportion of open area, 60%–75%; pitch of holes, 10–20 mm; a total of four sieve plates, with plate spacing of 600–800 mm.

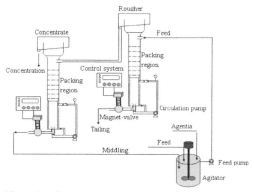

Figure 5. One-to-one experimental equipment.

The honeycomb was a B acrylic copolymer, with a diameter 40 mm; pipe thickness, 1.0 mm, height 500 mm; proportion of open area, > 90%.

During mixed packing, the honeycomb was placed close to the jetting mouth, and the sieve-plate layers were placed upwards.

3.4 *Results and discussion*

Table 2 shows the results of sieve packing and mixed packing, respectively, of the cyclonic static micro-bubble flotation column, processing copper ore in Yunnan, China. It is clear from Tables 2 and 3 that sieve packing has the potential to yield a product with a copper concentration of approximately 25.83%, with 91.78% recovery, from a feed concentration of approximately 0.727%. If mixed packing is used, the product concentration is further increased to 27.60% from a feed concentration of approximately 0.739%, with 93.38% recovery.

Table 2. Index comparison between sieve plate packing and honeycomb tube packing ()/%).

No.% %	mixed packing					sieve packing				
	Feed	Concentration	Tailing	Productivity	Recovery	Feed	Concentration	Tailing	Productivity	Recovery
1	0.650	25.08	0.044	2.48	95.69	0.724	25.15	0.063	2.63	91.53
2	0.852	29.93	0.055	2.67	93.79	0.752	25.72	0.050	2.73	93.53
3	0.720	29.07	0.056	2.29	92.46	0.672	24.33	0.060	2.52	91.30
4	0.745	25.40	0.046	2.76	94.10	0.794	27.25	0.069	2.67	91.54
5	0.728	28.69	0.050	2.37	93.40	0.693	26.72	0.064	2.36	90.98
Arithmetic Mean	0.739	27.60	0.050	2.50	93.38	0.727	25.83	0.061	2.58	91.78

From the view of fluidization, the purpose of sieve-plate packing is to rectify the fluidized bed layer and to form an even bed layer of homogeneous density, with a gap from the top down. The application of sieve-plate packing facilitates contact mineralization of the bubbles and mineral particles, and thus achieves highly effective separation (high recovery and high grade) of the micro-particles.

From the view of dynamic analysis, the flotation fluid was distributed reasonably by the sieve plate. The collision probability of the micro-bubbles and the particles was maximal, so that the mineralization effect and separation selectivity were improved.

4 CONCLUSIONS

In this paper, sieve-plate packing and mixed packing were investigated, which proved the packing medium were an effective way in increasing flotation efficiency.

The sieve plate normalizes the flow pattern and the mineralization environment of the flotation column, shears the flotation bubbles, reduces bubble merger, decreases the interference of the lower cyclone by the upper foam layer in the flotation column, and achieves the purposes of a stable flotation process and an increase in flotation efficiency.

The honeycomb tube forms smaller spaces and enhances the mass transfer process. Mixed packing further weakens the negative impact of the cyclone field and also allows high-grade flotation to concentrate the overflow quickly, so that a fast floating mechanism for mineral separation is achieved.

ACKNOWLEDGMENTS

The authors would like to acknowledge the financial support from the National Youth Foundation of China (Project No. 51404076) and the financial support from Guangxi Education Department of China (Project No. ZD2014005).

REFERENCES

[1] Liu, 2000 Liu, J.T. The separation method of cyclonic-static micro-bubble column flotation and practical application: Part 2: Stabilization and packing patterns of the separation process in the flotation column. Coal Preparation Technology, 2000, 2: 1–5.

[2] Yang, 1991 Yang, J .L. New Type Packed Column Flotation. Metallic Ore Dressing Abroad. 1991, 2: 8.

[3] Zhou, 2001 Zhou, K. The Research and Application of Packed Column Flotation.Chemical Engineering.

[4] 2001, 29(6):57. Aisaier, 2002 Aisaier.T.C. The stability of screen plate in flotation column. Metallic Ore Dressing Abroad. 2002,6:24-26.Liu, 1998 Liu, J.T. Study on cyclonic static micro-bubble column flotation and clean coal preparation, Doctor thesis. China University of Mining & Technology (Beijing), 1998, 27–36.

[5] Zhang, 2007 Zhang, M. Column Flotation Packed-Separation Zone and Study on the Hydrodynamics Doctor thesis. China University of Mining & Technology (XuZhou),2007,68–80.

Testing and Measurement: Techniques and Applications – Chan (Ed.)
© 2015 Taylor & Francis Group, London, ISBN: 978-1-138-02812-8

The research on using wood flour hydrolysate as carbon source to produce bacterial cellulose

Z.S. Ma, Y. Liu & Z. Wang
Key Laboratory of Pulp and Paper Science Technology of Ministry of Education, Shandong University of Technology, Jinan, Shandong, P. R. China

ABSTRACT: Bacterial Cellulose (BC), a kind of nano-biomaterial, is produced by some microorganism. It is of particular network structure which is composed of ultra-fine ribbon-shaped fibers. In comparison with plant cellulose, BC displays high purity, high crystallinity, high water-holding capacity, biocompatibility, high mechanical strength and controllability during the course of biosynthesis and so on. Therefore, BC can be used in areas where plant cellulose can hardly be used. Nowadays BC has been one of the most active topics in the materials field. At present, there are some problems about the production and application of bacterial cellulose in China. This paper regarded acetobacter xylinum selected by the laboratory as the experimental strains, and explored the fermentation conditions that it used wood flour hydrolysate of poplar as carbon source to produce bacterial cellulose. The optimal culturing conditions for acetobacter xylinum statically producing bacterial cellulose under cultivation and fermentation are: the age of the seed solution is 24h; the cultivation temperature is 30°C; the inoculum size is 8%; the fermentation period should be controlled within 6 days; the initial pH value is 6.0. The wood flour of poplar needs pretreatment of ball-milling, when the enzyme concentration is 30mg/ml, the temperature is 50°C, pH value is 6.0 and the time is 5 hours. It has a maximum absorbance at the place of 540nm, and by calculation the acid hydrolysis degree is 38%. When the content of wood flour is 18g/L and the content of glucose is 14g/L, it can produce the maximum bacterial cellulose.

KEYWORDS: Bacterial cellulose, Hydrolysis, Acetobacter Xylinum, Wood, Fermentation.

1 INTRODUCTION

Cellulose produced by some kinds of microbial cells, such as Aceto-bacter, Agrobacteria, and Sarcina strains is free from lignin and hemi celluloses[1].The degree of polymerization(DP) and fiber thickness of bacterial cellulose (BC) are 2,000-6,000 and 0.1μm, respectively[2,3]. Its high water holding capacity, high mechanical strength, elasticity, high crystallinity, etc. have increased its demand in various fields[4,5].

Environmental pollution and energy pinch have gradually become the hot concerns for the human beings of the 21st century. Cellulose, as a kind of environmental-friendly material without any pollution, not only is a kind of renewable energy, but also it can be gained from the broad resource of nature. It is the main component of such plants as cotton, sugarcane and trees, etc.. At the same time, it also is the main component of the cell wall of many funguses such as alga (e) and so on. In addition, some microorganisms also have the ability of compounding cellulose, in order to distinguish the cellulose from the plant source, and name the cellulose produced by such microorganism as Mierohial Cellulose or Bacterial Cellulose.

At present, the strains that can compound bacterial cellulose mainly include nine chondromyces such as achromobacter, hizobium, Pseudomonas, acetobacter, alcaligenes, sarcina, azorobaczer, Aerobacter and agrobaclerium, in which, acetobacter xylinum is the type strain of studying the fermentation of bacterial cellulose. It not only has a higher cellulose production rate, but has the potential of large-scale fermentation and production [6].

Bacterial cellulose and plant cellulose have the same basic chemical composition. As a kind of polymer compound, bacterial cellulose has been found by people for more than 100 years, but its use value has been fully recognized for recent decade. The application of bacterial cellulose involves many kinds of industries such as food industry, biological medicine, membrane material, acoustic equipment vibration of membrane, fuel cell, papermaking and the study of immobilized enzyme and modified materials, etc.

2 EXPERIMENTAL MATERIAL

2.1 *Acetobacter xylinum*

Strain M12. reserved in key microbiological laboratory of Shandong Province

2.2 Cant bacteria culture medium of acetobacter xylinum

20g glucose, 10g yeast powder, 10g calcium carbonate, 20g agar, 1000mL distilled water, pH value is 6, 20 minutes' sterilization under high temperature and high pressure.

2.3 Seed culture medium of acetobacter xylinum

2% glucose and wood flour hydrolysate with different volume, 0.5% peptone, 0.5% yeast powder, 0.27% disodium hydrogen phosphate, 0.115% citric acid, the pH value is 6.0, 20 minutes' sterilization under high temperature and high pressure.

2.4 Fermentation medium of acetobacter xylinum

2% glucose, 0.5% peptone, 0.5% yeast powder, 0.27% disodium hydrogen phosphate, 0.115% citric acid, pH value is 6.0, 20 minutes' sterilization under high temperature and high pressure.

3 THE EXPERIMENT CONTENT

3.1 The hydrolysis of wood chips

Pretreatment for wood chips: use vibration ball mill to grind the wood chips into wood flour, and make the standard curve of glucose; add the wood flour into 3% hydrochloric acid solution, and heat in the 95°water for 2 hours. Get the hydrolysate. Use spectrophotometer to measure the absorbancy of each group of experiment, and then make the graph and find the optimal condition for enzymolysis; Measure the degree of hydrolysis under the optimal condition.

Use hydrolysate to make different fermentation culture medium, according to the needs, add 5%, 10%, 20%, 30%, 40%, 50%, 60% glucose, and calculate the volume of the needed wood flour hydrolysate as per the degree of hydrolysis. Add the hydrolysate with corresponding volume to fermentation culture medium.

3.2 The development of bacterial cellulose

Pick up a ring of slantly cultured acetobacter xylinum, put it into the triangular flask(250mL) with 50mL fluid medium, and culture it under the constant temperature of 30°C for 24 hours. Use the above culture solution to the line on the solid culture medium plate, and then put it in the incubator with the constant temperature of 30°C to culture for 36 hours. Select the single colony with better growth state to inoculate into the fluid medium for

pure culture. Use inoculating needle to puncture and inoculate bacterial suspension that is gained after purification into the tube solid medium. After putting it in the incubator with the temperature of 30°C to culture for 36 hours, it will be reserved under 4°C in the refrigerator.

3.3 Culture method

L) with 50mL fluid medium, carry out shake cultivation under the appropriate temperature, the revolving Under aseptic condition, use inoculating needle to pick up a ring of seeds with good activation, and put it in the triangular flask (25mpeed of shaker is 160r/min. After 24 hours, as per different inoculum size, inoculate them into fermentation culture medium, and it needs fully shaking after inoculation, and then culture them under the constant temperature of 30°C and static state for 7 days,

3.4 The extraction and treatment method for bacterial cellulose

After cultivation under constant temperature and static state for several days, take out the membrane on the surface of the medium, and wash it with distilled water for several times. Then, keep the clean bacterial cellulose membrane under the condition of 90°C and use 0.1mol/L NaOH solution to keep them warm for 2 hours. This process aims at getting rid of the fermentation bacteria in bacterial cellulose membrane and the culture medium left in the bacterial cellulose membrane. Finally, wash it with distilled water for several times, use pH test strips to gently press the surface of bacterial cellulose membrane until it turns neutral, and its pH value is about 7.0.

4 RESULT ANALYSIS

4.1 The hydrolysis of wood powder

Best experiment under optimum reaction conditions as follows: reaction temperature is 90 °C, pH value of 6, reaction time for 2 hours, Hydrochloric acid concentration of Three percent

$$A(\%) = \frac{P \times \dfrac{V1}{V2}}{g} \times 100\% \qquad (1)$$

A: glucose P: Glucose mg number check curve obtained; VI; Extract total volume ; V2; The volume of determination of access; g; The sample quality
Reducing sugar content (%) = 37%

4.2 Variation of bacterial cellulose following with the wood flour hydrolysate

Make fermentation culture medium according to the sugar content that accounts for 5%, 10%, 20%, 30%, 40%, 50% of glucose of wood flour hydrolysate of poplar. Take out the frozen strain, after slantly lining it and culturing for 24 hours, use inoculating needle to pick a ring of seed culture medium, culture it on the shaker for 24 hours, and inoculate it into the fermentation medium as per different inoculum size. It needs fully shaking after inoculation. Culture it under the constant temperature of 30°C and static state for 6 days. Take out bacterial cellulose, clean up. Freeze and dry it, and weigh it. The results are showed in Figuer 1.

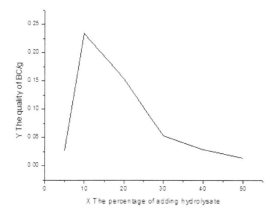

Figure 1. Variation of bacterial cellulose following with the wood flour hydrolysate.

4.3 Infrared spectroscopy

It is indicated by the given information that at 3395cm-1, the strong, wide absorption peak on both sides are caused by a variety of ways hydrogen bond formed by sugar O -H hydroxyl; the absorption peak at 2924cm-1 is caused by carbohydrate C - H stretching vibration; the absorption peak at 1685cm-1 is caused by crystal water; the absorption peak at 1563cm-1 is caused by 4 - the sugar hemiacetal group; the absorption peak at 1405cm-1 is caused by C - H variable angle vibration; the absorption peak at 1051cm-1 is caused by Sugar ring C - O - C and C - O - H stretching vibration. These peak position and peak intensities correspond to standard cellulose spectra. The results are showed in Figure 2.

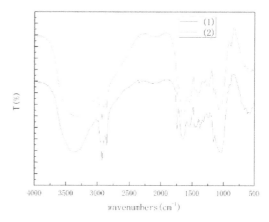

Figure 2.   The infrared spectra of bacterial cellulose.

5 CONCLUSION

Medium costs limit commercial use of bacterial cellulose: molasses from the sugar cane process effluents such as potato effluents and sugar beet as substrates for bacterial cellulose production. In addition, since the bacterial cellulose production has been demonstrated from d-xylose in this work, lignocellulosic materials will be an attractive feedstock for bacterial cellulose production because of their availability at low costs and large quantities. In this experiment, acid hydrolysis degree is 38%. When the content of wood flour is 18g/L and the content of glucose is 14g/L, it can produce the maximum bacterial cellulose.

REFERENCES

[1] J.Y. Jung, J.K. Park and H.N. Chang, Enzym. Microb. Technol.,37,347 (2005).

[2] T.Khan,H.Khanand, J.K.Park, ProcessBiochem.,42,252 (2007).

[3] S.H.Moon,   J.M.Park,   H.W.Chunand,   S.K.Kim, Biotechnol.Bio-processEng.,11,26 (2006).

[4] M. Phisalaphong, T. Suwanmajo and P. Tammarate, J. Appl. Poly.Sci.,107,3419 (2008).

[5] H.Backdahl,   G.Helenius,   A.Bodin,   U.Nannmark, B.R.Johansson,   B.Risbergand   P.Gatenholm, Biomaterials,27,2141(2006).

[6] Keshk S, Sameshima K. Influence of lingo sulfonate on crystal structure and productivity of bacterial cellulose in a static culture[J]. Enzyme and Microbial Technology, 2006, 40(1):4–8.

Laser, optics fiber and sensor

Testing and Measurement: Techniques and Applications – Chan (Ed.)
© *2015 Taylor & Francis Group, London, ISBN: 978-1-138-02812-8*

Volumetric accuracy management of multi-axis systems based on laser measurements

V.I. Teleshevskii & V.A. Sokolov
Moscow State University for Technology 'Stankin', Moscow, Russia

ABSTRACT: The paper describes a newly developed approach to CNC-controlled multi-axis systems geometric errors compensation based on optimal error correction strategy. Multi-axis CNC-controlled systems – machine-tools and CMM's are the basis of modern engineering industry. Similar design principles of both technological and measurement equipment allow usage of similar approaches to precision management. The approach based on geometric errors compensation is widely used at present time. The paper describes a system for compensation of geometric errors of multi-axis equipment based on the new approach. The hardware basis of the developed system is a multi-function laser interferometer. The principles of system's implementation, results of measurements and system's functioning simulation are described. The effectiveness of application of described principles of multi-axis equipment of different sizes and purposes for different machining directions and zones within workspace is presented. The concepts of optimal correction strategy are introduced and dynamic accuracy control is proposed.

1 INTRODUCTION

Multi-axis systems with CNC-controlled displacements of moving parts are the basis of modern engineering industry. These systems include various technological equipment – machine-tools, shaping the parts by determined relative movement and interaction of the part and a tool – edge cutting, abrasive, ultrasonic, laser, electro-erosion etc, as shown in Grigoriev and Teleshevskii (2011), Teleshevskii, et al. (2012), Grigoriev et al. (2013), Kosinskii et al. (2011), and Grishin (2012). Measuring devices – coordinate-measuring machines (CMMs) and other coordinate devices – also relate to these systems, with a contact or non-contact scanning probe interacting with the measured part, as shown in Schwenke H. et al. (2008).

Currently the technological reserves for precision machine-tools and CMMs accuracy improvement are almost used up, and further progress in this field is either impossible or quite expensive. A new method of accuracy improvement arose, which was based on measuring of machine-tool's or CMM's geometric errors measurement and compensation, as shown in Grigoriev and Teleshevskii (2011).

The paper describes a new approach to multi-axis equipment geometric errors correction. First, a multi-axis equipment geometric errors model is developed. Next, the components of geometric errors are measured by multi-function Laser Measuring Information System (LMIS). After that,

the error distribution within the machine's workspace (error mapping) is restored. Finally, corrections are introduced into the machine's CNC system based on optimal correction strategy. Approaches to errors correction will be presented in section 2. Implementation of the system will be introduced in section 3. The results of measurements and simulation will be demonstrated in section 4. Some conclusions will be presented in section 5.

2 METHODS OF GEOMETRIC ERRORS CORRECTION

Within recent 10-15 years, the method of geometric errors correction has developed significantly, as shown in Grigoriev and Teleshevskii (2011). Evolution of errors models, development of error mapping methods– see Schwenke et al. (2005), progress of NC and CNC systems, development of precision methods of machine-tools' and CMM's errors measurement – see Teleshevskii V. I. and Grishin (2006), Kosinskii et al. (2011), ANSI ASME B 5.54-2005, constant improvement of standards regarding error corrections are the most significant advances in this field.

Hence, the geometric errors correction method is currently widely used. In 2012, up to 30-50% of all new machine-tools will undergo the software correction; the usage of software correction for CMM's is even wider.

There are two main approaches to geometric errors compensation:

- The traditional approach;
- The volumetric errors based approach.

2.1 Traditional approach

The traditional approach based on axes positioning errors compensation is widely used. The approach is based on measuring positioning errors for each of machine's axes (i.e., X, Y and Z for 3-axis machine) and introducing corrections for these errors into the control system for each axis separately. Results of positioning accuracy measurement of a bridge-mill machine-tool are presented in Teleshevskii and Sokolov (2012). The two bridges of the machine tool move along the table, machining large parts (over 20 meters long) with velocity of up to 1000 mm per minute. The required machining accuracy limit was ±0.05 mm.

The bridges positioning accuracy of 18900 mm in length was examined via LMIS HP-5528A by Hewlett Packard (USA), with maximal error of 1.7 micron at length up to 30 meters. Figure 1 represents the results of geometric errors measurement after the machine-tool's assembly.

These results determined the corrections introduced into CNC system. Figure 2 represents the results of geometric error measurements after correction.

Thus laser correction of geometric errors is quite efficient and significantly increases the positioning accuracy (up to 10 times). This compensation made it possible to equip the machine-tool with a measuring probe and certify it as a CMM.

2.2 The approach based on volumetric error compensation

Volumetric error is defined as ability of a machine to produce accurate three-dimensional parts. The volumetric error is determined as difference between position vectors of actual and nominal position of tool relative to machined or measured part within the selected coordinate system. This article uses the geometric errors measurement approach based on measuring elementary components of error by means of LMIS and following calculation of total error.

As an illustration of this approach, a multi-axis machine is considered as a set of rigid components moving along the axes (for three-axis machine – X, Y and Z). Since a rigid body has six degrees of freedom – three displacements and three rotations, six errors appear per axis: three displacement errors: δxa, δya, δza, and three rotation errors: εxa, εya, εza , where a – one of three machine–tools axes. With three axes squareness errors, this makes a total of 21 parametric

error functions to be measured within the workspace. A multi-function LMIS allowing measurement of these error functions is the primary measuring device for this approach.

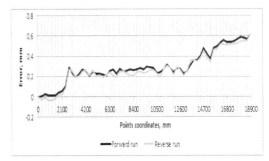

Figure 1. Positioning errors before correction.

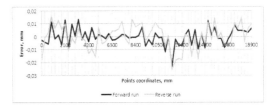

Figure 2. Positioning errors after correction.

The volumetric error is calculated using different formulae for different multi-axis systems configuration. This article considers as an example of a machine-tool with ANSI code 02-11-21-303-406 (see ANSI ASME B 5.54-2005); the volumetric error is determined by solution of matrix equation - see Donmez et al. (1986).

$$\mathbf{XYZ} = R_X\left[{}_Y\left(R_Z^{-1}\mathbf{T} + \mathbf{Z} - \mathbf{Y}\right) - \mathbf{X}\right], \qquad (1)$$

Here, $\mathbf{X,Y,Z}$ – linear and squareness error vectors; R_X, R_Y, R_Z – rotation matrices; $\mathbf{T}$ – tool offset vector.

Solving (1) gives following errors formulae:
X errors:

$$\Delta X = \delta_{XX}(X) + \delta_{XZ}(Z) + \delta_{XY}(Y) + Y\left[\epsilon_{ZY}(Y) + \epsilon_{ZX}(X)\right] + Y\alpha_{YX} - \\ -Z\left[\epsilon_{YY}(Y) + \epsilon_{YX}(X)\right] - -Z\alpha_{ZX} + X_T - \\ -Y_T\left[\epsilon_{ZX}(X) + \epsilon_{ZY}(Y)\right] - Z_T\left[\epsilon_{YZ}(Z) + \epsilon_{YY}(Y) + \epsilon_{YX}(X)\right] \qquad (2)$$

Y errors:

$$\Delta Y = \delta_{YY}(Y) + \delta_{YX}(X) + \delta_{YZ}(Z) - X\epsilon_{ZX}(X) - Z\left[\epsilon_{XY}(Y) + \epsilon_{XX}(X)\right] - \\ -Z\alpha_{ZY} + X_T\left[\epsilon_{ZY}(Y) + \epsilon_{ZX}(X)\right] + Y_T - \\ -Z_T\left[\epsilon_{XY}(Y) + \epsilon_{XX}(X) + \epsilon_{YZ}(Z)\right] \qquad (3)$$

Z errors:

$$\Delta Z = \delta_{ZZ}(Z) + \delta_{ZX}(X) + \delta_{ZY}(Y) - X \in_{YY}(X) - Y[\in_{XY}(Y) + \in_{XX}(X)] + \\ + X_T[\in_{YX}(X) + \in_{YZ}(Z) + \in_{YY}(Y)] + Y_T[\in_{XY}(Y) + \in_{XZ}(Z) + \in_{XX}(X)] + Z_T \qquad (4)$$

where X, Y, Z are the current coordinates of the tool center in the machine's workspace;

X_T, Y_T, Z_T – coordinate components of tool displacement vector **T**;

δ_{a1a2} - linear displacement error of axis a_2 in the direction of axis a_1

ε_{a1a2} - angular error of axis a_2 in the around axis a_1;

α_{a1a2} - squareness error of axes a_1 and a_2 as stated in Donmez et al. (1986).

After the volumetric error mapping, the most critical section of volumetric error, is selected and corrections are introduced into machine's CNC system in order to achieve the least possible error within the largest workspace zone, thus, we introduce the optimal error correction strategy.

3 IMPLEMENTATION OF GEOMETRIC ERROR COMPENSATION SYSTEM

The authors implemented a system for compensation of volumetric geometric errors. As an example, the system was based on: a three-axis coordinate boring machine, ANSI code 02-11-21-303-406, equipped with CNC system Flex-NC by Stankocenter, Russia (http://flexnc. com/); LMIS XL-80 by Renishaw, Great Britain (http:// www.renishaw.com); software for connection between LMIS and CNC system and error correction.

Figure 3 shows the scheme of implemented system. The movement of carriages is controlled by CNC system and driven by displacement drives. The elementary errors are measured during the movement of carriages by XL 80 system. Next, the total geometric errors are calculated and error correction is introduced into CNC system.

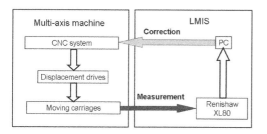

Figure 3. Scheme of implemented system.

3.1 *Hardware*

A three-axis coordinate boring machine, ANSI code 02-11-21-303-406, with vertical spindle alignment and horizontal two-axis table and workspace dimensions 640X400X500 mm was measured. A zone of workspace was selected for current investigation with dimensions of 300X200X100 mm.

The machine-tool is equipped with FlexNC CNC system, which is chosen for high reliability, flexibility and versatility.

The elementary machine-tool's geometric errors were measured by Renishaw XL-80 laser interferometer.

3.2 *Software*

A control program was written in order to compensate the machine-tool's errors. The program input is a set of workspace points used to measure error components. The program generates CNC commands and transfers them into the CNC system via serial port. Next, the PC is set in data acquiring and mode is set, and then the CNC program starts.

After the data are automatically acquired by means of laser interferometer, the coordinate components of volumetric error are calculated using the developed software.

Next, a decision is made if the error correction is necessary within a certain error correction strategy. The compensation zone limits and correction points are set. Next, corrections values are calculated via software. The corrections are input into software manually using CNC keyboard or automatically by means of serial port connection.

4 MEASUREMENTS AND SIMULATION RESULTS

The error correction is introduced as follows:

1 The X, Y and Z error values are calculated for every measured point of machine workspace according to (2)-(4);
2 The calculated error values are subtracted from current tool position by means of CNC system firmware, thus compensating the measured volumetric errors.

Figure 4 presents the results of software correction simulation for the developed system. The point's colors represent the volumetric error value in each workspace point. Black means error under 20 micron, dark-grey – under 40 micron, light-grey – under 60 micron; thus the darker, the better. The low-error points (under 20 micron) make up of less than 8% of the whole workspace.

Figure 5, which is similar to Figure 4, represents the results of software correction simulation. Compared with Figure 4, the low error area increased significantly, occupying more than 30% of machine's workspace.

The implemented system makes it possible to correct errors in real-time mode, so that machine-tool accuracy can be controlled dynamically by electronically changing the values of introduced corrections. As a result, different error distribution can be achieved for different machining stages and depending on multi-axis system status.

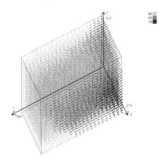

Figure 4. Error distribution before correction.

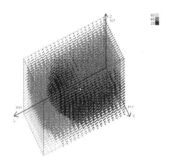

Figure 5. Error distribution after correction.

5 CONCLUSIONS

The authors' experimental work demonstrates the high effectiveness of CNC systems corrections even for long-sized machines (up to 30 m large), so that their precision level is close to that of CMMs.

A system is built for automatic error distribution measurement, being an integrated complex, including, for example, multi-function LMIS Renishaw XL-80 and a machine-tool CNC system.

Measuring elementary error components by means of this complex allows analytical acquiring of error distribution within workspace of multi-axis systems (error mapping).

Selecting a workspace section for error compensation allows changing this distribution for minimal resulting error within maximal workspace zone, i.e. optimal accuracy control.

Optimal software sections assure dynamic control and optimization of resulting volumetric error for different machining directions within workspace.

The exact algorithm of optimal section selection is the subject of further study.

ACKNOWLEDGMENTS

The results described in the paper were received as result of scientific research performed at The Moscow State University for Technology 'Stankin' requested and sponsored by The Ministry of Education and Science of Russia.

REFERENCES

[1] Grigoriev S. N., Teleshevskii V. I., 2011. Measurement problems in technological shaping processes. Measurement Techniques. V. 54. № 7. P. 744–749.
[2] Teleshevskii V. I., Sokolov V. A., 2012. Laser correction of geometric errors of multi-axis programmed-controlled systems. Measurement Techniques. V. 49, N 5, P 535–541.
[3] Grigoriev S. N., Teleshevsky V. I., Sokolov V. A., 2013. Volumetric Geometric Accuracy Improvement for Multi-Axis Systems Based on Laser Software Error Correction. International Conference on Competitive Manufacturing «COMA'13» 30 January - 1 February Stellenbosch, South Africa. Organized by the Department of Industrial Engineering Stellenbosch University. P. 301–306.
[4] Kosinskii D. V., Teleshevskii V. I., Sokolov V. A., 2011. Heterodyne laser interferometric techniques based on Fresnel diffraction. Measurement Techniques. –.V. 54, № 8. P. 859–864.
[5] Grishin S.G. 2012. An analysis of the polarization component of the measurement error in heterodyne laser interferometer measurement systems. Measurement Techniques. V. 54, № 12. – P. 1378–1387.
[6] Schwenke H. e. a., 2008. Geometric error management and compensation of machines – an update , Ann. CIRP. N 57. P. 660–675.
[7] Schwenke H., Franke M., Hannaford J., 2005. Error mapping of CMMs and Machine tools by a single tracking interferometer. Ann. CIRP. N 54(1). P. 475–478.
[8] Teleshevskii V. I., Grishin S. G., 2006. A heterodyne laser interferometer with digital phase conversion Measurement Techniques, V. 49. N 6. P. 545–551.
[9] Kosinskii D. V., Teleshevskii V. I., Sokolov V. A., 2011 Heterodyne laser interferometric techniques based on Fresnel diffraction// Measurement Techniques, V. 54. N8. P. 859–864.
[10] ANSI ASME B 5.54-2005 Methods for Performance Evaluation of Computer Numerically Controlled Machining Centers.
[11] Teleshevskii V.I., Sokolov V.A., 2012. Laser correction of geometric errors of multi-axis CNC machines, Measurement Techniques, V. 55 N6.
[12] Donmez M.A., Blomquist D.S., Hocken R.J., Liu C.R., Barash M.M., 1986 A General Methodology for Machine Tool Accuracy Enhancement by Error Compensation. Precision Engineering 8:187–196.

Testing and Measurement: Techniques and Applications – Chan (Ed.)
© 2015 Taylor & Francis Group, London, ISBN: 978-1-138-02812-8

Turn-on fluorescent Fe^{3+} sensor based on tris(rhodamine) ligand

X.M. Wang & H. Yan
Department of Chemistry and Material Engineering, Logistic Engineering University, Chongqing, China

X.C. Guo
Department of Military Oil Application & Management Engineering, Logistic Engineering University, Chongqing, China

S. Wang, D.W. Wang & M. Lei
Department of Chemistry and Material Engineering, Logistic Engineering University, Chongqing, China

ABSTRACT: The well-known Hg^{2+}-selective chemodosimetric behavior of tris(rhodamine) ligands was successfully swithced selectivity for Fe^{3+}. The fluorescence signaling of tris(rhodamine)-based fluorescent chemosensors **2** is remarkably selective fluorescence enhancement toward Fe^{3+} ions over other metal ions at 580 nm, under the optimized conditions (CH_3CN / Tris-HCl buffer (0.01 M, pH = 7.0) (75/25, v/v)). It is shown to be due mainly to the spirolactam ring-opening power of Fe^{3+}.

1   INTRODUCTION

Fe^{3+} is the most essential metal ion in biological systems and plays a crucial role in many biochemical processes, such as in celleular metabolism, in many enzymatic reactions as a cofactor, and in carrying oxygen by heme.[1-2] Detection of Fe^{3+} ions in solution is very difficult using fluorescence techniques because the paramagnetic nature leads to their fluorescence quenching ability. Fluorescence enhancement through chelation of this metal ion with any fluorophore is a challenging task in vitro as well as in vivo.[3]

In the recent past, several attempts have been made to develop chemosensors to detect paramagnetic species.[4-6] Among different sensors, rhodamine-B dericatives and their ring-open reaction have received greater attention following the report of a rhodamine-B hydrazine sensor for Cu^{2+} by Czarnik et al.[7] Ring-closed rhodamine-B derivatives are colorless and nonfluorescent, whereas the ring-open forms are strongly fluorescent and pink in color.[8] Several successful attempts have been made to develop selective fluorescent sensors based on rhodamine B, for Hg^{2+},[9-10] Cu^{2+},[11-12] Pb^{2+},[13] Cr^{3+},[14] and Fe^{3+}.[15] Moreover, they have a longer emission wavelength (about 550 nm), which is often preferred to serve as reporting group for analytes to avoid the influence of the background fluorescence (below 500 nm). Thus, in the present investigation, novel tris(rhodamine)-based fluorescent chemosensor **2** (Scheme 1) have been reported and developed to

increase the selectivity and sensitivity for Fe^{3+} and Hg^{2+} over other metal ions at 580 nm in CH_3CN/ H_2O (3/1, v/v) solution. During the optimized experiment condition by adding different buffers, we also have found that Tris-HCl effectively inhibits the interference of Hg^{2+} and Cu^{2+} ions during the detection of Fe^{3+}.[16]

In this paper, we report the prominent Fe^{3+}-selective and sensitive fluorescence signaling behavior of tris(rhodamine)-based fluorescent chemosensor **2** in 75% CH_3CN, 25% 0.01 M Tris-HCl buffer system (pH = 7.00).

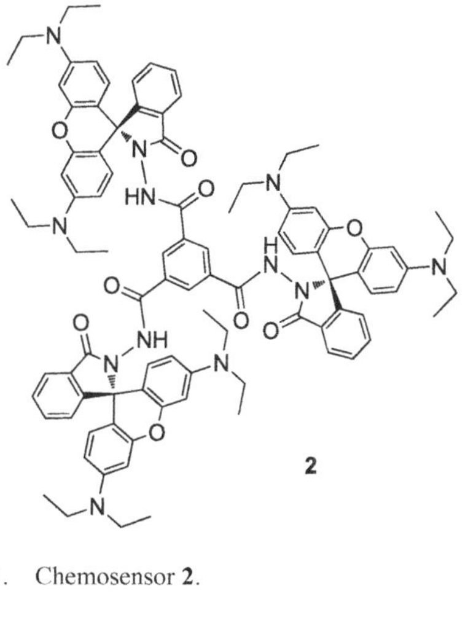

Figure 1.   Chemosensor **2**.

2 RESULTS AND DISCUSSION

2.1 *Selectivity measurements*

All the spectroscopic studies for chemosensor **2** were performed in 75% CH_3CN, 25% 0.01 M Tris-HCl buffer system with a pH of 7 to keep the dye molecules in their ring closed form. The fluorescence titrations of chemosensor **2** with various metal ions (Na^+, K^+, Pb^{2+}, Co^{2+}, Cd^{2+}, Cs^+, Ag^+, Cu^{2+}, Mg^{2+}, Zn^{2+}, Hg^{2+}, Fe^{2+}, and Fe^{3+}) were conducted to examine the selectivity. Among these metal ions (80 equiv), chemosensor **2** both showed large chelation enhanced fluorescence (CHEF) effects with Fe^{3+} and smaller CHEF effects with Cu^{2+} under optimized conditions (CH_3CN / Tris-HCl buffer (0.01 M, pH = 7.0) (75/25, v/v)) (Figure 1). The addition of 400 µM (80 equiv) of Fe^{3+} immediately yielded a pink solution with a strong fluorescence signal at 580 nm due to the strong affinity with chemosensor **2**. The addition of other metal ions especially Hg^{2+}, which has been found a trong enhanced fluorescence in CH_3CN / H_2O (3/1, v/v) solution,[16] produced almost no fluorescence change at 580 nm. For chemosensor **2**, there was 15-fold enhancement with Fe^{3+} (Figure 2). Under the optimzied experiment condition, the interference of Hg^{2+} ion towards chemosensor **2** effectively was inhibited by Tris-HCl buffer.

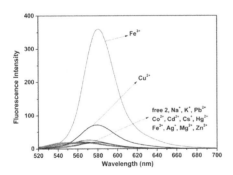

Figure 1. Fluorescence spectra of chemosensor **2** (5 µM, λ_{ex} = 510 nm) in 75% CH_3CN, 25% 0.01 M Tris HCl buffer with different metal ions (400µM).

In addition, a very weak fluorescence signal for free **2** was observed at 580 nm upon excitation at 510 nm, confirming the presence of a ring-closed spirolactam structure, whereas with the addition of Fe^{3+} ion, ring-opening of the spirolactam occurs and gives rise to a strong fluorescence emission at 580 nm. It is interesting to note that it is very easy to differntiate Fe^{2+} from Fe^{3+} with chemosensor **2**. Though Cu^{2+} gave a small color change and a very small fluorescence enhancement, the spectroscopy and interaction of chemosensor **2** with Cu^{2+} is completely different form of **2** with Fe^{3+} as recently reported by others.[15]

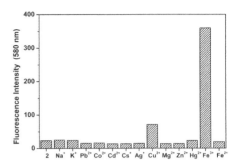

Figure 2. The bar profile of chemosensor **2** (5 µM, λ_{ex} = 510 nm) in 75% CH_3CN, 25% 0.01 M Tris HCl buffer with different metal ions (400 µM).

2.2 *Sensitivity measurements*

Concentration dependent fluorescence measurements were carried out to monitor the sensitivity for Fe^{3+} ion. Shown in Figure 3 are the fluorescence titration experiments carried out for chemosensor **2** with Fe^{3+} in 75% CH_3CN, 25% 0.01 M Tris-HCl buffer system (pH = 7.00) after excitation at 510 nm. It is observed that the sensitivities of chemosensor **2** for Fe^{3+} are arround 140 µM. Sensitivity is calculated by determining the concentration where the enhancement of fluorescence is three times that of the background.[15]

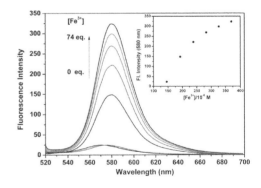

Figure 3. Fluourescence titration of chemosensor **2** (5 µM, λ_{ex} = 510 nm) in 75% CH_3CN, 25% 0.01 M Tris HCl buffer with Fe^{3+} ions (0 – 370 µM).

The fluorescence intensity at 580 nm is sensitivie to the structure of chemosensor **2** binding Fe^{3+} ion which indicates the constrution changes of chemosensor **2** and the formation of a complex between chemosensor **2** and Fe^{3+} ion. The plot of measured fluorescence (I_0/I-I_0) at 580 nm against 1/[Fe^{3+}] showed a linear relationship confirming the formation of a 1:1 complex between chemosensor **2** and Fe^{3+} (Figure 4).

The association constant (K_s) was calculated using modified Benesi-Hildebrand equation (1).[17]

$$\frac{I_0}{I-I_0} = \left(\frac{a}{b-a}\right) \cdot \left(\frac{1}{Ks \cdot [Fe^{3+}]} + 1\right) \qquad (1)$$

Where I_0 is the fluorescence intensity of the chemosensors **2** at 580 nm in the absence of Fe^{3+} and I is the fluorescence intensity upon the addition of Fe^{3+} at 580 nm. The association constant K_s is the ratio of intercept/slope which was obtained from the linear fit (Figure 4) and was found to be 1.48×10^2 M^{-1}.

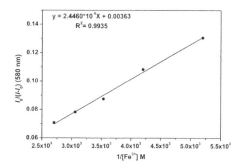

Figure 4.   Spectrofluorimetric titration linear for a complex 1:1 according to Eq. (1). The data are fitted to the linear with a correlation to coefficient of $R^2 = 0.9935$.

2.3   Mechanism of fluorescence sensing

As discussed above, chemosensor **2** showed excellent selectivity toward Fe^{3+}. The complex formation for chemosensor **2** with Fe^{3+} is ascribed to the ring-opening mechanism shown in Scheme 2. In chemosensor **2**, three carbonyl oxygens as well as three amide oxygens can provide a stable binding pocket for Fe^{3+} ions. It is reasonable to suggest that the Fe^{3+} can bind with the amide oxygen that causes the ring-opening, while other metal ions do not have the size advantage or the relative bongding strength to open up the spirolactone ring.

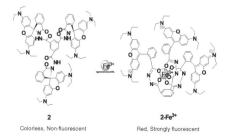

2
Colorless, Non-fluorescent

2-Fe³⁺
Red, Strongly fluorescent

Figure 6.   Turn-on fluorescence of chemosensor **2** in the presence of Fe³+ with the opening of the three spirolactam rings.

Furthermore, binding of Fe^{3+} with chemosensor **2** is reversible by the test using EDTA/Fe^{3+} (Figure 5). When 5 µM chemosensor **2** in 75% CH_3CN, 25% 0.01 M Tris HCl buffer was exposed to 5 mM Fe^{3+}, the pink color faded and fluorescence disappeared upon addition of 1 equiv of EDTA to Fe^{3+}, and when Fe^{3+} was added to the system again, the fluorescence could be reproduced. The results showed that the spectral sensing is reversible.

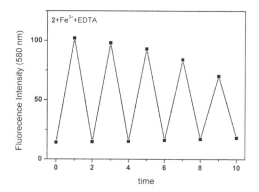

Figure 5.   Fluorescence intensity changes during the titration of chemosensor **2** (5µM, λex = 510 nm and λem = 580 nm) in 75% CH_3CN, 25% 0.01 M Tris HCl buffer with Fe^{3+} (5 mM) and EDTA (5 mM), respectively.

3   CONCLUSIONS

In conclusion, we have found that the well-known Hg^{2+}-selective chemodosimetric behavior of tris(rhodamine) ligands was successfully swithced selectivity for Fe^{3+}. Under the optimized experiment condition (CH_3CN / Tris-HCl buffer (0.01 M, pH = 7.0) (75/25, v/v)), the fluorescence signaling of tris(rhodamine)-based fluorescent chemosensors **2** is selective and also sensitive toward Fe^{3+} ions over other common interfering metal ions. Mechanistically, tris(rhodamine) derivative has a good assosiation constant with Fe^{3+} as its complex with three carbonyl oxygens and three amide oxygens, thereby assisting the ring-opening of spirolactam, which increase the fluorescence instensity. By adding excess EDTA, the binding process is reversible.

ACKNOWLEDGMENT

The authors would like to thank Prof. Jurriaan Huskens, Prof. Willem Verboom and acknowledge the support of MESA+ Institute for Nanotechnology of the Twente University in the Netherlands.

REFERENCES

[1] Brugnara, C. Iron deficiency and erythropoiesis: new diagnostic approaches. *Clin. Chem.* 2003(49): 1573–1578.

[2] Zhang, X. B.; Cheng, G.; Zhang, W. J.; Shen, G.; Yu, R. Q. A fluorescent chemical sensor for Fe3+ based on blocking of intramolecular proton transfer of a quinazolinone derivative. Talanta 2007(71): 171–177.

[3] Chen, X. Q.; Rradhan T.; Wang, F.; Kim, J. S.; Yoon, J. Fluorescent chemosensors based on sprioring-opening of xanthenes and related drivatives. Chem. Rev. 2012(112): 1910–1956.

[4] Zhang, J. F.; Kim J. S. Small-molecule fluorescent chemosensors for Hg2+ ion. Analytical Sciences 2009(25): 1271–1281.

[5] Quang, D. T.; Kim, J. S. Fluoro- and chromogenic chemodosimeters for heavy metal ion detection in solution and biospecimens. Chem. Rev. 2010(110): 6280–6301.

[6] Kim, J. S.; Lee, S. Y.; Yoon, J.; Vicens, J. Hyperbranched calixarenes: synthesis and applications as fluorescent probes. Chem. Commun. 2009: 4791–4802.

[7] Kim, H. N.; Lee, M. H.; Kim, H. J.; Kim, J. S. A new trend in rhodamine-based chemosensors: application of spirolactam ring-opening to sensing ions. Chem. Soc. Rev. 2008(37): 1465–1472.

[8] Valeur, B. Molecular Fluorescence: Principles and Applications. New York: Wiley-VCH.

[9] Huang, J.; Xu, Y.; Qian, X. A rhodamine-based Hg2+ sensor with high selectivity and sensitivity in aqueous solution: a NS2-containing receptor. J. Org. Chem.2009(74): 2167–2170.

[10] Wu, J. S.; Hwang, I. C.; Kim, K. S.; Kim, J. S. Rhodamine-based Hg2+ selective chemodosimeter in aqueous solution: fluorescent off-on. Org. Lett .2007(9): 907–910.

[11] Duan, Y.L.; Shi, Y.G.; Chen, J.H.; Wu, X.H.; Wang, G.K.; Zhou, Y.; Zhang, J.F. 1,8-Naphthyridine modified rhodamine B derivative and Cu2+ complex: colorimetric sensing of thiols in aqueous media. Tetrahedron Lett. 2012(53): 6544–6547.

[12] Kim, Y.-R.; Kim, H.J.; Kim, J.S.; Kim, H. Rhodamine-based "turn-on" fluorescent chemodosimeter for Cu(II) on ultrathin platinum films as molecular switches. Adv. Mater. 2008(20): 4428–4432.

[13] Kwon, J.Y.; Jang, Y.J.; Lee, Y.J.; Kim, K.M.; Seo, M.S.; Nam, W.; Yoon, J. A highly selective fluorescent chemosensor for Pb2+. J. Am. Chem. Soc. 2005(127): 10107–10111.

[14] Weerasinghe, A.J.; Schmiesing, C.; Sinn, E. Highly sensitive and selective reversible sensor for the detection of Cr3+. Tetrahedron Lett. 2009(50): 6407–6410.

[15] Weerasinghe, A.J.; Schmiesing, C.; Varaganti, S.; Ramakrishna, G.; Sinn, E. Single- and multiphoton turn-on fluorescent Fe3+ sensors based on bis(rhodamine). J. Phys. Chem. B 2010(114): 9413–9419.

[16] Wang, X. M.; Iqbal, M.; Huskens, J.; Verboom, W. Turn-on fluorescent chemosensor for Hg2+ based on multivalent rhodamine ligands. Int. J. Mol. Sci. 2012(13): 16822–16832.

[17] Connors, K. A. Binding Constants-The Measurement of Molecular Complex Stability. New York: Wiley & Sons.

Testing and Measurement: Techniques and Applications – Chan (Ed.)
© 2015 Taylor & Francis Group, London, ISBN: 978-1-138-02812-8

Precision frequency meter for basic metrology and displacement measurements

V. Zhmud
Novosibirsk State Technical University, Novosibirsk, Russia

A. Goncharenko
Institute of Laser Physics SB RAN, Novosibirsk, Russia

A.V. Liapidevskiy
Novosibirsk Institute of Program Systems, Novosibirsk, Russia

ABSTRACT: The paper describes precision frequency meter for investigation and testing of laser and atomic frequency standards. Action bases and technical parameters are given.

1 INTRODUCTION

Relevance of high-precision frequency measurements for different time intervals is a consequence of the need to develop a global satellite navigation system GLONASS, where high-precision measurements of distances require time or frequency standards of according precision. Available frequency standards do not agree to the requirements because their relative uncertainty is the unit of the 14-th sign, whereas this system requires frequency standards with the accuracy of the order of one of the 18-th sign. Therefore, the leading research teams carry out extensive research to create a prototype of a new generation of the frequency standard, which would provide the required accuracy. Certification of such standards requires precision frequency meters. The validation procedure uses two or more standard, and this frequency difference is formed by the heterodyne mixing which by mixing or directly is transferred to a carrier frequency, sufficiently stable for the subsequent measurement. The accuracy requirements to the frequency counters used for this measuring are in this case not so high, namely, the error of frequency counters should be not more than one unit of the 10-th sign. But the following additional requirements arise.

1 Measuring intervals must vary from ultra-small (less than 0.001 *s*) to very large-scale (more than 1000 *s*) values.
2 Dead time between the measurement intervals is unacceptable.
3 The possibility of simultaneous measurement of multiple frequencies a single device is desirable

(which guarantees a uniform time base and common measuring intervals).

Substantiation of the benefits of our "method of integer periods" for solving this problem is given in [Zhmud, V. A. 2014]. This measurement principle is as follows.

1 Initial measurement intervals are chosen sufficiently small, typically 1 *ms*.
2 The device counts the number of pulses of the measured frequency in this interval, and measures the exact time of the front of the first pulse, which fall into the interval, as well as the front of the first pulse, which fall into the next interval. The difference between these times gives the exact duration of a number of periods of the measured frequency, stacked on the selected measuring range.
3 There is no dead time: the end of each measurement interval is the beginning of the next one.

When the method of integer periods is used, the increase in duration of the interval can be done on the base of calculating of the earlier results of the measurements of the smaller intervals. In this case, the calculation result will be the same as if this interval is used directly as the time of fronts of intermediate boundary pulses shall not enter into the result. Therefore, if an increase in the averaging time use a conventional frequency (having dead time), the error will decrease in the inverse proportion to the square root of the magnification of the measurement interval, and if the frequency counter uses the described principle, this component of error will decrease in inverse proportion to this ratio without square root.

2 HARDWARE AND SOFTWARE PART OF THE FREQUENCY METER

The hardware part of the meter consists of two parts: the primary data collection device (PDCD) and a PC for secondary processing of the collected data and display the measurement results. PDCD is implemented on commercially manufactured evaluation board, which is a microcontroller with built-in clock, digital and analog inputs. Such devices have recently received widespread: their purpose is to provide developers with a tool to explore all possibilities of new microprocessor-based devices offered for sale. Such evaluation boards are inexpensive and versatile enough; they can easily be purchased and programmed under the given functions. Proposed evaluation is programmed so that eventually it comprises a reference oscillator, counters and the former of the counted pulses. PDCD provides three channels continuous count of the number of the leading edges of the input signal and sync signal once per millisecond to read the meter. It also measures the time from the first clock edge at the beginning of the measured interval to the first clock edge after this interval finish (with an error of less than 1 ns). These data are collected in packages, and passed to the program FreqAndAllanLabNSTU via a standard serial port (RS-232). Packages are complemented with checksum, so if the reception fails, the packet is retransmitted.

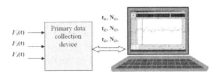

Figure 1. Simplified diagram of the hardware of the frequency counter.

The software part of the device is implemented as self-made program *FreqAndAllanLabNSTU* [Goncharenko, A.M., 2013]. This program collects and processes data from the PDCD, including the calculation of statistical parameters (arithmetic mean, standard deviation, two-sample Allan variance). This program develops the principles implemented in the program described in [Bugrov S.V., 2006]. From the obtained data we can calculate the frequency at any time interval measurement in increments of 1 *ms*. Dead time which has the characteristic of the digital frequency counters (i.e. the time when the frequency counter is restarted) is completely absent. The program in real time simultaneously calculates various measurement intervals which compute the statistical parameters of signals and provide them to display on the screen with the accumulation of

information. Short intervals contain 600 points from 1 *ms* to 600 *ms* in increments of 1 *ms*, the average intervals contains 600 points from 50 *ms* to 30 *ms* with increments of 50 *ms*, long intervals contains 600 points from 2.5 *s* to 1500 *s*, with increments of 2.5 *s*. Logarithmic scale for the measurement intervals with 600 points is realized, the beginning and the end is defined by the user. Regimes of measuring frequency, period, phase difference, the transfer function of the calculated statistical parameters in *Microsoft Excel* to display the data in charts are provided.

The program writes the received data to a file thus realizes the possibility to carry out post-processing and analysis of data from a file, a comparison with the data obtained earlier. The time of the collecting of information and continuous monitoring is not limited to (it is determined by the size of disk space of PC, the data rate is 6 *KB / s*, or 518.4 *MB per day*). In decimation mode, the amount of information data to record into the file is significantly reduced. For example, when data is recorded once per second, the volume of information file is 518.4 *kilobytes per day*. Type of computer: IBM PC-compatible personal computer. Programming language C++, development environment "Microsoft Visual Studio 2008". Operating system: Windows XP. Scope of the program: 174 *KB*.

3 SIMPLIFIED PRINCIPLES OF ACTION OF THE DEVICE

The proposed frequency meter has increased the accuracy and dead time in that it is eliminated due to the continuous count of the counters, which determines the number of reference pulses during the measuring interval. This device realizes the developed ideas and principles described in publications [Goncharenko, A.M., 2013; Bugrov S.V., 2006; Goncharenko A. M., 2007; Zhmud V.A., 2005a; Zhmud V.A., 2005b; Vasiliev, V.A., 2003; Goncharenko A.M., 2006; Zhmud V.A., 2000; Borisov B.D., 2002; Zhmud V.A. 2002]. The device has three measuring channels, but the principle of his actions could be considered at a single channel. Every channel has three counters, with one of them is common for all channels, so that the addition of two channels demands the addition of two pairs of counters more. Thus, in this device the first counter performs continuous measurement of the time by counting the number of pulses of the standard frequency, the second counter counts the number of pulses of the measured frequency, the third counter performs correction measurement durations of the pulses generated by a special circuit and stretched in time by 1000 times with high accuracy. In the multichannel device the time is measured by the common counter and the remaining counters are individually for each channel. Time counters and counters of the number of pulses are working non-stop. Reading their registrations is carried non-stop too. It is provided by special synchronization circuit of the counting pulses, as well as by the appropriate choice of low digit capacity of these counters. Small digit capacity provides quick installation of the final code value due to a short circuit path for the digits transfer. Typically, a small capacity in the conventional scheme is undesirable, since this limits the length of the smallness of the averaging frequency. In our technical solutions, high orders of the counter are restored by software. For example, the counter capacity is $Q = 128$. If the previous reading is equal $M_K = 37$, a new reading is $M_{K+1} = 86$, hence the number of received power pulses is $\Delta M_K = M_{K+1} - M_K = 86 - 37 = 49$. If, for example, the previous reading is equal $M_K = 37$, a new reading power $M_{K+1} = 26$, hence the number of received pulses is $\Delta M_K = Q + M_{K+1} - M_K = 128 + 26 - 37 = 117$.

Another technical solution that allows reading of the counters "on the fly" is the synchronization of the read command, for example, by the rising edge of the clock, provided that the count is carried out by counter on the falling edge of the same pulse. Thus, between the time of the count and time of the reading, a fixed time passes equal to half the period of clock pulses. Since the clock frequency is set in advance, hence the response time of the counters is also known. One can easily calculate the number of bits of the counter, where the reading is correct. For example, if the time of establishment of one-bit counter is t_0, and the number of bits in the counter is m, then the estimated time of the establishment of a stable output of the counter code is $t_m = m \cdot t_0$. If the half of the clock period is more than at least 10%, one can be assured that the reading "on the fly" is performed correctly. Thus, the timing circuit generates from each counting pulse new pulse, delayed to the time of arrival of the next rising edge of the standard frequency. So the time of counting always coincides with the moment of receipt of one of these fronts. Code reading into the register (snap) is synchronized with the falling edge of the same pulse. Therefore, the snapping is spaced in time relatively counting, and the reading into the computer is carried out from the register and does not require stopping the bill.

An additional advantage of such a measurement scheme is unambiguous assignment of each pulse of the measured frequency to only one of the two adjacent measurement intervals. The boundaries of the measurement intervals are synchronized with the end edge of the reference pulses, and counting pulses are synchronized with leading edge of them. Therefore, even if the front of the initial pulse coincides with the boundary of the measurement intervals, the synchronized pulse will be assigned strictly to the next interval, rather than the previous one, it will not be lost and regarded twice. The resulting delay of the pulse does not change the number of these pulses at a predetermined interval. Loss or distortion of information does not occur. The delayed pulse is only once used for counting of the number of whole periods of the measured frequency in the measurement interval, and refinement of the fractional part is carried out on the base of the time of the initial impulse and time of the refined final pulse. This provides reduction of measurement error together with the elimination of the dead time.

Clarification of discreteness error is achieved through supplemental evaluation of the duration between the leading edge of the pulse, controlling the reading, and the nearest edge of the pulse of the measured frequency [Vasiliev, V.A., 2003; Goncharenko A.M., 2006]. The measurement of these intervals of time is made by applying of the pulse stretching circuit and additional counting meter interval [Goncharenko A.M., 2006]. Thus, together with the forming of a given duration of the measurement interval τ_i, and counting the number of counting pulses N_i, caught in this interval, an additional channel measures the amendment $\Delta \tau_i$ which is the duration from the start of the measurement interval to the nearest edge of the next frequency to be measured. Since the measured intervals are followed strictly one after another, without interruption, the next amendment to the beginning of i-th interval is adjusted at the same time for the end of $(i-1)$-th interval. The resulting integer number of the pulses of the measured frequency F_X therefore relates not to an exemplary interval τ_i but to the adjusted interval $\tau_i^* = \tau_i - \Delta \tau_i + \Delta \tau_{i+1}$.

From the obtained by the program data from the device the following variables are formed: 1) the number of leading edges of the signal from the start of measurement; 2) the exact time of the last of the front from the start of the measurement; 3) the number of measurement. Using this data, the program calculates the real-time statistical properties of the signal, Allan function and standard deviation at the same time to 600 different times. The averaging of the obtained results is displayed on the monitor screen. The main parameter of the devise is the resolution which is 50 ps. The device is designed on the basis of *FPGA Cyclone IVE* of company *Altera*. For the measuring of the time intervals 128-bit adder is used. In the summation of the two codes in the adder transferred from junior to senior rank take place, the transfer time is about 50 ps. Thus, 6.4 ns is a full time of the addition of two 128-bit binary codes. With the front edge of the pulse adding the two numbers starts and the back edge snaps the result in the 128-bit register. Numbers for adding is selected so that the traveling wave appears at the output of the adder during the measurement. *Fig.* 2 shows a traveling wave at the output of the adder.

The result at the output of the adder is decoded, the decoder determines the location of the wavefront, and

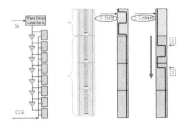

Figure 2. Traveling wave at the output of the adder.

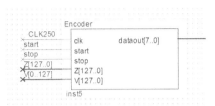

Figure 3. Project of the module Encoder.

outputs eight-bit number corresponding to the measured pulse time.

Module *Encoder*, a project for programming of which is shown in *Fig*. 3, performs the function of measuring of short intervals.

Input signals module *Encoder* are:

CLK250 - the reference frequency of 250 *MHz*.
Z [127..0] - 128-bit variable, all the bits are zero.
V [0..127] - 128-bit variable, all bits are equal to unity.
Z [127..0] and V [0..127] are used to generate codes in the result of adding of which, a traveling wave arises.
START- on the leading edge of the pulse measurement is started.
STOP - ends on the rising pulse measurement.
Exit DATAOUT [7..0] – 8-bit measurement result.

The module is written in hardware description language VERILOG.

Such technique for pulse width conversion into a digital code has the significant disadvantage that the time of delay of one cell depends on temperature and supply voltage. Therefore, an automatic calibration is used. Framework for the application of this calibration is shown in Fig. 4.

On the rising edge of the signal *LatchMeandr* the meander pulse of a frequency of 1 *kHz* defines the measurement frequency which startes every the next measurement, On the falling edge of the signal *LatchMeandr* it starts the calibration process. Thus calibration is performed continuously. After passing of the trailing edge of the signal *LatchMeandr* 1024 measurement are accomplised for the calibration of the signal *CALIBR* with a frequency equal to (CLK250 × 64) / 1023 = 15.64 *MHz*. On the rising edge pulse of signal *CALIBR*, signal *START* of the module *Encoder*, on the rising edge of the reference frequency signal CLK250 reads the cosed from the module *Encoder*. Frequencies 250 *MHz* and (CLK250 × 64) / 1023 are selected in such a way that for the 1024 measurements all phase relations between signals and the time between the start and stop evenly occupied all possible values. Next, the program builds a histogram.

Memory module *ram1port* consistin of 256 cells of 16-bit memory collects information about the frequency of occurrence of any code at the input *in* data[7.0] of module *Histogram* (which gets the information from the output of the module *Encoder*).

Duration of time delay *delay*[i] of transfering of *i*-th digit of the adder is proportional to the accumulated number in the *i*-th memory cell *ram1port*. After accumulating of the histogram, module *Histogram* generates a signal *meas done* on which using the processor module *FreqLab_SOPC* the memory is read and the information is transfered to a personal computer. With all this, input-output ports *CNTRL*[15..0] and *data* [31..0], respectively, are connected to the input module *Histogram CNTRL*[15..0] and to exit *data*[15..0] of module *ram1port*.

FreqAndAllanLabNSTU program on a personal computer recalculates the histogram (array of numbers) *Histogram*[256] received from the processor *FreqLab SOPC* so that the total length of the delay corresponds to the number of the resulting output *Encoder*. Measured time period in the terms of the reference clock period in the getting by the module *Encoder* of *EncoderData* value is calculated as follows:

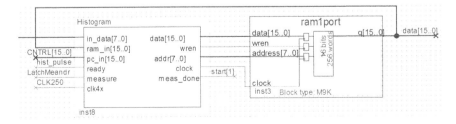

Figure 4. Structure for use automatic calibration.

IntervalTime = (1/2048) × Histogram [EncoderData].

Then, to get the time in seconds, it is nesessary to divide *IntervalTime* to the frequency of the reference signal.

Module *Histogram* accumulates data in the memory during 512 cycles of 1024 measurements after each falling edge of the *LatchMeandr* signal. In the result in a memory module *ram1port* 512 × 1024 = 524288 calibration results are accumulated to produce a histogram for the time 512 *ms*.

The number 2048 in the denominator of the formula for *IntervalTime* obtained from considerations that 524288 calibration results are divided into number 256 of memory cells of the histogram (524288/256 = 2048).

Structure of the module *FreqLab_SOPC* is shown at *Fig. 5*.

FreqLab_SOPC module uses interrupt, input signal *start*[0], generated by the falling edge of the signal *LatchMeandr* for the reading of the results of measurements (data from the counters of the measured signals, the data from the module *Encoder*, and data from the module of counetr of the reference frequency).

On the rising edge of the signal *meas done* from the module ,Histogram interruption occurs to read the histogram (input *start*[1]).

The program for module *FreqLab SOPC* buffers the data and sends the data on request to the computer. Buffer size corresponds to 1280 measurements, equivalent to 1.28 *s*. It is sufficient for *Windows*-measurements. Data packets are sent by amount 2048 B, used USB port. The package consists of 128 measurements and the current histogram.

Module *FreqCount* which counts the number of pulses of the measured signal, achives during the time from the start of the measurements is written on the hardware description language VERILOG [VERILOG site].

Module *IntervalCount* continuously counts the number of pulses of the reference frequency signal, and forms signal of the exact time of the passage of the leading edge of pulses *LatchPulse*.

According program fragment is the following:

module IntervalCount(input LatchPulse , input clk, output reg [31:0] count);

reg [31:0] cnt;
  always @ (posedge clk)
    cnt <= cnt + 1;
  always @ (posedge LatchPulse)
    count <= cnt;
endmodule

Processor *FreqLab_SOPC*, on the falling edge of *LatchMeandr* reads data from all registers by the multiplexer *CountValue*, asking address counter *CNTRL*[11..10], and reading the data from the output of the multiplexer *data*[31..0].

Module *LatchPulse* generates a signal of the same name *LatchPulse*. This signal plays a pivotal role in the measurement. Signal with frequency of 1 *kHz* from *LatchMeandr* is synchronized with the selected signal *Signal*[i], the selection is made by the multiplexer processor *Gate123*, and port *CNTRL*, signal *LatchChoice* is obtained.

Essentially the time of the positive edge of the signal *LatchChoice* with a resolution of 50 ps is measured, and the number of pulses of *Signal*[i] corresponding to *LatchChoice* since the start of the measurements. This scheme makes it easy to increase the number of simultaneously and continuously measured channels. *LatchChoice* signal in the module *LatchPulse* is synchronized by the reference frequency signal *CLK250* doubly: first by the positive edge of *CLK250*, and second by the negative edge of *CLK250*, the result is a signal *LatchPulse*, in which all counters in the respective registers become snapped. The time delay between signals and *LatchChoice LatchPulse*, is measured by module *Encoder* with a resolution of 50 *ps*.

4 TECHNICAL CHARACTERISTICS OF THE FREQUENCY METER

Technical characteristics of the frequency meter have been determined by the calculation, simulation and testing. The time resolution $\Delta t = 47$ ps, the calculation is made by Alan function [10], when the averaging time $\tau = 1$ ms.

Measuring of the frequency F = 250 × 64/1023 = 15.64 *MHz* gives the error $\Delta F = 0,73$ *Hz* in the function of Alan, which is $\Delta F / F = 0,73 / 15,640,000 =$

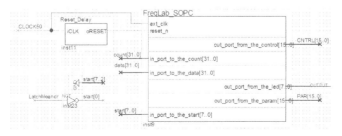

Figure 5.   Structure of the module FreqLab_SOPC.

0.00000004667. It mean in time part, that temporary resolution is $\Delta t = 46,7\ ps$.

Also providing of unique opportunity at relatively high frequency of new data measurements is worth mentioning. For example, if every new measurement is done and every $T = 1\ ms$, then the frequency of taking samples is $f = 1\ kHz$. Frequency averaging time can be of any value more then $1\ ms$. Moreover, measurement results with the full range of values of the measurement time is achieved, for example, with $\tau = 0.001\ s$; $0.01\ s$; $0.1\ s$, $1\ s$; $10\ s$; $100\ s$, and so forth. These values τ can be taken into the increments of discreteness $T = 1\ ms$. That is, the average frequency values are available (as well as functions of Alan) with values of $1\ ms$; $2\ ms$; $999\ ms$, $1000\ ms$; $1001\ ms$; $1002\ ms$;and so on. Another important feature is the ability to display the results of measuring, including those derived characteristics (function Alan) in real time on a computer screen. Due to the high accuracy of the instrument (ten orders of magnitude) even in experiments associated with the investigation or certification of laser or atomic frequency standards in the operation of such standards on the graph even very small changes in frequency are seen.

In particular, it is the study of the laser frequency standard response of the standard vibration which is observed. You can also see the relationship between the frequency of the supply voltage ripple, as well as the excitation systems of phase-locked loop used in these standards.

5 CONCLUSION

Thus, the developed frequency meter is a valuable tool for research and certification of laser and atomic frequency standards. We should note that the relative error on the level of a few tenth order should be attributed to the difference frequency of two or more standards, according to the conventional method of testing [Borisov et al, 2002]. When two close highly stable frequency are mixed (for example, the mixing of two laser beams of light on a common photodetector) their instability is added statistically (i.e., the total variance is the sum of the variances, the standard deviation is the square root of the sum of the standard deviations), and the average value obtained frequency equal to the difference between miscible frequencies, i.e. it is considerably smaller than the frequency of the study.

For convenience of measurement, this value, without loss of accuracy is transferred onto a carrier frequency, which is convenient for measurement. For example, the frequency standard is $F_S = 10^{14}\ Hz$, and instability in the 18-th sign of the corresponding increment of the frequency of the order of $F = 10^{-4}\ Hz$. If the difference frequency is transferred to the frequency F_0

$= 10^5\ Hz$, the measurement of this instability in the ideal is sufficient with frequency error not greater then $\sigma F = 10^{-5}\ Hz$. In this case, the measured value is at least an order of magnitude larger than the error of the meter. This is the error which is the feature of the described frequency meter. Any frequency counetr with the worse parameters is not fit to this task.

This work was financially supported by the Ministry of Education and Science of the Russian Federation on the state task №2014/138, theme: "New structures, models and algorithms for the management of breakthrough technical systems based on high technology of intellectual property".

REFERENCES

[1] Zhmud, V. A. 2014. Precision measurements of frequency for testing of laser standards. Automatics and Software Engineering, 2014. 1(7). P.104–109. (*In Russian*). http://jurnal.nips.ru/en/taxonomy/term/3.

[2] Goncharenko, A.M., Zhmud, V.A. 2013.Registered computer program FreqAndAllanLabNSTU, № 2013611160 (In Russian).

[3] Bugrov S.V., Goncharenko A.M., Zhmud V.A., Voevoda A.A. 2006. Software and possibilities of multi-channel precision fast frequency counter. Science Bulletin of NSTU, 2006, N 3(24), P.171–178. (In Russian).

[4] Goncharenko A. M., Zhmud V.A., Voevoda A.A., Avilov S.A. 2007. Microprocessor precision three-channel high-speed frequency counter without dear time. Instruments and Experimental Techniques. http://www.maik.ru/cgi-perl/journal.pl?lang=eng&name=instr (In Russian). http://elibrary.ru/item.asp?id=9495285)

[5] Zhmud V.A., Voevoda A.A. Goncharenko, A.M., 2005. Multi-channel precision frequency counter. Science Bulletin of NSTU, 2005, N 1(19), P.73–82. (In Russian).

[6] Zhmud V.A., Voevoda A.A. Goncharenko, A.M., 2005. Attestation of three-channel precision frequency counter. Science Bulletin of NSTU, 2005, N 2(20), P.175–178. (In Russian).

[7] Vasiliev, V.A., Zhmud V.A., Goncharenko A.M. 2003. Russian Patent N 2210785 Frequency meter. (In Russian).

[8] Goncharenko A.M., Zhmud V.A., 2006. Russian Patent N 2278390. Frequency meter. (In Russian).

[9] Zhmud V.A., Goncharenko A.M., 2000. Russian Patent N 2210783. Transformer of the time scale. (In Russian).

[10] Borisov B.D., Vasiliev V.A., Goncharenko A.M., Zhmud V.A. 2002. Methodic of the estimation of the stability of frequency standards. Optoelectronics, Instrumentation and Data Processing. http://www.springer.com/physics/optics+%26+lasers/journal/11974 (In Russian).

[11] Zhmud V.A. 2002. Frequency measurings in the laser systems. Science Bulletin of NSTU, 2002, N 1(13), P.127–136. (In Russian).

[12] Verlog language. Official web-site of the developer. URL: https://ru.wikipedia.org/wiki/Verilog.

Two oxygen sensors based on the fluorescence quenching of pyrene bonding on side-chain polysiloxanes

H. Yan, X.M. Wang, D.W. Wang & R.S. Yu
Department of Chemistry and Material Engineering, Logistic Engineering University, Chongqing, China

ABSTRACT: The two novel side-chain polysiloxanes containing different pyrene groups (–COO– and –OOC–) were synthesized by the hydrosilylation of polymethylhydrogensiloxane (PMHS) respectively with two pyrene derivatives, 1-decenyl–4–(1–pyrenyl) butanoate (DCPB) and 4–(1–pyrenyl) butyl–9–decenoate (PBDE). The polymers structures were confirmed by FTIR and ^{1}HNMR. The photophysical properties of the monomers and polymers in toluene were studied by using Hitachi F-2500 Fluorescence Spectrometer. It was found that the spectrum of the polymer was different from that of its monomer. The fluorescence intensity and the oxygen queching sensity (OQS) of the polymers are due to the side-chain construction. The OQS of the polymer with –COO– groups was as high as 59% at the monomer emission (ME) and 54% at the excimer emission (EE). With –OOC– groups, it was about 37% at ME and 56% at EE. It was found that the side-chain polysiloxane with –COO– groups was higher oxygen quenching performance than that with –OOC– groups. With the decreasing of the concentration, the changes of fluorescence intensity and OQS were both increasing at ME and decreasing at EE.

1 INTRODUCTION

Pressure sensitive paint (PSP) which is based on the luminescence quenching of some material by oxygen has been increasingly used for the surface pressure measurement in aerodynamic testing[1-6]. Since pyrene and some of its derivatives have long excited state lifetime, relatively high luminescence quantum yield and oxygen quenching efficiency, they are often used as active luminophor in PSP[6-8]. The PSPs with pyrene or its derivatives as the pressure sensor and silicone polymer as the binder often lack stability due to the loss of active luminophors from the matrix by evaporation or sublimation.

Basu et al. have carried out a detailed study of the mechanism of degradation of pyrene-based PSPs and identified that the cause of the degradation is the high diffusion coefficient of pyrene in silicone polymers, which assists evaporation of pyrene from the matrix[9]. It is found that the side group rotation and segmental motion of polymer chains of silicone polymers are due to the high diffusion rate of pyrene in silicone coatings. They had introduced the long-side chain into pyrene groups for improving the stability of active luminophores, and synthesized a new pyrene derivative for this purpose. It had been found that the coating which used pyrene derivative as active luminophores was better stability than pyrene-based PSP[10-12]. But its stability is still a problem. More over, all the PSPs were prepared by dispersing the fluorescent probe molecules in the silicone polymer matrix. The preparation of PSP easily results in generating the heterogeneous and unstable system. It wasn't the radical method for improving the stability of oxygen-sensitive coatings by modifying constitution of fluorescent probe molecules.

We have designed the different preparation of PSP, which is not only composed of luminescent molecules embedded in a polymeric binder, but also is synthesized by the chemical reaction with luminescent molecules and silicone polymers. The two novel side-chain polysiloxanes containing different pyrene groups (–OOC– and –COO–) were synthesized by the hydrosilylation of polymethylhydrogensiloxane (PMHS) respectively with two pyrene derivatives, 1–decenyl–4–(1–pyrenyl) butanoate (DCPB) and 4–(1–pyrenyl)butyl–9–decenoate (PBDE). Their structures were confirmed by FTIR and ^{1}HNMR. In the present study, the photophysical properties of the monomers and polymers in toluene are compared for attaining the better active luminophor used for PSPs and the results are studied here.

2 EXPERIMENTAL

2.1 *Materials*

9–Decen–1–ol and 9–decenoic acid were obtained from Tokyo Chemical Industry Co., Ltd.. 1–Pyrenyl butyric acid (PBA), 4–pyrenyl butanol (PBO) and polymethylhydrogensiloxane (PMHS) were procured from Aldrich Co.. Chloroplatinic acid (H$_2$PtCl$_6$) was purchased from Shanghai Shiyi Chemicals Reagent Co., Ltd. and p–toluene sulfonic acid from Chongqing Medicines Co., Ltd.. Toluene, n–hexane, cyclohexane, ethyl acetate and column-layer chromatographic silica gel were obtained from Chongqing Medicines Co., Ltd.

2.2 Methods for characterizations

^{1}HNMR spectroscopy was performed on a Bruker AV-300 with CDCl$_3$. FTIR spectra were recorded on a nicolet avatar 360 FTIR spectrometer, and samples were prepared by compressing disks of KBr in case of solid or scribbling on KBr disks in case of liquid.

Fluorescence emission spectra were recorded by using a fiber optic spectrometer, model F-2500 from Hitachi, Ltd.. The specimens were prepared by solving DCPB, PBDE, Polymer-A and Polymer-B in toluene respectively. The fluorescence emission spectra of the solution were recorded before deaeration and after deaeration.The oxygen quenching sensitivity (OQS)was calculated by using the equation

$$\frac{I_{N_2} - I_{air}}{I_{N_2}} \times 100$$

where I$_{N2}$ and I$_{air}$ are the fluorescence emission intensity values before deaeration and after deaeration, respectively.

2.3 Syntheses

2.3.1 Syntheses of 1–decenyl–4–(1–pyrenyl) butanoate (DCPB) and 4–(1–pyrene) butyl–9–decenoate (PBDE)

A direct esterification reaction between PBA and 9–decen–1–ol was used for the preparation of DCPB. The mixture of 1.15 g (0.4 mmol) PBA and 1.25 g (0.8 mmol) 9–decen–1–ol was added to 50 ml toluene in a 250 ml round bottom flask. 0.05 g p–Toluene sulphonic acid was used as catalyst. The above solution was refluxed for 6 h under nitrogen atmosphere. During the reaction, water was removed via a Dean-Stark trap. For there are different separating degrees of the reactants and the product in toluene, thin-layer chromatography was used to confirm the optimum terminal time for the reflux reaction. The brownish yellow reaction mixture was cooled to room temperature. After being filtered, the filtrate containing the product in toluene was concentrated to about 3 ml by reduced pressure evaporation. The product was

purified by using silica gel column chromatography. A mixture of n–hexane and ethyl acetate (2:1) was used for eluting the product. The resulting light yellow solid was dried under vacuum at 60℃ for 24 h (yield 82%). ^{1}HNMR (CDCl$_3$, δ in ppm): 8.30~7.82 (9H in pyrene), 4.07 (–COOCH$_2$–), 3.37 (pyrene–CH$_2$–), 2.18 (pyrene–CH$_2$–CH$_2$–), 2.44 (–CH$_2$–COO–), 1.60 (–COOCH$_2$–CH$_2$–), 1.27 (–CH$_2$–(CH$_2$)$_5$–CH$_2$–), 2.02 (–CH$_2$–CH=CH$_2$), 5.80 (–CH=CH$_2$), 4.90~5.00 (–CH=CH$_2$). The infrared spectrum of the product shows signals at 1730.0 and 1180.0cm^{-1} (due to C=O group), which clearly indicates that the product is ester. The absence of signal around 3500 cm^{-1} shows the absence of –OH group of 9–decen–1–ol and the produce is pure and free from 9–decen–1–ol and PBA.

PBDE was synthesized by the same esterification reaction between PBO and 9–decenoic acid, and its yield was 75%. ^{1}HNMR (CDCl$_3$, δ in ppm): 8.19~7.76 (9H in pyrene), 4.06 (–CH$_2$OOC–), 3.29 (pyrene–CH$_2$–(CH$_2$)$_2$–), 1.67~1.77 (pyrene–CH$_2$–CH$_2$CH$_2$–), 2.19 (–OOC–CH$_2$–), 1.19 (–OOCCH$_2$–(CH$_2$)$_5$–), 1.92 (–CH$_2$–CH=CH$_2$), 5.69 (–CH=CH$_2$), 4.85 (–CH=CH$_2$). IR: 1730.02 and 1180.02 cm^{-1} (due to C=O group). The signal around 3340 cm^{-1} disappeared due to the absence of –OH group in the pure product.

2.3.2 Hydrosilylations of PMHS with DCPB and with PBDE

The method for the syntheses of the two side-chain polysiloxanes containing different pyrene groups (–COO– and –OOC–) is shown in Scheme 1. The two polysiloxanes are called polymer-A and polymer-B respectively. The process to synthesize polymer A is detailed in the next text.

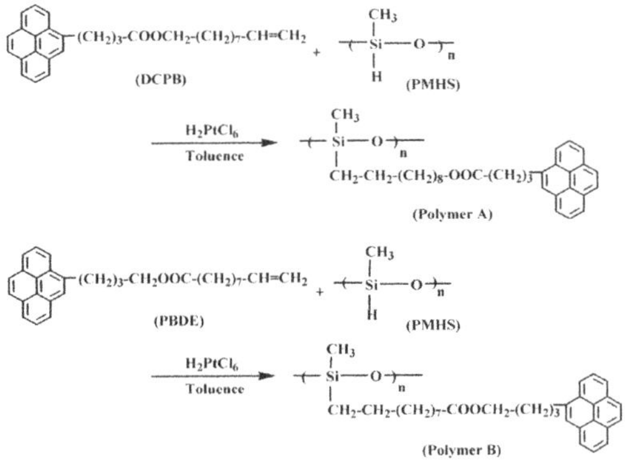

Figure 1. The syntheses of polymers.

Polymer-A with -COO- groups was synthesized by a hydrosilylation of PHMS and DCBP[13]. To a 100 ml three-necked flask equipped with a reflux condenser, magnetic stirrer was added 0.31 g (2.0 mmol of Si-H units) PMHS, 0.94 (2.19 mmol)DCPB and 50 ml of dry, freshly distilled toluene. The reaction mixture was heated under nitrogen for 1h. H_2PtCl_6 (2 mg), in 4ml dry, freshly distilled toluene ,was injected and stirred under nitrogen for about 60 h. The reaction was monitored by thin-layer chromatography. After reaction, 1.0 g active carbon was added and then filtered. The filtrate containing the product was added to 500 ml toluene for removing the unreacted monomers. The final product is a light yellow viscosity liquid (yield 80%). ^{1}HNMR (CDCl$_3$, δ in ppm): 8.32~7.74 (9H in pyrene), 4.08 (–COOCH_2–), 3.40 (pyrene–CH_2–), 2.20 (pyrene–CH$_2$–CH_2–), 2.43 (–CH_2–COO–), 1.52 (–COOCH$_2$–CH_2–), 1.20 (–COOCH$_2$CH$_2$–$(CH_2)_7$–), 0.50(–CH_2–Si(CH$_3$)–O–), 0.06 (–Si(CH_3)–O–). IR: 2926.0 cm^{-1} (–(CH$_2$)$_7$–), 1735.0 cm^{-1} (C=O).

3 RESULTS AND DISCUSSION

3.1 Absorption characteristics of monomers and polymer in solutions

The solutions of DCPB and PBDE (2 mM) and polymers (0.165g/10ml) in toluene were prepared, respectively, which are illuminated with UV radiation at the wavelength of 400 nm and shown in Figure 1. For both containing the founctional group of -COO-, the comparison of the absorption characteristics between DCPB and Polymer-A in toluene is shown in Figure 1A. The comparison of that between PBDE and Polymer-B for the fouctional group of -OOC- is shown in Figure 1B.

The spectra of monomers and polymers are very similar to that of pyrene and pyrene derivatives at the same concentration. In Figure 1A, the absorption peaks of DCPB and Polymer-A are shifted to the longer wavelength side by about 10 nm compared to that of pyrene and to the shorter wavelength side by about 2.5 nm compared to that of PBA. The changes are similar to that of PBDE and Polymer-B in Figure 1B. The absorption maximum is around 364 nm. This is in agreement with the results reported earlier for other probes based on PBA and PBO[14]. It has been reported that pyrene is connected to other functional groups though bridges of -COO- and -OOC- groups. Then the absorption spectra of the resulting probe exhibited well-resolved spectra similar to that of unsubstituted pyrene. The -COO- and -OOC- groups as bridges for attached pyrene groups attached to

main-chain of polymers have almost no effect on absorption characterisics.

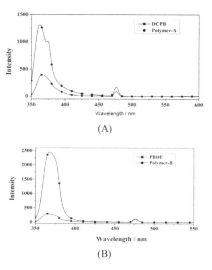

(A)

(B)

Figure 2. Absorption spectra of DCPB and Polymer-A (A) and PBDE and Polymer-B (B) in toluene.

3.2 Fluorescence emission characteristics of monomers and polymers in solutions

Figure 2 presents the results of the fluorescence emission spectra of solutions of monomers and polymers with the excitation wavelength at 364 nm. The solutions were prepared respectively. For comparison, the fluorescence emission spectra of DCPB and Polymer-A in toluene are shown in Figure 2A. And that of PBDE and Polymer-B in toluene are shown in Figure 2B.

It can be seen that monomers and polymers in solutions had high UV stability and there were no decrease in the fluorescence intensity on continuous UV exposure of the solutions for several minutes. In Figure 2A, the fluorescence emission spectra of DCPB and Polymer-A in toluene exhibited the same characteristics peaks of unsubstituted pyrene. The broad emission peak at 475 nm is due to the excimer emission (EE) of pryene group and the narrow peak at 395 nm is due to the monomer emission (ME). So we could confirm that pyrene group had been introduced into the side-chains of polysiloxanes further. The fluorescence monomer emission (ME) peaks of DCPB and Polymer-A are shifted to the longer wavelength side by about 3 nm and the fluorescence intensity of the excimer emission (EE) decreases prominently compared to that of pyrene and pyrene derivatives at the same concentration. The changes are similar to that of PBDE and Polymer-B in Figure 2B.

The spectrum of the polymer existed some different changes compared to that of its monomer. Whether Polymer-A or Polymer-B, the excimer emission band (EE) at about 475 nm was more prominent than the monomer emission peaks (ME). It can be seen that the excimer emission of polymers has higher quantum efficiency than that of monomers.

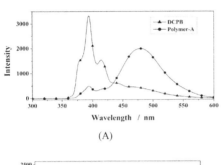

(A)

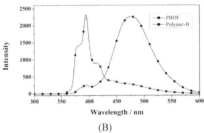

(B)

Figure 3. Fluorescence emission spectra of DCPB and Polymer-A (A) and PBDE and Polymer-B (B) in toluene.

For PSP applications, the excimer emission should be higher oxygen quenching. The fluorescence emission spectra of DCPB and PBDE in air and in absence of oxygen are shown in Figure 3. In Figure 3A, the emission intensities of DCPB in toluene were measured at 395 nm and at 475 nm respectively in air and in the presence of nitrogen. By using the equation $(I_{N2}-I_{air})\times100/I_{N2}$, the oxygen quenching sensitivity (OQS) of DCPB were about 66% at ME (395 nm) and 78% at EE (475 nm). The spectra of PBDE shown in Figure 3B are very similar to that of DCPB. The OQS of PBDE were about 73% at ME (395 nm) and 67% at EE (475 nm). It can be seen that the fluorescence emissions of both solutions of DCBP and PBDE have high quantum efficiency and high oxygen sensitivity, and the oxygen quenching sensitivity (OQS) of DCBP solution is higher than that of PBDE at EE.

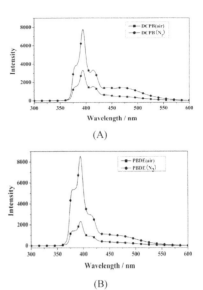

(A)

(B)

Figure 4. Fluorescence emission spectra of DCPB (A) and PBDE (B) before deaeration and after deaeration.

3.3 *Effect of polymer concentration in solution on fluorescent intensity and oxygen quenching sensitivity*

Since pyrene group was attached to the main chains of polymers by the functional groups, clusters of pyrene groups could be formed easily. Fluorescence studies on pyrene crystals have shown the formation of short lived excited dimers from ground state dimers 14-16. If the small clusters in the solutions are in the form of small crystals, then this would lead to quenching of dynamic excimer intensity by formation of ground state dimmers.

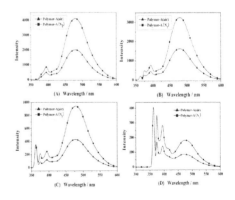

Figure 5. Fluorescence emission spectra of solutions of Polymer-A at different concentration, (A) 1.67mg/ml, (B) 0.33mg/ml, (C) 0.07mg/ml, (D) 0.01mg/ml, before deaeration and after deaeration.

With the decreasing of the concentration, the excimer intensity decreased from the trend at lower concentrations[12, 14, 17]. Influences of Polymer-A and Polymer-B concentration in solutions were studied. The solutions of Polymer-A and Polymer-B with different concentrations (1.67 mg/ml, 0.33 mg/ml, 0.07 mg/ml and 0.01mg/ml) in toluene were prepared, respectively. The emission intensities of solutions were recorded in air and in absence of oxygen. The results of Polymer-A solutions are shown in Figure 4, the same as that of Polymer-B. It was found that the fluorescence characteristics of polymer solutions exhibited the regular changes similar to that of pyrene and pyrene derivatives. With the decreasing of its concentration, the emission intensity at ME (395 nm) was increasing and was the opposite at EE (475 nm). By using the equation which was mentioned above, the oxygen quenching sensitivities (OQS) of Polymer-A and Polymer-B in toluene with different concentrations were calculated. The results are shown in Table 1.

Table 1. The oxygen quenching performances of Polymer-A and Polymer-B in toluene with different concentrations.

Concentration (g/10ml toluene)	The OQS (%) of Polymer-A solution		The OQS (%) of Polymer-B solution	
	(ME)	(EE)	(ME)	(EE)
1.67	43	51	17	41
0.33	43	50	13	40
0.07	59	54	37	56
0.01	50	53	33	51

The OQS of Polymer-A with -COO- groups was as high as 59% at the monomer emission (ME) and 54% at the excimer emission (EE). That of Polymer-B with -OOC- groups was about 37% at ME and 56% at EE. It was found that the side-chain polysiloxane with -COO- groups is higher oxygen quenching performance than that with -OOC- groups.

4 CONCLUSIONS

The photophysical properties of two new pyrene alkenyl eaters and side-chain polysiloxanes containing different pyrene groups were found to be similar to that of pyrene with the excitation wavelength at 364 nm. The fluorescence emission intensities have high quantum efficiency and effectively quenching by oxygen.

The results have shown -COO- and -OOC- groups bridges attached pyrene groups to main chains of polysiloxanes affected the fluorescence quenching.

1–decenyl–4–(1–pyrenyl) butanoate (DCPB) and polysiloxane with -COO- groups have higher oxygen quenching performance than 4–(1–pyrenyl) butyl–9–decenoate (PBDE) and polysiloxane with -OOC- groups. The OQS of the polymer with -COO- groups was as high as 59% at the monomer emission (ME) and 54% at the excimer emission (EE). With -OOC- groups, it was about 37% at ME and 56% at EE.

Whether containing -COO- or -OOC- groups bridges, the emission and oxygen quenching sensitivity all were affected by the concentration of pyrene groups in solutions. With the decreasing of the concentration, the changes of fluorescence intensity and OQS were both increasing at ME and decreasing at EE.

REFERENCES

[1] M. Gouterman, Oxygen quenching of luminescence of pressure sensitive paint for wind tunnel research. *J. Chem. Educ.* 1997, 74, 697–702.

[2] C. Klein, Application of pressure sensitive paint (PSP) for the determination of the instantaneous pressure field of models in a wind tunnel. Aerospace Science Technology, 2000, 4, 103–109.

[3] R. H. Engler; M. Merlenne; C. Klein, Application of PSP in low speed flows. Aerospace Science Technology, 2002, 6, 313–322.

[4] L. M. Coyle; M. Gouterman, Dual luminophor pressure sensitive paint:II. Lifetime based measurement of pressure and temperature. Sens. Actuators B, 2003, 96, 304–314.

[5] E. Purlin; B. Carlson; S. Gouin; et al., Ideality of pressure-sensitive paint. I. Platinum tetra(pentafluorophenyl)porphine in fluoroacrylic polymer. J. Appl. Polym. Sci., 2000, 77, 2795–2804.

[6] Y. Fujiwara; Y. Amao, An oxygen sensor based on the fluorescence quenching of pyrene chemisorbed layer onto alumina plates. Sens. Actuators B, 2003, 89, 187–191.

[7] T. Ishiji; M. Kaneko, Photoluminescence of pyrenebutyric acid incorporated into silicone film as a technique in luminescent oxygen sensing. Analyst, 1995, 120, 1633–1638.

[8] W. Xu; R. Schmidt; M. Whaley; J. N. Demas; B. A. Degraff; E. K. Karikari; B. L. Farmer, Oxygen sensors based on luminescence quenching: interactions of pyrene with the polymer supports. Anal. Chem., 1995, 67, 3172–3180.

[9] B. J. Basu; C. Anandan; K. S. Rajam, Study of the mechanism of degradation of pyrene-based pressure sensitive paints. Sens. Actuators B, 2003, 94, 257–266.

[10] C. Anandan; B. J. Basu; K. S. Rajaml, Investigations of the effect of viscosity of resin on the diffusion of pyrene in silicone polymer coatings using steady state fluorescence technique. Eur. Polym. J., 2004, 40, 335–342.

[11] B. J. Basu; K. S. Rajam, Comparison of the oxygen sensor performance of some pyrene derivatives in silicone polymer matrix. Sens. Actuators B, 2004, 99, 459–467.

[12] B. J. Basu; A. Thirumurugan; C. Anadan; et al., Optical oxygen sensor coating based on the fluorescence

quenching of a new pyrene derivative. Sens. Actuators B, 2005, 104, 15–22.

[13] H. Yan; W. L. Chan; Y. S. Szeto, Electrorheological behavior of side-chain polysiloxane containing 3-(4-amidophenyl)sydnone moieties. J. Appl. Polym. Sci., 2003, 91, 2523–2528.

[14] H. Yan; X. M. Wang; R. S. Yu; B. Zhang; S. L. Chen, Synthesis and oxygen quenching of pyrene derivatives containing long chain alkyl. J. Functional Materials, 2007, 38, 795–797.

[15] T. Kobayashi, The observation of the excimer formation process in pyrene and perylene crystals using a picosecond ruby laser and streak camera. J. Chem. Phys., 1978, 69, 3570–3574.

[16] J. G. Xu; Z. B. Wang, Flourimetry, Science Press, Beijing, 2006.

[17] C. Anandan; B. J. Basu; K. S. Rajaml, Study of the diffusion of pyrene in silicone polymer coatings by steady state fluorescence technique: effects of pyrene concentration. Eur. Polym. J., 2004, 40, 1833–1840.

Testing and Measurement: Techniques and Applications – Chan (Ed.)
© 2015 Taylor & Francis Group, London, ISBN: 978-1-138-02812-8

Measurement of thermal neutrons with fiber-optic radiation sensors using Cerenkov effect

K.W. Jang, J.S. Kim, S.H. Shin, J.S. Jang, H. Jeon, G. Kwon, W.J. Yoo & B. Lee
School of Biomedical Engineering, College of Biomedical & Health Science, BK21 Plus Research Institute of Biomedical Engineering, Konkuk University, Chungju, Korea

T. Yagi, C.H. Pyeon & T. Misawa
Nuclear Engineering Science Division, Research Reactor Institute, Kyoto University, Osaka, Japan

ABSTRACT: The purpose of this research is to measure the neutrons in mixed radiation fields by using Cerenkov effect. In this research, we fabricated a fiber-optic Cerenkov radiation sensor using a Gd-foil, a rutile crystal and an optical fiber. A reference sensor was also fabricated with the rutile crystal and an optical fiber to remove noise signals induced by background gamma-rays. To clarify the relationship between electron fluxes and intensities of Cerenkov radiation, the electron fluxes inducing Cerenkov radiation in the fiber-optic Cerenkov radiation sensor were calculated by using the Monte Carlo N-particle transport code simulations. Finally, relative fluxes of thermal neutrons in a polyethylene generated by a Cf-252 neutron source were obtained by using the fiber-optic Cerenkov radiation sensor. The results obtained by the fiber-optic Cerenkov radiation sensor were close to calculated thermal neutron fluxes by using the Monte Carlo N-particle transport code simulations.

1 INTRODUCTION

Generally, fiber-optic radiation sensors (FORSs) for detecting neutrons consist of neutron converters, optical fibers and scintillators (Yagi et al. 2011). In order to detect neutrons, converting materials having high cross-sections for neutrons are required; for example, Li-6 and Gd-157 having cross-sections of 940 and 255,000 barns, respectively, are used as converters for thermal neutrons (Abdushukurov et al. 1994; Crow et al. 2004). As a waveguide, optical fibers are widely used to transmit scintillation signals to a photodetector. Optical fibers have many advantages in hazardous environments of nuclear facilities. Their most favorable capability is remote transmission without significant diminution of a light signal. In this process, the light signals transmitted by optical fibers are immune to environmental influences such as pressure, humidity, and electromagnetic field (Lee et al. 2000). In addition, their small radii or thicknesses make it possible to detect neutrons with a high spatial resolution in a narrow space (Yagi et al. 2011). Moreover, specific optical fibers such as metal coated optical fibers can be used in a high temperature condition. Scintillators have been usually exploited to convert the charged particles or gamma-rays emitted from neutron converters into measureable light signals (Mori et al. 1999). Although conventional scintillators have high scintillation efficiencies and short decay

times, these types of materials have some defects in specific conditions. In high ionization density, scintillation molecules can be temporarily damaged by high energy charged particles; this phenomenon, which is known as the ionization quenching, causes non-proportionality between energy losses of charged particles and scintillation outputs (Mouatassim et al. 1995). In addition, light yields of scintillators vary with an ambient temperature and this characteristic restricts the use of scintillator in high temperature conditions (Boivin et al. 1992). Therefore, a novel technique is required to detect neutrons in extremely harsh environments such as molten-salts in pyroprocessing or cores of nuclear reactors.

Meanwhile, a charged particle cannot travel with a velocity greater than the phase velocity of light in a vacuum. However, in some dielectric media including water, silica, and polymethyl methacrylate (PMMA), a high energy particle can pass through the media with a velocity greater than the phase velocity of light. When the charged particle does this, the electromagnetic field close to the particle polarizes the media along its path, and then the electrons in the atoms follow the waveform of the pulse. Here, the waveform and the medium are called Cerenkov radiation and Cerenkov radiator, respectively (Jelley 1955). In contrast to scintillation generated in a scintillator, Cerenkov radiation generated from the radiator can be used in extremely harsh conditions.

In a high temperature tokamak, electron fluxes can be obtained by measuring Cerenkov radiation generated from some crystals. Also, in a radiotherapy dosimetry, it is possible to measure relative depth doses for heavy charged particle beams without the ionization quenching by measuring the intensities of Cerenkov radiation (Jang et al. 2012). Therefore, Cerenkov radiation generated in the radiators can be a significant signal in hazardous radiation conditions.

In order to measure the neutrons using Cerenkov effect, we have fabricated a fiber-optic Cerenkov radiation sensor (FOCRS) using a Gd-foil, a rutile crystal and an optical fiber, and have measured pure thermal neutrons successfully (Jang et al. 2013)

The purpose of this research is to measure the neutrons in mixed radiation fields by using the FOCRS. Thus, in this study, we fabricated a reference sensor consisting of only a rutile crystal and an optical fiber to remove noise signals induced by background gamma-rays. Also, we calculated energies and fluxes of electrons inducing Cerenkov radiation in the FOCRS by using the Monte Carlo N-particle transport code (MCNPX) simulations. Finally, relative fluxes of thermal neutrons in a polyethylene (PE) generated by a Cf-252 neutron source were obtained by using the FOCRS.

2 MATERIALS AND METHODS

In our experiments, a Gd-foil (GD-143220, Nilaco Co., Tokyo, Japan) with 99.9% Gd-157 composition is used as a neutron converter. The Gd-157 has a cross section for thermal neutron as 255,000 barns, which is among the highest nuclear cross sections found in any material. After interaction with the neutrons, the Gd-157 emits gamma-rays with energies up to 7.8 MeV and 72 keV conversion electrons. The gamma-rays and electrons emitted from the Gd-157 induced by the neutrons have sufficient energies to produce Cerenkov radiation in a dielectric medium having a high refractive index. The dimensions and density of the Gd-foil used in this research are $5\times5\times0.025$ mm^3 and 7.9 g/cm^3, respectively.

As Cerenkov radiators, rutile crystals (TiO$_2$, Shinkosha Co. Ltd., Yokohama, Japan) were employed throughout this study. In general, Cerenkov radiators should have high refractive indices to produce Cerenkov radiation for low energy electrons. The rutile crystals have a refractive index of 2.87 for 430 nm wavelength. The CTE of electrons in the crystals to produce Cerenkov radiation was calculated as 34 keV by using Eq. (1); since this energy is smaller than the conversion electrons and gamma-rays emitted from the Gd-157, Cerenkov radiation can be produced in the rutile crystals. The density and dimensions of rutile crystals used in this research are 4.23 g/cm^3 and $10\times10\times0.1$ mm^3, respectively.

Optical fibers used to transmit Cerenkov radiation are step-index multimode plastic optical fibers (SH6001, Mitsubishi Rayon Co. Ltd., Tokyo, Japan) having core/cladding structures. The outer diameter of plastic optical fibers (POFs) is 1.5 mm and the cladding thickness is 0.015 mm. The refractive indices of the core and the cladding are 1.492 and 1.402, respectively, and the numerical aperture (NA) is 0.510. The NA denotes the light-gathering power, and more light can be guided by an optical fiber with a higher NA. The materials of core and cladding are polymethyl methacrylate (PMMA) and fluorinated polymer, respectively, and the jacket is made of PE.

A photomultiplier tube (PMT; R1635, Hamamatsu Photonics K.K., Hamamatsu, Japan) with a multichannel analyzer (MCA; 2100C/MCA, Laboratory Equipment Corporation Co. Ltd., Tsuchiura, Japan) was used to measure the intensities of Cerenkov radiation. The measurable wavelength range of the PMT is from 300 nm to 650 nm and its peak sensitive wavelength is 420 nm.

As a neutron source, a Cf-252 having a half-life of 2.65 years was used. The Cf-252 emits neutrons with a mean energy of about 2 MeV. To increase the ratio of thermal neutrons, we used PE blocks as a moderator. The activity of Cf-252 used in our experiments was about 10 µCi.

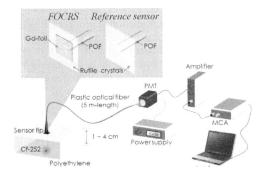

Figure 1. The structure of FOCRS and experimental setup.

The structure of FOCRS and experimental setup can be found in figure 1. The sensor probe of FOCRS consists of a rutile crystal with a Gd-foil. In case of the reference sensor for measuring background gamma-rays, only the rutile crystal was used without the Gd-foil. By subtracting signals of the reference sensor from those of the FOCRS, therefore, the intensities of Cerenkov radiation induced by only the Gd-foil can be obtained. When the sensor probes are irradiated by the neutrons, the electrons and gamma-rays emitted from the Gd-foil pass through the rutile crystals. In this process, Cerenkov radiation is generated in the crystals and is then transmitted by a 5 m POF to the

PMT. The amplified electric signals are then measured by using the MCA and the supporting software (MCAWinUC, Laboratory Equipment Corporation Co. Ltd., Tsuchiura, Japan).

3 RESULTS AND DISCUSSIONS

Figure 3 shows calculated energy distributions of electrons generated in rutile crystals by using MCNPX code. As mentioned above, because the CTE of electrons in the crystal is about 34 keV, the electrons in this range cannot contribute to producing Cerenkov radiation in the rutile crystal. Most of these electrons are derived from conversion electrons with energy of 72 keV emitted from the Gd-157 and background gamma-rays; the conversion electrons can lose their energy to a level below the CTE by passing through the Gd-157 and rutile crystal. On the other hand, gamma-rays generated from the Gd-157 have relatively high energies of up to 7.8 MeV, thus subsequent electrons induced by interactions between the gamma-rays and the rutile crystals have sufficient energies to produce Cerenkov radiation in the crystals. Therefore, we can estimate that almost Cerenkov radiation generated in the FOCRS is induced by the gamma-rays generated from the Gd-157.

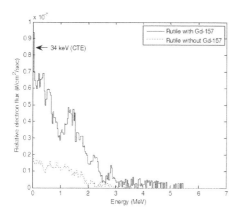

Figure 3. Calculated electron fluxes for energy in rutile crystals with and without a Gd-157.

The result of rutile crystal without a Gd-157 presents the electron fluxes induced by background gamma-rays. Typically, background gamma-rays are produced by interactions between neutrons and a PE. In our simulation, total electron flux of the rutile crystal with Gd-157 was 4.2 times higher than that of the rutile crystal without Gd-157. Since the purpose of this study is to detect thermal neutrons by measuring intensities of Cerenkov radiation induced by

electrons and gamma-rays emitted from the Gd-foil, the intensities caused by the background gamma-rays should be eliminated. Throughout this study, therefore, Cerenkov radiation that is obtained by using the reference sensor was subtracted from that using the FOCRS.

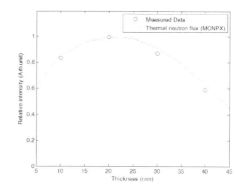

Figure 4. Measured intensities of Cerenkov radiation using the FOCRS and calculated electron fluxes over CTE for thicknesses of PE.

Figure 4 shows measured intensities of Cerenkov radiation by using the FOCRS and calculated electron fluxes over CTE for thicknesses of PE. In the prior simulation, the thermal neutron fluxes have a peak at 20 mm thickness of PE. In this result, the calculated electron fluxes did not perfectly match with the thermal neutron fluxes, but had a similar trend. The difference between the calculated fluxes of electrons and thermal neutrons was about 4.3%. In case of measured Cerenkov radiation, although the intensities obtained by the FOCRS have a relatively large instrumental error (10%), the intensities were in good agreement with the calculated thermal neutron fluxes. The difference between the measured intensities of Cerenkov radiation and the calculated fluxes of thermal neutrons was about 2.5%.

4 CONCLUSIONS

By measuring the intensities of Cerenkov radiation generated in a dielectric medium, it is possible to detect some radiations without any scintillation material. In this research, we proposed a novel method for detecting thermal neutrons with a FOCRS. The FOCRS for detecting thermal neutrons was fabricated by using a Gd-foil, a rutile crystal, and an optical fiber. A reference sensor was also fabricated with the rutile crystal and the optical fiber for measuring background gamma-rays. To clarify the relationship between electron fluxes and intensities of Cerenkov radiation, electron fluxes inducing Cerenkov radiation in the FOCRS were calculated by using the MCNPX simulations. Finally, relative fluxes of thermal neutrons in a PE generated by a Cf-252 neutron source were obtained by using the FOCRS. The results obtained by the FOCRS were close to calculated thermal neutron fluxes using the MCNPX.

Further study will be carried out to exploit FOCRS for detecting the neutrons in high temperature conditions. It is anticipated that the novel and simple FOCRS using the Cerenkov effect proposed here can be effectively used to measure radiations in hazardous nuclear facilities.

ACKNOWLEDGMENTS

This research was supported by Basic Science Research Program through the National Research Foundation of Korea (NRF) funded by the Ministry of Science, ICT & Future Planning (No. 2013R1A1A1061647 and No. 2014002620). Also, this research was supported by National Nuclear R&D Program through the National Research Foundation of Korea (NRF) funded by the Ministry of Science, ICT and future planning (No. 2014000457).

REFERENCES

[1] Yagi, T., Unesaki, H., Misawa, T., Pyeon, C. H., Shiroya, S., Matsumoto, T., and Harano, H. 2011. Development of a small scintillation detector with an optical fiber for fast neutrons. *Appl. Radiat. Isotopes* 69(2): 539–544.

[2] Abdushukurov, D.A., Dzhuraev, A.A., Evteeva, S.S., Kovalenko, P.P., Leskin, V.A., Nikolaev, V.A., Sirodzhi, R.F., and Umarov, F.B. 1994. Model calculation of efficiency of gadolinium based converters of thermal neutrons. Nucl. Instrum. Meth. B 84(3): 400–404.

[3] Crow, M.L., Hodges, J.P., and Cooper, R.G. 2004. Shifting scintillator prototype large pixel wavelength-shifting fiber detector for the POWGEN3 powder diffractometer. Nucl. Instrum. Meth. A 529(1-3): 287–292.

[4] Lee, B., Choi, W.Y., and Walker, J.K. 2000. Polymer-polymer miscibility study for plastic gradient index optical fiber. Polym. Eng. Sci. 40(9): 1996–1999.

[5] Mori, C., Uritani, A., Miyahara, H., Iguchi, T., Shiroya, S., Kobayashi, K., Takada, E., Fleming, R.F., Dewaraja, Y.K., Stuenkel, D., and Knoll, G.F. 1999. Measurement of neutron and γ-ray intensity distributions with an optical fiber-scintillator detector. Nucl. Instrum. Meth. A 422(1-3): 129–132.

[6] Mouatassim, S., Costa, G.J., Guillaume, G., Heusch, B., Huck, A., and Moszynski, M. 1995. The light yield response of NE213 organic scintillators to charged particles resulting from neutron interactions. Nucl. Instrum. Meth. A 359(3): 530–536.

[7] Boivin, R.L., Lin, Z., Roquemore, A.L., and Zweben, S.J. 1992. Calibration of the TFTR lost alpha diagnostic. Rev. Sci. Instrum. 63(10): 4418–4426.

[8] Jelly, J.V. 1955. Cerenkov radiation and its applications. J. Appl. Phys. 6: 227–232.

[9] Jang, K.W., Yoo, W.J., Shin, S.H., Shin, D., and Lee, B. 2012. Fiber-optic Cerenkov radiation sensor for proton therapy dosimetry. Opt. Express 20(13): 13907–13914.

[10] Jang, K.W. Yagi, T., Pyeon, C.H., Yoo, W.J., Shin, S.H. Misawa, T., Lee, B. 2013. Feasibility of fiber-optic radiation sensor using Cerenkov effect for detecting thermal neutrons. Opt. Express 21(12): 14573–14582.

Testing and Measurement: Techniques and Applications – Chan (Ed.)
© 2015 Taylor & Francis Group, London, ISBN: 978-1-138-02812-8

Fabrication and characterization of a fiber-optic temperature sensor based on NaCl solution

H.I. Sim, W.J. Yoo, S.H. Shin, S.G. Kim, S. Hong, K.W. Jang & B. Lee[*]
School of Biomedical Engineering, College of Biomedical & Health Science, BK21 Plus Research Institute of Biomedical Engineering, Konkuk University, Chungju, Korea

S. Cho
Department of Organic Materials & Fiber Engineering, College of Engineering, Soongsil University, Seoul, Korea

J.H. Moon
Department of Nuclear & Energy Engineering, College of Energy & Environment, Dongguk University, Gyeongju, Korea

ABSTRACT: In this study, a fiber optic temperature sensor was developed using an aqueous solution of sodium chloride (NaCl solution) and an Optical Time-Domain Reflectometer (OTDR) to measure temperature at a long distance. By changing temperature, the refractive index of the NaCl solution is varied and then Fresnel reflection arising at the interface between the distal end of optical fiber and the NaCl solution is also changed. Therefore, we measured the optical power of the light reflected from the sensing probe according to the temperature of water and also obtained the relationship between the temperature of water and the optical power.

1 INTRODUCTION

To measure real-time temperature, many different types of temperature sensors, including thermocouple, resistance temperature detector (RTD), thermopile, and thermistor, have been developed and commercialized (Crunelle et al. 2009, Pan et al. 2012, Yoo et al. 2011a). However, existing electric and electronic temperature sensors have some disadvantages to be used at inaccessible regions or harsh environments, such as a deep sea, a nuclear power plant, and medical or chemical facilities, due to their limit of measurable distance, power supply, and high electromagnetic interference (EMI) or radiofrequency interference (RFI). To overcome these problems, various fiber-optic temperature sensors have been investigated as a promising candidate (Yoo et al. 2011a, Fernandez-Valdivielso et al. 2002, Han et al. 2014, Eom et al. 2005). Generally optical fiber-based sensors have many attractive advantages, such as small sensing volume, real-time monitoring, remote sensing, and immunity to EMI or RFI (Bolognini et al. 2007, Mendonca et al. 2007, Yoo et al. 2011b, Yoo et al. 2010, Sade& Katzir 2001).

In this study, a fiber-optic temperature sensor was fabricated using an optical time-domain reflectometer (OTDR) which can measure specific signals at a long distance. We measured the reflected light in order to obtain temperature using the fabricated fiber-optic temperature sensor and then compared the results with temperature obtained using a conventional thermometer.

2 MATERIALS AND METHODS

When light transmitted through an optical fiber, a portion of the light is reflected at the end of the optical fiber by the refractive index (RI) difference between two media. This phenomenon is called Fresnel reflection and the fraction of optical power (R) is expressed by the following equation 1 (Pedrotti et al. 2007, Khare 2004).

$$R = ((n_1 - n_2) / (n_1 + n_2))^2 \qquad (1)$$

Where n1 is the RI of the core of optical fiber and n2 is the RI of the contact sensitive material. As delineated in equation 1, the optical power of the reflected light is determined by the RI of the two media. When n2 approached to n1, the optical power of the reflected light is decreased, as shown in figure 1.

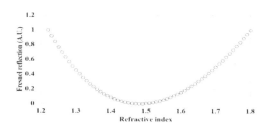

Figure 1. Relationship between the RI of contact material and the Fresnel reflection.

*Corresponding author

A sensing probe of the fiber-optic temperature sensor is composed of an optical fiber, a stainless-steel case, a rubber sealing-ring, and an aqueous solution of sodium chloride (NaCl solution), as shown in figure 2. In general, NaCl exists as a white solid at room temperature and can be easily dissolved in water because of its hydrophilic property. Specially, the RI of a NaCl solution is changed according to the temperature, and thus we used this phenomenon for measuring temperature of water. A single-mode optical fiber (980HP, Thorlabs) was employed to transmit modulated light from the sensing probe to an OTDR (AQ7275, Yokogawa) device. Fiber-optic FC connectors were installed on the both ends of the optical fiber in order to easily and exactly connect with the sensing probe and the OTDR device.

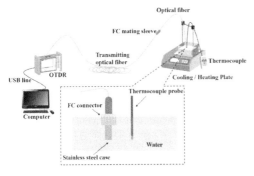

Figure 2.   Internal structure of a sensing probe.

In fabricating sensing probe, first, the NaCl solution was filled in the stainless-steel case. Second, an FC connector was coupled with the stainless-steel case. Finally, the gap at the coupling interface between the stainless-steel case and the ferrule of FC connector was sealed by the rubber sealing-ring. By changing temperature or concentration of NaCl, the NaCl solution has different RI. For example, as increasing temperature or decreasing concentration of NaCl, the RI of the NaCl solution decreases.

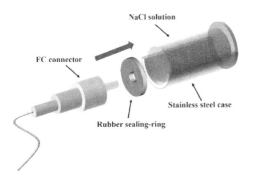

Figure 3.   Experimental setup using the fiber-optic temperature sensor based on a NaCl solution.

Figure 3 shows the experimental setup for measuring temperature using the fiber-optic temperature sensor based on the NaCl solution. The fiber-optic temperature sensor system consists of a sensing probe, a transmitting optical fiber, an OTDR, and a laptop computer. The sensing probe was located in a water tank and the temperature of water was controlled by a cooling/heating plate (CP-7200GT, Intec) from 5 to 65°C. At the same time, the temperature of water was monitored using a commercialized thermocouple (54II thermometer, Fluke). The light of near-infrared (IR) wavelength range emitted from the light source in the OTDR was transmitted to the sensing probe through the transmitting optical fiber with a length of 100 m. Then, the reflected light occurred which resulted from Fresnel reflection at the coupling interface between the optical fiber and the NaCl solution in the sensing probe. The optical power of the reflected light depends on the RI change of the NaCl solution in the sensing probe at the different temperature. In this study, we measured the optical power of the reflected light according to the changes in the temperature of water or the concentration of NaCl.

3   RESULTS

To evaluate Fresnel reflection in the sensing probe, we measured the optical power using the distilled water with a RI of 1.33 and the index-matching oils with five different RI, such as 1.35, 1.4, 1.45, 1.5, and 1.6. Figure 4 shows the variation of the optical power according to the RI change. With the decreasing of RI difference between the core of optical fiber (n = 1.464) and the index-matching oil, the Fresnel reflection also decreases, as can be seen in figures 1 and 4.

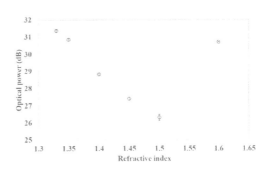

Figure 4.   Variation of the optical power according to the RI change of the index-matching oils.

Figure 5 shows the output signals of the fiber-optic temperature sensor according to the temperature variation of the water when the concentration of NaCl solution in the sensing probe was changed from 5 to 25%. As the temperature of water increased from 10 to 50°C, the optical power of the reflected light increased because the RI of the NaCl solution decreased. When the concentration of the NaCl solution was 25%, the gradient of curve increased more steeper than those of other concentrations according to the temperature of water. In this study, we selected the NaCl solution with a concentration of 25% and the light source with a wavelength of 1550 nm.

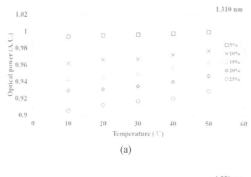

(a)

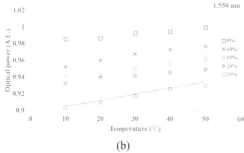

(b)

Figure 5. Output signals of the fiber-optic temperature sensor with two wavelengths of the light source (1310 and 1550 nm) and five concentrations of NaCl solution (5, 10, 15, 20, and 25%) according to the temperature of the water.

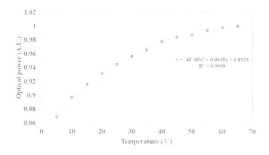

Figure 6. Relationship between the optical power of the fiber-optic temperature sensor and the temperature of water.

Figure 6 shows the relationship between the optical power of the fiber-optic temperature sensor and the temperature of water. When the temperature of water increased, the RI of NaCl solution decreased and thus, the optical power of the reflected light increased. In this experiment, as the temperature of water increased from 5 to 65°C, the optical power of the reflected light also increased, as shown in figure 6.

4 CONCLUSIONS

We developed a fiber-optic temperature sensor using a NaCl solution for real-time thermometry at a long distance. The performance of the fabricated sensing probe was evaluated by measuring optical power of the reflected light according to the temperature of water. In this study, we selected the optimum concentration of the NaCl solution and the optimum wavelength of the light source and then obtained the relationship between the optical power of the fiber-optic temperature sensor and the temperature of water. Based on the results of this study, it is anticipated that a fiber-optic temperature sensor using a NaCl solution can be developed to monitor temperature of water at a very long distance.

ACKNOWLEDGMENT

This research was supported by National Nuclear R&D Program through the National Research Foundation of Korea (NRF) funded by the Ministry of Science, ICT and future Planning (No. 2014040951 and No. 2014031841). Also this research was supported by Basic Science Research Program through the National Research Foundation of Korea (NRF) funded by the Ministry of Science, ICT and future Planning (No. 2014002620).

REFERENCES

[1] Bolognini, G., Park, J., Soto, M.A., Park, N.,Pasquale, F.D. 2007. Analysis of distributed temperature sensing based on Raman scattering using OTDR coding and discrete Raman amplification. *Measurement Science and Technololgy* 18(10): 3211–3218.
[2] Crunelle, C., Legre, M., Wuilpart, M., Megret, P.,Gisin, N. 2009. Distributed temperature sensor interrogator based on polarization-sensitive reflectometry. IEEE Sensors Journal 9(9): 1125–1129.
[3] Eom, T.J., Kim, M.J., Lee, B.H.,Park, I.C. 2005. Temperature monitoring system based on fiber Bragg grating arrays with a wavelength tunable OTDR. IEICE Transaction on Electronics E88-C: 933–937.
[4] Fernandez-Valdivielso, C., Matias, I.R., Arregui, F.J. 2002. Simultaneous measurement of strain and temperature using a fiber Bragg grating and a thermochromic material. Sensors and Actuators A: Physical 101: 107–116.

[5] Han, W., Tong, Z., Cao, Y. 2014. Simultaneous measurement of temperature and liquid level base on core-offset singlemode-multimode-singlemode interferometer. Optics Communications 321: 134–137.

[6] Khare R.P. 2004. Fiber optics and optoelectronics. New Delhi: Oxford University Press.

[7] Mendonca, S., Frazao, O., Baptista, J.M., Santos, J.L. 2007. Fiber optic displacement sensing monitored by an OTDR and referenced by Fresnel reflection and by fiber Bragg gratings. Microwave and Optical Technology Letter 49(4): 768–770.

[8] Pan, J., Huang, X., He, Y., Huang, B. 2012. Fresnel-reflection-based fiber sensor for high-temperature measurement. Review of Scientific Instruments 83(3): 035004.

[9] Pedrotti, F.L., Pedrotti, L.M., Pedrotti, L.S. 2007. Introduction to Optics 3rd Edition. San Francisco: Pearson Education.

[10] Sade, S. & Katzir, A. 2001. Fiberoptic infrared radiometer for real time in situ thermometry inside an MRI system. Magnetic Resonance Imaging 19: 287–290.

[11] Yoo, W.J., Jang, K.W., Seo, J.K., Heo, J.Y., Moon, J., Park, J.-Y.,Lee, B. 2010. Development of respiration sensors using plastic optical fiber for respiratory monitoring inside MRI system. Journal of the Optical Society of Korea 14(3): 235–239.

[12] Yoo, W.J., Jang, K.W., Seo, J.K., Moon, J., Han, K.-T., Park, J.-Y., Park, B.G., Lee, B. 2011a. Development of a 2-channel embedded infrared fiber-optic temperature sensor using silver halide optical fibers. Sensors 11(10): 9549–9559.

[13] Yoo, W.J., Seo, J.K., Jang, K.W., Heo, J.Y., Moon, J., Park, J.-Y., Park, B.G., Lee, B. 2011b. Fabrication and comparison of thermochromic material-based fiber-optic sensors for monitoring the temperature of water. Optical Review 18(1): 144–148.

Testing and Measurement: Techniques and Applications – Chan (Ed.)
© 2015 Taylor & Francis Group, London, ISBN: 978-1-138-02812-8

A novel Sagnac-based fiber-optic acoustic sensor using two laser diodes with external optical injections

L.T. Wang, N. Fang & Z.M. Huang

Key Laboratory of Specialty Fiber Optics and Optical Access Networks, School of Communication and Information Engineering, Shanghai University, Shanghai, China

ABSTRACT: A novel fiber-optic acoustic sensor consisting of two laser diodes and a fiber Sagnac interferometer is presented. Two laser diodes operate under external optical injections, including self and mutual injections. The Sagnac interferometer acts as a reflectivity-variable loop mirror. When external acoustic waves perturb the Sagnac fiber loop, its reflectivity will change with the intensity of external perturbation, which alters the amount of injection lights of two laser diodes and induces the output powers of laser diodes fluctuating. The acoustic signals can be detected by monitoring the changes of the laser output power with the photodiode mounted inside the laser diode. The basic principles and the proposed concepts will be introduced. The experimental results on the investigations of sensor performances as well as the actual measurements of partial discharges and power-frequency electric field intensities will be presented. The experimental results demonstrate that the proposed sensing concept is practicable and the sensor has very high sensitivity and good frequency response property in detection of high-frequency and weak ultrasonic waves produced by the discharges.

1 INTRODUCTION

In recent years, the fiber-optic interferometer-based acoustic sensors with the property of immunity from electromagnetic interferences have been employed within the power industry for partial discharge (PD) monitoring [1, 2]. In fiber-optic acoustic sensors, the fiber Sagnac interferometer can provide ideal sensing performances [3, 4], in terms of the sensitivity and frequency response, to detect the faint, high-frequency, PD-induced ultrasonic waves.

On the other hand, the utilization of external optical injection technology, such as the self-mixing (SM) interferometry, for sensing applications has attracted considerable interests [5, 6]. In the SM-type sensor configuration, a fraction of the light back-scattered by a remote target is allowed to re-enter into the laser cavity, and perturb the lasing field, which modulates the frequency and output power of the laser diode. SM interferometry based laser sensors with very high sensitivity and simple system scheme have been widely employed in various measurement fields for the measurements of different physical quantities [6], such as for the measurements of displacement, vibration, velocity, angle and absolute distance. To our knowledge, until now there is not any such sensor configuration by combining the external optical injection scheme in the Sagnac interferometer. In this paper, we will propose a novel fiber-optic sensor configuration,

by using a fiber Sagnac interferometer as the sensor and two laser diodes as light sources as well as receivers in order to enhance the sensor sensitivity, especially in high frequency range for sensing weak, high-frequency ultrasonic waves.

In residual parts of this paper, we will simply introduce the proposed concepts and the operation principles and demonstrate some experimental results with respect of evaluating sensor characteristics as well as the actual measurements of PD discharges.

2 SENSING PRINCIPLES

2.1 *Sensor system configuration*

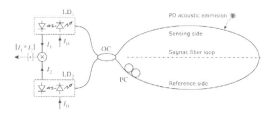

Figure 1. The scheme of proposed fiber acoustic sensor system, where LD_i represents the laser diode, OC, optical fiber coupler, PC, polarization controller. I_D is laser driving current and I_i is detection current.

The scheme of the proposed sensor system is presented in Figure 1, in which a fiber Sagnac interferometer with a 3-dB, 2 × 2 fiber coupler (OC) and two laser diodes (LD₁ & LD₂) without optical isolators are used. Two detection current signals (I_1 & I_2) are obtained individually from the corresponding output of the photodiode mounted within the laser package. The final detection signal is formed by multiplying I_1 with I_2 and then taking its absolute value. Two laser diodes are similar in their spectral characteristics with close lasing wavelengths and output powers. The PC used in the Sagnac fiber loop is to balance the intensities of two interference signals at two output ports of the OC.

2.2 Concept of reflectivity-variable loop mirror

The fiber Sagnac interferometer used in this sensor configuration is taken as a wide-band, reflectivity-ariable loop mirror. The reflectivity of loop mirror for the light input port can be expressed as $R = 1 - |\cos(\theta_b + \delta)|$ and its transmissivity to another port is $T = 1 - R = |\cos(\theta_b + \delta)|$. Here θ_b represents a constant nonreciprocal phase shift generated by the PC and fiber loop birefringence, and generally θ_b is set at $\theta_b = \pi/2$. δ is a time-dependent phase shift induced by external acoustic fields. When the sensor is in the static state, that is without the external influences, having $\delta = 0$. In this case, the reflectivity R will achieve at its maximum value, having $R = 1$. As external acoustic fields perturb the loop mirror, δ will change and results in a decrease of R and an increase of T. In this way, we can detect the external acoustic fields by measuring the changes of the reflectivity of loop mirror. This concept is schematically demonstrated in Figure 2.

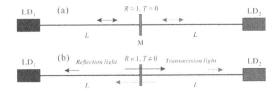

Figure 2. The scheme of proposed sensor concept, here (a) and (b) are the case without and with external perturbations, respectively.

2.3 Detection method

In the SM interferometry based laser sensor, the laser output power under optical feedbacks is a function of the intensity of light back-scattering from the target, also of the reflectivity of the target surface [5, 6]. Therefore, as an intensity modulation factor, as the reflectivity of the target changes, the laser output power will consequently change. For the detection of the external perturbation, we can monitor changes in the laser output power.

In this sensor configuration, besides feedback process, there exists a mutual injection process. As the reflectivity of the loop mirror decreases and the transmissivity increases, the mutual optical injection existing between two laser diodes will occur. It is a more completed nonlinear process and will induce a four-wave mixing effect within both laser diodes, which obviously modifies the spectral characteristics and output powers of the laser diodes [7-9]. According to our experimental observations, in the static state, output powers in two laser diodes are lowest owing to the optical feedback effect, and increase with the decrease of the reflectivity of loop mirror as well as the increase of mutual injection power. Compared to the traditional detection method used in the ordinary Sagnac sensor with a light source and a detector, this nonlinearity-based detection method can greatly enhance the sensitivity of the sensor system. From the view of the laser diode with the gain, the amplitude of the detection signal actually is enlarged by laser diode as an optical amplifier.

2.4 Experimental observations

We experimentally investigated the spectral characteristics of the laser diodes under the external optical injections, including SM and mutual injections with different injection powers, as well as disturbing the loop mirror by imposing an external pressure on the fiber, as a technical support of the proposed sensor concept. Figure 3 is a schematic of the setup used in this experiment. The injection optical power was selected in a way by exchanging the ports of the fiber coupler (80:20) or by directly connecting the loop mirror with the laser diode, thus having three injection powers, 20%, 80% and 100%, respectively. Additional parameters chosen in this experimental setup are shown in Figure 3.

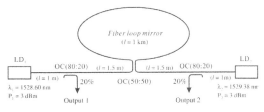

Figure 3. Schematic of setup for spectrum measurements.

The spectrums of two laser diodes (LD₁ & LD₂) under different experimental conditions were measured with the optical spectrum analyzer and are presented in Figure 4 and Figure 5, respectively.

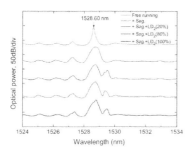

Figure 4. Output spectrums of LD_1 under different experimental conditions. Here Sag. is an abbreviation of "Sagnac" word.

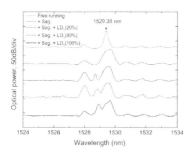

Figure 5. Output spectrum of LD_2 under different experimental conditions.

The output spectrums of two laser diodes (LD_1 & LD_2) measured at different time under disturbing the fiber loop with a variable external pressure are shown in Figure 6(a) and Figure 6(b), respectively.

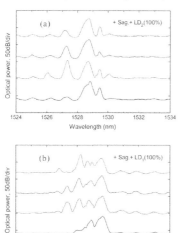

Figure 6. Output spectrums of LD_1 (a) and LD_2 (b) at different time, 1when the fiber loop was disturbed by external pressures.

From the investigation results as shown in Figure 4 and 5, it is clear that the spectral characteristic of laser diode under optical injections had considerable changes. The spectral line was broadened and multiple peaks arose as a product of four-wave mixing process. Also it is clear, as shown in Figure 6, that the spectrums of two laser diodes changed obviously, following the changes of external pressures, which actually modified the reflectivity of the fiber loop mirror. In addition, in the output waveforms detected by the power monitor of the laser diode, the large fluctuations in the signal amplitude also can be observed.

2.5 Signal processing

In this sensor system, we adopted a new signal processing method for enhancing the signal-to-noise ratio (SNR) of the sensor system. It is done by multiplying two detection signals, I_1, I_2 and then taking an absolute operation to constitute a new output signal $| I_1 \cdot I_2 |$. The reason for doing it is that as the sensor system is in the static state (balance state), having $T = 0$, two detection outputs are uncorrelated, and as the sensor system is in the dynamic state (sensing state), having $T \neq 0$, two detection signals become correlated. Therefore by multiplying two uncorrelated output signals in the static state, the noise floor can be effectively suppressed, and by multiplying two correlated output signals in the dynamic state, the signal amplitude can be greatly enhanced. Therefore, as a result, the SNR of sensor system will be improved obviously. A set of measured signal waveforms are presented in Figure 7 for explaining this method. From these results, one can see that after taking this processing method, the noise floor in the new signal, $| I_1 \cdot I_2 |$, had been reduced significantly.

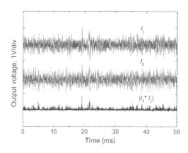

Figure 7. Detected signal waveforms.

With the experimental system as shown in Figure 1, we carried out numerous experiments for investigating the performances of proposed sensor system and its actual detection ability. The experimental results are presented in the following section.

3 EXPERIMENTAL RESULTS

3.1 *Performance evaluation of sensor system*

The performances of sensor system, in terms of sensitivity as well as frequency response, were evaluated with an experimental setup as shown in Figure 8, in which the fiber Sagnac loop with a diameter of 60 mm was made with the 1-km long single-mode fiber, as sensor head was placed on an aluminum plate where the silicone grease was applied prior. A corner of aluminum plate, located on the same side, separated by 50 cm from the sensor was utilized as the test area, in which a piezoelectric transducer (PZT) and a short piece of pencil lead were used to generate the required ultrasonic signals with different oscillating frequencies and tiny impacts for testing the performances of sensor system. The measured results for evaluating the frequency response property of sensor system are shown in Figure 9.

Figure 8. A setup used for evaluating sensor performances.

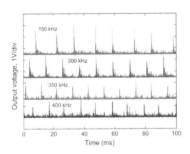

Figure 9. Measured waveforms of ultrasonic pulses generated, respectively, by a PZT with different oscillating frequencies.

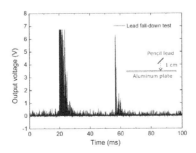

Figure 10. Measured impact signal generated through a pencil lead falling action for testing sensor sensitivity.

A 0.5-cm long pencil lead was used to generate a tiny impact on the aluminum plate by falling it from 1-cm height for testing the sensor sensitivity. The measured impact signal waveform is presented in Figure 10, in which, multiple pulses can be seen, which indicates that actually there existed a springing process after the lead hitting the aluminum plate.

The experimental results presented above demonstrate that this sensor system has an excellent frequency response especially in detecting the high frequency ultrasonic signals and fairly high sensitivity in the detections of very faint impacts with a considerable SNR.

3.2 *Detections of partial discharges*

Based on this experimental system, we tried to detect the partial discharges generated between two high-voltage electrodes. The experimental setup is illustrated in Figure 11, in which two electrodes were immersed into the transformer oil and imposed with an AC voltage of up to 5 kV. The gap of two electrodes was fixed at about 8 mm. The sensor was attached to the shell of the oil tank, facing on the electrodes. In this experiment, the room temperature was set at 25 °C and the humidity was kept in 65%RH. Several measurement results are shown in Figure 12 and Figure 13, respectively. The PD pulses shown in Figure 12 were obtained at the beginning stage of experiment, when the oil was clear and no obvious discharge around the electrodes could be observed. The results shown in Figure 13 were obtained after the discharge process lasted for about 10 minutes, when the transformer oil became dirty and some bubbles around the electrodes appeared.

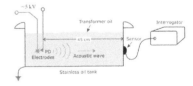

Figure 11. Experimental setup used for detections of partial discharges occurred in oil tank filled with the transformer oil.

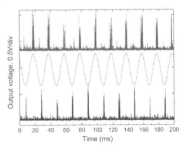

Figure 12. Two PD pulse traces (upper and lower ones), measured at different time. Sine wave signal (middle trace) is a power frequency signal here adopted as a phase reference.

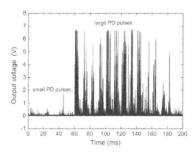

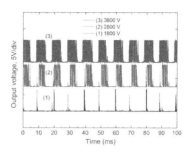

Figure 13. Measured PD pulses with different discharge intensities when the transformer oil became dirty.

Figure 15. Recorded ultrasonic pulses at different applied HV voltages, which were induced by air ionizations.

We measured the RMS voltages of recorded pulses at different HV voltages and plot them against the imposed HV voltage. This result is illustrated in Figure 16, in which the RMS voltage of detection pulses increases proportionally with the applied HV voltage with a sloop $K = 0.00119$. The corresponding electric field intensity can be calculated out with a relation, $E = V/D$, and here V is voltage imposed and D is the spacing distance between two plate electrodes.

3.3 Detections of electric field strength

With this experimental system, we tried to detect the intensity of power frequency electric field. The experimental setup is illustrated in Figure 14, in which two copper plates with same size (20×20 cm^2) as the plate electrodes were used and fixed in parallel with a spacing of 10 cm. The fiber Sagnac loop was placed in the middle of two plate electrodes as illustrated in Figure 14. The plate electrodes were connected to a high-voltage (HV) generator, which can generate the AC (50 Hz) voltage of up to 5 kV.

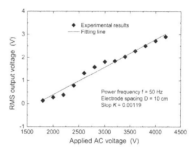

Figure 16. Experimental results on detections of electric field intensity. Marks are measured values and solid line is a fitting line.

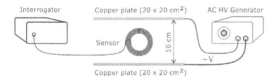

Figure 14. Experimental setup used for detections of electric field existed between two parallel plate electrodes.

4 CONCLUSIONS

In the experiment, the AC voltage imposed on the electrodes was increased gradually and the sensor outputs were recorded at each measurement point. Figure 15 are three groups of recorded pulse waveforms, obtained at 1800 V, 2800 V and 3800 V, respectively. As shown in these experimental results, the number of induced pulses within a cycle of AC voltage or the pulse density increases with the applied AC voltage. The pulses induced by the air ionization under a strong electric filed environment will generate high-frequency ultrasonic pulses in air which act on the fiber Sagnac loop. Therefore with the increase of the applied AC voltage, the ultrasonic field was enhanced. Consequently, the induced pulse density increased.

In this paper, we presented a novel fiber-optic acoustic sensor by using two laser diodes and a fiber Sagnac interferometer. The basic principles of the sensor system were introduced, including the system configuration, the concept on the reflectivity-variable loop mirror and the way of detecting the reflectivity changes in the loop mirror by utilizing the effects of external optical injections in the laser diodes. As a technical background, we carried out an experiment for investigating the output spectrums of laser sources under optical injections, including self as well as mutual injections. We also evaluated the performances of an experimental system, in terms of frequency response property and sensitivity. The measured results showed that this experimental system had considerably high sensitivity and better high-frequency response properties in detecting high-frequency, faint ultrasonic waves of up to 400 kHz. For a concept proving, numerous experiments were carried out with the proposed sensor system in detections of the partial discharge as well as the power-frequency, HV electric field intensity. The experimental results prove that the proposed sensing concept is practicable and the sensor system has an excellent sensing ability in detections of high frequency ultrasonic waves.

ACKNOWLEDGMENTS

This work is supported by the National Natural Science Foundations of China (Grant No. 61108004 as well as Grant No. 61377082), the Science and Technology Commission of Shanghai Municipality (Grant No. 11ZR1413300) and Shanghai Leading Academic Discipline Project (S30108).

REFERENCES

[1] Culshaw, B. 2000. Fiber optics in sensing and measurement. *IEEE Journal of Selected Topics in Quantum Electronics,* 6: 1014–1021.

[2] Udd, E. 1983. Fiber-optic acoustic sensor based on the Sagnac interferometer. *In Single Mode Optical Fibers 90, Proc. of SPIE*, 0425: 90–95.

[3] Jang, T.S. & Lee, S.S. et al. 2002. Noncontact detection of ultrasonic waves using fiber optic Sagnac interferometer. *IEEE Trans. Ultrason., Ferroelect., Freq. Contr.,* 49: 767–775.

[4] Wang, L. & Fang, N. et al. 2014. A fiber optic PD sensor using a balanced Sagnac interferometer and an EDFA-based DOP tunable fiber ring laser. *Sensors,* 14(5): 8398–8422.

[5] Miles, R.O. & Dandridge, A. 1983. An external cavity diode laser sensor. *J. Lightwave Technol.,* LT-1(1): 81–93.

[6] Bosch, T. 2004. An overview of self-mixing sensing applications. *IEEE Conference on Optoelectronic and Microelectronic Materials and Devices,* 385–392.

[7] Kong, Z.M. & Wu, J.G. et al. 2008. Experimental observations on the nonlinear behaviors of DFB semiconductor lasers under external optical injection. *Chaos Soliton Fractals,* 36(1): 18–24.

[8] Lin, R.Y. & Liu, J.M. 2003. Nonlinear dynamical characteristics of an optically injected semiconductor laser subject to optoelectronic feedback. *Opt. Commun.,* 221(1): 173–180.

[9] Lang, R. & Kobayashi, K. 1980. External optical feedback effects on semiconductor injection laser properties. *IEEE Journal of Quantum Electronics,* 16: 347–355.

Filterless Vacuum Ultraviolet detector based on YF_3 thin films grown by Pulsed Laser Deposition

M. Yanagihara, H. Ishikawa & S. Ono
Nagoya Institute of Technology, Gokiso-cho, Showa-ku, Nagoya, Aichi, Japan

H. Ohtake
Aishin Seiki Co., Ltd., Kariya, Aichi, Japan

ABSTRACT: We report on the development of Vacuum Ultraviolet (VUV) photoconductive detector based on YF_3 thin films. YF_3 thin films were grown by Pulsed Laser Deposition (PLD) method. The detector showed high sensitivity only below 180 nm in wavelength without any UV-visible cut filter. The current value of the detector increased 3-digit before and after VUV irradiation at an applied bias of 300 V. The size of the YF_3 particles, which constituted the thin films, was controlled by laser fluence and substrate heating temperature during PLD. Sensitivity of the detector was improved with smaller particle size. Such filterless VUV photoconductive detector is expected to be a strong substitution of conventional detector and play important role in monitoring of the VUV light sources.

1 INTRODUCTION

Vacuum ultraviolet (VUV) light sources, such as a xenon excimer lamp (172 nm in wavelength) and a low-pressure mercury lamp (185 nm in wavelength), are used in various applications such as surface treatment, optical cleaning of semiconductor substrates, and sterilization. For example, Viktor et al. (2004) demonstrated surface treatment of polyethylene. Accordingly, demand for the VUV detectors, which are capable of monitoring these light sources, is increasing.

The photodetector using diamond (Hayashi et al. 2005), nitrides such as AlGaN (Saito et al. 2009), and oxides such as ZnMgO (Yang et al. 2003) is a well established research area. However, these detectors response deep-UV range not only VUV range because of limitation of the band gap. To detect only VUV light without any filters, we need materials with wider band gaps. Some of the most prominent candidates are fluorides. Some fluorides have relatively wide band gaps in contrast to these materials as it was shown by Ouenzerfi et al. 2004. Until now, we have achieved filterless UV and VUV detector by applying multiple fluoride thin films such as CeF_3 (Ieda et al. 2012) and NdF_3 (Ishimaru et al. 2013). In this work, we applied YF_3 thin film for the detector expecting for the realization of different response wavelength.

We applied pulsed laser deposition (PLD) for the thin film growth method. PLD has a prominent advantage that there are few composition gaps between a target and a thin film. Thus, we can obtain the fluoride thin film without the inflow of toxic fluoride assist gasses as shown by Yanagihara et al. 2014. PLD also has other features such as vaporizing the target with high melting points, not being polluted by a melting pot and pulsingly controllable process which is as shown by Rajiv et al. 1990.

2 EXPERIMENTAL METHOD

2.1 *Growth of YF_3 thin films*

YF_3 thin films were grown by PLD on the quartz glass substrate. YF_3 ceramic target was irradiated with the focused femtosecond laser pulses (wavelength: 790 nm, repetition rate: 1 kHz, pulse width: 180 fs) in the vacuum chamber. The quartz glass substrate was placed on the rotating holder located towards the target and attached to the heater. Growth was carried out 3 h under high vacuum condition (2×10^{-4} Pa). In this experiment, we valued laser fluence and substrate heating temperature as the dominant parameters to control morphological characteristics. The laser fluence and substrate temperature were adjusted within the range from 13.5 to 23.0 mJ/cm^2 and from room temperature (R.T.) to 600 °C, respectively.

2.2 *Fabrication of photoconductive detector*

Subsequently, a pair of interdigitated aluminum electrodes was fabricated onto the thin film by vacuum deposition. The patterned area and the gap of the electrodes were 8×10 mm^2 and 0.4 mm, respectively.

3 RESULTS AND DISCUSSION

3.1 *Surface morphology of YF₃ thin films*

Surface morphology of YF$_3$ thin films was evaluated by using scanning electron microscope (SEM). Size distributions of YF$_3$ particles which constituted thin film was statically analyzed from the SEM images by measuring randomly 300 particles as shown in Figure 1. SEM image of each thin film is shown in the inset. The symbol "d$_a$" in the figure indicates the average diameter of YF$_3$ particles. As a result, average particle diameter increased with higher laser fluence. In addition to that, substrate temperature also affected the particle size of YF$_3$. We evaluated photoconductivity of 5 thin films

3.2 *V-I characteristics*

Photoconductivity was evaluated by measuring the current value under non-VUV irradiation (dark current) and under VUV irradiation (photo current) by a deuterium lamp (Hamamatsu L10366).

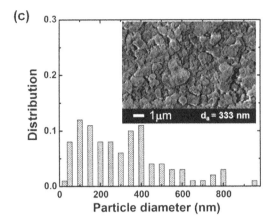

Figure 1. Size distribution of YF$_3$ particles. Laser fluence were (a) 13.5 mJ/cm^2, (b) 18.2 mJ/cm^2 and (c) 23.0 mJ/cm^2. SEM images of surface of YF3 thin film are shown in the inset. Measured average particle diameter is shown as "d$_a$".

Figure 2 shows V-I characteristic of YF$_3$ detector using the thin film prepared by laser fluence of 13.5 mJ/cm^2. Under non-VUV irradiation, detector showed high insulation with dark current below 1 pA. Meanwhile, the current increased 3-digit in response to VUV irradiation at an applied bias of 300 V.

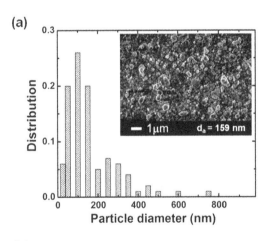

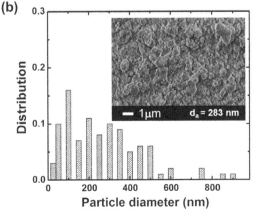

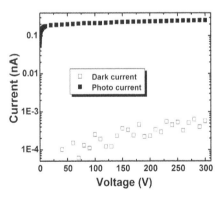

Figure 2. V-I characteristic of YF$_3$ thin film.

In addition to such high sensitivity and photoconductivity of YF$_3$, we also confirmed the correlation between the current and particle size of thin films. We calculated electric resistances from V-I characteristics. Figure 3 shows the dependence of resistances on the average particle diameter. R$_d$ and R$_p$ indicate resistances for dark and photo current, respectively. The line (a) shows the dependence of

R_d on d_a. The Y axis of line (b) indicates decreasing rate of resistance before and after VUV irradiation. Value of R_p greatly decreases with higher photoconductivity, and accordingly this value becomes close to 1. These results revealed that small d_a brought low dark current and high photoconductivity. Thus, we can obtain higher sensitivity for the detector with the thin film consisting of smaller particles.

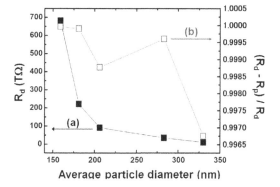

Figure 3. Dependences of electric resistances on particle size of thin films. R_d and R_p indicate resistances for dark and photo current, respectively.

3.3 *Spectral sensitivity*

Transmission spectrum and spectral sensitivity of the detector were measured as shown in Figure 4. Spectral sensitivity was measured at each wavelength of synchrotron radiation (UVSOR BL-7B) from 130 to 300 nm at an applied bias of 300 V. As the radiation wavelength decreased, the transmittance of thin film gradually decreased by the light scattering on the surface of thin film. Transmittance edge appeared at around 160 nm. The sensitivity of the detector extremely increased below 180 nm, which was roughly corresponding to the transmission edge of YF_3 thin film.

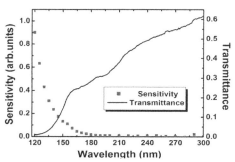

Figure 4. Transmission spectrum and spectral sensitivity of the detector.

4 CONCLUSION

We achieved filterless VUV photoconductive detector by applying YF_3 thin film grown by PLD. Particle size of YF_3 thin film, which was controlled by the parameter of PLD, affected the sensitivity of the detector. This detector showed high sensitivity only below 180 nm without any UV-visible cut filter. Such filterless detector could contribute to the simplification of fabrication processes. This VUV photoconductive detector was expected to be a strong substitution of conventional detector and expanded further application not only monitoring of the VUV lamp but also radiation detection.

REFERENCES

[1] Hayashi, K., Tachibana, T., Kawakami, N., Yokota, Y., Kobashi, K., Ishihara, H., Uchida, K., Nippashi, K., & Matsuoka, M. 2005. Diamond Sensors Durable for Continuously Monitoring Intense Vacuum Ultraviolet Radiation. *Japanese Journal of Applied Physics* 44: 7301–7304.

[2] Ieda, M., Ishimaru, T., Ono, S., Kawaguchi, N., Fukuda, K., Suyama, T., Yokota, Y., Yanagida, T. & Yoshikawa, A. 2012. Filterless Ultraviolet Detector Based on Cerium Fluoride Thin Film Grown by Pulsed Laser Deposition. *Japanese Journal of Applied Physics* 51: 062202.

[3] Ishimaru, T., Ieda, M., Ono, S., Yokota, Y., Yanagida, T. & Yoshikawa, A. 2013. Vacuum ultraviolet photoconductive detector based on pulse laser deposition-grown neodymium fluoride thin film. *Thin Solid Films* 534: 12–14.

[4] Ouenzerfi, E. R., Ono, S., Quema, A., Goto, M., Sakai, M. Sarukura, M. 2005. Design of wide-gap fluoride heterostructures for deep ultraviolet optical devices. *Journal of Applied Physics* 96: 7655–7659.

[5] Rajiv, K. S. & Narayan, J. 1990. Pulsed-laser evaporation technique for deposition of thin films: Physics and theoretical model. *Physical Review B* 41: 8843–8859.

[6] Saito, T., Hitora, T., Ishihara, H., Matsuoka, M., Hitora, H., Kawai, H., Saito, I. & Yamaguchi, E. 2009. Group III-nitride semiconductor Schottky barrier photodiodes for radiometric use in the UV and VUV regions. *Metrologia* 46: S272–S276.

[7] Viktor, N. V., Artem, V. K. & Viktor, I. S. 2004. Vacuum ultraviolet treatment of polyethylene to change surface properties and characteristics of protein adsorption. *Journal of Biomedical Material research* 69A: 428–435.

[8] Yanagihara, M., Yusop, M. Z., Tanemura, M., Ono, S., Nagami, T., Fukuda, K., Suyama, T., Yokota, Y., Yanagida, T & Yoshikawa, A. 2014. Vacuum ultraviolet field emission lamp utilizing $KMgF_3$ thin film phosphor. *APL Materials* 2: 046110.

[9] Yang, W., Hullavarad, S. S., Nagaraj, B., Takeuchi, I., Sharma, P. R. & Venkatesan, T. 2003. Compositionally-tuned epitaxial cubic $Mg_xZn_{1-x}O$ on Si (100) for deep ultraviolet photodetectors. *Applied Physics Letters* 82: 3424–3426.

Testing and Measurement: Techniques and Applications – Chan (Ed.)
© 2015 Taylor & Francis Group, London, ISBN: 978-1-138-02812-8

A high sensitive of an optical Raman sensor system to detect bisphenol A

N.A. Bakar, M.M. Salleh & A.A. Umar
Institute of Microengineering and Nanoelectronics (IMEN), Universiti Kebangsaan Malaysia, Bangi, Malaysia

ABSTRACT: An optical system has attracted researchers to develop a highly sensitive sensor since the system can produce low cost and portable equipment. In this study, we report an attempt to develop an optical Raman sensor system to detect Bisphenol A (BPA) particles on quartz surface. The Raman sensor system was set up, comprising a butterfly laser diode, Raman fiber optic probe, spectrophotometer, and sensor chamber. BPA samples of surface were prepared by drop-casting technique with seven concentrations of BPA from 0.001M to 1M. The detection of a Raman scattering effect from BPA molecule was studied by analyzing the interaction between BPA film and 785 nm of laser light in order to sense the presence of BPA on the surface. Hence, Raman intensity is directly proportional to the concentrations when Raman intensity of BPA peaks has been enhanced with the increasing of BPA concentrations into surface.

1   INTRODUCTION

Bisphenol A (BPA) is an organic chemical extensively utilized in our daily life to make polycarbonate plastics and broadly used in food and drinking packaging purpose (Borrell, 2010). However, the BPA functions as a protective coating to drink containers which might be leached and transferred from polycarbonate into food after heated and repeat—use containers (Bakar et al. 2013). Thus, consumers would be highly exposed to BPA contamination in food and harmful to consumer's health and any living species (Staples et al. 2000).

The peoples' awareness of the exposure to BPA is poor due to deficiency of data monitoring as well as inadequate sophisticated equipments and proficient technicians. Therefore, this is critical to develop a simple, cheaper and portable system to monitor BPA residue. In this decade, there were a few approaches which were widely used to scrutinizing BPA, such as liquid chromatography (Rezaee et al. 2009), fluorescence (Ballestreros-Gómez et al. 2009), electrochemical (Ntsendwana et al. 2012), quartz crystal microbalance (Tsuru et al. 2006) and biosensor (Marchesini et al. 2005). Lately, sensor based optical sensing technique has attracted a curiosity among researchers to develop a portable optical technology because the system was able to produce low cost and portable equipment (Massie et al. 2006). A high sensitivity Raman optical sensing system might be developed to detect BPA.

This research paper explains the development of Raman sensor used to detect the presence of BPA on the quartz surface with different concentrations. The result was interpreted in Raman shift and the intensity change was learning to observe the sensitivity of Raman sensor.

2   EXPERIMENTAL

A Raman sensor was successfully self-developed by combining a few optical components in order to design an optical Raman system. This sensor consists of butterfly diode laser with 785 nm of wavelength (PD-LD Inc.), diode laser driver (Arroyo Instruments), temperature controller (Arroyo Instruments), fiber optic Raman probe (Inphotonic), thermoelectric-cooler (TEC) spectrophotometer QE-65000 Raman (Ocean Optics) and sensor chamber. The butterfly diode laser was acted as light source, fiber optic probe as an excitation light and a collection light path, and finally a spectrophotometer as a detector and an analyzer for the Raman sensor. Meanwhile, bisphenol A (BPA) sample was located inside the sensor chamber.

This sensor was invented to sense BPA with a variety of concentrations from 0.001M to 1M. The BPA sample was prepared for surface via drop-casting approach by dropping 0.01 ml of BPA solution on the cleaned quartz surface. The surface was then dried under room temperature prior to its utilization of the Raman sensor system. To prepare the seven concentrations of BPA solution, BPA powder (Sigma Aldrich) was diluted in ethanol absolute (HmbG Chemicals) for 0.001M, 0.005M, 0.05M, 0.1M, 0.25M, 0.5M and 1M of concentration. After that, each sample in solution was prepared on surface to learn a Raman effect of BPA molecule using Raman sensor.

The Raman effect study was understood by irradiating 785 nm wavelength of laser light and 5 mWatt laser power onto the BPA film surface. An obtained of Raman scattering was detected via interaction between laser photon and the presence of BPA molecules on the surface. This interaction was interpreted by spectrophotometer and turning out a precise spectral fingerprint in the Raman shift spectrum. The BPA sample was scanned for 1 second until 50 seconds of integration time and five times of scan for average per spectrum before the data of the spectrum was saved to avoid surrounding vibrational noises.

3 RESULTS AND DISCUSSION

The development of the Raman sensor system was simply self-designed based on an optical system. This sensor consists of a butterfly diode laser, fiber optic Raman probe, spectrophotometer, and sensor chamber as illustrated in Figure 1.

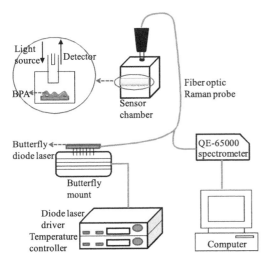

Figure 1. Schematic picture of Raman sensor system.

A competent laser system was developed from a few components in order to fabricate a high sensitivity of Raman sensor. The system was comprised of butterfly diode laser, butterfly mount, diode laser driver and temperature controller. Butterfly diode laser, worked as an electromagnetic light source, was purchased from PD-LD Inc., with 785 nm of wavelength. This laser was positioned on butterfly mount which was equipped with a cooling system for laser safety. At the same time, a temperature controller was connected to butterfly mount to monitor the laser diode temperature and power a fan of cooling system. Meanwhile, a diode laser driver was functioned as power supply for diode laser by providing a maximum current up to 500mA. Thus, this laser is able to

provide an output power up to 100mWatt. However, the maximum temperature of temperature controller was fixed at 25°C to ensure the diode is not damaged during the diode laser driver was supplying higher power to diode laser.

The Raman sensor was operated by emitting the laser light via fiber optic Raman probe to BPA sample which was located in the sensor chamber. This fiber probe has two path arms, namely excitation and collection arms. The laser light emitted the sample from excitation path and the formed scattering light was congregated by collection arm path. After that, the scattering light was transmitted to spectrophotometer to interpret in the Raman intensity mode. This Raman QE 65000 spectrophotometer was provided by 1200 grating and high resolution with 0.05 mm slit. This spectrometer is capable to detect a Raman shift ranging from 0–2000 cm^{-1}.

Inphotonic Raman fiber optic has been specifically built for signal Raman detection and provided with one optical body holding one probe that has a focus lens at the end of the probe. This probe was attached by 90° with an optical body that functioned to separate the excitation and collection light inside the optical body system. With this geometry, the excitation light source from excitation arm will not enter the collection arm fiber and vice versa.

The sensor chamber body was built using teflon material and the inside and outside of the body was wholly painted in black to avoid the fluorescence from surrounding light and teflon material. Owing to this model, the result of Raman signal spectrum has shown that the spectrum will be free of noise and high sensitivity as we will be discussing in Figure 2. However, the BPA sample position in the chamber is considered as a crucial role in order to get a high Raman scattering effect. Therefore, the sample was placed at 90° with irradiated laser light. Meanwhile, the working distance between fiber probe and BPA sample is 7.5 mm. Based on this configuration, we were expecting the high Raman signal will be able to collect by spectrophotometer for low concentration of BPA in order to learn the limit of detection of this Raman sensor.

This work was carried out to sense the seven variations of BPA concentrations namely 0.001M, 0.005M, 0.05M, 0.1M, 0.25M, 0.5M and 1M using the Raman sensor system. Each tested sample of BPA was prepared by solving BPA powder in absolute ethanol and was then dropped of 0.01 ml of BPA solution into quartz surface. After 5 minutes, ethanol was evaporated under ambient air and BPA film was formed on the surface. In order to evaluate the sensor sensitivity towards different concentration of BPA, we learnt the interaction between the vibration of BPA molecule and a photon of laser light when irradiating the laser light into the BPA film surface. The change of photon energy was interpreted by spectrophotometer into a Raman shift spectrum that revealed the characteristic of the assignment of BPA molecule bonding. This detection was repeated a few times for each sample to observe the sensitivity and repeatability of the sensor. Figure 2 shows the Raman scattering effect for seven concentrations of BPA in quartz surface in Raman intensity mode.

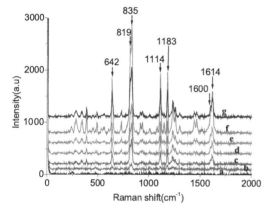

Figure 2. Raman shift spectra for seven concentrations of BPA a) 0.001M, b) 0.005M, c) 0.05M d) 0.1M, e) 0.25M, f) 0.5M, g) 1M.

Based on Raman spectra in Figure 2, the peaks of Raman shift for seven concentrations were positioned in same Raman shift referring the peaks for BPA molecule. The existence of the peaks was known when a spectral fingerprint was revealed, because this Raman sensor system is highly sensitive towards the BPA presence. The sensor was able to detect 0.001M of BPA, a relatively low BPA concentration and 1.0M, the highest concentration.

As we can see from Raman spectra, 0.001M gave the smallest Raman intensity of all of the peaks and the Raman intensity was kept enhancing with the increasing of the concentration of BPA on the surface. This result shows the Raman intensity is directly proportional to the concentrations. However, the intensity of Raman was decreasing for the high concentration of 1M. The decreasing intensity might because of the high percentage of the bigger size of BPA formed on the surface and the interaction between the light photon and area surface of BPA coverage was not yet effective (Yan et al. 2009). The uniform distribution is easier to form for lower BPA concentration in order to get a good cross section for Raman photon that scattered elastically. Based on overall Raman spectrum analysis, it was found that 0.5M concentration is optimum concentration which the sensor was able to detect and to give the optimum Raman intensity signals.

Each Raman spectrum exhibited almost nineteen similar peaks at vathe same Raman shift, however, only seven peaks have shown the significant change towards the increasing of BPA concentrations and other twelve peaks remained unchanged (Yao et al. 2012). The Raman intensity of the seven peaks at 642 cm^{-1}, 819 cm^{-1}, 835 cm^{-1}, 1114 cm^{-1}, 1183 cm^{-1}, 1600 cm^{-1} and 1614 cm^{-1} has been enhanced when the concentration of BPA solution into surface was increased excluding 1M of concentration. The increasing of BPA concentrations directly increases the presence of BPA particles on the surface which reacted with laser photon. Table 1 shows the rising value of Raman intensity at their Raman shift peaks for each concentration. In this table, we have compared the value of normal Raman spectrum that has been taken using Raman spectroscopy and our experimental value using the Raman sensor system. The comparison showed that our Raman system has successfully detected BPA particles only with a small percentage of uncertainty.

Besides, Raman scattering signal from BPA molecule was highly sensitive which can be detected by the sensor as it has been proved by the existence of minor peaks of BPA at 340 cm^{-1}, 457 cm^{-1}, 488 cm^{-1}, 560 cm^{-1}, 732 cm^{-1}, 918 cm^{-1}, 936 cm^{-1}, 1148 cm^{-1}, 1232 cm^{-1}, 1256 cm^{-1}, 1440 cm^{-1}, 1464 cm^{-1}. These Raman shift peaks have revealed the particular bond in BPA molecule (Yao et al. 2012). However, it is not the purpose of this paper aiming to discuss on the BPA assignment bonding.

Table 1. The Raman intensity of certain peaks for each spectrum a) 0.001M, b) 0.005M, c) 0.05M d) 0.1M, e) 0.25M, f) 0.5M, g) 1M.

*NRS (cm^{-1})	*ERS (cm^{-1})	Intensity (a.u) of Raman shift towards BPA concentrations						
		a	b	c	d	e	f	g
640	642	31	68	249	310	398	1057	507
820	819	40	94	376	465	612	1624	748
830	835	27	112	487	590	761	2072	960
1112	1114	−11	44	246	338	425	1141	555
1180	1183	20	87	297	340	437	1169	532
1600	1600	−4	15	88	134	179	501	226
1614	1614	26	53	176	241	305	782	373

*NRS: normal Raman spectrum
*ERS: experimental Raman spectrum

To verify the performance of the sensor towards the variation of BPA concentrations, we studied the Raman intensity for the entire Raman shift peaks. Figure 3 showed the plot of linear relationship between BPA concentrations and Raman intensity for seven major Raman shift peaks. It can be seen in Figure 3 and Table 2, the calibration curves have revealed the value of r for each peaks which are >0.9 as its good linear correlation coefficient relationship. Hence, these results are worth to apply in monitoring the BPA residue in the environment and food. However, this relationship is merely satisfied from 0.005 M to 0.5M of BPA concentrations.

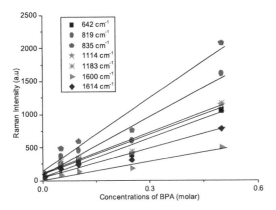

Figure 3. Linear relationship for Raman intensity towards concentrations of BPA at each Raman shift peak.

Table 2. Linear relationship for the spectra in Figure 3.

Peaks of Raman shift (cm^{-1})	Linear relationship
642	y = 1830.8x + 85.049; r = 0.972
819	y = 2833.9x + 121.24; r = 0.973
835	y = 3611.2x + 150.94; r = 0.970
1114	y = 2021.9x + 72.839; r = 0.973
1183	y = 1992.4x + 105.48; r = 0.968
1600	y = 1356.6x + 65.947; r = 0.976
1614	y = 913.11x + 18.382; r = 0.977

4 CONCLUSION

The Raman sensor system was successfully developed to detect bisphenol A (BPA) particles on the quartz surface from 0.001M to 1M of concentrations. Hence, Raman intensity of BPA peaks has been enhanced when the concentration of BPA solution into surface was increased excluding 1M of concentration. This is possible because the sensor is able to detect the highest concentration of BPA up to 0.5M due to the

effective area surface of BPA film. Moreover, there is a linear relationship between BPA concentrations and the Raman intensity with good linear correlation coefficient (r) of >0.9. Therefore, this Raman sensor is a promising tool to use in toxic materials detection.

ACKNOWLEDGMENTS

This research has been supported by the Malaysian Ministry of Higher Education and Universiti Kebangsaan Malaysia under research grants of ERGS/1/2012/STG02/UKM/01/1.

REFERENCES

[1] Bakar, N.A., Salleh, M.M., Umar, A.A. & Yahaya, M. 2013. Localized surface plasmon resonance sensor using gold nanoparticles for detection of bisphenol A. *Key Engineering Materials* 543: 342–345.
[2] Ballesteros-Gómez, A., Rubio, S. & Pérez-Bendito, D. 2009. Analytical methods for the determination of bisphenol A in food. *Journal of Chromatography A* 1216: 449–469.
[3] Borell, B. 2010. The big test for bisphenol A. *Nature* 464: 1122–1124.
[4] Marchesini, G.R., Meulenberg, E., Haasnoot, W. & Irth, H. 2005. Biosensor immunoassays for the detection of bisphenol A. *Analytica Chimica Acta* 528: 37–45.
[5] Massie, C., Stewart, G., McGregor, G. & Gilchrist, J.R. 2006. Design of a portable optical sensor for methane gas detection. *Sensor and Actuator B: Chemical* 113: 830–836.
[6] Ntsendwana, B., Mamba, B.B., Sampath, S. & Arotiba, O.A. 2012. Electrochemical detection of bisphenol A using grapheme-modified glassy carbon electrode. *International Journal of electrochemical Science* 7: 3501–3512.
[7] Rezaee, M., Yamini, Y., Shariati, S., Esrafili, A. & Shamsipur, M. 2009. Dispersive liquid-liquid microextraction combined with high-performance liquid chromatography-UV detection as a very simple, rapid and sensitive method for the determination of bisphenol A in water samples. *Journal of Chromatography A* 1216: 1511–1514.
[8] Staples, C.A., Dorn, P.B., Klecka, G.M., O'Block, S.T., Branson, D.R. & Harris, L.R. 2000. Bisphenol A concentrations in receiving waters near US manufacturing and processing facilities. *Chemosphere* 40: 521–525.
[9] Tsuru, N., Kikuchi, M., Kawaguchi, H. & Shiratori, S. 2006. A quartz crystal microbalance sensor coated with MIP for "Bisphenol A" and its properties. *Thin Solid Films* (1-2)499: 380–385.
[10] Yan, B., Thubagere, A., Premasiri, W.R., Ziegler, L.D., Negro, L.D. & Reinhard, B.M. 2009. Engineered SERS substrates with multiscale signal enhancement: nanoparticles cluster arrays. *American Chemical Society Nano* (5)3: 1190–1202.
[11] Yao, W., Wang, S. & Wang, H. 2012. *Method for detecting BPA by surface enhanced Raman spectroscopy*. United States: Patent Application Publication.

Testing and Measurement: Techniques and Applications – Chan (Ed.)
© 2015 Taylor & Francis Group, London, ISBN: 978-1-138-02812-8

Eu^{3+} doped Gd$_2$Ti$_2$O$_7$ nanoparticles as a luminescence thermometry probes

S. Ćulubrk, V. Lojpur, M. Medić & M.D. Dramićanin

Vinca Institute of Nuclear Sciences, University of Belgrade, Belgrade, Serbia

ABSTRACT: In this report, we analyzed the potential of Eu^{3+} doped Gd$_2$Ti$_2$O$_7$ nanopowders for the luminescence thermometry. Gd$_2$Ti$_2$O$_7$ powders doped with 5 at% Eu^{3+} ions were prepared using Pechini-type polymerized complex route. Sample was investigated for luminescent thermometry over the temperature ranges from 303 - 423 K with fluorescence intensity ratio (FIR) method. Luminescent spectra recorded in range from 400 - 650 nm discovered two distinct spectral regions: the high-energy region associated with the trap emission of the Gd$_2$Ti$_2$O$_7$ host and the low-energy region with well-resolved emission peaks of the Eu^{3+} ions. FIR is determined using trap emission from host and one emission band from Eu^{3+} ion in Gd$_2$Ti$_2$O$_7$. The relative sensitivity varies over the 303-423 K temperature range, and has maximum value of 0.558 % K^{-1} at 303 K.

1 INTRODUCTION

Pyrochlores are compounds with the general formula A$_2$B$_2$O$_7$, where A element is rare earth metal (RE = Ln, Sc, or Y) or an element with an inert lone-pair of electrons while B element can be either a transition or a post-transition metal. Recently, pyrochlores have been extensively explored due to their broad range of physical, chemical and magnetic properties that the substitution of ions depends on the A and B cation sites. Specifically, high ionic conductivity, superconductivity, luminescence and ferromagnetism are found with pyrochlores (Porat et al. 1997, Heremans et al. 1995, Moon et al. 1993). Titanate pyrochlores with the formula A$_2$Ti$_2$O$_7$ have been latterly investigated and they also showed excellent physical and chemical characteristics. They have applications as solid electrolytes and mixed ionic/electronic conducting electrodes, catalysts, and ferroelectric/dielectric device components. Among them, rare earth titanates (RE$_2$Ti$_2$O$_7$, RE = Ln^{3+}, Sc^{3+}, Y^{3+} or Gd^{3+}) provide a good candidate for phosphor matrices since they have ability to accept rare earth elements in high concentrations. Investigations of gadolinium-titanate, Gd$_2$Ti$_2$O$_7$, have a great interest since luminescent nanoparticles with Gd ions can be used for magnetic resonance imaging (MRI) (Gavrilovic et al 2014). Solid-state reactions as well as different wet chemical methods were used for obtaining Gd$_2$Ti$_2$O$_7$. In order to reduce temperature of producing material (below 1400°C) and particle size, wet chemical methods such as sol-gel, molten salt, combustion, hydrothermal, and stearic acid methods have become more frequent (Heredia et al. 2010, Lin et al. 2007, Dharuman et al. 2013, Joseph et al. 2008, Zhang et al. 2010). Although, many different pyrochlore compounds have been synesized, a very few papers has focused on the photoluminescent properties of rare earth doped Gd$_2$Ti$_2$O$_7$ powders (Zhang et al. 2009, Pang et al. 2004).

New materials and methods for the measurements of temperature have become a necessity in most experimental and applied fields of science, engineering and medicine. Phosphor thermometry represents an optical technique for surface temperature measurements. Temperature evolution can be remotely determined by measuring changes in the luminescent properties of the probe, such as absolute and relative emission intensities, lifetime values of excited states, peak position, and the emission bandwidth (Khalid et al. 2008, Brubach et al. 2011). Luminescent thermometry enables a spatial distribution of temperature with sub-micrometer resolution, which is a not possible to use conventional temperature sensor. Two types of temperature sensing measurements are commonly used: luminescent lifetime measurements method and fluorescence intensity ratio method (FIR). Lifetime method is based on the determination of emission decays over the observed temperature range (Brites et al. 2012). With the increasing of temperature, lifetime decay decreases due to the thermal quenching effect so temperature is measurable only at the point and within a relatively limited range. On the other hand, FIR method is based on the intensity ratio between two emission lines in photoluminescence spectrum. This method has a number of advantages like wider temperature range compared to the emission decay method and, in contrast to single emission measurements, is insensitive to fluctuations of the excitation light or other changes in measurement conditions (Lojpur et al. 2013).

In this paper, the temperature dependence of photoluminescence of $Gd_2Ti_2O_7$ nanopowders doped with Eu^{3+} ions is presented. We have recently show with TiO_2 nanoparticles (nas rad Sensors and Actuators B) that trap emission of the host material can serve as an excellent internal reference for the FIR thermometry method. Here, we aimed to test the sampe principle using $Gd_2Ti_2O_7$ nanoparticle over a wide temperature range (293-423 K), and to assess sensitivity of the thermometry method.

2   EXPERIMENTAL

The sample investigated in luminescence thermometry is prepared by Pechini-type polymerized complex route. Detailed procedure was given in our previous paper (Ćulubrk et al. In press). For luminescence measurements, round pellets were prepared from $Gd_2Ti_2O_7$: Eu^{3+} powders under a load of 5 tons. Photoluminescence spectra were collected using a Fluorolog-3 Model FL3-221 spectrofluorometer system (Horiba Jobin-Yvon), shown in figure 1, over the temperature range from 293 to 400 K. The photoluminescence measurements were performed under continuous excitation from a 450 W Xenon lamp at a wavelength of 393 nm. The samples were placed in a custom-made, temperature controlled furnace, and emission spectra were collected via an optical fiber bundle. The temperature of the samples was controlled within the accuracy of ±0.5°C by a temperature control system utilizing proportional-integral-derivative feedback loop equipped with T-type thermocouple for temperature monitoring. Experimental set-up consists of the following parts: 1. Lamp, 2. Detector, 3. Monochromator, 4. Optical fiber, 5. Lense, 6. Sample, 7. Oven, 8. T-regulator, 9. Controler and 10. PC.

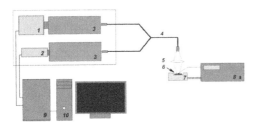

Figure 1.   Experimental set-up for temperature dependent photoluminescent measurement.

3   RESULTS AND DISCUSSION

Luminescence spectra of $Gd_2Ti_2O_7$:5 at% Eu^{3+} powders measured as a function of temperature (293-423 K) are shown in the Figure 2. Two distinct

spectral regions can be observed : high energy broad band spectral region (400-550 nm) that belongs to the trap emission of the $Gd_2Ti_2O_7$ host and the low energy spectral region that belongs to emission of Eu^{3+} ions. Emission of dopant is composed of distinctive bands peaking at 589 nm, 597 nm ($^5D_0 \rightarrow ^7F_1$) 611 nm and 627 nm ($^5D_0 \rightarrow ^7F_2$) that arise from spin-forbidden electron transition. It is clearly seen from the figure 2 that dopant emission is extremely sensitive to temperature, showing a rapidly decreasing intensity with the temperature increase while changes of trap emission are minimal with the change of temperature in the measured range. The ideal case for FIR measurement is to have one emission whose properties don't change with the temperature, so the trap emission can serve as an excellent internal standard for FIR measurements. Small changes in trap emission intensities are a consequence of slight changes in optical properties of the host with temperature such as band gap energy, reflectivity and changes of experimental parameters during the measurements. On the other hand, sharp decline of dopant emission is due to the non-radiative relaxation processes that increase with increasing of temperature.

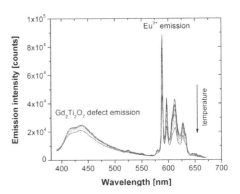

Figure 2.   Photoluminescence spectra of 5.0 at.% Eu^{3+} doped $Gd_2Ti_2O_7$ powder over the temperature range of 293-563 K (temperature increment was 20 K).

FIR data were calculated as the ratio between trap emission of host at 438 nm and Eu^{3+} emission intensity 589 nm, and their temperature dependence (303- 423 K) is presented with symbols in Figure 3. The temperature dependence of FIR can be described by the Arrhenius-type equation. The solid line represents the FIR obtained by fitting experimental data to Eq. (2).

$$FIR = \frac{\text{Intensity of trap emission}}{\text{Intensity of Eu emission}} = C \exp\left(\frac{\Delta E}{kT}\right), \quad (1)$$

$$\ln FIR = \ln C + \frac{\Delta E}{k}\frac{1}{T}, \qquad (2)$$

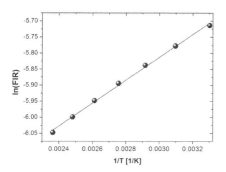

Figure 3. Intensity ratio of emission at 436 nm and emission at 589 nm. The experimental data are represented as symbols, whereas the theoretical curves were obtained using Eq.(2) and are represented with solid line.

where C is constant, ΔE is the energy gap between the excited and ground state. The solid line in Figure 3 obtained after fitting with Eq. (2) is in excellent agreement with the experimental data (R-Square indicator of fit is 0.996). The fitting procedure provides the following values: C= 8.515, and $\Delta E = 287.69$ cm^{-1}.

A useful parameter that describes the performance of the temperature sensor is the relative sensitivity (S_r), given by the following equation:

$$S_r = 100\% \times \left| \frac{1}{FIR}\frac{\partial FIR}{\partial T} \right| \text{ in } [\,\%K^{-1}\,]. \qquad (3)$$

The relative sensor sensitivity values are presented in figure 4. One can see that sensitivity is temperature dependent. The maximal value is 0.558 %K^{-1} obtained at 303 K while minimum is 0.286 %K^{-1} at 423 K.

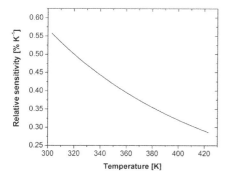

Figure 4. Relative sensor sensitivity of 5.0 at.% Eu^{3+} doped Gd$_2$Ti$_2$O$_7$ powder as a function of temperature.

4 CONCLUSION

To conclude, Gd$_2$Ti$_2$O$_7$ nanopowders doped with 5 at% Eu^{3+} ions were prepared using Pechini-type polymerized complex route and investigated for high-temperature luminescent thermometry over the temperature ranges from 303 - 423 K. FIR method was used and data were calculated using trap emission from host and emission form Eu^{3+} ion in host. The maximal relative sensor sensitivity is 0.558 %K^{-1} at 303 K.

ACKNOWLEDGMENT

This work is supported by the Ministry of Education, Science and Technological Development of the Republic of Serbia (grant No. 45020) and the APV Provincial Secretetariat for Science and Technlogocal Development of the Republic of Serbia, through project no. 114-451-4787.

REFERENCES

[1] Brites,C.D.S Lima, P.P. Silva, N.J.O.Millan, A. Amaral, V.S. Palacio, F. & Carlos, D.L. 2012. Thermometry at the nanoscale. *Nanoscale*. 4:4799–4829.
[2] Brubach, J. Kissel, A. Frotsher, M. Euler, M. Albert, B. & Dreizler, A.. 2011. A survey of phosphors novel for thermography. *Journal of luminescence*. 131:559–564.
[3] Ćulubrk, S. Antić, Ž. Marinović-Cincović, M. Ahrenkiel, S.P. & Dramićanin, D.M. 2014. Synthesis and luminescent properties of rare-earth (Sm^{3+} and Eu^{3+}) doped Gd$_2$Ti$_2$O$_7$ pyrochlore nanopowders. *Optical Materials*. DOI:10.1016/j.optmat.2014.08.001
[4] Dharuman, N. & Berchmans, L.J. 2013. Low temperature synthesis of nano-crystalline gadolinium titanate by molten salt route. *Ceramics International*. 3:8767–8771.
[5] Gavrilović, T. Jovanović, D. Lojpur V. & Dramićanin, D.M. 2014. Multifunctional Eu^{3+} and Er^{3+}/Yb^{3+}-doped GdVO$_4$ nanoparticles synthesized by reverse micelle method. *Scientific Reports*. 4:4209–4216.
[6] Heredia, R.A. García, Q.M. Mazariego, P.L.J. & Escamilla, R. 2010. X-ray diffraction and raman spectroscopy on Gd$_2$(Ti$_{2-y}$Te$_y$)O$_7$ prepared at high pressure and high temperature, *Journal of Alloys and Compdounds*. 504:446–451.
[7] Heremans, C. Wuensch, J. B. Stalick J.K. & Prince, E. 1995. Fast-ion conducting Y$_2$(Zr$_y$Ti$_{1-y}$)$_2$O$_7$ pyrochlores: Neutron rietveld analysis of disorder induced by Zr substitution. *Journal of Solid State Chemistry*. 117:108–121.
[8] Joseph, K.L. Dayas, R. K. Damodar, S. Krishnan, B. Krishnankutty, K. Nampoori, N. P.V. & Radhakrishnan, P. 2008. Photoluminescence studies on rare earth titanates prepared by self-propagating high temperature synthesis method. *Spectrochimica Acta Part A*. 71:1281–1285.

[9] Khalid, A.H. & Kontis, K. 2008. Thermographic phosphors for high temperature measurements: principals, current state of the art and recent applications. *Sensors*. 8:5673–5744

[10] Lin, L & Yan, B. 2011. Rare earth titanates ceramics $Na_2La_2Ti_3O_{10}$: Pr^{3+} and $RE_2Ti_2O_7$: Pr^{3+} (R = Gd, Y): sol–gel synthesis, characterization and luminescence. *Journal of Material Science – Materials in Medicine*. El. 22:672–678.

[11] Lojpur, V. Nikolic, M. Mancic, L. Milosevic, O. & Dramicanin, M.D. 2013. Y_2O_3:Yb,Tm and Y_2O_3:Yb,Ho powders for low-temperature thermometry based on up-conversion fluorescence. *Ceramics International*. 39:1129–1134.

[12] Moon, K.P. & Tuller, L.H. Ionic conduction in the $Gd_2Ti_2O_7$–$Gd_2Zr_2O_7$ system. 1988. *Solid State Ionics* 28–30:470–474.

[13] Pang, L.M. Lin, J. Fu, J. & Cheng, Y.Z. 2004. Luminescent properties of $Gd_2Ti_2O_7$: Eu^{3+} phosphor films prepared by sol–gel process. *Materials Research Bulletin*. 39:1607–1614.

[14] Porat, O. Heremans, C. Tuller & L.H. 1997. Stability and mixed ionic electronic conduction in $Gd_2(Ti_{1-x}Mo_x)_2O_7$ under anodic conditions. *Solid State Ionics*. 94:75–83.

[15] Zhang, W. Zhang, L. Zhong, H. Lu, L. Yang, X. & Wang, X. 2010. Synthesis and characterization of ultrafine $Ln_2Ti_2O_7$ (Ln = Sm, Gd, Dy, Er) pyrochlore oxides by stearic acid method. *Materials Characterization*. 61:154–158.

[16] Zhang, Y. Ding, L. Pang, X. & Zhang, W. 2009. Influence of annealing temperature on luminescent properties of Eu^{3+}/V^{5+} co-doped nanocrystalline $Gd_2Ti_2O_7$ powders. *Journal of Rare Earths*. 27:900–904.

Testing and Measurement: Techniques and Applications – Chan (Ed.)
© 2015 Taylor & Francis Group, London, ISBN: 978-1-138-02812-8

Sensitivity analysis of aerosol based on MODIS remote sensing data

C.F. Li, Y.Y. Dai, S.Q. Zhou, J.J. Zhao & J.Y. Yin
School of Computer Engineering and Science, Shanghai University, Shanghai, China

ABSTRACT: The selection of aerosol types is the key factor of haze pollution monitoring based on remote sensing data. Aiming at the problem, the sensitivity analyses of common aerosol types are carried out and have reached a conclusion that the continental aerosol type is roughly identical to the V5.2 algorithm of aerosol optical depth inversion. In combination with the severe haze pollution on December 5, 2013 in Shanghai, China, the Moderate Resolution Imaging Spectroradiometer (MODIS) remote sensing images are taken as data sources, this paper tries to use V5.2 algorithm to monitor the severe haze pollution. The result shows that it can select the suitable aerosol types by aerosol sensitivity analysis for different types of remote sensing data, and it is helpful to improve monitoring precision of the haze pollution based on remote sensing data.

1 INTRODUCTION

The haze weather is one kind of the atmosphere pollution. Generally speaking, the suspended particulate matter less than 2.5 microns in diameter (PM2.5) are responsible for the haze pollution (Du et al. 2009, Guo et al. 2010). It is easy to be attached with poisonous and endanger not only people's health and traffic safety, but also the balance of the ecosystem. Traditional haze measurement mainly takes advantage of ground observation stations and has some limited effectiveness outside the scope of station monitoring (Fu et al. 2008, Guo & Ma 2005, Ramanathan et al. 2007). Fortunately, the appearance of remote sensing makes up for the shortage of traditional methods in PM2.5 monitoring (Lei et al. 2013). Remote sensing has the characteristics of macro and objectivity, and can obtain particulates (aerosol) parameters of the whole layer of the atmosphere rapidly as well as automatically. It has become an important means of observing the spatial distribution of aerosols and the key is to select the aerosol type (Lee et al. 2006). Because of the complexity of the aerosol type and various data types of remote sensing, there is still a lot of uncertainty in haze monitoring using remote sensing data, it is necessary to do sensitivity analysis of the aerosol types so as to find out the aerosol type appropriate to the characteristics of remote sensing data and the surrounding environment.

According to different types of aerosol, this paper firstly put forward sensitivity analysis on the most commonly used aerosol types including continental, marine, dust and city type aerosols. Then taking the severe haze pollution occurred December 5, 2013 in Shanghai area as an example, the V5.2 algorithm has been used to monitor the haze pollution based on MODIS remote sensing data. The first part of this article is an introduction, the second part is sensitivity analysis of the aerosol type, the third part is monitoring process of this severe haze pollution in the Shanghai area, and the last part is our conclusion and discussion.

2 SENSITIVITY ANALYSIS

The MODIS remote sensor is carried on the NASA/EOS U. S. satellite. It covers the globe once a day and has 36 spectral bands with ranges between 0.4 µm—14 µm. Its scanning width is 2330 km with the ground resolution of 250 m, 500 m and 1000 m respectively.

Regarding the MODIS remote sensing image obtained on October 1, 2012 in Shanghai area as the data source, this study conducts a sensitivity analysis, respectively for the continental, marine, dust and city type aerosols. Thereinto, according to the geographical distribution of research area, in this study, the atmospheric model is the mid-latitude winter model, the aerosol optical depth within 0.1–1.2, and the step size is 0.1, the surface type is uniform without direction, the surface reflectance within 0.05–0.5, respectively. For each type of aerosol, the sensitivity analysis will be carried out in terms of the red light, blue light and intermediate infrared spectral bands in the following research.

2.1 *Continental aerosol*

As can be seen from Figure 1a, b, the threshold value of red light and blue light spectrum wavelengths are 0.2 and 0.25, respectively. When the surface reflectance is less than the threshold, the apparent reflectance aggrandizes with the increase of the aerosol optical depth. However, when the surface reflectance is greater than the threshold, the apparent reflectance reduces with the increase of the aerosol optical depth. As can be seen from Figure 1c, when the surface reflectance in the intermediate infrared spectral bands is very low (approximately between 0.05–0.15), the surface reflectance is seldom affected by the aerosol optical depth. Therefore, the apparent reflectance can be approximately equal to the surface reflectance; this is also identical with the fundamentals of the V5.2 algorithm.

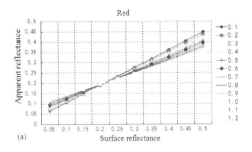

(a)

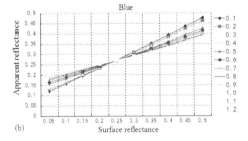

(b)

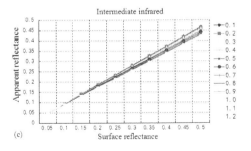

(c)

Figure 1.   Sensitivity analysis of continental aerosol.

2.2   Marine aerosol

As can be seen from Figure 2a, b, there is no threshold in surface reflectance from both red light and blue light spectrum wavelengths, respectively. When the surface reflectance is between 0.05 and 0.4, the apparent reflectance aggrandizes with the increase of the aerosol optical depth, but the increase rate gradually

diminishes, and the trend is becoming more and more vague. When the surface reflectance is between 0.4 and 0.5, the apparent reflectance becomes no longer affected by the aerosol optical depth. However, there is an apparent threshold value of 0.35 in the intermediate infrared spectral bands (see Fig. 2c). When the surface reflectance is less than 0.35, the apparent reflectance rises with the increase of the aerosol optical depth, and when the surface reflectance is greater than 0.35, the apparent reflectance gradually reduces with the increase of the aerosol optical depth. Thus, the marine aerosol cannot match the fundamentals of the V5.2 algorithm.

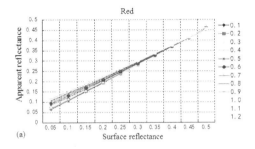

(a)

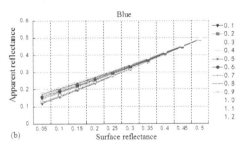

(b)

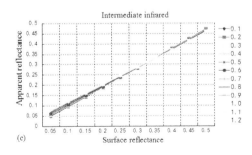

(c)

Figure 2.   Sensitivity analysis of marine aerosol.

2.3   Dust aerosol

As can be seen from Figure 3a, b, the thresholds in both red light and blue light spectrum wavelengths are similar to those of the continental aerosol (see Fig. 1). But in dust aerosol, there is an apparent threshold value of 0.3 in the intermediate infrared spectral

164

bands (see Fig. 3c). When the surface reflectance is less than 0.3, the apparent reflectance rises with the increase of the aerosol optical depth, and when the surface reflectance is greater than 0.3, the apparent reflectance reduces with the increase of the aerosol optical depth. Therefore, the dust aerosol cannot match the fundamentals of the V5.2 algorithm.

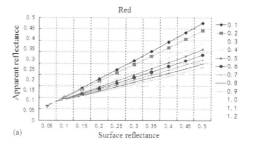

(a)

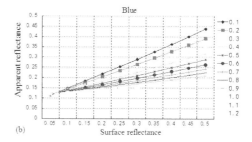

(b)

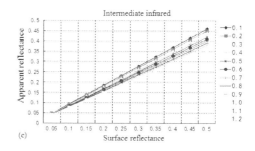

(c)

Figure 4. Sensitivity analysis of city type aerosol.

The sensitivity analysis result of continental aerosols, marine aerosols, dust aerosols and city type aerosols show that there is only continental aerosol model that corresponds to the fundamentals of the V5.2 algorithm for MODIS remote sensing data. So next we will adopt the continental aerosol model to establish the 6S look-up table and monitor the severe haze pollution practical case in December 2013.

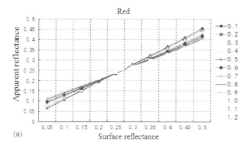

(a)

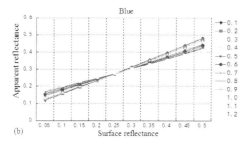

(b)

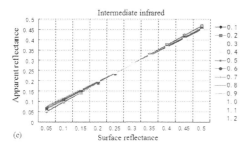

(c)

Figure 3. Sensitivity analysis of dust aerosol.

2.4 City type aerosol

As can be seen from Figure 4a, b, there is a small threshold near the value of 0.07 in both red light and blue light spectrum wavelengths in the city type aerosol. When the surface reflectance is greater than 0.07, the apparent reflectance aggrandizes fast with the increase of the aerosol optical depth. As can be seen from Figure 4c, the city type aerosol band has been consistently affected by the aerosol optical depth, the apparent reflectance aggrandizes gradually with the increase of the aerosol optical depth. Therefore, the city type aerosol cannot match the fundamentals of the V5.2 algorithm either.

3 REMOTE SENSING MONITORING CASE

3.1 6S look-up table and V5.2 algorithm

6S (Second Simulation of Satellite Signal in the Solar Spectrum) model is an atmospheric radiative transfer model, which has been widely used at present. It is developed on the basis of 5S model. Compared with 5S model, 6S model can ensure that the input parameter is more in conformity with the actual situation due to the adoption of advanced approximation algorithm and successive scattering algorithm. The main parameter settings of 6S model are as shown below: the atmospheric model is 3 (mid-latitude winter atmosphere mode), the aerosol type is continental aerosols, the aerosol content parameter is aerosol optical depth with 550 nm spectrum wavelengths, the target height parameter is 0, and it doesn't need to activate the atmospheric correction mode.

Although V5.2 algorithm is developed on the basis of the dark pixel algorithm, there are still differences between the two. In the dark pixel algorithm, the reflectance ratio of the intermediate infrared wavelengths with red light and blue light wavelengths is constant. However, the V5.2 algorithm puts forward that it is not a constant, but a functional relation, and also considers the effects of normalized difference vegetation index (NDVI).

3.2 Remote sensing monitoring results

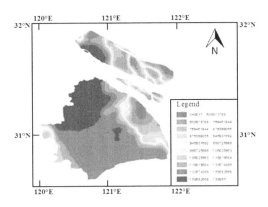

Figure 5. Aerosol distribution on December 5, 2013 in Shanghai area.

As can be seen from Figure 5, the aerosol optical depth of Shanghai area was very large on December 5, 2013. The maximum aerosol optical depth lies in the Baoshan District and Jiading District, the minimum aerosol optical depth lies in the Chongming County, which is near the Yangtze River estuary and coastal areas of Pudong New Area. From the overview of haze pollution distribution in the Shanghai area, the aerosol optical depth gradually decreased mainly from northwest to southeast, and it is also coincided with the ground observation stations of the same area. In addition, this study proposes to use V5.2 algorithm to monitor the aerosol optical depth in the Shanghai area from MODIS remote sensing data, the errors of the inversion aerosol optical depth is within the scope of permission and the distribution is consistent with the actual haze pollution distribution.

4 CONCLUSION AND DISCUSSION

Remote sensing technology can obtain the aerosol optical depth without destroying the atmospheric particulates, and realize three-dimensional monitoring of regional haze pollution in cities. On the basis of the sensitivity analysis of various aerosol types in terms of MODIS remote sensing data, the continental aerosol type was more identical with the characteristic of MODIS remote sensing data and the actual situation in the Shanghai area. And then the V5.2 algorithm is used to build 6S look-up table and further realize the inversion of aerosol optical depth. Finally, it is validated to combine the case of severe haze pollution case on December 5, 2013 in the Shanghai area. The results of this study could not only make up for the deficiency of the traditional ground monitoring of haze pollution, but also provide the technology reference for monitoring, early warning and governance haze for the relevant departments.

ACKNOWLEDGMENTS

This work was supported by the Project of National Natural Science Foundation of China (41404024), and partially by the Young Teachers Training and Supporting Plan in Shanghai Universities (2014–2016), and Laboratory Technician Team Building Program in Shanghai Universities (B.60-E108-14-101), and Higher Education Connotation Constructions 085 Project in Shanghai.

REFERENCES

[1] Du, Q.S., Liu, Z.P., Wang, X.S., et al. 2009. Method for MODIS data pre-processing based on ENVI. Geospatial information 7(4): 98–100.
[2] Fu, Q., Zhuang, G., Wang, J., et al. 2008. Mechanism of formation of the heaviest pollution episode ever recorded in the Yangtze River Delta, China. Atmospheric Environment 42(1): 2023–2036.
[3] Guo, G.M. & Ma, L. 2005. Urban aerosol optical thickness retrieval from MODIS data. Remote Sensing Technology and Application 20(3): 343–345.
[4] Guo, J., Zhang, X., Cao, C., et al. 2010. Monitoring haze episodes over the Yellow Sea by combining multi-sensor measurements. International Journal of Remote Sensing 31(17): 4743–4755.
[5] Lee, K., Kim, Y. & Kim, M. 2006. Characteristics of aerosol observed during two severe haze events over Korea in June and October 2004. Atmospheric Environment 40(27): 5146–5155.
[6] Lei, M., Xue, D., Li C.F. et al. 2013. Study on AOD of Shanghai Based on MODIS data. Geomatics & Spatial Information Technology 36(10): 40–43.
[7] Ramanathan, V., Ramana, M.V., Roberts, G., et al. 2007. Warming trends in Asia amplified by brown cloud solar absorption. Nature 448 (7153): 575–578.

Testing and Measurement: Techniques and Applications – Chan (Ed.)
© *2015 Taylor & Francis Group, London, ISBN: 978-1-138-02812-8*

Hilbert diagnostics of phase disturbances of the light fields in air

V.A. Arbuzov & Y.N. Dubnishchev
Kutateladze Institute of Thermophysics Siberian Branch of the Russian Academy of Sciences (SB RAS),
Novosibirsk, Russia
Technological Design Institute of Scientific Instrument Engineering SB RAS, Novosibirsk,Russia

V.G. Nechaev & E.O. Shlapakova
Novosibirsk State Technical University, Novosibirsk, Russia

ABSTRACT: Methods of optical Hilbert-filtration of phase disturbances of the light field in the spectral range of source emission are discussed, such as, the evolution of complimentary vortex rings induced in an air medium by a pressure pulse on the hole in the combuster wall with application of the Hilbert-optics methods.

1 INTRODUCTION

Optical diagnostics of flows has been successfully applied in experimental hydrodynamics and gas dynamics for a long time. It offers a large variety of modern methods and techniques: shadowgraphy and schlieren methods (Settles 2001, Belozerov 2007, Dubnishchev et al. 2013), laser Doppler anemometry (LDA) Doppler technologies of velocity field measurements (Dubnishchev et al. 2009), and particle image velocimetry (PIV). The Hilbert diagnostics of flow based on visualization of space-time phase disturbances induced in the light field by the flow passing through the examined medium has a high potential of development and application. Though the theoretical and experimental details of the Hilbert-optics have been discussed in many publications (Arbuzov & Dubnishchev 2007; Dubnishchev 2011), its potential is far from being exhausted. This refers, in particular, to the analysis of the amplitude-phase characteristics of filters and the dynamic range of phase disturbances during the Hilbert-diagnostics in the spectral range of the probing field. The present study was motivated by the necessity of considering these issues. For example, the application of Hilbert-optic method for on the hole in the combuster wall is discussed.

2 METHOD OF RESEARCH AND RESULTS

Among the Hilbert optics methods based on the use of filters having a quadrant structure, the Hilbert transform is not isotropic. In the case of an isotropic Hilbert transform, filters with axial or screw symmetry are used (Eu & Lohmann 1973; Arbuzov&Dubnishchev 2007; Dubnishchev 2011).

For zov & Dubnishchev 2007; Dubnishchev 2011). For the filter with a quadrant structure, the coherent transfer function $H(K_x, K_y)$ generally looks like

$$H(K_x, K_y) = \left[e^{i\varphi}\sigma(K_x) + e^{-i\varphi}\sigma(-K_x)\right]\sigma(K_y) +$$

$$+\left[e^{-i\varphi}\sigma(K_x) + e^{i\varphi}\sigma(-K_x)\right]\sigma(-K_y) =$$

$$= \left[\cos\varphi + i\sin\varphi\,\mathrm{sgn}\,K_y\right]\sigma(K_x) +$$

$$+\left[\cos\varphi - i\sin\varphi\,\mathrm{sgn}\,K_y\right]\sigma(-K_x) =$$

$$= \cos\varphi + i\sin\varphi\left[\mathrm{sgn}\,K_x\,\mathrm{sgn}\,K_y\right], \tag{1}$$

where φ is the phase shift, and $\sigma(\pm K_x)$ and $\sigma(\mp K_y)$ are the Heaviside functions: $\sigma(\pm K_x) = \frac{1}{2}(1 \pm \mathrm{sgn}\,K_x)$ and $\sigma(\pm K_y) = \frac{1}{2}(1 \pm \mathrm{sgn}\,K_y)$. K_x and K_y are the spatial frequencies in the Cartesian coordinate system, and sgn K is the sign function.

As the phase shift j depends on the wave length, $\varphi = \varphi(\lambda)$, the filter with the coherent transfer function (1) performs the Foucault–Hilbert polychromatic filtration in a spectral band of the light source. At $\varphi = \frac{\pi}{2}$, the coherent transfer function (1) becomes $H(K_x, K_y) = i\,\mathrm{sgn}\,K_x\,\mathrm{sgn}\,K_y$.

Let us address the Hilbert-diagnostics of the optical phase density in air or liquid. We assume that phase disturbances of the light field induced by the investigated fluid change harmonically in the *x* direction and are described by the expression

$$s(x, y) = e^{i\eta\sin(K_{x_0}x)}, \tag{2}$$

where, K_{x0} is the spatial frequency, and η is the amplitude of phase disturbances.

Let us present the function $s(x, y)$ by the Fourier series

$$s(x, y) = e^{i\xi \sin(K_{x0}x)} = \sum_{n=-\infty}^{\infty} J_n(\xi) e^{inK_{x0}x} =$$

$$= J_n(\xi) + 2 \sum_{n=-1}^{\infty} J_{2n}(\xi) \cos(2nK_{x0}x) +$$

$$+2 \sum_{n=-0}^{\infty} J_{2n+1}(\xi) \sin[(2n+1)K_{x0}x].$$

Here $J_n(\xi)$ is the Bessel function. From here, for the Hilbert-conjugated signal, we obtain

$$\hat{s}_x(x, y) = 2\sum_{n=1}^{\infty} J_{2n}(\xi) \sin(2nK_{x0}x) -$$

$$-2\sum_{n=0}^{\infty} J_{2n+1}(\xi) \cos[(2n+1)K_{x0}x] \approx -2J_1(\xi)\cos(K_{x0}x) \quad (3)$$

Here we neglect terms with harmonics of the higher order. As it may be seen from Eqs. (3) and (2), phase disturbances with all amplitudes are visualized by the Hilbert-diagnostics in the plane of registration of the Hilbert-conjugated signal. The Hilbert-filtration allows one to visualize the structure of the phase optical density in a wide dynamic range.

The optical measuring complex was created on the basis of the IAB–463M shadowgraph (www.ckb-photon)in which the optical Hilbert-filtration module and a source of illumination were specially developed for the experiment. A sketch of the optical measuring system is shown in Figure 1a. The system consecutively includes a light source 1a collimation objective 2, and a Fourier-objective 3 in the frequency plane K_x, K_y where the Hilbert–filter 4 is placed. There is a phase disturbance $s(x,y)$. The Fourier-objective 5 brings back the Fourier-transformation and restores the filtered image $s(x,y)$, which is registered by a CCD camera 6. Vortex rings are formed inside and outside the chamber 7.

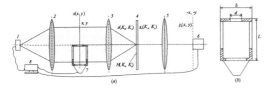

Figure 1. Sketch of the experimental measuring complex (a) and generator of vortex rings (b): 1 – light source; 2 – collimation objective; 3 – Fourier–objective; 4 – quadrant Hilbert–filter in the frequency plane K_x, K_y; 5 – Fourier–objective; 6 – CCD camera; 7 – generator of vortex rings; 8 – computer control system.

The design of the chamber (Figure 1b) is similar to that described in (Arbuzov et al. 2008; Dubnishchev

et al. 2010). Its internal sizes are 0.19×0.19 wall of the chamber is formed by the diffuser 0.38 m. The back wall of the chamber is formed by the diffuser of an electrodynamic loudspeaker initiated by electric pulses acting through the amplifier from the output of the sound card of a computer. Windows of optical quality were inserted into lateral walls for visualization of phase disturbances inside the chamber. A computer control system of the experiment 8 provided the control of the shape and on/off time ratio of pulses acting on the electrodynamic generator, measurement of pressure pulses in the chamber, and also synchronization of operation of the electrodynamic generator and the system of registration of vortex structure images. The experiments were performed with exhaust outlet diameters of 20 and 24 mm. Visualization of phase disturbances induced by vortex rings in the light field was consecutively performed by the optical Hilbert-filtration of the Fourier-spectrum of the disturbed optical field, by the inverse Fourier-transformation of the filtered Fourier-spectrum, and by registration of the Hilbert-conjugated image. In the case of the Foucault-Hilbert transformation, the result is an analytical signal as a superposition of the original signal and its Hilbert-conjugated image (Dubnishchev 2011). The Hilbert-conjugated image contains the information about the structure of the phase optical density disturbance in the fluid.

Figure 2 shows examples of the positive pressure pulse (a) and negative pressure pulse (b) on the hole formed by the electrodynamic loudspeaker.

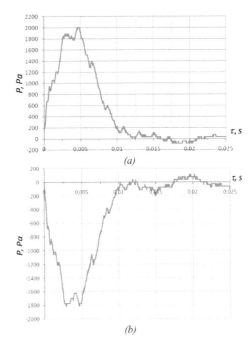

Figure 2. Positive pressure pulse (a) and negative pulse (b) on the hole.

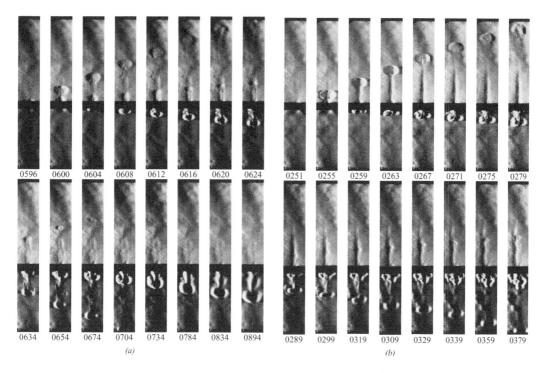

(a)

0596 0600 0604 0608 0612 0616 0620 0624

0634 0654 0674 0704 0734 0784 0834 0894

0251 0255 0259 0263 0267 0271 0275 0279

0289 0299 0319 0309 0329 0339 0359 0379

(b)

Figure 3. Vortex structures induced by the positive pressure pulse on the round hole in diameter: (*a*) – 20 mm, (*b*) – 24 mm.

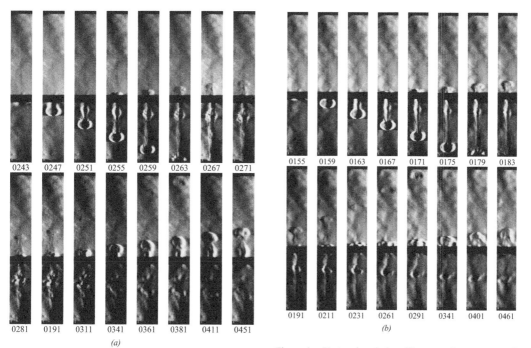

0243 0247 0251 0255 0259 0263 0267 0271

0281 0191 0311 0341 0361 0381 0411 0451

(a)

0155 0159 0163 0167 0171 0175 0179 0183

0191 0211 0231 0261 0291 0341 0401 0461

(b)

Figure 4. Vortex rings induced by a negative pressure pulse on the round hole in diameter: (a) – 20 mm, (b) – 24 mm.

169

Figure 3 shows the Hilbert-images of the vortex structures induced by the positive pressure pulse (Figure 2*a*) on the hole. The vortex structures were visualized by the Hilbert-optics methods. The visualization was performed on the experimental setup whose sketch is shown in Figure 1.The sequence of the Hilbert-images illustrates the evolution of complementary vortex rings. The difference of the serial numbers by unity corresponds to the interval between the frames equal to 1.3 ms. Each frame is a rectangular area 52 mm wide and 330 mm high.

The vortex ring induced by the positive pressure pulse inside the chamber appears with a delay relative to the external ring; this delay approximately corresponds to the pressure pulse duration.

As is evident from Figure 3*a*, the velocity of propagation of the ring outside the chamber is greater than the velocity of propagation of the complementary ring inside the chamber. Figure 3*b* shows the vortex rings induced by the same positive pressure pulse on the hole 24 mm in diameter. As is seen in Figures 3*a* and 3*b*, the velocity of propagation of vortex rings induced on the hole 24 mm in diameter is higher.

For example, Figure 4*a* shows the evolution of vortex rings induced on the hole by the negative pressure pulse presented in Figure 2*b*.

The difference of the serial numbers of the frames by unity corresponds to the interval between the frames equal to 1.3 ms. A negative pressure pulse induces the external vortex ring with the certain delay relative to the internal ring. For comparison, Figure 4*b* shows the evolution of vortex rings induced by the same negative pressure pulse on the hole 24 mm in diameter.

The developed methods of the optical Hilbert-filtration allowed us to detect the existence of structures in the form of vortex rings induced by the pressure pulse in the chamber with a nozzle. For example, Figure 5 shows the photographs of the

Hilbert images which illustrate the evolution of the vortex rings induced by the pressure pulse in the side the chamber with the cylindrical nozzle whose internal diameter is 20 mm and whose length is 70 mm. The shape of the pressure pulse is identical to that shown in Figure 2*b*. The structure of the vortex rings induced in air by the pressure pulse is clearly seen as an image of the phase optical density fields visualized by the methods of the Hilbert-optics.

The evolution of complementary vortex rings induced by a pressure pulse and propagating outside the chamber is depended from boundary conditions (diameter of the diaphragm) and the initial conditions (the amplitude, sign and shape of the pressure pulse).

Detection of vortex rings induced by the pressure jump inside the chamber with the nozzle illustrates the effective application of the developed methods of the optical Hilbert-diagnostics of the phase optical density fields in problems of experimental hydrodynamics and gas-dynamics.

3 CONCLUSION

The amplitude-phase structure of optical filters in systems of the Hilbert-diagnostics of phase optical density fields in gaseous and condensed systems was studied in this work. A possibility of visualization of disturbances in phase optical density fields with an arbitrary amplitude was demonstrated. The effectiveness of the methods of the optical Hilbert-diagnostics was illustrated by an example of the pioneering detection of the evolution of complementary vortex rings induced by the pressure pulse inside the chamber with the hole and the cylindrical nozzle.

ACKNOWLEDGMENTS

This work was supported by the Russian Foundation for Basic Research (RFBR), No. 14-08-00818.

REFERENCES

[1] Arbuzov V.A. & Dubnishchev Yu.N. 2007. Methods of the Hilbert Optics in Measurement Technologies Novosiirsk: Izd. Nov. Gos. Tech. Univ. [in Russian].
[2] Arbuzov V.A. & Dvornikov N.A. & DubnishchevYu.N. 2008. Detecting Opposite Vortex Rings Formated During Pressure Front Diffraction on a Hole. *Techn. Phys. Lett.* 34 (5), 394–396.
[3] Belozerov A.F. 2007. Optical Methods of Gas Flow Visualization. Kazan': Izd. Kazan. Gos. Tekh. Univ. [in Russian].

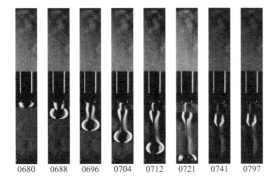

0680   0688   0696   0704   0712   0721   0741   0797

Figure 5. Vortices induced by the negative pressure pulse on the input hole of a cylindrical nozzle 20 mm in diameter and 70 mm long.

[4] Dubnishchev Yu.N. & V.A. Arbuzov, & P.P. Belousov, & P.Ya. Belousov 2003.Optical Methods of Flow Investigation. Novosibirsk: Sib. Univ. Izd., [in Russian].

[5] Dubnishchev Yu.N. & Chugui Yu.V. & Kompenhans Ju. 2009. Laser Doppler Visualization of the Velocity Field with Elimination of the Influence of Multiparticle Scattering Quant. Elektron. 39 (10), 962–966.

[6] Dubnishchev Yu.N. 2011. Theory and Transformation of Signals in Optical Systems. St. Petersburg: Izd. Lan', [in Russian].

[7] Dubnishchev Yu.N. & Dvornikov N.A. & Nechaev V.G. et al. 2012. *Polychromatic Hilbert Diagnostics of the Evoluyion of* Vortex Ring Induced by a Pressure Jump

on a Hole. *Optoelectron., Instrum. Data Process.* 48 (3), 227–234.

[8] Eu J.K.T. & Lohmann A.V. 1973 Isotropic Hilbert Spatial Filtering, *Opt. Commun.* 9 (3), 257–262.

[9] Raffel M & Willert C.T. & Wereley S.T. & Kompenhans Ju. 2007. Particle Image Velocimetry. A Practical Guide:Berlin, Springer.

[10] Settles G.S. 2001. Schlieren and Shadowgraph Techniques: Visializing Phenomena in Transparent Media: Springer, 376p.

[11] Soroko L.M. 1981 Hilbert Optics (Moscow: Nauka) [in Russian].

[12] www. ckb-proton.

Testing and Measurement: Techniques and Applications – Chan (Ed.)
© *2015 Taylor & Francis Group, London, ISBN: 978-1-138-02812-8*

Simultaneous measurements of two optical fiber lengths using chaotic ring laser with optical feedback

N. Fang, H.J. Qin, L.T. Wang & Z.M. Huang
Key Laboratory of Specialty Fiber Optics and Optical Access Networks, School of Communication and Information Engineering, Shanghai University, Shanghai, China

ABSTRACT: A method of simultaneous measurements of two optical fiber lengths is proposed. A semiconductor optical amplifier based fiber ring laser with an external optical feedback was constituted which can work in chaotic state. The short fiber was measured as a part of the fiber ring, and the long one as optical feedback. By using two different time intervals of autocorrelation peaks or two different frequency intervals of resonant peaks in the spectrum function of the chaotic signal, the two fiber lengths were simultaneously obtained in a measurement process. The experimental results verify that the proposed method is fast and efficient. The measurement range free of the dead zone is from less than 1 m to a few tens of kilometers, based on the sampling rate of 1GS/s and the gain of the optical amplifier used. The length measurement with autocorrelation method is better than that with the spectrum analysis method in the measurement stability.

1 INTRODUCTION

The measurement of physical lengths, involving optical fiber length, is needed in laboratories and industry applications. There exist some typical optical systems for fiber length measurements, such as optical time domain reflectometry (OTDR) (Barnoski et al. 1977), optical frequency domain reflectometry (OFDR) (Eickhoff et al. 1981) and optical low-coherence reflectometry (OLCR) (Gottesman et al. 2002). However OTDR has the low precision, unavoidable inherent error and no capacity for measuring short optical fiber. OFDR and OLCR are expensive in system configurations and low in the measurement stability. Moreover the available measuring range of OLCR is very limited, only within a few centimeters. Other optical fiber length measurement methods or systems also have been reported, for example, an all-fiber interferometer (Jia et al. 2002), which has no dead zone; nevertheless, the measuring precision is low. Another method by using a mode-locked fiber laser configuration (Hu et al. 2007) has very high measurement resolution. However, it cannot directly measure the fiber length shorter than 500 m. The measurement system based on a phase modulation optical link (Ye et al. 2013) is only applicable in the measurements of the short-to-medium length fibers. A semiconductor optical amplifier (SOA) based fiber ring laser can be used to measure the length of optical fiber within the fiber ring (Fang et al. 2014). All the systems mentioned above, however, only can obtain one optical fiber length in a measurement process.

In this paper, we propose a new fiber optic measurement system which uses a SOA-based fiber ring laser with a modified structure. This system employs two optical fibers with different lengths to constitute two optical resonant cavities: an inherent ring cavity and an external cavity as feedback. Thus this laser system can simultaneously measure the lengths of two optical fibers. Usually, a short fiber to be measured is inserted into the ring as a part of the ring cavity, while another long fiber to be measured is used as the feedback fiber in the external cavity. By calculating the autocorrelation or the frequency spectrum of detection signals, the two optical fiber lengths can be obtained simultaneously in a measurement process. In the following parts, the structure of the proposed measurement system will be introduced and the experimental results are demonstrated.

2 STRUCTURE OF MEASUREMENT SYSTEM AND OUTPUT WAVEFORM

The structure of the proposed system for fiber length measurements is shown in Figure 1, which is based on the fiber ring laser with an optical feedback. The fiber laser is composed of a SOA, a polarization controller (PC), a fiber coupler (FC), and a single mode optical fiber (Fiber 1). The SOA is the main nonlinear device. The PC controls the polarization states of the light travelling in the ring. Some light outputs through the FC. The external feedback is formed by using a single mode optical fiber (Fiber 2) with a mirror, at another end connected with the FC. The mirror can be a normal reflection mirror or a Faraday rotation mirror (FRM). By adjusting the PC and the driving current of the SOA, the laser system can be operated in a chaotic state. The output chaotic light is converted to an electrical chaotic signal by the photodetector (PD). Through the data acquiring and signal processing, two optical fiber lengths can be obtained. The whole system is consisted of the measured fibers (Fiber 1 and Fiber 2) and other inherent parts contained in the red dash line rectangle of Figure 1.

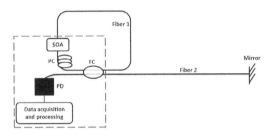

Figure 1. Structure of the measurement system based on the fiber ring laser with an external optical feedback.

Figure 2 presents the chaotic waveform output from the measurement system. The left inlet is a zoomed part of the waveform. The right inlet is the frequency spectrum of the left one. From Figure 2 and the left inlet, we can see that the waveform is like a stochastic signal. The spectrum in the right inlet is wide-band and continuous one. These characteristics demonstrate that the output signal waveform is a chaotic one.

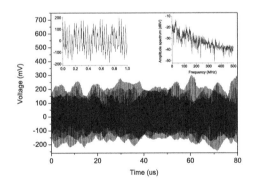

Figure 2. Chaotic waveform output from the measurement system.

3 MEASUREMENT METHODS

3.1 Autocorrelation method

The measurement principles are similar to the one reported in 2014 by Fang (Fang et al. 2014). Unlike the former autocorrelation function, every autocorrelation peak is not a single peak, but a cluster of multiple peaks with small and equal time intervals. Figure 3 shows the autocorrelation function curve of the measured chaotic signal from the ring laser with the optical feedback. The lengths of Fiber 1 and Fiber 2 are about 50 m and 6 km, respectively. In Figure 3 (a), the multiple equally spaced autocorrelation peak clusters can be seen clearly. The time interval τ_L is equal to the time that the light propagates twice in Fiber 2. Hence the length of Fiber 2, L can be obtained with $L = \tau_L / 2 \times v$, here v is the light speed in the fiber.

By expanding the autocorrelation peak cluster at zero delay time, as shown in Figure 3 (b), we can see that it actually consists of multiple autocorrelation peaks with small and equal time intervals. Here, the small time interval τ_l corresponds to the frame length of the system output waveform (Fang et al. 2011), also to the fiber ring length with a relation $\tau_l \times v$ (Fang et al. 2014). The length l of Fiber 1 can be obtained by subtracting the inherent length l_h from the total fiber ring length, which is measured previously (Fang et al. 2014). That is, $l = \tau_l \times v - l_h$.

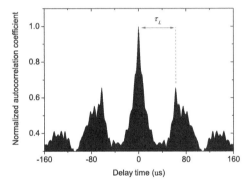

(a) Whole autocorrelation function curve.

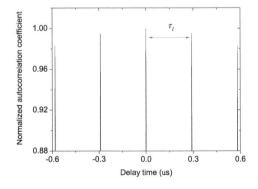

(b) The zoomed autocorrelation peak cluster at zero delay time.

Figure 3. Autocorrelation function of the chaotic signal from the fiber ring laser with the optical feedback.

3.2 *Spectrum analysis method*

For a chaotic laser with optical feedback, the time-delay signature is not only embodied in the autocorrelation function but also in the spectrum function of the chaotic waveform (Guo et al. 2012). We made a Fourier transform of the collected signal data and the obtained frequency spectrum is shown in Figure 4. From Figure 4 (a) we can clearly see multiple equally spaced resonance peak clusters. The frequency interval Δf_1 is equal to the resonant frequency of the fiber ring which is a reciprocal of the frame length. Therefore we can calculate the fiber ring length with a relation $v / D\Delta f_1$ and finally obtain the length of Fiber 1 with $l = v / \Delta f_1 - l_h$.

In the same way, by expanding the biggest resonance peak cluster, we can see that it actually consists of multiple resonance peaks with small and equal frequency intervals as shown in Figure 4 (b). The small frequency interval Δf_2 corresponds to the reciprocal of the time that the light propagates twice in Fiber 2. Hence the length of Fiber 2 can be determined by a relation, $L = v / (2\Delta f_2)$.

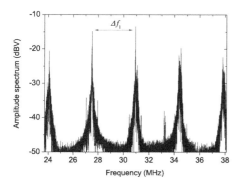

(a) Whole frequency spectrum.

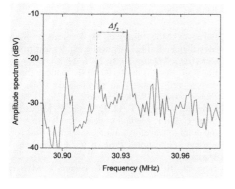

(b) The zoomed biggest resonant peak cluster.

Figure 4. Frequency spectrum of the chaotic signal from the ring laser with the optical feedback.

4 EXPERIMENTAL RESULTS AND DISCUSSIONS

According to Figure 1, a SOA-based fiber ring laser with an external optical feedback was constructed to measure the fiber lengths. The SOA used is a SOA-S-C-14-FCA module from CIP Technologies Corporation. The PC is an optical fiber squeezer PLC-001 of General Photonics Corporation. The coupling ratio of the FC is 50:50. The bandwidth of the PD is 1.25 GHz. The fiber-type mirror is a normal one made by ourselves. The output of the PD was input to a digital oscilloscope (Picoscope 5203) which was controlled by a computer via the USB port. The sampling rate was set at 1 GS/s.

At first, the inherent length l_h of the fiber ring was measured previously. It is about 9.4 m. Then we inserted Fiber 1 and Fiber 2 with different lengths into the system for length measurements. Two measured results on Fiber 1 and Fiber 2 are shown in Figure 5 and Figure 6, respectively. By setting the length of Fiber 1 at 1 m, we can measure Fiber 2 length from 1 km to 30 km. The measurement results of Fiber 2 length obtained with the proposed method and with OTDR, respectively, are shown in Figure 5 as a comparison. The red solid cycles represent the results obtained with the autocorrelation method, and blue hollow cycles are ones obtained with the spectrum analysis method. The maximum relative error compared with OTDR is less than 2% for the autocorrelation method, 6.3% for the spectrum analysis method. It is clear that the average length measured with the autocorrelation method is more accurate and stable than the spectrum analysis method. Figure 5 also gives the results on Fiber 1 by the proposed method in the same measurement process. Red solid triangles correspond to the autocorrelation method, and blue hollow triangles to the spectrum analysis method. The black solid squares are results of the length of Fiber 1 obtained by using a ruler, here as a reference length. The lengths obtained with the spectrum analysis method are closer to the reference one than that with the autocorrelation method. However, the measurements are influenced easily by the feedback optical fiber. Moreover, the lengths of Fiber 1 and Fiber 2 measured with two methods are both less than the reference lengths obtained with OTDR as well as with the ruler.

Next we set the length of Fiber 2 at about 6 km and measured the length of Fiber 1 from 1 m to 300 m. Figure 6 shows the measurement results. The meanings of the symbols in Figure 6 are basically the same as those shown in Figure 5. Except at the 1 m, the relative errors of the two proposed measurement methods, compared with OTDR, are nearly the same. Similarly, Figure 6 also gives the measured results of the length of Fiber 2 by the proposed methods in the same measurement process. The black solid square is the measurement result of Fiber 2 with OTDR, as a reference. In the condition of using a short fiber as Fiber 1, for example, less than 100 m in length, the measured lengths of Fiber 2 with two methods, are both less than the reference lengths. However, when the Fiber 1 was set at middle length long, for example, at 200 m or 300 m, the measured lengths of Fiber 2 with two methods, are both larger than the reference lengths. The measurement results of Fiber 2 are different from the different fiber ring length or using Fiber 1 with different length. Therefore there is an optimal ring length. For example, Fiber 1 of 50 m long is an optimal length for a length measurement of Fiber 2 with a length around 6 km.

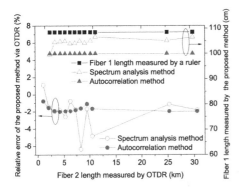

Figure 5. Measurement results as Fiber 1 was selected at 1 m.

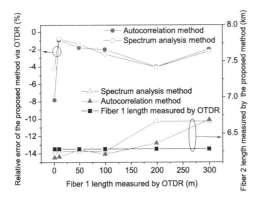

Figure 6. Measurement results in the condition of fixing the length of Fiber 2 at 6 km.

The measurement errors mainly arose from the insufficient sampling rate and the data recording length. The chaotic signal is wide-band and the fluctuations in intensity are higher than 1.25 GHz, however the maximum sampling rate of the measurement system available only is 1 GS/s. The maximum data length recordable is 1,000,005 at a sampling rate of 1 GS/s, recorded by the digital oscilloscope used in our measurements. On the other hand, the birefringence in the fiber also is one error source which will reduce the accuracy in the measurement of the long-scale optical fiber.

5 CONCLUSIONS

We proposed a novel method for simultaneously measuring two optical fiber lengths in an optical measurement process. This system is based on a chaotic ring laser with an external optical feedback. By utilizing two different time intervals of autocorrelation peaks or two different frequency intervals of resonant peaks in the spectrum function of the chaotic signal, the two fiber lengths can be simultaneously obtained in a measurement process. The experimental results verify that the proposed method is fast and efficient. The measurement range free of the dead zone is changeable from less than 1 m to a few tens of kilometers, depending on the gain in the chaotic ring laser as well as the sampling rate available in the data acquisition. The autocorrelation method is better than the spectrum analysis method in respect to the measurement stability. In addition, in measurements, for getting an accurate result, the shorter fiber should be adopted as a part of the optical fiber ring, while the longer one is taken as an external feedback fiber.

ACKNOWLEDGMENTS

This work is supported by the National Natural Science Foundation of China (Grant No. 61108004), the Science and Technology Commission of Shanghai Municipality (Grant No. 11ZR1413300) and Shanghai Leading Academic Discipline Project (S30108).

REFERENCES

[1] Barnoski, M.K., Rourke, M. D., Jensen, S. M. & Melville, R. T. 1977. Optical time domain reflectometer. *Applied Optics* 16(9): 2375–2379.
[2] Eickhoff, W. & Ulrich, R. 1981. Optical frequency-domain reflectometry in single-mode fiber. *Applied Physics Letters* 39(9): 693–695.
[3] Fang, N., Wang, L. T., Li, J., Huang, Z. M., Lu, M. M., Shan, C. 2011. Method to determine length and starting point of waveform frame from chaos time series in sensing applications. *Proc. ICICTA Shenzhen, 28–29 March 2011.*
[4] Fang, N., Ji W., Wang, L. T., Huang, Z. M. accepted on July 2014. Real-time measurement of optical fiber length with chaotic ring laser. *WIT Transactions on Engineering Sciences.*
[5] Gottesman, Y., Rao, E. V. K. Sillard, H. & Jacquet, J. 2002. Modeling of optical low coherence reflectometry

recorded Bragg reflectograms: evidence to a decisive role of Bragg spectral selectivity. *Journal of Lightwave Technology* 20(3): 489–493.

[6] Guo Y. Y., Wu Y. & Wang Y. C. 2012. Method to identify time delay of chaotic semiconductor laser with optical feedback. *Chinese Optical Letters* 10(6): 000000–1-000000–5.

[7] Hu, Y. L., Zhan, L., Zhang, Z. X., Luo, S. Y. & Xia, Y. X. 2007. High-resolution measurement of fiber length by using a mode-locked fiber laser configuration. *Optics Letters* 32(12): 1605–1607.

[8] Jia, B., Qian, S. R., Hua, Z. Y., Hu, L. & Ye, K. Z. 2002. Optic fiber length measurement using all fiber interferometer. *Chinese Journal of Lasers* A29(1): 73–75.

[9] Ye, Q. Y. & Yang, C. 2013. Fiber length measurement system based on phase modulation optical link. *Chinese Journal of Lasers* 40(5): 0505003–1-0505003–5.

Testing and Measurement: Techniques and Applications – Chan (Ed.)
© 2015 Taylor & Francis Group, London, ISBN: 978-1-138-02812-8

A high dynamic range all-fiber acoustic pressure sensing system

Z. Ye & C. Wang
Department of Materials Science, Fudan University, Shanghai, China

ABSTRACT: A new all-fiber system is introduced to realize the acoustic pressure sensing. Acoustic wave is detected by the photoelastic effect of fiber and demodulated by all-fiber interferometer. Phase Generated Carrier (PGC) techniques and phase retrieval algorithm are used in this sensor system to analyze the acoustic signals. Through the experiments of the sinusoidal signals, the system proved its broad dynamic range and high sensitivity.

1 INTRODUCTION

Nowadays, acoustic sensors have been widely applied in our world and played a very important role in military, health care and our daily life. There are three basic parameters in acoustic measurement: acoustic pressure, acoustic intensity and acoustic power. For the reason that acoustic intensity and acoustic power can be derived from acoustic pressure under the condition of free acoustic field, we can choose the acoustic pressure as the measurement goal.

The traditional acoustic sensor usually utilizes the acoustic pressure to make the film electrets in microphone vibrate, leading to the change of capacity, which would cause the corresponding small variation in voltage. Though amplification, AD converter and data collection, the small variation in voltage would be sent to the computer. Although this kind of acoustic sensor is simple and cheap, with the development of technology and society, the requirements for the properties of acoustic sensor to keep growing. The traditional acoustic sensor's sound-electronic feature dooms the limitation of applications and properties. In order to meet society's needs, the technology of fiber acoustic sensor emerged. Different from the traditional acoustic sensor, the fiber acoustic sensor can change the acoustic signals into optical signals. As a passive system, it has the advantages of low loss, anti-electric magnetic field interference, anti-corrosion, high security, flexibility, etc.

Since the first fiber sensor—endoscope appeared, In the late 1970s, the invention of low-loss optical fiber set up a new climate for the application and blossom of fiber technology. Until today, the technology of fiber sensor has kept developing for several decades. Fiber vibration sensor's history has a history of more than 30 years. The most common types include intensity-modulated, phase-modulated, polarization-modulated and fiber Bragg grating-modulated.

In this article, a new all-fiber acoustic pressure sensing system with high dynamic range and high sensitivity is presented.

2 THEORETICAL ANALYSIS

As a kind of mechanical wave, sound can cause tiny stress on optical fiber. Because of the photoelastic effect, the tiny stress would lead to the change of optical fiber refractive index and axial length, and cause the phase changes of light transferring in fiber: $\Delta\phi = \gamma\Delta l$. γ is constantly related to light property. So the phase change $\Delta\phi$ is in proportion to the axial strain Δl, making the conversion from acoustic pressure to optical properties.

To convert phase change into observable quantity, here we use the structure of single-mode interferometric fiber to transfer phase signal into the intensity signal. This kind of structure can guarantee the sensitivity and stability of the system. The optical structure is shown in Figure 1.

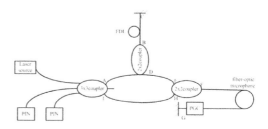

Figure 1. Single fiber distributed senor.

In the structure of the single fiber distributed sensor, the fiber microphone is made from the hundreds of meters of bare fiber winding into the coil form. The light beams follow the transmission path A→B→C→B→D→E→F→G→F→H→I and another transmission path I→H→F→G→F→E→D→B→C→B→A produce interference at the 3*3 coupler. Since the two light beams have the same length of the optical path, the produced interference term is $P = p_1 \cos\left[\phi_0 + \Delta\phi_1(t) + \Delta\phi_c(t)\right]$. p_1 is a constant related to the system, ϕ_0 is the initial phase, $\Delta\phi_1(t)$ is the interference phase difference caused by disturbance, $\Delta\phi_c(t)$ is the interference phase difference caused by PGC.

PGC is to make the measured signal bandwidth beyond a certain band into the large phase modulation so that the measured signal would be located in the sideband of modulation signal, which can ensure the interference signal always in in great condition and convert the effects of external disturbance into the effects to modulation signal. By separating the signal and random drift, we can remove the effects of random drift. Besides, the structure takes advantage of 3*3 coupler's property that the two output have a phase difference to realize the broad dynamic range. The combination of PGC and 3*3 coupler realizes the broad dynamic range and high sensitivity sensor.

On the base of traditional PGC technology, our system made the improvement that set the PGC between sensing optical fiber end and reflection mirror to subtract effects caused by parasitic interference of backscatter light because the backscatter light won't be modulated by PGC. If the carrier produced by PGC is sinusoidal signals with frequency f_m, then we can get:

$$P = p_1 \cos[\phi_0 + \nabla\phi_1 + \phi_m + \cos(2\pi f_m t)]$$
$$= p_1 \cos(\phi_0 + \nabla\phi_1) [J_0(\phi_m)$$
$$+ p_1 \sin[\phi_0 + \Delta\phi_1] 2[J_1(\phi_m) \cos(4\pi f_m t) + \dots]$$
$$2J_3(\phi_m) \cos(6\pi f_m t) + \dots]$$

The phase change caused by acoustic disturbance is on the sideband of carrier fundamental frequency and double frequency. But as the carrier has nothing to do with the interference caused by backscatter light, the sideband won't get affected. Using this method of modulation, we can extract pure phase information caused by acoustic disturbance to get more accurate information about the acoustic signal.

3 TECHNICAL PROPOSAL

The interferometric fiber acoustic sensor is as Figure2.

The laser source in this system is a super luminescent diode. The light comes out from SLD, goes

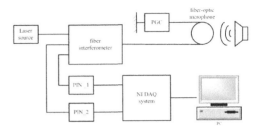

Figure 2.   Interferometric fiber acoustic sensor.

through FI(all-fiber interferometer) to arrive at microphone made by bare fiber, then gets modulated by PGC, reaches the reflection mirror. After that, the light come back to FI following the initial path and produce interference in FI. The light intensity at the output pin change with the phase difference. When the fiber microphone gets the acoustic signal, the photoelastic effect causes the phase changes of light transferring in fiber. The interference signals are exported to PIN_1 and PIN_2, through the photovoltaic conversion, then collected by NI-DAQ and be sent to PC for further data processing as Figure3.

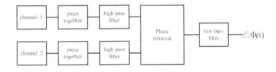

Figure 3.   Data processing.

The two signals channel_1 and channel_2 from PIN_1 and PIN_2 get piece together first. Since f_m is big(250kHz), all the desired signals distribute in the high frequency part. 100kHz 10order high-pass filter can filter the unmodulated parasitic interference signal or other noise signals. With the phase retrieval algorithm, we can get $\Delta\phi(t) = \Delta\phi_1(t) + \phi_m \cos(2\pi f_m t)$. Finally, the correlation and FFT present the power spectral information of the acoustic signal.

4 EXPERIMENTAL RESULTS

In the experiments, the sinusoidal signals are used to produce acoustic signals in order to make the detection of results comparison much more obvious.

Under the large signal, the waveform of channel_1 and channel_2 is as Figure 4. Phase difference can be observed in this figure.

The result of phase retrieval and power spectral information are shown in Figure 5. We can see that the frequency is about 1 kHz and the amplitude is about 18.6V. In the power spectral information, there is a peak at the frequency of 1 kHz. Besides, there are four troughs at the end of power spectral information caused by filters, but they have nothing to do with our analysis so that we can ignore them in analysis.

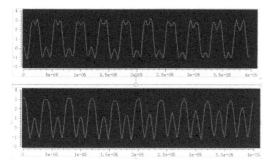

Figure 4. Waveform of channel_1 and channel_2.

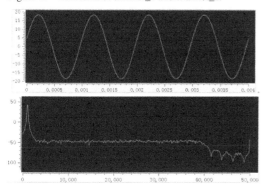

Figure 5. Phase retrieval and power spectral information.

Under the small signal, the waveform of channel_1 and channel_2 is as Figure6.

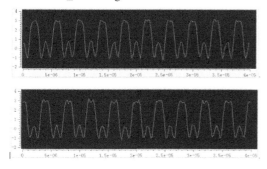

Figure 6. Waveform of channel_1 and channel_2.

The result of phase retrieval and power spectral information are shown in Figure 7. We can see that the frequency is about 1 kHz and the amplitude is about 0.1 V. In the power spectral information, there is a peak at the frequency of 1 kHz.

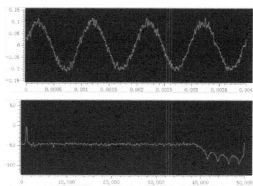

Figure 7. Phase retrieval and power spectral information.

5 CONCLUSION

With the improvement and innovation on the basis of traditional fiber sensors, the new all-fiber acoustic pressure sensing system can be much less affected by scattered light interference. Through experiments, the new sensing system can reach a greater dynamic range more than 45dB. Further work will be done to improve its properties in the future.

ACKNOWLEDGMENT

This work was supported in part by the National Key Support Program of China under Grant 2013BAK02B03 and the National Key Scientific Instrument and Equipment Development Project under Grant 2012YQ150213. (Corresponding author: C. Wang.)

REFERENCES

[1] Ke S, Bo J. A novel fiber audio transmission system for secure communication [J]. Journal of Optoelectronics Laser, 2005, 1(3).
[2] SUN Y, JIA B, ZHANG T. Position determination sensing system based on all-fiber interferometer with feedback loop [J][J]. Transducer and Microsystem Technologies, 2006, 1: 015.
[3] JIA Bo. Audio transmission system P. CN ZL 0Z Z83475 .3 Z 003-11-1Z.
[4] S.C. Huang et al. Fiber optic in-line distributed sensor for detection and localization of the pipeline leaks [J]. Sensors and Actuators A, 2007, 135:570–579.
[5] Qian X, Bo J. The characteristic phase generated carrier technology for the single-core fiber optic sensing [A]. Chinese Journal of Scientific Instrument.2014, 35(1): 36–42.
[6] Lee B. Review of the present status of optical fiber sensors [J]. Optical Fiber Technology, 2003, 9(2):57–79.
[7] Culshaw B, Kersey A. Fiber-Optic Sensing: A Historieal Perspective [J]. Journal of Lightwave Technology, 2008, 26(9):1064–1078.

Testing and Measurement: Techniques and Applications – Chan (Ed.)
© 2015 Taylor & Francis Group, London, ISBN: 978-1-138-02812-8

Active hydrothermal and non-active massive sulfide mound investigation using a new multi-parameter chemical sensor

C.H. Han & G.H. Wu
Second Institute of Oceanography, State Oceanic Administration, Hangzhou, China

Y. Ye
Zhejiang University, Hangzhou, China

H.W. Qin
Hangzhou Dianzi University, Hangzhou, China

ABSTRACT: A new multi-parameter chemical sensor method to study the active hydrothermal mound of the seafloor as well as non-active massive sulfide mound was proposed. And this method has been used for investigation and study during the first two legs of the 26th COMRA cruise on the Carlsberg Ridge of the Northwest Indian Ocean and the TAG hydrothermal area. During 1st Leg, we found one hydrothermal field located at 3.5~3.8°N on the Carlsberg Ridge in which few investigation has been carried out for hydrothermal activities. The known TAG area was also investigated for comparing and evidencing the ready instruments, which is helpful to find a new method to survey the massive sulfide mound, and the chemical sensor method can provide the similar result as the Transient ElectroMagnetic and Electric Self-potential methods do.

KEYWORDS: Carlsberg Ridge, Indian Ocean, North Atlantic Ridge, Seafloor hydrothermal activity, New multi-parameter chemical sensor method.

1 INTRODUCTION

Investigation of active hydrothermal mound as well as non-active massive sulfide mound has been studied recently. And the survey methods and instruments were diversified and convenient to study the active hydrothermal field. However, there is still lack of in-situ detection method for the non-active massive sulfide mound. Even though Transient ElectroMagnetic (TEM) and Electric Self-potential (SP) methods are good, they both are laborious, and time and money consuming. We proposed a new multi-parameter chemical sensor method to study the seafloor active hydrothermal mound as well as non-active massive sulfide mound. This sensor integrates with Eh and H_2S concentration chemical electrodes together, and can find chemical change in water column caused by the active hydrothermal vent, even weak chemical abnormalities by non-active massive sulfide hydrothermal mound which MARP and CTD sometimes cannot detect.

2 GEOLOGICAL BACKGROUND

2.1 Carlsberg Ridge

Carlsberg Ridge (CR) in the northwest Indian Ocean is a typical slow-spreading ridge with a half-spreading rate of about 11–16 mm/a (Semp and Klein, 1995) that lies between Indian and Somalian plates. The CR begins at the Owen fracture zone near 10°N and extends to the Central Indian Ocean Ridge at 2°S. Its depth ranges from 4700 m at ridge bottom to 1800 m on the ridge crest. The northern part (10°–3°N) and southern part (3°N–2°S) of CR section are nearly the NNW trench (~345°), while the ridge segments of the southern part (78 km) are longer than the northern part (70 km)(Mercuriev et al., 1996; Mudholkar et al., 2002; Raju, 2008; Ray et al., 2012).

2.2 TAG hydrothermal area

TAG (short for Trans-Atlantic Geotraverse) hydrothermal field extends over an area of at least 5×5 km and is located between Kane and the Atlantis Fracture Zones at the base and slope of the eastern wall of the Mid-Atlantic Ridge at 26ºN. It has a long history that the massive sulfides started to form about 50-20 ka ago with high-temperature pulses every 5000-6000 years (Lalou et al., 1993).The active TAG mound is mainly the Black Smoker Complex which is located on a 10- to 15-m high, 20- to 30-m diameter cone located in the northwest of the center, sulfide deposit.

3 INSTRUMENTS AND METHODS

3.1 *Instruments*

The H_2S and Eh sensors (ECS, short for Electrochemical sensor) were integrated in one sensor tube and utilized for observing and surveying hydrothermal diffusion flow field surrounding the seafloor active hydrothermal mound as well as non-active massive sulfide mound. All the sensors were self-made in the lab. The H_2S electrode was made by using the fabrication and calibration method of Ye et al. (2008). Considering the H_2S electrode was inconvenient to be calibrated on board when it was bound with pH and Eh sensor together, we generally calibrated it in the lab before and after the sea trial. The Eh oxidation-reduction electrode was simply a clean platinum wire (Gillespie, 1920; Connelly et al., 2012) and it didn't need to be calibrated because we used the primitive electromotive force (EMF) data of the Eh electrode. And these data were relatively potential with respect to an Ag/AgCl reference electrode. Ag/AgCl electrodes were made with the silver chloride melt quenching method (Han et al., 2009). Both sensors were connected to an underwater multi-channel data acquisition controller (DAC) developed (Qin et al., 2012).

3.2 *Sensors exploration*

The ECS was carried out by Deep-sea Towed Camera System (DTCS) in the 1st Leg and carried by transient electromagnetic (TEM) and electric self-potential (SP) in the 2nd Leg. It collected continuously the chemical parameters usually every 5 s. We output data from the data logger or directly receive data from the optic or electrical cable. During the sensor exploration, a GPS (Kongsberg DPS132) was mounted on board, and its time was set to the output time by the DAC. Additionally, the DACs were synchronized to the Greenwich Mean Time (GMT) before ready-to-use.

4 RESULTS AND DISCUSSION

4.1 *Sensors exploration on the Carlsberg Ridge*

The RV Dayangyihao of 26th cruise was used to conduct a 5-day plume survey on the CR. During the 1st Leg, we deployed the DTCS and CTD sampler which carried the Miniature Autonomous Plume Recorder (MAPR), CTD and ECS to find the hydrothermal field. However, no turbidity and temperature anomalies were detected by MAPR and CTD. The ECS

showed an obvious negative Eh anomalies and positive H_2S concentration anomalies at 3.69°N/63.66°E and 3.66°N /63.8°E on the CR. The ECS investigation results of the survey lines of L1 and L4 on the CR were shown in Figs 1-2. Figure 1 showed a 15 min-long obvious negative Eh abnormality (amplitude 20~40 mV) and positive H_2S abnormality during survey line L1. And the similar chemical abnormality (EMF amplitude of Eh range 15~20 mV) lasted for more than 1 hour during survey line L4, which indicated that massive sulfide mound may exist around this site (Fig.2). While these abnormalities existed, the DTCS observed that it was merely distributed widely and uniformly sediment in the deep seafloor, occasionally fractured basalt. And no rock alteration or sulfide chimney was found. Consequently, the methane weak anomaly was also detected on the board from the bottom seawater column sample.

(a)

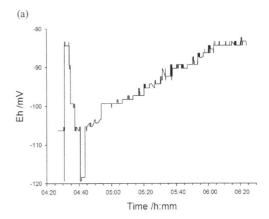

(b)

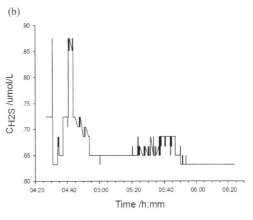

Figure 1. Detection results of the survey line L1 on CR (a) Eh; (b) H_2S.

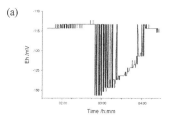

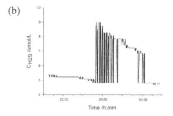

(a)

(b)

Figure 2. Detection results of the survey line L4 on CR. (a) Eh; (b) H₂S

Figure 3. Rock and sediment samples from TV-grab on the CR. (a) alteration basalt; (b) poly-metallic sediment deposit; (c) poly-metallic sediment deposit on board.

We also deployed the TV-grab to collect the rock and sediment samples in the vicinity of survey lines L1 and L4. The result evidenced chemical abnormalities Hydrothermal poly-metallic sediment deposit and the alteration basalt was successfully collected (Fig.3). And the surface of the basalt was overlaid with a thin liquid iron-manganese oxide layer, which initially concluded that this site was located in the periphery of the sulfide deposit.

Besides, chemical sensor also had detected the chemical abnormality at the seafloor in CTD03 station which was near the survey line L4. All results evidenced that the buried non-active sulfide mound existed in this field. The detailed survey line distribution and the results of the sulfide deposit field detected by the ECS on the CR were shown in Fig.4. We could see that the relative low and high value parts were distributed along each survey line

even though the absolute EMF value was not accurate due to the instability of the Ag/AgCl reference electrode. We used the relative EMF value to determine the changing trend during this survey process.

4.2 Sensors exploration on TAG hydrothermal field

Here we showed two survey lines L04TEM03 and L05TEM04 results on TAG field in the northern Atlantic Ocean Ridge, from which we can see two type chemical abnormalities (Figs.5-6). 30 umol/L H₂S positive abnormality and as high as nearly 50 mV Eh negative abnormality can be observed in survey line L04TEM03, while less than 5 umol/L H₂S positive and 5 mV Eh negative abnormalities were detected in survey line L05TEM04. As we have known, survey line L04TEM03 passed through a known active hydrothermal vent, and the massive sulfide deposits were collected near this site by TV-grab. Therefore, the abnormalities were larger than the results of survey line L05TEM04 where the altered basalt was sampled.

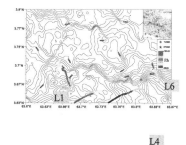

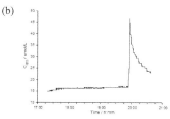

Figure 4. Distribution of ECS survey line, TV-grab and CTD sampler located on the CR.

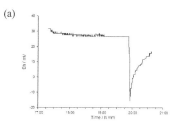

(a)

(b)

Figure 5. Detection results of the survey line L04TEM03 at the TAG field (a) Eh; (b) H₂S.

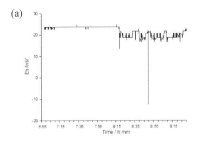

(a)

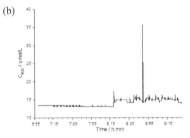

(b)

Figure 6.   Detection results of the survey line L05TEM04 at the TAG field (a) Eh; (b) H₂S.

5   CONCLUSION

During the 1ˢᵗ Leg of the Chinese 26ᵗʰ cruise, we found one hydrothermal field at 3.68°N/63.8°E on the CR by the multi-parameter chemical sensor during a 5-day survey. And the small negative Eh and positive H₂S concentration abnormalities were used to infer the surrounding area to be a non-active sulfide mound field and this was evidenced by the TV grab. During the 2ⁿᵈ Leg, this chemical sensor was carried out with TEM and SP survey system in the TAG field, and detected the chemical abnormalities which were matched very well with both TEM and SP survey analysis. The results showed that the multi-parameter chemical sensor method not only can detect active hydrothermal mound, but also can help find and study the non-active massive sulfide hydrothermal mound.

ACKNOWLEDGMENT

This study was supported by the National Basic Research Program of China under contract No. 2012CB417305; China Ocean Mineral Resources Research & Development Association Project under contract No. DY125-11-R-04; Endowment Fund of International Seabed Authority (the International Cooperative Study on Hydrothermal System at Ultraslow Spreading SWIR). And we are grateful to the captain and the crew members of RV Dayangyihao.

REFERRENCES

[1] Connelly, D.P. Jonathan, T. C. Bramley, J.M. et al. 2012. Hydrothermal vent fields and chemosynthetic biota on the world's deepest seafloor spreading centre. *Nature Communication* 3:620. doi: 10.1038/ncomms1636.

[2] Gillespie, L.J. 1920. Reduction potentials of bacterial cultures and of waterlogged soils. *Soil Science* 9: 199–216.

[3] Han, C. H. Pan, Y. W. Ye, Y. 2009. CO₂ microelectrode based on Zn-Al-LDH-ion carrier and its characterization. *Journal of Tropical Oceanography* 28(4): 35–41. (in Chinese).

[4] Herzig, P.M. Humphris, S.E. Miller, D.J. and Zierenberg, R.A. (Eds.), 1998. Proceedings of the Ocean Drilling Program, *Scientific Results* 158: 47–70.

[5] Lalou, C. Reyss, J.L. Brichet, E. et al. 1993. New age data for Mid-Atlantic Ridge hydrothermal sites: TAG and Snakepit geochronology revisited. *Journal of Geophysical Research* 98: 9705–9713.

[6] Mercuriev, S. Patriat, P. Sochevanova, N. 1996. Evolution de la dorsal de Carlsberg: évidence pour une phase d' expansion très lente entre 40 et 25 Ma (A18 à A7). *Oceanologica Acta* 19: 1–13.

[7] Mudholkar, A.V. Kodagali, V.N. Raju, K.A.K. et al. 2002. Geomorphological and petrological observations along a segment of slow spreading Carlsberg Ridge. *Current Science* 82(8): 982–989.

[8] Qin, H.W. and Ying, Q.W. 2012. Integration of chemical sensors chain and its practical application around the Kueishan Tao hydrothermal diffusion flow. *Applied Technology* 4: 171–172. (in Chinese).

[9] Raju, K.A.K. 2008. Recent cruise onboard R/V Sonne to the Carlsberg Ridge and the Andaman Sea. *InterRidge News* 17: 34–35.

[10] Ray, D. Raju, K.A.K. Baker, E.T. et al. 2012. Hydrothermal plumes over the Carlsberg Ridge, Indian Ocean. *Geochemistry Geophysics Geosystems* 13(null): Q1009.

[11] Semp, J.C and Klein, E.M. 1995. New insights in crustal accretion expected from Indian Ocean spreading centres. *Eos* 76(11): 113–116.

[12] Ye, Y, Huang, X. Pan, Y.W. et al. 2008. In-situ measurement of the dissolved S²⁻ in seafloor diffuse flow system: sensor preparation and calibration. *Journal of Zhejiang University: Science A* 9(3): 423–428.

Testing and Measurement: Techniques and Applications – Chan (Ed.)
© 2015 Taylor & Francis Group, London, ISBN: 978-1-138-02812-8

The research based on the level four-axis laser diameter measuring instrument

Q.R. Tian
A.A. Kunming University of Science and Technology, Kunming, China

B.G. He
Yunnan Packaging and Printing Engineering Technology Center, Kunming, China

ABSTRACT: Considering the existing disadvantage of detection technology of the filter rod circle, we apply the Laser scanning method to measure the diameter in order to realize the high accuracy of the filter rod line circle. On the basis of statistics, we analyzed the off-line measurement data of 87 groups of samples of filter rod, which was chosen randomly, and we arranged diameter gauge in four directions preliminarily. And then we applied the one-way dividing measuring system to test the diameter measuring of different types of 97 filter rod which were different from the machine tool. Accordingly, we determined the specific layout angle. In the end, we manufactured the diameter measuring instrument and put it into production. The results show that the proposed scheme can realize the accuracy of ±0.1mm.

1 INTRODUCTION

With the development of the tobacco industry, the competition is more and more intense, and the quality requirements of the cigarette are also getting higher. Filter rod is an important part of cigarette, and the circumferential size directly affects the quality of cigarette tipping. The circular of filter rod is too large or too small, which will lead to leakage, off the mouth of cigarettes, bad appearance and other major quality defects. This will cause a lot of harm. It not only damages the product image of enterprise and the interests of consumers seriously, but also increases the consumption of materials. Therefore, ensuring the stability of filter rod circle size is the key to improve the quality of cigarette products.

The current measurement methods of filter rod circular are mainly online pneumatic testing methods and off-line integrated test bench detection method.

The change in air pressure caused by the circumferential variation of nozzle in the filter rod, which is measured through the way of online pressure detection is very weak, and the detection signal is susceptible to interference. The relationship between the pressure and the circumference of the change is nonlinear, and the gas pressure fluctuations provided by the site are great[1], and low detection accuracy and poor stability cannot meet the requirement of production. And its measurement nozzle is easily set dust and glue scale to cause inaccurate measurement, and the downtime to maintain measuring tube frequently can lead to low efficiency of equipment operation and the great consumption of raw materials, which can constraint the improvement of the stability of the filter rod circle seriously. Because of many existing defects, the factory doesn't use the air pressure detection device in the production. In the actual production, the factory mainly relies on the experienced operator to estimate the circumferential size of filter rod by the amount of smoke plate. This detection method has the problem of large labor intensity and low accuracy.

Offline detection way is to extract a certain number of filter rod by manual or automatic sampling device and to measure the circumference in use of filter rod integrated test bench. According to the off-line measurement data, adjust the diameter motor left-right rotate artificially, and reduce or increase the height of Cooling bar to increase or decrease the circumference[2]. The device extracts samples to detect automatically after a period of time, and it cannot reach online real-time monitoring level of the filter rod circumference. The time at which the detected data is fed back to the equipment and the operator is lag. And this offline detection method has very limited sampling number, and the measurement of filter rod is through the part of the data to determine the overall circumference. It is a rough estimation of measurement and cannot fully reflect the real situation of the on-line production of the filter rod circle.

Based on the defects of the existing technology, this paper presents a planar four axis laser on-line measuring method. It can realize online real-time and high precision detection of the filter rod circle.

2 THE DIFFICULTIES OF THE STUDY

1 The standard circle of filter rod is 24mm, and the diameter is 7.64mm. The circumference is limited in 24±0.1mm in production. So the sensor we choose must comply with the measurement range of filter rod circumference and the measurement accuracy requirements.

2 KDF2/3 filter rod forming machine is the core equipment of the existing filter rod production, and its technical level and ability is the latest. KDF2/3 filter rod forming machine's structure and system is very complete and compact. Under the conditions that the existing functions of the equipment cannot be affected, its structure is analyzed. The detection space of filter rod is extremely limited (only 270mm*250mm*70mm), along the length direction of the filter rod with only 70mm. This will limit the installation dimension of measuring device, so we can only choose the appropriate volume of the sensor.

3 The filter rod is measured and it is a deformation and irregular shape flexible material, and the forming paper and external shape of a filter rod cannot be damaged in the detection process, so we need to choose a non-contact measurement to avoid measuring force interfering with filter rod.

4 The forward speed of filter rod is 300m/min and 600m/min two types. The speed is so high, so we can only use the dynamic measurement method. This will limit the scanning speed of the sensor.

5 The small jitter of filter stick, the vibration of equipment and other interfering factors will all cause fluctuation of circumferential precision, and this requires that the designed online detection device has high reliability and strong anti-interference ability to adapt to the scene dust, vibration, noise and other interference and meet the line production rhythm of filter rod.

These difficulties all need to be made a breakthrough.

3 THE RESEARCH CONTENT

3.1 *The reasons for choosing the laser scanning measuring method*

Based on the difficulty in this research, we choose the laser measuring method to achieve the non-contacting and long-distance measure by making use of the characteristics of high directivity, monochromaticity and brightness[3]. This method has the advantages of high speed, precision, measurement range, and good ability to resist interference[4]. As the tobacco industry speeds up the pace of "reduce harm reducing tar", the high transparency and composite filter rods are widely used in cigarette manufacturing processes, and the non-contacting measure of laser measuring method will not be affected by the air permeability of filter rod, so it can be used in all kinds of filter rod circumference detection.

At present, the laser sensor used in diameter measurement can be divided into laser scanning caliper sensor, CCD caliper sensor and laser diffraction caliper sensor. The previous two sensors use the geometrical principle of light, and the last one depends on the wave theory of light. By analyzing the theories of all the major methods, we could get to know that the laser diffraction caliper method is not very suitable, and CCD caliper sensor is too complex and expensive[5] to realize. Though laser scanning caliper sensor has system error because of the unstable scanning speed, it can be reduced by optimizing the algorithm. So we choose the laser scanning caliper method as the final detection method to measure the filter rod.

3.2 *Theoretical study of the four direction arrangement*

We have done off-line circumference test to a random sample of 50 filter rods of specific production equipment using one-way laser measuring diameter instrument and bidirectional laser diameter. We find that a group or two groups of diameter values that are gained exist a great chance and has a big difference between the standard value and the measured value. It is difficult to meet the requirement of testing accuracy.

Then we select offline measurement data of 87 groups of filter rod samples to analyze from the point of view of the data itself and overall judgment[6]. We find that the average value of four points and the whole value is the closest. The concrete analysis is as follows:

1 The data point

At present, KC instrument and QMU instrument are just an approximate estimation of measurement on off-line measurements of the filter rod circumference. In theory, the same filter rod circumference measurement value on 0° and 180° should be equal because of their symmetry, but the actual data show that there are some differences. There are some methods to find the rules of data, such as the mean, standard deviation and skewness. Select 4 sets of data in the 87 groups of data, and the error is relatively small. At the ±0.05 range, the 4 groups of measured data values are the closest to the measured value of 87 groups. If you take 7 points, the data values will be more precise.

2 The overall sample judgment

Only the 4 points of circumferential data can determine the qualified rate of a batch of product by hypothesis testing. The 4 points must be vertical and sharing each other. The vertical point detection can be more precise, Sharing can reduce the error. And we find that the 4-point and 9-point detection error is the same through the analysis of measurement data. So by comparing the value, the set value and the error of measurement data can judge the qualified rate of a batch of product. The probability of error is 6/10000.

So we determine the laser diameter measuring instrument with four vertical layouts. They are arranged in the same plane. It is conducive to the structural design of measuring diameter instrument and data processing.

3.3 Experimental study on layout angle

We apply the one-way dividing measuring system to test the diameter measuring of different types of 97 filter rod, which were different from machine tool in order to determine the specific layout angle. We measure a diameter value interval 1° for each filter rod, get a total of 180 diameter values, calculate the average diameter and the circumference value of each filter rod and draw the deviation map of circular 180 dividing value and the average circumference value of each filter rod making use of Excel.

We analyze the experimental data and deviation map of the 97 root filter rod, and the results are as follows:

1 The average diameter and average circumference value of 15°, 60°, 105°, 150° four directions and 45°, 90°, 135°, 180° four directions of each filter rod have below ±0.1mm deviation with the average diameter and the average circumference of 180 degrees of each filter rod, which can meet the predetermined requirement.

2 We record the average circumferential deviation of 45°, 90°, 135°, 180°four directions and 180° of filter rod as circular deviation 1. The average circumferential deviation of 15°, 60°, 105°, 150° four directions and 180° of filter rod is recorded as circular deviation 2. We can see that the distribution of circular deviation 2 is more uniform than circular deviation 1 and the absolute value of the maximum and minimum deviation of circumference deviation 2 is smaller than circular deviation 1 from Fig. 1. Consequently, 15°, 60°, 105°, 150° four directions are more reasonable.

4 APPLICATION

We develop a planar four-axis laser measuring diameter instrument on the basis of the above theoretical and experimental research. We install the diameter gauget in the exit for KDF3 filter rod forming machine to detect filter rod diameter of on-line production. The standard circle value of the filter rod of the equipment in the production is 24.000mm. We compare 100 groups of circumference values read randomly with the 24.000mm and finds that the deviations are below ±0. 1mm. Table 1 lists part of the circumference value of a filter rod read randomly, and the unit is mm.

Table 1. Part of the filter rod circumference values.

24.015	23.974	23.982	23.955	24.083
24.063	24.058	24.003	23.988	23.992
24.027	24.009	24.011	23.931	23.940
24.013	24.055	23.932	23.929	23.931
24.058	23.989	23.975	24.014	23.987
23.968	24.081	24.030	24.027	24.015
24.027	24.033	23.997	24.009	24.012
24.009	23.977	23.951	23.942	23.990

5 CONCLUSION

The device can realize the high accuracy detection of online real-time of filter rod in production. The device can collect four directions diameter values of filter rod at the same section. It has high detection precision and convenient data processing. The scanning speed of the gauge is larger than >1200/sec, the measuring precision is ±(0.001+0.02%D) mm (D is the filter rod diameter), the measurement range is 0.100~10.000mm and the resolution is 0.001mm. It can meet the production requirements. The measurement precision of the device is not affected by the normal range of environmental temperature changes and vibration of the machine, and laser detection method is non-contact detection method, We don't worry about glue adhesion because filter rod doesn't contact with measuring element absolutely.

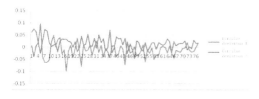

Figure 1. The contrast diagram circular deviation 1 and 2.

REFERENCES

[1] Huang, W.H. 2012. Comparative analysis of pressure type and optical type filter stick peripheral control system. Tobacco science and technology(12):26–29.
[2] Li, X.Q. & Li, Z.W. & Liu D.J. 2009. Design of Dual Axis Laser Scanning Diameter Measuring Gauge System with PID Co. Microcomputer Information(7):5–6.

[3] Gao, Y. 2007. Analysis on the development status and application prospect of laser. China science and technology information(19):270–271.

[4] Hu, Y. 2012. Non-contact detection system research and design of on-line roll. Xiamen: Xiamen University.

[5] Le, S.Y. 2010. The realization of the wire diameter measurement using CCD. Wuhan: Wuhan University.

[6] Zhao, Y.H. 2013. Mathematical statistics. Beijing: Science Press.

Testing and Measurement: Techniques and Applications – Chan (Ed.)
© 2015 Taylor & Francis Group, London, ISBN: 978-1-138-02812-8

A study on rhodamine 6G – based fiber-optic sensing system for detection of mercury ion in aqueous environment

A.R. Lee, B.Y. Jung, H.J. Han, Y.I. Kim, B.K. Kim & B.G. Park
Department of Energy & Environmental Engineering, College of Engineering, Soonchunhyang University, Asan, Republic of Korea

ABSTRACT: A fiber-optic sensor for the detection of mercury ions in an aqueous environment was studied with rhodamine derivatives as sensing materials, namely Rhodamine-Ethylenediamine (RE) and Rhodaminederived Schiff base (RS). The rhodamine derivatives were synthesized with rhodamine 6G, and characterized by nuclear magnetic resonance spectrometer. The absorption intensity of RE dissolved in water with various metal ions was measured by UV spectrometer. The experimental results showed that the absorption intensity was about two times higher in solution with mercury ion, than that in solutions with other metal ions. Experimental results performed with RS showed that it selectively reacted with mercury ions. In order to evaluate the compatibility of the RS as a thin film sensor, it was installed in a fiber-optic sensing system, which was composed of plastic optical fiber, a Y-coupler, and a spectrometer. The thin film to immobilize RS was fabricated by a sol-gel process, and was evaluated with the fiber-optic sensing system. The absorption spectra of RS-immobilized thin film were measured under different mercury ion concentrations from 0.3ppm to 100ppm in water solution. Peak intensities of RS-immobilized thin film and RS-dissolved water solution were observed at 530nm. The RS-immobilized thin film showed a linear correlation between the absorption intensity and mercury ion concentration. This result reveals that RS is a good sensing material for mercury ion, and could be utilized to develop fiber-optic sensors for monitoring mercury ion in an aqueous environment.

1 INTRODUCTION

Mercury is one of the most deleterious pollutants that is released through natural events or human activities. When exposed to the environment in a variety of forms, Mercury is diffused through the air, water, industrial, and food chains (Renzoni et al. 1998). When absorbed in the human body, it causes paralysis symptom and language disorder, and fatal damage to the brain, central nervous system, kidneys, and liver (Wang et al. 2012). Therefore, prompt detection for toxic mercury ion is a very important topic in various chemical and environmental systems, including living systems (Zhao et al. 2010).

Concerns about toxic exposures to mercury provide the motivation to explore new methods to monitor Hg^{2+} ions from environmental systems. Several methods, including atomic absorption spectroscopy and inductively coupled plasma atomic emission spectrometry, make it possible to detect low limits of Hg^{2+} ions (Butler et al. 2006) However, these methods require expensive equipment, and are time-intensive. Alternatively, various types of sensors have been developed for the detection of mercury in the environment. Mercury detection sensors based on various organic compounds have been reported (Wu et al.

2007). However, most of these sensors are irreversible, and cannot be used to monitor the variation of Hg^{2+} concentration in the environment at one time. Therefore, the development of new types of sensors to detect mercury is still required.

Many kinds of physicochemical sensors employing fiber-optics have been developed, and commercialized. Fiber-optic sensors have many advantages, such as small size, easy processing, and remote sensing. Also, fiber-optic sensors can offer sensitivity over previous technology, and have geometric versatility in arbitrary shapes (Yoo et al. 2011b, Patil et al. 2013). Recently, the development of fiber-optic sensors for the detection of environmentally important species has become of the major concern.

Noelting and Dziewonsky first reported Rhodamine derivatives for the detection of mercury in 1905 (Noelting et al. 1906). Rhodamine derivatives having high light-sensitive properties are often used to detect a variety of metal ions. Rhodamine derivatives are non-fluorescent and colorless; whereas, ring-opening of the corresponding spirolactam gives rise to strong fluorescence emission, and a pink color (Kim et al. 2008, Zhao et al. 2010). Rhodamine derivatives have been studied in various fields, such as their biological aspects (Kwon et al. 2010) and fluorescent chemical sensors (He et al. 2010), to detect various heavy metal ions in water using fluorescent properties; and optical sensors (Kim et al. 2008), to determine the concentration of analysis ions, through absorption spectrum measurement.

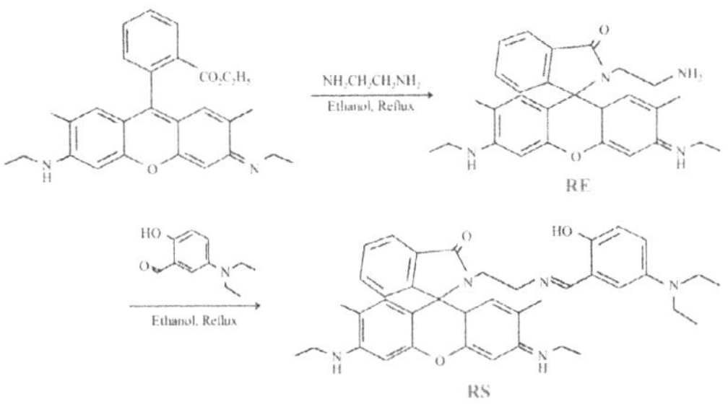

Figure 1. Synthesis procedure of RE and RS.

In this study, Rhodamine-Ethylenediamine (RE) and Rhodamine-derived Schiff base (RS) were synthesized with Rhodamine 6G, and their optical absorption properties were evaluated in water solution with various metal ions. Also, the possibility of RS as a fiber-optic sensor was investigated by fabrication of an RS-immobilized thin film, and by characterization of the absorption property of the thin film under different mercury ion concentrations in water solution.

2 EXPERIMENTAL

2.1 Apparatus and reagents

All reagents and solvents for the experiment were purchased from Aldrich, and used without further purification. Absorption spectra were measured by an Ocean Optics UV/Vis Spectrometer QE65000. A JEOL ECS-400 NMR spectrometer was used to analyze the structure of the synthetic products, for the sensing material of mercury ion in water solution.

2.2 Synthesis

N-(rhodamine 6G)-lactam ethylenediamine (RE) and Rhodamine-derived Schiff base (RS) were synthesized from Rhodamine 6G. The preparation of derivatives was based on the reported procedure (Wu et al. 2007, Quang et al. 2011). Figure 1 shows the molecular structural change according to the synthetic procedure of RE and RS. RE is synthesized from rhodamine 6G, and then RE is used to synthesize RS. The detailed procedures are as follows.

2.2.1 Synthesis of N-(Rhodamine 6G) lactam ethylenediamine (RE)

Rhodamine 6G (958mg, 2mmol) was dissolved in 20mL of Ethanol at 70°C, and then followed by the addition of ethylenediamine (0.67mL, 10mmol). A prepared solution was refluxed for 4 hours, until the fluorescence of the solution disappeared. After the solution without fluorescence was cooled to room temperature, the precipitate was collected, and washed three times with absolute ethanol. Crude product was purified by recrystallization in 20ml of acetonitrile, for about 10 minutes. After completion of this procedure, N-(Rhodamine 6G)-lactam ethylenediamine (RE) was obtained.

2.2.2 Synthesis of Rhodamine-derived Schiff base (RS)

Portions of N-(rhodamine-6G)-lactam ethylenediamine (456mg, 1.0mmol) and 4-diethylamino-salicylaldehyde (212 mg, 1.1mmol) were added to absolute ethanol (30mL). The solution was refluxed for 6hr under N_2 atmosphere at 60°C, and then stirred for another 2 hours at room temperature, to obtain precipitate. The precipitate in the solution was collected with filtration, and washed three times with ethanol. The crude product was purified by recrystallization in absolute ethanol 30mL, as the solution was stirred. The solid was collected with filtration. This procedure showed a 95.37% yield.

2.3 Fabrication of thin film

In order to measure the optical properties of the synthetic derivative, the derivative was immobilized by a sol-gel process. We prepared tetramethylorthosilicate (TMOS) as a silica precursor, for the synthesis of a colloid; trimethoxymethylsilane (MTMS) to prevent crack formation in the sol–gel film, and improve the chemical bonding between silica and the polymer; ethanol as a solvent, to uniformly mix the sol–gel materials, and distilled water (H_2O), to promote hydrolysis (Yoo et al. 2011a). The prepared Sol-Gel materials, which were composed of 24ml of TMOS, 6ml of MTMS, 7.5ml of deionized H_2O, 60ml of Ethanol, and 0.1g of synthetic derivative, were mixed together at 60°C for 2 hours. A thin film was prepared with a dip coater (E-Flex EF-4100), and then indurated at 80°C for 6 hours. After aging, the thin film was dried at room temperature for 6 hours.

2.4 Fiber-optic sensing system

To determine the reactivity of the synthetic derivatives RE and RS with the mercury ion in water solution, the same amounts of synthetic derivatives were added to water solutions with different metal ions, namely Hg²⁺, K⁺, Ca²⁺, Mn²⁺, Mg²⁺, Zn²⁺, Cd²⁺, Cu²⁺, Bi²⁺, and Li⁺. The absorption spectrum of each solution with a different metal ion was measured by UV-VIS spectrometer.

The absorption spectra of the derivative that was immobilized at the thin film was also measured by the fiber-optic sensing system, as Fig. 2 shows. The fiber-optic sensing system was composed of a light source, a mercury-sensing thin film, plastic optical fibers (GH 4001, Mitsubishi Rayon), a Y-coupler, and a spectrometer. The light emitted by the light source was guided by plastic optical fibers into the thin film in an aqueous solution, and the optical characteristic of reflected light was analyzed by spectrometer. The water solutions with different mercury ion concentrations were used to determine the relationship between absorption intensity and mercury ion concentration.

Figure 2. The set-up of the fiber-optic sensor system.

3 RESULT AND DISCUSSION

3.1 *Absorption property of the RE and the RS*

In the experiment, the same amount of synthetic derivative RE was added to the water solutions with different metal ions, namely Ca²⁺, K⁺, Li⁺, Mg²⁺, Zn²⁺, and Hg²⁺. The prepared solution was transferred into a quartz cell, and each individual absorption spectrum was then measured by UV-VIS spectrometer. Figure 3 shows the absorption spectrum of each solution with a different metal ion. The RE derivatives showed that absorbance of the solution with mercury was 2 times higher than that of the solutions with other metal ion.

The synthesized RS was characterized by NMR spectrometer, and the results were: ¹H NMR (CDCl₃) 7.94 (d, 1H), 7.72 (s, 1H), 7.46 (s, 2H), 7.26 (m, 1H), 7.05 (m, 1H), 6.87 (s, 2H), 6.36 (m, 3H), 3.48–3.45 (m, 4H), 3.37–3.22 (m, 4H), 3.15 (m, 4H), 1.87 (s, 6H), 1.60–1.34 (t, 6H), and 1.16 (t, 6H). FAB-MS (M+H⁺): m/z = 632. The results were similar to the NMR results of Quang et al. 2007.

The absorption spectra of the RS were obtained with the same procedure as the experiment with the RE. Figure 4 shows the experimental results. Higher absorbance of the RE for mercury

ion was observed. It is well known that the attached 4-diethylamino-salicylaldehyde of the RE provides additional binding groups to induce the ring-opening of spirolactam, upon complexation with Hg²⁺ ions (Wu et al. 2007, Quang et al. 2011). The experiment confirmed the selective reaction of RS with mercury ion in water solution.

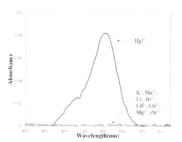

Figure 3. Results of the absorbance spectrum of Rhodamine-Ethylenediamine (RE) for various cation solutions (Hg²⁺, Ca²⁺, Mg²⁺, Mn²⁺, Zn²⁺, Li⁺, and K⁺).

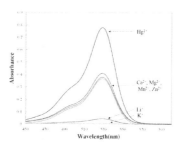

Figure 4. Absorption spectrum of Rhodamine-derived Schiff base (RS) dissolved in the water solutions with different metal ions of Hg²⁺, K⁺, Mn²⁺, Li⁺, Bi²⁺, Cd²⁺, Cu²⁺, Mg²⁺, and Zn²⁺.

3.2 *Relationship of the absorbance with the Hg²⁺ ion concentration*

Figure 5 shows the absorption spectra of the RS in the water with various Hg²⁺ ion concentrations, which is determined by UV-VIS spectrometer. The absorption peak of each concentration was observed at about 530nm. The addition of Hg²⁺ ion in the water solution resulted in the increase of absorbance. The experimental results show that the peak absorbance at about 530 nm is in linear correlation with the Hg²⁺ ion concentration (R²=0.9974), as Figure 5 shows.

The same experiment was conducted with the RS immobilized in thin film, using the fiber-optic sensing system. Figure 6 shows the absorption spectra of the RS immobilized in thin film. A peak absorbance was observed at about 531 nm, regardless of the Hg²⁺ ion concentration, and the peak absorbance at about 531 nm was in linear correlation with the

Hg^{2+} ion concentration, as Fig. 6 shows. This result indicates that the RS has the same optical property, regardless of the conditions in the water solution. However, the absorbance of the RS immobilized at the thin film is lower, than that of the RS dissolved in the water.

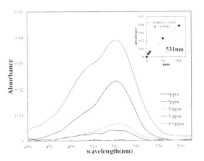

Figure 5. Absorption spectrum of Rhodamine-derived Schiff base (RS) dissolved in the water solutions with the various Hg^{2+} concentrations.

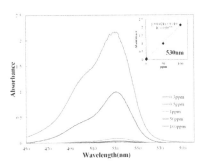

Figure 6. Absorption spectrum of Rhodamine-derived Schiff base (RS) immobilized within the thin film in the water solutions with the various Hg^{2+} concentrations.

4   CONCLUSIONS

Rhodamine-Ethylenediamine (RE) and Rhodamine-derived Schiff base (RS) were synthesized from rhodamine 6G. The RS was characterized by NMR spectrometer, which confirmed that the synthesis procedure of RS was adequate. The experimental results showed that the absorption intensity was about two times higher in solution with mercury ion, than that in solutions with other metal ions. The experimental results performed with RS showed that RS selectively reacted with mercury ion. The compatibility of the RS as a thin film sensor was evaluated by the fiber-optic sensing system. The absorption spectra of RS-immobilized thin film were measured under different mercury ion concentrations. Peak intensities of the RS-immobilized thin film and RS-dissolved water solution were observed at 530nm. RS-immobilized thin film showed a linear correlation between the absorption intensity and mercury ion concentration. This result shows that RS is a good sensing material for mercury ions, and could be utilized to develop fiber-optic sensors to monitor mercury ions in an aqueous environment.

REFERENCES

[1] Butler, O. T., Cook, J. M., Harrington C. F., Hill, S. J, Rieuwerts, J. and Miles, D. L. 2006. Atomic spectrometry update. Environmental analysis. *Journal of Analytical Atomic Spectrometry.* 21: 217–243.
[2] He, G., Zhang, X., He, C., Zhao, X., Duan, C. 2010. Ratiometric fluorescence chemosensors for copper (II) and mercury (II) based on FRET systems: *Tetrahedron.* 66: 9762–9768.
[3] Kim, H.N., Lee, M.H., Kim, H.J., Kim, J.S. and Yoon, J.Y. 2008. A new trend in rhodamine-based chemosensors: application of spirolactam ring-opening to sensing ions". *Chemical Society Reviews.* 37: 1465–1472.
[4] Kwon, P.S., Kim, J.K., Kim, J.W. 2010. Heavy metal Ion Detection in Living Cell Using Fluorescent Chemosensor. Journal of the Korean Chemical Society. 54(4): 451–459.
[5] Noeling, E. and Dziewonski, K. 1906. Zur Kenntniss der Rhodamine. European Journal of Inorganic Chemistry. 39(3): 2744–2749.
[6] Patil, S.S., Shaligram, A.D. 2013. Refractometric Fiber Optic Sensor for Detecting Salinity of Water. *Journal of Sensor Technology.* 3: 70–74.
[7] Renzoni, A., Zino, F. and Franchi, E. 1998. Mercury Levels along the Food Chain and Risk for Exposed Populations. *Environmental Research*, Section A. 77: 68–72.
[8] Wang F, Nam SW, Guo Z, Park S, Yoon, J. 2012. A new rhodamine derivative bearing benzothiazole and thio-carbonyl moieties as a highly selective fluorescent and colorimetric chemodosimeter for Hg^{2+}. *Sens Actuators B*. 161: 948–953.
[9] Wu, J.S., Hwang, I.C., Kim, K.S. and Kim, J.S. 2007. Rhodamine-Based Hg^{2+}-Selective Chemodosimeter in Aqueous Solution: Fluorescent OFF-ON. *Organic Letters.* 9(5): 907–910.
[10] Yoo, W.J., Jang, K.W., Seo, J.K., Moon, J., Han, K.T., Park, J.Y., Park, B.G., Lee, B. 2011a. Development of a 2-channel embedded infrared fiber-optic temperature sensor using silver halide optical fibers. *Sensors* 11(10): 9549–9559.
[11] Yoo, W.J., Seo, J.K., Jang, K.W., Heo, J.Y., Moon, J., Park, J.Y., Park, B.G., Lee, B. 2011b. Fabrication and comparison of thermochromic material-based fiber-optic sensors for monitoring the temperature of water. *Optical Review* 18(1): 144–148.
[12] Zhao, Y., Sun, Y., Lv, X., Liu, Y., Chen, M. and Guo, W. 2010. Rhodamine-based chemosensor for Hg2+ in aqueous solution with a broad pH range and it application in live cell imaging. *Organic & Biomolecular Chemistry.* 8: 4143–4147.

A new current-based technique for discriminating between internal faults and inrush current within power transformers

M. F. El-Naggar
Electrical Engineering Department, Salman bin Abdulaziz University, Saudi Arabia
Electrical Power and Machines Engineering Department, Helwan University, Cairo, Egypt

A. Abu-Siada
Department of Electrical and Computer Engineering, Curtin University, WA, Australia

A.M. Mahmoud
Electrical Power and Machines Engineering Department, Helwan University, Cairo, Egypt

Khaled M. Gad El Mola
Department of Mechanical Engineering, College of Engineering, Salman Bin Abdul-Aziz University, Saudi Arabia

ABSTRACT: In rush current may lead to a false operation of the power transformer differential protection relay. Hence, it is very essential to distinguish in rush current from various internal transformer faults in order to improve the reliability of the differential protection scheme. This paper presents a new current-based technique to discriminate transformer inrush current from transformer internal faults. The proposed technique relies on converting the three phase transformer currents on both sides into modal current components using Clarke's transformation to produce three current modes, namely, ground mode, aerial mode 1 and aerial mode 2. The features extracted from the three modes are then applied to the inrush discrimination equation proposed in this paper. The robustness of the proposed technique is evaluated through extensive simulation studies on a 132/15 KV, 155MVA power transformer using an Electromagnetic Transient Program (EMTP-ATP). Simulation results show the effectiveness of the new proposed technique in discriminating magnetizing in rush current from other transformer internal faults.

1 INTRODUCTION

Power transformers are vital links and one of the most critical and expensive assets in electrical power systems. The day-by-day increase in load demand, global trend to develop smart grids and the growing number of nonlinear loads such as smart appliances and electric vehicles will further increase the likelihood of non-sinusoidal and transient transformer operating conditions [1].

Therefore, it is essential to adopt simple, reliable and accurate transformer protection schemes that can precisely discriminate between internal transformer faults and normal operating condition in order to avoid any unwanted transformer outages due to mal-operation of its protection system.

Differential protective relays based on the comparison of transformer input and output currents are widely accepted as the principal protection scheme for power transformers.

However, power transformer differential protection exhibits some shortcomings, of which false relay tripping during transient magnetizing inrush current that flows in the energized winding during transient and switching operation is about the great concern [2, 3]. As transformer inrush current comprises a large second order harmonic component when compared to that of the typical fault current, conventional transformer protection systems are designed to restrain during inrush transient phenomenon by sensing this large second harmonic [4]. This technique is, however, has a slow relay operating speed [5]. Moreover, in certain cases the magnitude of the second harmonic component of some internal fault currents may be close to or greater than that is presented in the magnetizing in rush current waveform, leading to relay mal-operation.

Furthermore, because of the rapid improvement in the transformer core material, magnetizing inrush current second harmonic component tends to be relatively small. Consequently harmonic restraint relay based on second harmonic restraint is not a reliable discrimination tool for internal fault current and an inrush current [6, 7]. There are many papers in the literature investigating various inrush current identification techniques [8–16]. However, none of these proposed techniques has been widely adopted as an effective discrimination tool for power transformer magnetizing inrush current and internal fault current. This is mainly due to the slowness, complexity, requirement of substantial training data (for artificial intelligence-based approaches), and sensitivity to substation noise (in case of high frequency analysis-based methods) [7].

This paper presents a new current-based technique to discriminate transformer inrush current from transformer internal faults. The proposed technique relies on converting the three phase transformer currents into three modal current components using Clarke's transformation. The features extracted from the three modes are then applied to a proposed inrush discrimination equation as will be elaborated below.

2 PROPOSED TECHNIQUE

In this technique, the three phase transformer input and output currents are measured and converted to its modal currents using Clarke's transformation as below [17].

$$
\begin{bmatrix} I_0 \\ I_1 \\ I_2 \end{bmatrix} = \frac{1}{3} \begin{bmatrix} 1 & 1 & 1 \\ 2 & -1 & -1 \\ 0 & \sqrt{3} & -\sqrt{3} \end{bmatrix} \begin{bmatrix} I_a \\ I_b \\ I_c \end{bmatrix} \tag{1}
$$

Where Ia, Ib, and Ic represent the transformer input or output three phase currents, I0 is the ground mode current component and I1 and I2 are known as the aerial mode current components.

The ground mode current component I0 is defined as the zero sequence symmetrical component of the three phase currents while aerial mode current component I1 passes in phase a and returns back through phase b and phase c while aerial mode current component I2 is circulating in phases b and c [17]. For phase to phase faults and inrush current cases, the ground mode I0 is zero while it exceeds zero for phase to ground faults.

In this case, the level of the ground mode component can be used to differentiate between phase to ground faults and inrush current. To show the main idea of the proposed technique, the system shown in Figure 1 is simulated using electromagnetic transient program alternative transient program (EMTP-ATP).

The system consists of a synchronous generator connected to an infinite bus that is represented by a three phase source of constant frequency through two transmission lines and a 132/15 kV, 155 MVA, Ynd11 3-phase power transformer connection.

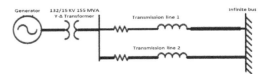

Figure 1. System under study.

Two fault internal scenarios namely, single line to ground and double line faults have been investigated and compared with the transformer magnetizing in rush current that is established due to switching on the transformer to the grid.

Figure 2(a) shows the three phase transformer input currentst during internal line to ground fault on phase a applied at t=0.04s and is assumed to 0.12s. As is shown in Figure 2(a), the current in phase a reaches a crest value of 8 pu after the fault application. Figure2(b) shows the modal current components I0, I1 and I2 calculated using Clarke's matrix applied to the three phase currents shown in Figure 2(a) which reveals a value of the ground mode exceeding zero.

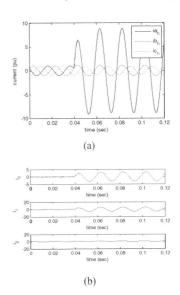

Figure 2. Phase a to ground fault; (a) Three phase currents, (b) Modal currents.

A double line short circuit fault to ground fault is simulated on phases (a and b). The three phase transformer input currents during this fault are shown in Figure 3(a) while the corresponding modal current components are shown in Figure 3(b).

Results in Figure 3 show that the current in the faulty phases has significantly increased and the ground modal current component is zero.

Figure 4 shows the transformer magnetizing inrush three phase currents and their corresponding modal current components of which the ground mode component is equal to zero.

Figures 2 through 4 show that the value of the ground mode I0can be used to differentiate between the internal transformer faults to ground and the inrush current.

However, phase to phase faults and inrush current cannot be distinguished through their ground mode components as it equals zero in both cases.

An inrush discrimination equation is proposed below to accurately differentiate between in rush current and non-ground fault conditions.

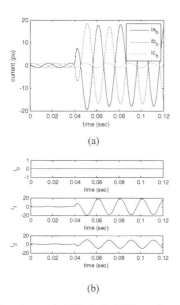

(a)

(b)

Figure 3. Phase a to b SC fault; (a) Three phase currents, (b) Modal currents.

The proposed inrush discrimination equation is based on the average of the absolute value of aerial modal currents I1 multiplied by aerial modal current I2 as given below:

Inrush Discrimination equation = average($|I1 * I2|$) (2)

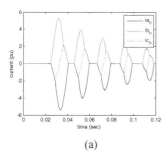

(a)

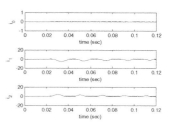

Figure 4. (a) Magnetizing inrush current and (b) Modal currents.

For inrush current, slope of the inrush discrimination equation is negative, whereas, it is constant for non-ground phase faults.

3 SIMULATION RESULTS

To evaluate the effectiveness of the proposed technique, various fault scenarios within the power transformer shown in Figure 1 are simulated and the corresponding modal current components are calculated, from which the slope of the inrush discrimination equation given in (2) can be obtained and compared with the inrush current.

3.1 *Response to internal double line fault on phases a and c*

An internal double line short circuit fault between phases a and c is simulated at t=0.04s. The three phase current and the corresponding modal current components are shown in Figure 5 (at) and 5 (b) respectively.

By applying the in rush discrimination equation, the slope shown in Figure 5(c) can be obtained. As is shown in Figure 5(c), the slope of the proposed discrimination equation is constant throughout the fault duration.

3.2 *Response to internal double line fault on phases a and b*

Figure 6 shows the three phase current, modal components and the slope of the proposed discrimination equation during such fault. Similar to case A, the slope of the proposed equation is constant during the double line fault.

3.3 *Response to internal double line fault on phases b and c*

For internal double line short circuit fault on phases b and c, the three phase current and the corresponding calculated modal current components and the slope

197

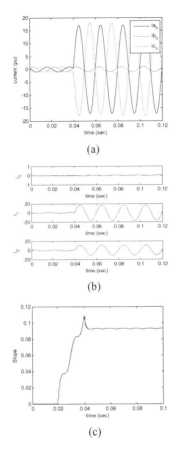

(a)

(b)

(c)

Figure 5. Phases a and c double line fault; (a) Three phase currents, (b) Modal currents and (c) Inrush discrimination equation output slope.

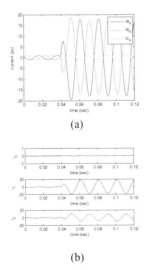

(a)

(b)

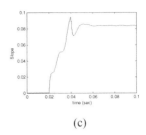

(c)

Figure 6. Phases a and b double line fault; (a) Three phase currents, (b) Modal currents and (c) Inrush discrimination equation output slope.

of the proposed discrimination equation are shown in Figure 7. Again, the slope of the proposed discrimination equation is constant during the internal line to line short circuit fault.

3.4 Response to internal 3-phase short circuit fault

For internal three phase short circuit fault the three phase fault currents and modal currents are shown in Figure 8(a) and (b), respectively. As no ground fault is involved, the slope of the inrush discrimination equation will be constant as shown in Figure 8(c).

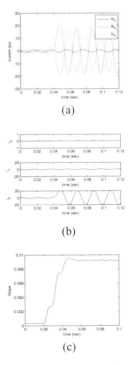

(a)

(b)

(c)

Figure 7. Phases a and b double line fault; (a) Three phase currents, (b) Modal currents and (c) Inrush discrimination equation output slope.

3.5 Response to inrush current

The three phase inrush currents due to energizing the transformer primary windings at no-load at t=0.02s are shown in Figure 9(a) with its modal currents and the output slope of the proposed inrush discrimination equation as shown in Figure 9(b) and 9(c), respectively. In contrary, with line to line faults, the slope of the proposed inrush discrimination equation is negative during the energization periods revealed from Figure 9(c).

3.6 Response to switching on fault

The proposed technique is assessed against switching on the transformer to an external short circuit fault as shown in Figure (10). The proposed technique reveals constant slope during the fault which indicates the effectiveness of the proposed technique in discriminating external fault conditions from inrush currents.

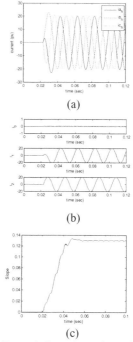

(b)

(c)

Figure 9. (a) Three phase inrush currents, (b) Modal currents and (c) Inrush discrimination equation output slope.

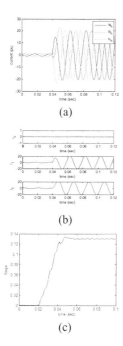

(a)

(b)

(c)

Figure 8. Three-phase short circuit fault; (a) Three phase currents, (b) Modal currents and (c) Inrush discrimination equation output slope.

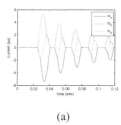

(a)

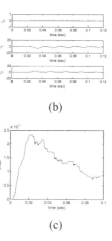

(a)

(b)

(c)

Figure 10. External three-phase short circuit fault; (a) Three phase currents, (b) Modal currents and (c) Inrush discrimination equation output slope.

4 CONCLUSION

This paper presents a new current-based technique to precisely distinguish between transformer magnetizing inrush currents and various transformer internal fault currents. The proposed technique is based on converting transformer three phase currents into modal current components namely, ground mode, a real mode 1 and a real mode 2 using Clarke's transformation. Simulation results show that the ground mode for short circuit to ground faults has a value greater than zero while it is equal to zero for phase to phase short circuit faults and for transformer inrush current. A real mode 1 and a real mode 2 are used to compose a proposed inrush discrimination equation to precisely distinguish inrush current from another line to ground faults. The slope of the proposed discrimination equation is found to be constant in case of line to ground short circuit faults and negative in case of inrush current. The proposed technique is simple, reliable and easy to implement within digital relays to avoid any false tripping due to inrush current that may lead to costly outage of power transformers.

ACKNOWLEDGMENT

This paper was supported by the deanship of scientific research at the Salman binAbdulaziz University under the research project no. 2014/01/734.

REFERENCES

[1] A. Abu-siada S. Islam, "A Novel Online Technique to Detect Power Transformer Winding Faults," Power Delivery, IEEE Transactions on, vol. 27, pp. 849–857, 2012.

[2] Zhiqian Bo, Weller and G. Lomas, "A New Technique for Transformer Protection Based on Transient Detection ", IEEE Transactions on power delivery, vol. 15, no. 3, July 2000.

[3] M. F. El-Naggar, A. M. Hamdy, S. M. Moussa, and E. H. Shehab El-Din, "A Novel Image-Based Approach forDiscrimination between Internal Faults andMagnetizing Inrush Currents in PowerTransformers", AUPEC, Perth, Western Australia, pp. 347–352, 2007.

[4] S Sudha, A. Ebenezer Jeyakumar, "Wavelet and ANN Based Relaying for Power Transformer Protection", Journal of Computer Science 3 (6): 454–460, June 2007.

[5] M. A. Rahman and R. Jayasurya, "A state-of-art review of transformerprotection algorithm," IEEE Transaction on Power Delivery, vol. 3, pp.534–544, April 1988.

[6] R. Hamilton, "Analysis of Transformer Inrush Current and Comparison of Harmonic Restraint Methods in Transformer Protection," Industry Applications, IEEE Transactions on, vol. 49, pp. 1890–1899, 2013.

[7] A. Hooshyar, M. Sanaye-Pasand, S. Afsharnia, M. Davarpanah, and B. M. Ebrahimi, "Time-Domain Analysis of Differential Power Signal to Detect Magnetizing Inrush in Power Transformers," Power Delivery, IEEE Transactions on, vol. 27, pp. 1394–1404, 2012.

[8] D. Q. Bi et al., "Correlation analysis of waveforms in nonsaturationzone-based method to identify the magnetizing inrush in transformer,"IEEE Trans. Power Del., vol. 22, no. 3, pp. 1380–1385, Jul. 2007.

[9] Á. L. Orille-Fernández, N. K. I. Ghonaim, and J. A. Valencia, "A FIRANNas a differential relay for three phase power transformer protection,"IEEE Trans. Power Del., vol. 16, no. 2, pp. 215–218, Apr. 2001.

[10] M. C. Shin, C. W. Park, and J. H. Kim, "Fuzzy logic-based relaying forlarge power transformer protection," IEEE Trans. Power Del., vol. 18,no. 3, pp. 718 –724, Jul. 2003.

[11] E. C. Segatto and D. V. Coury, "A differential relay for power transformerusing intelligent tools," IEEE Trans. Power Syst., vol. 21, no.3, pp. 1154–1162, Aug. 2006.

[12] S. P. Valsan and K. S. Swarup, "Protective relaying for power transformerusing field programmable gate array," IET Elect. Power Appl.,vol. 2, no. 2, pp. 135–143, Mar. 2008.

[13] S. A. Saleh and M. A. Rahman, "Testing of a wavelet-packet-transform-based differential protection for resistance-grounded three-phasetransformers," IEEE Trans. Ind. Appl., vol. 46, no. 3, pp. 1109–1117,May/Jun. 2010.

[14] P. L. Mao and R. K. Aggarwal, "A novel approach to the classificationof the transient phenomena in power transformers using combinedwavelet transform and neural network," IEEE Trans. Power Del., vol.16, no. 4, pp. 654–660, Oct. 2001.

[15] X. N. Lin, P. Liu, and O. P. Malik, "Studies for identification of theinrush based on improved correlation algorithm," IEEE Trans. PowerDel., vol. 17, no. 4, pp. 901–907, Oct. 2002.

[16] A.M. Mahmoud, M.F. El-Naggar, E.H. Shehab_Eldin, "A New Technique for Power Transformer Protection Basedon Transient Components", ", ICAEE, Thailand, pp. 318–324, 2011.

[17] A. M. Elhaffar, "Power Transmission Line Fault Location Based on Current Travelling Waves", Doctoral Dissertation, Helsinki University of Technology (Espoo, Finland), March 2008.

Testing and Measurement: Techniques and Applications – Chan (Ed.)
© 2015 Taylor & Francis Group, London, ISBN: 978-1-138-02812-8

Development of Test Access Port (TAP) design for IEEE 1149.1 standard improvements

S.A. Jayousi & M.S. Muhammad

Department of Electrical and Electronic Engineering, Universiti Malaysia Sarawak, Malaysia

ABSTRACT: IEEE 1149.1 standard drew the way to a modern Integrated Circuits (ICs) testing techniques with unique characteristics such as low cost, high speed, high reliability, accuracy and non-intrusive properties. Exploiting Test Access Port (TAP) in controlling the testing system increases the diagnostic efficiency, reliability and flexibility. TAP resources waive the tester design from the need of the most costly parts such as memory and microcontroller. In this paper, we present a new design of TAP that provides a means to arbitrarily observe test results and source test stimulus. TAP is to be built in printed circuit board (PCB) and shared between several ICs. The design requires minimal on chip/board resources (pins/nets). Also, it is not limited by chip function or complexity. Moreover, test access is not limited by the board physical factors and the test generation is highly automated.

KEYWORDS: IEEE 1149.1 standard, Integrated circuits (ICs) testing, Test access port (TAP), Printed circuit board (PCB).

1 INTRODUCTION

During handling or even manufacturing PCBs and ICs, defects may develop (Tseng et al., 2004). Those defects such as open circuits or short circuits may appear in or between circuit pathways and electronic components. Effective testing system is necessary for maintenance purposes and also for manufacturing quality insurance. The rapid development of electronic module assembly manufacturing requires a parallel development in test procedures (Moganti et al., 1996, Moganti and Ercal, 1998, Kusiak and Kurasek, 2001, Tong et al., 2004).

PCB testing is becoming more expensive and difficult due to the complexity of PCBs design. The common methods for diagnosing PCBs are still suffering from many difficulties; it needs a long time, a lot of manual work, and direct contact with PCB, and it is so expensive (Sheen et al., 1997).

Testing has to be good (having high defect coverage), cheap and fast. The IC defect level in 1970 was 1000 defective chips per million (DPM) delivered, but nowadays it bellows 10 DPM. Experts are still pushing this number to reach 0 DPM level (Vermeulen et al., 2004).

Previously, the unique method to inspect PCBs was manual testing method; it involves using visual inspection, multi-meters, oscilloscopes and other testing equipments. This method is almost inapplicable for the recent PCBs due to the hugely mounted number of components installed on PCBs. Moreover, using ICs limits the ability of manual testing and makes it so difficult.

Manual testing takes a long time to be performed. The efficiency of such diagnose method depends on the repairer's knowledge and experience (Janóczki et al., 2013). In manual testing, the repairer needs to choose the suitable testing equipment according to the device to be tested (Maxim, 2001).

In this paper, we introduce new design architecture for TAP. The main objective is the development of on-chip design for testability (DFT) architecture to reduce the complexity of testing nets, resources and pins connection. Using the new architecture, it is possible to achieve high cost reduction for complex system on chip (SOC) designs with negligible design and test overheads.

The TAP design can efficiently utilize one test trace between TAP and each device under test (DUT). Hence, one test access pin is required for each DUT. As an advantage of the new design, each DUT is tested separately without depending on other DUTs. Also, unlike IEEE 1149.1 Std TAP, there is no need to embed any instruction, identification or memory registers in the TAP (Vermeulen et al., 2002).

2 TEST ACCESS PORT

Generally, PCB includes many ICs on a single board, each of which is separately designed and verified before use. Multiple test patterns may be applied to a single chip to test its functionality in accordance with IEEE 1149.1 Std, which states that three or four test input pins and one test output pin may be used for testing and debugging of the SOC cores (Kac et al., 2003).

In our design, TAP reads the input data to select which route should be used to forward the data packet, and through which physical interface connection. Routing is achieved by using internal pre-configured directive. Then, TAP forwards data packets between incoming and outgoing interface connections. It routes them to the correct DUT using information that data packet contains.

Each data packet consists of two kinds of data: control information and user data. The control information provides data the TAP circuit needs to deliver the pattern, for example: DUT address, test pattern length, and response pattern length. Typically, control information is found in packet headers, with payload data in trailers.

TAP receives a serial data from USB as a bit stream. The proposed design formats the information at a bit level. We can achieve two major results by using test data packets: response detection and single device addressing. A packet has the following components.

1 Transmission Counts

It represents the transmission time, which is a field that is decreased by one each time a bit goes through a test node. If the field reaches zero, the transmission ends and the data direction are turned out.

2 Reception Counts

It is simply the time required to receive the response pattern from a DUT. Same as "transmission counts" field, it is decreased by one each time a response bit come into the test port. If the field reaches zero, the reception ends and the data direction is turned out again to transmission mode.

3 DUT Address

PCBs usually have three or more ICs together; in such cases the test pattern header generally contains addressing information so that the packet is received by the correct device under test. In our design, one address is included, the destination address, which is where the test pattern is intended to go.

4 Test Pattern

In general, the test pattern is the data that is carried to be applied to DUT inputs. It is usually of variable length that depends on the sitting DUT. It is created and transmitted by computer according to predefined database about the board under test. Figure 1 illustrates the packet format.

Start	Transmission Counts	Reception Counts	DUT Address	Test Pattern

Figure 1.   Test data packet format.

For example, let us consider the data packet depicted in Figure 2. The first two bits (01) apply a rising edge to start the TAP circuit. The next bits (111) represent the transmission counts. The TAP processor counts down seven times, allowing test pattern bits depart TAP moving to the DUT. Following, the reception counts down (101). TAP processor counts down 5 times, letting captured response pattern come in TAP. Then, the DUT address (1001), which means that DUT9 is the intended device to perform the test. Finally, test pattern goes in TAP to be redirected toward the targeted device under test. The pattern length is 7 which is compatible with "transmission counts" value.

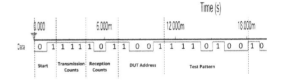

Figure 2.   Test data packet example.

Because of this, the proposed TAP is often considered more "intelligent" than the conventional TAP. Traditional TAPs don't recognize data packets of automated test equipments. This means that data packets have to be transmitted out to all devices under test, greatly degrading the efficiency of the connection.

Unlike traditional TAPs, the designed TAP has its own microprocessor that controls the overall test process. It can analyze test data packets, control transmission and reception counts, broadcast test patterns and receive the captured response signals. As described in Figure 3, the new TAP is composed of four main units, namely bypass register, transmission counts, reception counts and data direction control unit. Each of these units has its own function as discussed in the following sections.

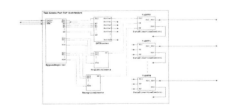

Figure 3.   TAP architecture.

2.1   *Bypass register*

Bypass register undertakes responsibility for intercepting and sorting data packet sections, and then redirecting each section to its proper destination or unit. It's composed of a shift register that loads the

202

serial input stream and expels it synchronously in a parallel mode. As mentioned earlier, the first two bits of the input steam start running the circuits.

Since the D+ input stream initially passes through a serial to parallel shift register, the "start section" is the first section that exits from the shift register; Hence, we made a considerable benefit from that section; It is ought to represent the first indicator for all other units to start running. Also, it halts the shift register waiting the first device testing to break-up.

The start section (01) performs a rising edge of the D Flip Flop clock pulse, hence a high signal results on the D Flip Flop output. Then the Q1 transistor is biased, allowing the following bit stream, to pass through the bypass register. See Figure 4.

DTC114EET1G is a digital transistor, which designed to replace a single device and its external resistor bias network. This BRT (Bias Resistor Transistor) contains a single transistor with a monolithic bias network consisting of two resistors; a series base resistor and a base–emitter resistor. The BRT eliminates these individual components by integrating them into a single device. The use of a BRT can reduce both system cost and board space. The device is housed in the SC–75/SOT–416 package which is designed for low power surface mount applications.

DTC114EET1G transistor has many features; it simplifies circuit design, reduces board space, reduces component count, can be soldered using wave or reflow, the modified Gull–Winged leads absorb thermal stress during soldering eliminating the possibility of damage to the Die and it is available in 8 mm, 7 inch/3000 unit tape & reel (Semiconductor, 2014).

According to the data packet in the previous example (Figure 2), D+ feeds bypass register with the following packet [01 111 101 1001 1110100]. As a result, the first '1' digit starts its journey in S0 till reaching START output of the shift register. 'START' output is connected with 'RESET' input of the D Flip Flop, so when the first '1' turns up "START", the Q output of D flip flop turns to low level. Then, register clock pulse discontinues imposing a halt status for the shift register.

The timing diagram of inputs and outputs of the shift register is captured in Figure 5. The operating frequency is 1 KHz. It's obviously seen that at time=11.6 ms, the first "1" digit reaches START output, concurrently the register outputs stabilized at the same values.

The challenge ambushes in the asynchronous release of register output data. To overcome this problem, we designed a "Signal Ejector" complementary circuit that ands every register output to "START", hence all outputs appear synchronously once START turns into high status as seen in Figure 6. At 1 KHz operating frequency, at time =11.6 ms, START turns into high level, all other outputs outcropped together synchronously.

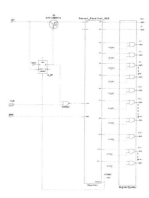

Figure 4.   Bypass register circuit.

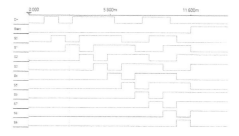

Figure 5.   I/O shift register timing diagram.

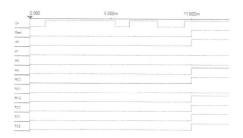

Figure 6.   I/O signal ejector timing diagram.

2.2 DUT router

A routing circuit based on demultiplexer routes the test data input signal to one of several different data outputs. That is, one of its outputs is identical to the data input signal, while the other output signals are 0. The output that matches the input signal is determined by address data signal on the output of shift register. The routing circuit can in principle have any number of data outputs. The address data has as many bits as needed for the selection. N address bits suffice for 2N data outputs. The routing circuit plays an important role in transmitting and also receiving test data and response data with DUT. By doing so, the routing circuit determines which DUT gets data written to it.

The test router consists of two major units; Routing unit and TDO receiver (See Figure 7). The routing unit receives the address from bypass register and accordingly opens the corresponding port to release TDI to reach its final destination.

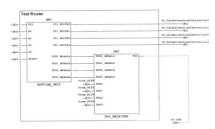

Figure 7. Test pattern router for 4 address inputs.

It's obviously seen that gates are totally controlled by address inputs A0A1A2A3. Also the same unit is in charge of enabling the related receiver port for the DUT. For example, suppose we have the address A0A1A2A3 = 1000. Therefore the AND gate U40 is enabled. It's still required permission from START input to enable U43. Biasing U43 is an enough indicator for the receiver port to start listening for any response pattern. While U43 is active, U44 state reflects the TDI signal. The test pattern is now ready to continue its journey to the targeted device under test.

TDI penetrates the corresponding route according to address inputs. While START input is off, no route selected. Also, receiver port listens to the corresponding DUT accordingly.

The TDO receiver is a logic circuit composed of N AND gates according to the number of the DUT. The circuit receives a unique enable signal which activates only one AND gate, the TDO signal which contributes on that gate is the only patter released from the circuit. Such combination avoids any prospected contention between signals, especially if there are unwanted signals from other out of test devices. The circuit is captured in Figure 8.

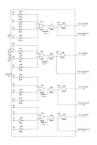

Figure 8. Routing unit architecture.

3 CONCLUSION

The newly proposed TAP overcomes the difficulties in the traditional TAPs. It reduces the reliability on ICs built-in testing circuits. Also, it's universal enough to test all devices regardless of its internal structure. The proposed TAP is considered as a code driven TAP because it can read, recognize, analyze and distribute the test patterns to its destination.

REFERENCES

[1] JANÓCZKI, M., BECKER, Á., JAKAB, L., GRÓF, R. & TAKÁCS, T. 2013. Automatic Optical Inspection of Soldering.
[2] KAC, U., NOVAK, F., AZAÏS, F., NOUET, P. & RENOVELL, M. 2003. Extending IEEE Std. 1149.4 analog boundary modules to enhance mixed-signal test. Design & Test of Computers, IEEE, 20, 32–39.
[3] KUSIAK, A. & KURASEK, C. 2001. Data mining of printed-circuit board defects. Robotics and Automation, IEEE Transactions on, 17, 191–196.
[4] MAXIM 2001. Automatic Test Equipment on a Budget. TUTORIAL. California: Maxim Integrated Co.
[5] MOGANTI, M. & ERCAL, F. 1998. A Subpattern Level Inspection System for Printed Circuit Boards. Computer Vision and Image Understanding, 70, 51–62.
[6] MOGANTI, M., ERCAL, F., DAGLI, C. H. & TSUNEKAWA, S. 1996. Automatic PCB Inspection Algorithms: A Survey. Computer Vision and Image Understanding, 63, 287–313.
[7] SEMICONDUCTOR, O. 2014. Product Overview DTC114EE: NPN Bipolar Digital Transistor (BRT). February, 2014 ed.: Semiconductor Components Industries, LLC, 2014.
[8] SHEEN, T. W., CHEN, J.-N., COHEN, S. A., BAGLINO, M. A. & WRINN, J. F. 1997. Printed Circuit Board Tester Using Magnetic Induction. Sheen et al., 5,631,572, 13.
[9] TONG, J., TSUNG, F. & YEN, B. 2004. A DMAIC approach to printed circuit board quality improvement. The International Journal of Advanced Manufacturing Technology, 23, 523–531.
[10] TSENG, T.-L. B., JOTHISHANKAR, M. C. & WU, T. T. 2004. Quality control problem in printed circuit board manufacturing—An extended rough set theory approach. Journal of Manufacturing Systems, 23, 56–72.
[11] VERMEULEN, B., HORA, C., KRUSEMAN, B., MARINISSEN, E. J. & VAN RIJSINGE, R. Year. Trends in testing integrated circuits. In: Test Conference, 2004. Proceedings. ITC 2004. International, 2004. IEEE, 688–697.
[12] VERMEULEN, B., WAAYERS, T. & BAKKER, S. Year. IEEE 1149.1-compliant access architecture for multiple core debug on digital system chips. In: Test Conference, 2002. Proceedings. International, 2002 2002. 55–63.

Testing and Measurement: Techniques and Applications – Chan (Ed.)
© *2015 Taylor & Francis Group, London, ISBN: 978-1-138-02812-8*

Au-Polypyrrole nanorod Volatile Organic Compounds gas sensor

S.W. Kim, J.S. Lee, S.W. Lee, S.H. Cha, K.D. Lee, J.H. Lee & S.W. Kang
School of Electronics Engineering, College of IT Engineering, Kyungpook National University, Daegu, South Korea

N. R. Yoon & H. J. Yun
Department of Sensor and Display Engineering, Kyungpook Nationatinal University, Daegu, South Korea

B. H. Kang
Center for Functional Devices Fusion Platform, Kyungpook Nationatinal University, Daegu, South Korea

D. H. Kwon
Department of Electronics Engineering, Kyungil University, Hayang-up, Gyeongsang buk-do, South Korea

ABSTRACT: In this work, we proposed a gas sensor for detection of Volatile Organic Compounds (VOCs) gas using Au-Polypyrrole (PPy) nanorods based on Localized Surface Plasmon Resonance (LSPR). Au-PPy nanorods array was fabricated by electrochemical deposition method and it exhibited LSPR phenomenon which can be the sensing properties of the Au-PPy nanorods with gas molecules. The sensing behavior of the gas sensor was evaluated by measuring the change in LSPR peak in a VOCs gas condition. We confirmed well-fabricated Au-PPy nanorods by scanning electron microscopy and the VOCs gas-sensing system performed under toluene gas. As a result, the sensitivity of the gas sensor is linearly related to the concentration of toluene gas in the range between 0 ppm and 60 ppm.

1 INTRODUCTION

Volatile organic compounds (VOCs) gas, which has a high vapor pressure at room temperature, can induce health problems and environmental risks. VOCs gas is generated in industrial environments such as solvents, paints, glues, and other products [1]. Accordingly, in order to monitor VOCs gas, a high efficiency gas sensor which involves sensitivity, selectivity, and real-time monitoring is essential. Recently, many types of VOCs gas sensor based on different sensing mechanisms have been developed such as oxide semiconductor, optical sensor, and chemical sensor [2]. In case of optical sensor, it has been proved to excellent properties which include high sensitivity, low power consumption, and real-time monitoring [3]. Among optical sensors, a sensor based on localized surface plasmon resonance (LSPR) phenomenon which is generated by light when it interacts with nanostructure has been studied. A LSPR sensor has high sensitivity and rapid response [4].

To detect for VOCs gas, variety sensing membranes have been used such as metal-oxide and conducting polymer. Semiconductor with metal-oxide gas sensor is difficult to operate at room temperature and needs high temperature to desorb gas molecules [5].

However, gas sensor with conducting polymers is sensitive to gas molecules in room temperature and has relatively simple process for fabrication of sensing membrane [6]. One of the representatives conducting polymers is polypyrrole (PPy) which includes –NH groups and π-conjugations that react VOCs gas molecules.

In this study, we proposed a gas sensor for detection of VOCs gas based on LSPR using Au-PPy nanorods. In order to evaluate for VOCs gas, the proposed gas sensor system was performed by measuring LSPR peak.

2 EXPERIMENTAL METHOD

2.1 *Preparing the sensing membrane*

The bilayer type of Au-PPy nanorods was synthesized by electrochemical deposition (ECD) by using anodized aluminum oxide (AAO) which used as a template (with a pore diameter of 200 nm and thickness 60 μm, Watman International Ltd.). Prior to the growth of Au-PPy nanorods, a thin layer of Au, which was used as working electrode, was deposited on the one side of the AAO template by electron-beam evaporation. After then, to synthesis for Au-PPy nanorods,

ECD was performed in a three electrode system with an Au plating solution and PPy electrolyte solution. A platinum wire was used as the counter electrode and an Ag/AgCl electrode was used as the reference electrode. Firstly, Au nanorods were electrochemically deposited by using Au plating solution (Ortemp 24 RTU, Technic Inc.) at -850 mV. Secondly, the PPy segments were grown by ECD using a pyrrole monomer on the Au nanorods at 950 MV. After that process, the synthesized Au-PPy naorods were immersed in 3 M NaOH for 1 h to dissolve for AAO template. The vertically Au-PP2y nanorods array was fixed on a glass and prepared for the detection of VOCs gas.

2.2 *Sensing system setup*

The layout of the measurement system is depicted in Fig. 1. The gas detection system consists of white-light source (DH-2000-BAL, Ocean optics), a reflectance optical probe (QR-400, Ocean optics), gas reaction chamber, a spectrometer (QE65000, Ocean optics), analysis software (Spectrasuit, Ocean optics) and Au-PPy nanorods. The optical response of the Au-PPy nanorods was measured by exposing toluene gas at a concentration varying from 0 ppm to 60 ppm into the gas chamber.

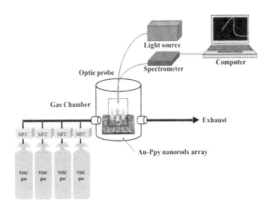

Figure 1. Schematic diagram of VOCs gas detection system.

3 RESULTS

3.1 *Must delete*

In order to confirm the structure of synthesized Au-PPy nanorods, we analyzed scanning electron microscopy (SEM) image which is shown in Figure 2. The aspect ratio of the Au-PPy nanorods was approximately 4 and their length can be controlled by the voltage and processing time. The morphology of the

fabricated Au-PPy nanorods had a good uniformity and the nanostructure of them can be enhanced their surface area which relates the sensitivity intimately.

The changes in the absorption spectrum of the Au-PPy nanorods were confirmed by the UV-Visible spectrometer to verify their sensitivity for toluene gas molecules which is one of the VOCs gas and it is shown in Figure 3. The results demonstrate that the absorption peak intensity relatively increases as the concentration of the toluene gas increase. As a result, the response of Au-PPy nanorods with the toluene gas molecules represented linearly from 0 ppm to 60 ppm at room temperature.

The represented responses between Au-PPy nanorods and toluene gas molecules relate to a transverse surface plasmon resonance of Au nanorods. On the other hand PPy does not affect the transverse mode of LSPR [7] and they relate to the sensing mechanism which is not yet clear but can be demonstrated that –NH groups and π-conjugations of PPy react with toluene molecules.

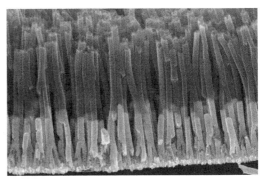

Figure 2. SEM image of Au-PPy nanorods.

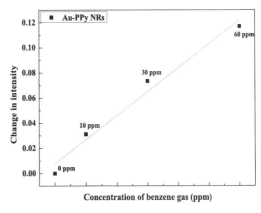

Figure 3. Changes in absorbance wavelength according to variations in toluene gas concentration.

4 CONCLUSION

In this study, we fabricated gas sensor based on Au-PPy nanorods and measured the sensitivity of this gas sensor according to toluene gas concentration.

ACKNOWLEDGMENT

This work was supported by the National Research Foundation of Korea (NRF) grant funded by the Korea Government (MSIP) (No.2008–0062617).

REFERENCES

[1] S. Stegmeier, M. Fleischere, and P. Hauptmann, Sens. Actuators B 148, 439–449 (2010).
[2] C. Elosua, C. Bariain, and I. R. Matias, The Open Construction and Building Technology Journal 4, 113–120 (2010).
[3] A. Cusano, M. Consales, A. Crescitelli, M. Penza, P. Aversa, C. D. Veneri, and M. Giordano, Carbon 47, 782–788 (2009).
[4] B. Sepulveda, P. C. Angelome, L. M. Lechuga, L. M. Liz-Marzan, Nano today 4, 244–251 (2009).
[5] J. J. Miasik, A. Hooper, and B. C. Tofield: J. Chem. Soc., Faraday Trans. 1, 82, 1117–1126 (1986).
[6] J. Huang, S. Virji, B. H. Weiller, and R. B. Kaner: Chem. Euro. J., 10, 1314–1319 (2004).
[7] S. Link, M. B. Mohamed, M. A. El-Sayed, J. Phys. Chem. B., 103, 3073–3077 (1999).

Testing and Measurement: Techniques and Applications – Chan (Ed.)
© *2015 Taylor & Francis Group, London, ISBN: 978-1-138-02812-8*

Study on the E_{vd} ranges and the interrelated relation with the bearing capacity of recently deposited layers in Huaibei Plain

F. Yu, S.X. Chen & Z.J. Dai
State Key Laboratory of Geo-mechanics and Geotechnical Engineering, Institute of Rock and Soil Mechanics, Chinese Academy of Sciences, Wuhan, China

ABSTRACT: In light of the geologic features of the newly deposited layers of Huaibei Plain such as substantial changes of soil properties and universal development of calcareous concretion, the thesis conducts a number of dynamic plate load tests on a highway under construction. The tests are designed to statistically analyze the ranges and variation of values with regard to different soil conditions and soil mass and to explore the distributive features of dynamic deformation modulus E_{vd} in different geologic parts of Huaibei Plain. Based on the above-mentioned tests, the paper further carries out cone penetration tests, spiral plate load tests and plate load tests to establish the correlation formula among dynamic deformation modulus E_{vd}, specific penetration resistance and the subsoil bearing capacity. Findings of the study show that the dynamic plate load test is a fast and convenient method for in situ detection and that the E_{vd} test values reflect geo-mechanic features and law of variation of the original sub-soils and correlates well with specific penetration resistance p_s and bearing capacity of sub-soils, promising a broad application as a mensurational index in highway engineering projects.

1 INTRODUCTION

As a result of the southward flow of the Huanghe River and the inundation as well as silting-up along the Huaihe River, the newly deposited layers of Huaibei Plain are mainly composed of the quaternary Huanghe River-flooded sediments. Silt and mucky soil are widely distributed on the ground surface of the depressions along the river and lowlands near the river. The inter-stream depressions derive mainly from the clay sedimentation of depression and limnetic facies or inter-bedding between paper clay and sandy soil. Besides, the lowland plain area of inter-stream terraces has witnessed the universal development of fluvio-lacustrine calcareous concretion. [1-2] Most of the recent sediments take the transitional form of mealy sand—sandy silt—silt—silty clay in contact relations and multiple inundations of the river has resulted in recurrences of this contact relation and consequently prominent stratification and inhomogeneity of the soil. [3] Calcareous concretion originates from the calcium carbonate of loess subject to the effects of leaching and deposition in formerly Huanghe River-flooded areas, which makes its engineering property differ from those of clayed soil and gravelly soil and hence contributes to its salient non-uniformity as well as unique geological engineering property.[4] All these jointly gave birth to the nowadays exceptional geological unit of the Huanghuai area.

Owing to the complexity and changeability of geological and soil conditions of Huaibei Plain, engineers are often confronted with the thorny problem of identifying the subsoil bearing capacity in a fast, easy and precise manner when it comes to highway constructions. Consequently, it is urgent to find an efficient and reliable technology for in situ detection to serve the construction of infrastructure engineering projects.

The dynamic plate load test boasts advantages such as easy operation, stable data and fast testing. To date, dynamic deformation modulus E_{vd}, an intensity indicator for filling quality of the roadbed, has been extensively harnessed in design and construction of railway sub-soil and the dynamic deformation modulus E_{vd} standard has become a development orientation of quality control standards for roadbed compaction.[5–8] However, there is a shortage of relevant study and standard directed at highway projects in China. Worse still, few relevant tests have been conducted in Huaibei Plain and systematic studies need to be done concerning the relationship between value features of dynamic deformation modulus E_{vd} and the sub-soil bearing capacity.

Taking the engineering projects in Xuzhou-Mingguang Highway and Sihong-Xuchang Highway, which is presently under construction in the Huaibei Plain area as examples, the paper launches extensive dynamic plate load tests and contrasts the results of cone penetration tests, spiral plate load tests and plate load tests. On the basis of this, it analyses the numerical features and variation laws of E_{vd} under typical soil conditions in the region and probes into the interrelation between E_{vd} and the subsoil bearing capacity, which presents theoretical proof for the use of dynamic deformation modulus E_{vd} to identify the bearing capacity of shallow subsoil in the Huaibei Plain area.

2 THE EXPERIMENTAL PRINCIPLE AND FEATURES OF DYNAMIC PLATE LOAD EVD TESTS

The dynamic plate load test is a new method designed to quickly, easily and accurately check and detect the bearing capacity index, which is used to show dynamic load characteristics of the roadbed. [9] The dynamic deformation modulus E_{vd} can be calculated as follows:

$$E_{vd} = 1.5 \times r \times \sigma / S \qquad (1)$$

Where, r refers to the radius (mm) of the bearing plate; σ refers to the maximum dynamic stress (MPa) of the roadbed; and S, the settlement (mm) of the bearing plate.

The paper uses the German-made LFG-K dynamic deformation modulus tester by the name of light falling deflectometer to carry out the dynamic plate load test. The gross weight of the block hammer is 10 kilograms, the maximum impact force 7.07kN (kilonewtons) and the diameter of the load plate 300 millimeters.

3 RANGES AND DISTRIBUTIVE FEATURES OF DYNAMIC DEFORMATION MODULUS EVD VALUES

In Huaibei Plain, clayed soils of the subcutaneous layer mainly consist of $Q_4{}^{al}$ and $Q_3{}^{al}$ clay and silty clay. $Fig.1$ shows the ranges and distributions of E_{vd} values of clay layer and silty clay layer.

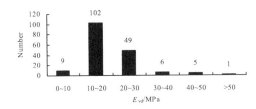

Figure 1. The range and distribution of E_{vd} values of clayed soil.

It can be seen from $Fig.1$ that the clay layer and the silty clay layer have a relatively wide range of dynamic deformation modulus E_{vd} values, which are specifically between 0 and 50MPa. E_{vd} values of the clayed soil are distributed concentratedly, with 88% from 10 to 30MPa and over 60% between 10 and 20MPa. Yet, owing to the uneven distribution of calcareous concretion in the soil and other factors, a fraction of the numerical values range from 30 to 50MPa or even beyond. Therefore, the dynamic deformation modulus E_{vd} values of clay layer and silty clay layer are relatively stable and few dispersedly-distributed data reflect the differences of calcareous concretion content in various geological areas.

Silt layer of Huaibei Plain is composed of holocene series $Q_4{}^{al}$ newly deposited silt that lies in a shallow depth and in a state of slight-to-medium compactness. The E_{vd} value range and distribution of the silt layer is shown in $Fig.2$.

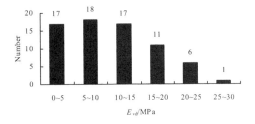

Figure 2. The range and distribution of E_{vd} values of silt.

According to $Fig.2$, the silt layer witnesses relatively smaller values and smaller range of the dynamic deformation modulus E_{vd}. The E_{vd} values of the silt layer all range from 0 to 30MPa. To be more specific, they chiefly range from 0 to 15MPa and 74% of all E_{vd} values of the silt layer are roughly evenly distributed in 3 sub-ranges from 0 to 5 MPa, 5 to 10MPa and 10 to 15MPa. By contrast, increasingly fewer E_{vd} values of the silt layer range from 15 to 20MPa, 20 to 25MPa and 25 to 30MPa, with barely one above 25 MPa. The value difference in dynamic deformation modulus E_{vd} embodies different degrees of compactness of silt in various geological sections. The greater the compactness is, the larger the E_{vd} values become. Compactness is the principal factor influencing the E_{vd} values of newly deposited silt.

A Mealy sand layer of Huaibei Plain is composed of Holocene series $Q_4{}^{al}$ mealy sand and upper Pleistocene Series $Q_3{}^{al}$ mealy sand as well as fine sand embedded deeply. $Q_4{}^{al}$ newly deposited mealy sand generally lies below 3.0 meters and is in a state of slight-to-medium compactness. $Q_3{}^{al}$ mealy and fine sand layer is embedded even deeper and most in a medium-to-high compactness. Dynamic deformation modulus E_{vd} values of mealy sand layer have a wider range generally between 10 to 50MPa. Its E_{vd} values are unstable and vary greatly, largely between 10 to 40MPa. Still, compactness remains the principal factor influencing the E_{vd} value of newly deposited mealy sand layer.

By illustrating the Anhui Segment of Xuzhou-Mingguang Highway, the paper carries out 433 sets of dynamic plate load tests at 81 testing points with average dynamic deformation modulus E_{vd} being 25.18MPa. The variation of dynamic deformation modulus E_{vd} values is consistent with the geological conditions and soil characteristics of Huaibei Plain. This also proves that dynamic deformation modulus E_{vd} gives a good reflection of the geological features and mechanical properties of all typical geological bodies in Huaibei Plain.

4 CORRELATION BETWEEN DYNAMIC DEFORMATION MODULUS EVD AND SPECIFIC PENETRATION RESISTANCE

Fig.3 presents the correlationships between E_{vd} values and p_s average values of clay layer, silty clay layer, silt layer and mealy sand layer in Huaibei Plain.

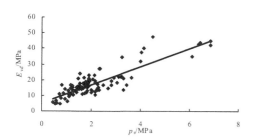

Figure 3. Correlation between Evd values and average specific penetration resistance in affected depths.

According to *Fig.3* dynamic deformation modulus E_{vd} shares a good correlation with the specific penetration resistance p_s within the affected depth. A fitting is conducted by an equation of linear regression, with the relevant coefficient R being 0.86. The Correlation equation is:

$$E_{vd} = 5.7898\overline{p_s} + 5.3735 \qquad (2)$$

Where, $\overline{P_s}$ refers to the average specific penetration resistance (MPa).

It can be seen from *Fig.6* that parts of E_{vd} values show a discrete feature. This is primarily attributed to the differences in calcareous concretion content of the

soil in Huaibei Plain. *Table 1* shows the E_{vd} test values and the values of specific penetration resistance p_s of soil layers with different calcareous concretion contents. *Table 1* suggests that on one hand, the rise of calcareous concretion content can contribute to a dramatic increase of E_{vd} test values; and that on the other hand, E_{vd} test results are evidently discrete because of the limited dimension of the bearing plate of the instrument and the uneven distribution of calcareous concretion in soil layers.

Table 1. A statistic table of E_{vd} values and p_s values of soil with calcareous concretion.

Content of Calcareous Concretion (%)	Dynamic Deformation Modulus E_{vd} (MPa)	Specific Penetration Resistance (MPa)	Content of Calcareous Concretion (%)	Dynamic Deformation Modulus E_{vd} (MPa)	Specific Penetration Resistance (MPa)
5	13.8	2.24	20	22.13	3.22
5	13.2	2.02	20	21.9	3.22
10	22.1	3.24	30	34.9	6.1
10	15.3	2.31	35	43.83	6.43
10	14.6	2.26	35	34.9	6.1
10	17.47	2.26	40	44.7	6.84
20	21.5	3.64	40	47.67	4.49

Fig.4 shows the correlation curves involving E_{vd} test values of all soil layers, specific penetration resistance p_s values and the contents of calcareous concretion. E_{vd} values enlarge as the increase of calcareous concretion content. Fitting is made by means of linear regression equation and the correlation equation is:

$$E_{vd} = 0.9188w + 7.2535 \qquad (3)$$

Where, w refers to the average content of calcareous concretion (%).

Meanwhile, linear fitting is also made between specific penetration resistance and the contents of calcareous concretion. According to *Fig.4*, the specific penetration resistance increases alongside the increase of the contents of calcareous concretion; the variation tendency of specific penetration resistance is basically similar to that of E_{vd} values as the contents of calcareous concretion change; the shapes of the two curves are basically consistent with each other. From *Fig.4*, we may draw the following conclusions: the mechanical properties of soil layers are relevant to the contents of calcareous concretion; the higher degree of calcareous concretion content, the better the mechanical property of the soil layer and the larger the bearing capacity of it; both E_{vd} values and the values of specific penetration resistance very well reflect the mechanical properties of soil layers with calcareous concretion and their variation features.

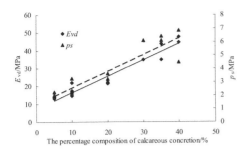

Figure 4. Correlation among E_{vd} values, p_s values and the percentage composition of calcareous concretion.

5 CORRELATION BETWEEN DYNAMIC DEFORMATION MODULUS EVD AND THE SUBSOIL BEARING CAPACITY

Though the dynamic deformation modulus E_{vd} has been adopted as an important indicator for roadbed quality control, nevertheless, as a result of the changeability of geological conditions, there is still a shortage of a uniform theoretical formula in order for it to be used to identify the subsoil bearing capacity. So far, the determination of the subsoil bearing capacity is done by securing an empirical relationship through various in situ tests. To establish the relationship between E_{vd} values of typical sub-soils in Huaibei Plain and the subsoil bearing capacity, spiral plate load tests and plate load tests are simultaneously launched near the E_{vd} testing locations to determine the eigenvalues of the subsoil bearing capacity. A WDL-type tester with its probe rated load of 1500kPa and a spiral plate diameter of 200 cm^2 is used in spiral plate load tests. The conventional slow method is used in plate load tests where the dimension of load plate is 0.5 m^2. Results of in situ tests are shown in *Table 2*.

Fig.5 shows the correlation between the dynamic deformation modulus E_{vd} and the eigenvalues of subsoil bearing capacity. It can be concluded that the two share a relatively good correlation where the fitting correlation coefficient of linear regression R equals 0.83 and that their correlation equation should be:

$$f_0 = 7.8108E_{vd} + 58.704 \quad (2.5 < E^{vd} < 50) \quad (4)$$

Where, f_0 stands for the characteristic value of sub-soil bearing capacity (kPa).

Table 2. A correlation table between dynamic deformation modulus E_{vd} and subsoil bearing capacity.

Names of Rock-Soil	Dynamic Deformation Modulus E_{vd} (MPa)	Eigenvalue of Subsoil Bearing Capacity f_0 (kPa)	Note
Q_4^{al} Clay	19.45	150	-
	13.65	142	-
	22.1	240	with calcareous concretion
Q_4^{al} Silty Clay	14	133	-
	15.91	162	-
	13.83	181	with calcareous concretion
	13.01	177	with calcareous concretion
	15.99	204	with calcareous concretion
	15.39	190	with calcareous concretion
	17.94	184	-
Q_4^{al} Silt	11.23	185	-
	6.6	130	-
	13.43	180	-
	11.93	160	-
	2.75	60	-
	7.31	100	-
	12.53	160	-
	3.79	70	-
Q_4^{al} Mealy Sand	15.26	190	-
	15.58	220	-

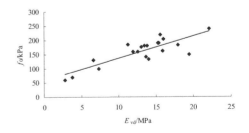

Figure 5. Correlation between E_{vd} and subsoil bearing capacity.

Formula 4 is applicable to all typical types of soil layers in Huaibei Plain area, including clay, silty clay, silt, mealy sand, and fine sand. When $E_{vd}>50$, because of the limited dimension of the bearing plate and the uneven sizes of soil particles, the subsoil bearing capacity varies in a wider range, particularly in areas where there are decomposed rocks. In the latter case, *Formula 4* can only be used for theoretical calculation.

The affected depths of impact load are 40 to 50cm during the dynamic deformation modulus tests. Thus, though the dynamic deformation modulus E_{vd} enables a sound reflection of the mechanical properties of a single soil layer, yet for layered foundation, the determination of subsoil bearing capacity need allow for associative effect and interface effect of the bearing capacity of all soil layers and integrate the dynamic deformation modulus tests with other in situ tests.

Table 3 presents the ranges of all typical subsoil bearing capacity of Huaibei Plain in accordance with the value ranges of dynamic deformation modulus E_{vd}. It can be known from *Table 3* that the subsoil bearing capacities of the newly deposited layers in Huaibei Plain have their average values mainly between 80 and 350kPa. The bearing capacities of clay and silt are more or less the same and universally low. Silty clay largely contains calcareous concretion and hence its subsoil bearing capacity is noticeably higher than general clayed soil. Mealy sand and silty clay with calcareous concretion have wider ranges of bearing capacity while clay and silt have comparatively more stable bearing capacities.

Table 3. Recommended ranges of bearing capacity of all typical types of soil in Huaibei Plain.

Types of soil	Dynamic Deformation Modulus E_{vd} (MPa)	eigenvalue of Subsoil Bearing Capacity f_0 (kPa)
Q_4^{al} general clayed soil	5~15	95~175
Q_4^{al} silty clay (with calcareous concretion)	15~30	175~290
Q_4^{al} silt	3~15	80~175
Q_4^{al} mealy sand	10~40	135~350

6 CONCLUSION

1 As the size of soil particles increases, the E_{vd} value range gets wider; the E_{vd} value range of clayed soil

is the narrowest, while that of silt wider and that of sandy soil the widest.

2 There is an incremental impact of specific penetration resistance on dynamic deformation modulus, meaning E_{vd} increases alongside with the increase of specific penetration resistance. Both elements are able to reflect the mechanical property and its variation features of soil layers containing calcareous concretion and the presence of calcareous concretion leads to larger E_{vd} values of clayed soil layers.

3 There is a significant linear correlation between dynamic deformation modulus E_{vd} and the subsoil bearing capacity. Different types of subsoil have distinctive features of the bearing capacities which are mutually diverse. E_{vd} values give a good reflection of the bearing capacity features of typical geological bodies in Huaibei Plain and hence can be used as an index for mensurating the subsoil bearing capacity in highway engineering projects.

REFERENCES

[1] Jin Quan. The Quaternary in Huaibei Plain of AnHui Province[M]. Beijing: Geological Publishing House, 1990.

[2] Zhang Keqian. Discussion on the Formation of the Huaibei Deposited Plain in Anhui and Its Tectonic Issues [J]. Geology in China, 1962, (6): 10–17.

[3] Wu Zhihai. Study on the Engineering Property of the Newly-Deposited Silt in Huaibei Plain of AnHui Province [Ms. D. Thesis][D]. Hefei: Hefei University of Technology, 2006.

[4] Cao Yajuan. Study on Distribution and Formation of the Calcareous Concretion Soils in Huaibei Plain of AnHui Province [Ms. D. Thesis][D]. Hefei: HeFei University of Technology, 2009.

[5] Claus Goebel, Klaus Lieberenz, Frank Richter. Der Eisenbahnunterbau [M]. Eisenbahn-Fachverlag, 1996.

[6] Li Nufang. Application and Prospect of Dynamic Deformation Modulus E_{vd} Standard [J]. Railway Standard Design, 2003, (6): 37–40.

[7] Dong Xiuwen. An Experimental Study on Detecting the Subgrade Compaction Quality Standard by E_{vd} [Ms. D. Thesis][D]. Chengdu: Southwest Jiaotong University, 2005.

[8] Cheng Yuanshui. Research on Correlativity among Relevant Parameters to Subgrade Compactness and the Control Indexes for Improved Soils [Ms. D. Thesis][D]. Beijing: China Academy of Railway Sciences, 2007.

[9] Li Nufang,Xiao Jinfeng, Li Shufeng. A Study on Dynamic Plate Load Testing Method and the Development of a Testing Device [J]. Railway Standard Design, 2002, (6):7–8.

Industrial autoimmunization and measurement

Testing and Measurement: Techniques and Applications – Chan (Ed.)
© 2015 Taylor & Francis Group, London, ISBN: 978-1-138-02812-8

Measuring of internal residual stress after machining using Eddy Current in dependence of technological parameters

D. Mital, J. Zajac, M. Hatala, J. Duplak & P. Michalik
Faculty of Manufacturing Technologies with the seat in Presov, Technical University of Kosice, Presov, Slovakia

J. Mihok
Faculty of mechanical engineering, Technical University of Kosice, Kosice, Slovakia

ABSTRACT: This paper deals with the identification of internal layers, where the influence of the load is often caused by the defect of the lattice, which should not cease even after the discontinuation of the presence of external force effect on the material. This paper focuses on the actual problem of the detection of residual stress after machining. The measuring is often complicated and expensive. Our study is focused on the experimental identification of residual stress after machining using a non-destructive method based on Eddy Current (EC). The results are verified by RTG diffraction method, which was used in the creation of mathematical equations of residual stress depend on technological cutting conditions for steel C45.

1 INTRODUCTION

1.1 Residual stress

Machining often causes deformation and defects of grid bonds in internal layers, which do not lapse even after the discontinuation of effects of external forces on the material. The measurement of it is often complicated and costly.

Residual stress or internal stresses are tensions in the internal layers of material, which are closely related to plastic deformation. In fact, this affects the functionality of machined surfaces. The formation of this stress can be caused by the heating of material during machining in the zone of cutting. Residual stress can be classified into three main classes:

Class I: Macroscopic residual stress, which arises in the process of machining, heat treatment, the forming of the material and others operations. Occur in deep layers of the material (more than 7 mm).

Class II: Operate in the area of grains and crystals in a small volume (from 0,2 mm to 3 mm).

Class III: Operate in atoms of material (under 0, 2 mm) (dislocations, gapping,…).

The Second and Third class are not characterized as significant classes that would significantly affect the mechanical and physical properties.

The formation of thermal residual stress also occurs during the cooling process of the material. With the increase of the temperature, the plasticity of the surface also increases. Cooling of the material causes the reduction of grains. The tensile stress is formed on the surface and comprehensive stress under the surface. Edges and surfaces are then susceptible to microscopic cranks

1.2 Actual state on the issue

Recent research indicates that eddy current conductivity measurements can be exploited for non-destructive evaluation of subsurface residual stresses in surface-treated nickel-base super alloy components. Most of the previous experimental studies were conducted on highly peened (Almen 10–16A) specimens that exhibited harmful cold work in excess of 30% plastic strain. Such high level of cold work causes thermo-mechanical relaxation at relatively modest operational temperatures; therefore the obtained results were not directly relevant to engine manufacturers and end users. The main reason for choosing peening intensities in excess of recommended normal levels was that in low-conductivity engine alloys, the eddy current penetration depth could not be forced below 0,2mm without expanding the measurements above 10MHz which is beyond the operational range of most commercial eddy current instruments. In this paper, we will report the development of a new high-frequency eddy current conductivity measuring system that offers an extended inspection frequency range up to 50MHz with a single spiral coil. In addition to its extended frequency range, the new system offers better reproducibility, accuracy, and measurement speed than the previously used conventional system.

Because of their frequency-dependent penetration depth, eddy current measurements are capable of mapping the near-surface depth profile of the electric conductivity. This technique can be used to nondestructively detect the characteristic of the subsurface residual stress distribution in certain types of shot-peened metals, e.g., in nickel-base superalloys. In this paper, a highly convergent iterative inversion procedure is presented to predict the frequency-independent intrinsic electric conductivity depth profile from the frequency-dependent apparent eddy current conductivity (AECC) spectrum. The proposed technique exploits three specific features of the subsurface electric conductivity variation caused by near-surface residual stresses in shot-peened metals. First, compressive residual stresses are limited to a shallow surface region of depth much less than typical probe coil diameters. Second, the change in electric conductivity due to residual stresses is always very small, typically less than 1%. Third, the electric conductivity depth profile is continuous and fairly smooth. The accuracy of the proposed iterative inversion procedure is one order of magnitude better than that of the previously developed simpler method (J Appl Phys 2004;96:1257).

2 MATERIALS AND METHODS

Eddy current can be defined as induced current in conductive materials, which is put near alternating magnetic field or moved in a constant magnetic field (intersecting the magnetic fields lines). Eddy current flow in the plane perpendicular to the primary magnetic field (f and their direction can be determined by Lenz's law, where eddy current flow in close loops.

Measuring methods based on eddy current are ranked as non-destructive methods of material testing. Electrical, conductive, magnetic permeability and certain geometrical dimension are initial conditions of measuring, due to the closing of magnetic field, where the whole measurement system is on AC power. Eddy current depends on the magnetic permeability of the material, which varies influenced of defects, cranks and others defect, what is reflected by changing of the magnetic field track. The penetration depth is one of the surface phenomena, where standard penetration depth can be evaluated:

$$S = \sqrt{\frac{1}{\pi.f.\mu_0.\mu_r.\ \sigma}} = \frac{503,3}{\sqrt{f.\mu_r.\sigma}} \qquad (1)$$

where:
s - standard penetration depth [m]
f - frequency [Hz],
μ - permeability of the material [–],
μ_r- relative permeability (4.π.10⁻⁷) [H.m⁻¹],
σ   - specific electric conductivity [s.m⁻¹].

Resulting magnetic fields is the vector sum of two magnetic fields. The first is an exciting magnetic field, which is caused by the primary winding; the second magnetic field is caused by induced voltage. The induced voltage is also dependent on the frequency of excitation current, which directly affects the depth of the test and also the penetration depth of eddy current.

A method based on annex coil consists of two coils (transmitting and receiving). It is attached to material in the desired places.

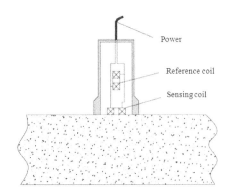

Figure 1.   Principle of attach sensor.

Working diagrams are used to determinate work area of the probe. Solid line in diagram represents point of impedance with sufficient thickness when changing angle β represent change of conductivity or excitation current. Curves on the left side from solid lines show impedance of the sensor in the distance α (distance between probe and surface). Parameter α is called the coefficient of performance. Parameters can be summarized as follows: direction ξ is thickness effect (thickness of the material), which is greater than the penetration depth of eddy current; direction α determines lift off effect and direction β represent conductance effect at constant permeability.

$$\beta = R.\sqrt{\omega.\mu_0.\mu_r.\sigma} \qquad (2)$$

where:
ω - angular velocity [rad.s⁻¹],
β - conductance effect,
R - diameter of probe [mm],
μ_0 - permeability of vacuum (4.p.10⁻⁷) [H.m⁻¹],
μ_r - relative permeability [–],
σ - specific electric conductivity [s.m⁻¹].

$$\alpha = \frac{2.h}{R} \qquad (3)$$

where:
  h - distance between probe and surface [mm],
  α - lift off effect,
  R - diameter of probe [mm].

$$\xi = \frac{2.d}{R} \qquad (4)$$

where:
  d - penetration depth of magnetic field [mm],
  ξ - thickness effect,
  R - diameter of probe [mm].

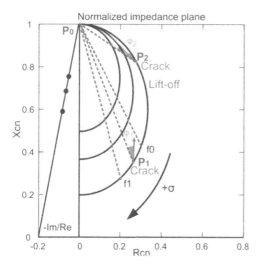

Figure 2. Working diagram of probe.

This paper focuses on monitoring internal residual stress after machining (milling with variant cutting technological parameters). Determination of residual stress is experimentally studied in this article as a new possible way of measuring residual stress using commercial eddy current measuring systems.

A material used in the experiment was structural steel 1205 by euro standard C45; En 10083–2-91. Steel C45 is intended for refining or normalizing annealing and is often used for the production of less stressed machine parts. The chemical composition of

Table 1. Chemical composition of steel C45.

Chemical element							
C	Mn	Si	Cr	Ni	Cu	P	S
0,51	0,69	0,25	0,15	0	0,12	0,023	0,017

(% ratio)

the steel C45 is shown in percentage ratios in the following table (Table 1).

3 EXPERIMENTAL MEASUREMENT

Samples were milled on CNC milling center Pinnacle VMC 650 S, which is available in the laboratories of Technical University of Kosice. CNC milling center is a 3 axes machining unit with control system Fanuc.

The surface was milled in two transitions with cooling, and because of the insufficient diameter of the mill, the process of machining was down milling. First of all, it was necessary to use finishing milling (end milling) to eliminate imperfections after cutting. An experiment was carried on by different technological parameters, where two parameters were set as a constant and one as a variable.

Figure 3. Milling of the sample on CNC milling center Pinnacle VMC 650S.

Cutting conditions were done in three steps, where spindle speed ranged from 1000min⁻¹ to 3000min⁻¹ at a constant feed rate and depth of cut.

The major aim of this presented study is to establish the impact of internal residual stress on changes of conductivity respectively permeability of material in the zone of penetration of eddy current.

Experimental measurement of residual stress was done on measuring device Zetec type designation MIZ – 22 Eddy Current Instrument (Fig4). A device based on eddy current was designed to test defects

Table 2. Cutting conditions of experiment.

N. sample	Spindle speed n [min⁻¹]	Feed rate vf [mm.min⁻¹]	Depth of cut ap [mm]
1.	1000	500	1
2.	1500		
3.	2000		
4.	2500		
5.	3000		

in materials, coating thickness, thickness of material, direct conductivity or detection of metallurgical inhomogeneities and to detect conductivity, permeability, changes of feriticity and other parameters.

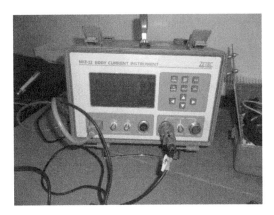

Figure 4. Milling of the sample on CNC milling center Pinnacle VMC 650 Measuring device ZETEC MIZ – 22 EDDY CURRENT INSTRUMENT.

The selecting the measuring probe was according to two basic criteria. The first criterion was the size to ensure the most accurate evaluation of residual stress in required point. The second step was necessary to choose the suitable frequency of the probe, according to the most significant changes in conductivity and permeability due to internal residual stresses. As the most appropriate variant was chosen probe with a frequency 150kHz, type Fersternf 58 (designation 2.830.012011, which means that the sensor is working absolutely). The diameter of the used probe is 12 mm (Fig. 5)

It must also be noted that the experiment was realized as a pilot test for material steel C45 for variant input parameters. Conductivity electromagnetic properties of the material depend on the type of material, its density, apparent density, specific electrical conductivity and so on (Fig.6). During the measuring, it is important to ensure the perfect contact of the probe to the measured material. Lift off effect causes displacement of the signal.

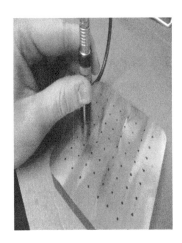

Figure 5. Measuring probe.

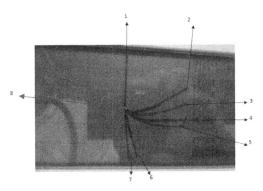

Figure 6. Position of signal for different materials (1 – ferity material, 2 – weld joint, 3 – TTO, 4 –steel C45, 5 – Unidentified steel, 6 – austenite SS 304, 7 – Aluminum Al 7075-T6, 8 – balance air.

Verification measurement of internal residual stress was done on the ground of Faculty of Mechanical Engineering of Zilina University. Difractometer type Proto XRD was used for this purposes which provides measuring of internal residual stress and also residual austenite in crystal materials. Anode of RTG lamp is made of Cr and Ka with cooling.

Measuring was carried on each sample automatically using CNC controlled positioning table, where 9 points to identifying internal residual stress were set.

Measuring points were determined for each sample at the same geometrical position to achieve comparability of values for each sample and displacement of the signal from eddy current.

Values of deviations created by eddy current are responding to changes in the material (macrostructures). Thus, changes in the macro structures caused by the influence of the forming process and the formation of residual stress. Deviations in the axes

Figure 7. RTG Difractometer type Proto XRD.

X and Y are readout from the screen of measuring device, where was recorded position of the point relative to the zero point (value of etalon for material C45 250Mpa). (Fig. 9).

Figure 9. Position of point from the screen (1 – position of measured point 2 – calibration point).

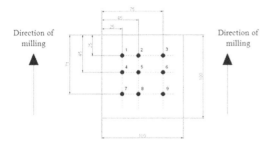

Figure 8. Geometrical disposition of measured points.

Measured points were recorded as ordered couples [X,Y], which were analyzed separately (Tab. 4).

Software Mathcad 14.0 was used for interpolation and creation of graphical dependences of deviation and input parameters such as spindle speed, feed rate and depth of cut. Interpolation was done using function interp, which is located in library of software

Table 3. Measured values of residual stress on difractometer.

	Measured point	1	2	3	4	5	6	7	8	9
	1	136,4	−18,2	101,5	151,3	−79,4	−73,3	256,7	207	320
	2	422,2	−7,4	192,9	388,3	308,1	−83,8	235,9	204,1	231,7
	3	415	−8,5	245	412,3	−152,4	−7,5	248,9	169,6	278,4
	4	264	−9,4	212,5	487,5	24	143,4	300,4	113	297,8
	5	362	−11,7	202,7	513	260,8	259,5	336,9	93,9	355,3
Residual stress [MPa]	6	375,3	−62	151,3	489,8	334,8	194,2	342,3	148,3	365,9
	7	396,4	25,8	241,2	326,5	45,5	202,1	436,5	−80,5	459,7
	8	292,1	−74,5	418	301,1	−11,1	344	403,5	108,1	293,6
	9	413,3	123,1	479,5	310,2	99	286,5	432,2	−36,1	442,9
	10	266,2	116,5	248,8	280,3	218,6	218,1	375,8	173	219,7
	11	422,1	182,3	86,2	321,6	156,9	142,4	246,1	−60,1	363,1
	12	472,3	165,6	160,6	398,4	69	65,1	170,7	172,7	264,2
	13	385	85,4	198,1	376,4	125	89,6	198,3	154,6	318,6
	14	298,3	−12,8	105,6	321,4	187,3	102,5	227,7	189,3	395,1
	15	274,1	−105	88	250,9	238,2	135,3	279,6	169,6	461,3

Table 4. Measured values of residual stress on difractometer.

Point	Stress versus calibration [MPa]	Position X [-]	Position Y [-]
1	125,3	5,3	4,3
2	−312	5	3,8
3	−98,7	4	1
4	239,8	4,9	3,5
5	84,8	5	4,8
6	55,8	5,2	4,2
7	92,3	5	3,2
8	101,7	5,2	3,4
9	115,9	5	3,2

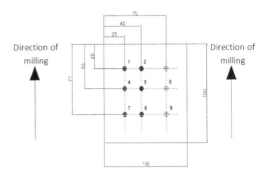

Figure 11. Define and geometrical identification of lines (Line 1 points 1, 4 and 7; line to points 2, 5 and 8; Line 3 points 3, 6 and 9).

Mathcad and is designed to complete values between two measured points, where measured point must be contained in set of interpolated values.

Interpolation of deviation X and Y sample 3 point 1 to 15

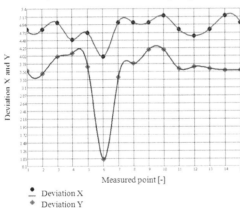

• — Deviation X
♦ — Deviation Y

Figure 10. Graphical dependences of deviation X and Y.

Graphical dependences in this phase have an informative character, because the initial phase of analyzing the experiment was devoted to identifying stress on individual samples. This phase of analyzing focuses on position of cutting tool and analyzing was done in three lines.

$$\sigma = V(X,Y).PK + KBN \qquad (5)$$

where :
σ - Evaluated residual stress [MPa],
V(X,Y) - deviation in direction X or Y [MPa],
PK - conversion coefficient [-],
KBN - value of calibration residual stress 250 MPa.

Conversion coefficients and constructed graphical dependencies shows, that internal residual stress has parabolic shape as deviation in the direction Y. Calculation of real values through conversion coefficients can also be stated, that coefficients for same positions obtain similar values:

– At the beginning at the distance of 25 mm from the edge of the sample are coefficients on line 1 31,03, on the line 2 is values of coefficient 29,5 and on the third line is 36,9
– In the middle length are coefficients -62,75, -52,3 and 87,2
– At the end from the edge of the samples are conversion coefficients -25,8, -37,1 and -42,8

Using average coefficients, we can receive informative character of size and type of residual stress. Graphical dependences confirm that maximal residual stress is on the edges of the samples. Values in the zone of leaving the tool is on average 30% lower than values of residual stress in start point of machining.

From average graphical, dependences can be assessed. Deviation in the direction X expresses decrease under value 4,5 and value of maximal proportional residual stress to the value of 100MPa. Deviation from the range from 4,5 to 5,2 is value of proportional residual stress 200MPa. General conversion coefficients are shown in table below (Tab. 5). Average values of conversion coefficients are similar at the deviation Y, where dispersion is 3. Value of average conversion coefficient for the line 1 is 30, line 2 is -67 and line 3 is -14. Determination of conversion coefficients for deviation X was for line 1 on value 19, -41 for line 2 and -7 for line 3.

Three lines at distance 25, 45 and 75 mm from the edge of the sample in the plane normal to the direction of milling were analyzed. Graphical dependences

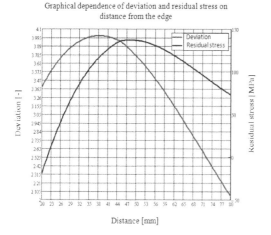

Figure 12. Graphical dependences of average deviations X and Y and evaluated residual stress sample 3.

were made based on measured values of internal residual stress and evaluated conversion coefficients, which were derived in previous chapters. Values of

conversion coefficients were for deviations in X and Y for lines different:

- Line 1 conversion coefficient X =39 a Y=31,17
- Line2conversion coefficient X= 40,2 a Y=50,5
- Line3conversion coefficient X= 40,2 a Y=41,25

As graphical dependence (Fig. 14) indicates, deviation in direction Y and its corresponding values of residual stress oscillating around the curve of real residual stress measured by verification method (RTG).

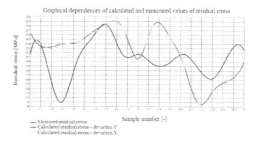

Figure 14. Graphical dependence of evaluated and measured values of internal residual stress.

Table 5. Table of conversion coefficients for deviation X and Y.

	Y			X		
	Line 1	Line 2	Line 3	Line 1	Line 2	Line 3
Sample 1–5	31,03361	−62,7389	−16,7889	19,68463	−48,1113	−8,92918
Sample 6–10	28,50383	−52,2866	−15,1669	18,8815	−25,7349	−23,4851
Sample 11–15	30,97851	−87,9095	−6,99293	18,28681	−50,3993	9,280142
Average	30,1719888	−67,645072	−13,654328	18,9510019	−41,415188	−7,7113962

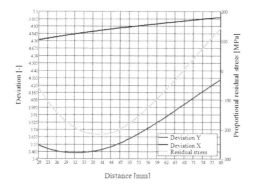

Figure 13. Graphical dependence of deviation X, Y and proportional residual stress.

4 CONCLUSION

By analyzing the values obtained from eddy current measurement, it is proved that deviation in X direction is more significant than deviation in Y direction. In fact, this deviation in Y direction is considered as the informative value of real evaluated internal residual stress, and the deviation in the X direction is considered as the primary indicator of residual stress. Conversion coefficients depending on distance from the surface in the direction of the axis normal to the direction of milling, where values of conversion coefficients in deviation Y evaluated on value 32 at distance 25, at distance 45 is value 50 and at distance 75 is calculated coefficient 42. Conversion coefficient calculated for deviation X direction is 40 for all analyzing lines and samples. Deviation in the X direction is considered as more significant and deviation in Y direction has an informative character.

REFERENCES

[1] LIPTÁK, O. et al. Technology of machining, Bratislava, 1979, 437 pages, 302 05 48.

[2] Abu-Nabah.A.High-frequency eddy current conductivity spectroscopy for residual stress profiling in surface-treated nickel-base superralloys In: NDT and E International, USA, pages 405–418, ISSN: 0963–8695.

[3] Bassam A. Abu-Nabah. Iterative inversion method for eddy current profiling of near-surface residual stress in surface-treated metals In: NDT and E International, USA, pages 641–651, ISSN: 0963–8695.

[4] REZ,J. Magnetic nondestructive testing of material by using feed through test oil and computer supported optimalization of its parameters In: Journal of electrical engineering, 1999, ISSN 1335–3632.

[5] KOPEC, B. at al. Non destructive testing of materials and constructions. Brno 2008, 384 pages, ISBN 978–80–7204–591–4.

[6] Hutyrová, Z., Harničarová, M., Zajac, J., Valíček, J., Mihok, J. Experimental study of surface roughness of wood plastic composites after turning (2014) Advanced Materials Research, 856, pp. 108–112.

[7] Semanco, P., Fedak, M., Rimar, M. Simulation study of two alternative workstations for pressure diecasting process (2012) Applied Mechanics and Materials,110–116, pp. 660–664.

[8] Čep, R., Janásek A., Martinický B., Sadílek M.Cutting Tool Life Tests of Ceramic Inserts for Car Engine Sleeves. Tehnički vjesnik – Technical Gazette, Strojarski fakultet, University of Osijek, 2011, s. 203–209.

Testing and Measurement: Techniques and Applications – Chan (Ed.)
© 2015 Taylor & Francis Group, London, ISBN: 978-1-138-02812-8

Quantitative prediction of residual stresses from 3D nanoindents formed on artificially bent samples

Y.H. Lee, Y.I. Kim, S.W. Baek, Y.H. Huh & H.M. Lee
Korea Research Institute of Standards and Science, Daejeon, South Korea

ABSTRACT: Three-sided pyramidal nanoindents formed on artificially bent stripes were observed with an atomic force microscope. 3D morphology of a remnant nanoindent including impression and pile-up was identified and analyzed for estimating the indentation strain which was expressed by the ratio of the remnant contact area and its projected area. By considering an elastic interaction between the indentation stress/strain with the in-plane stress applied on the stripe sample, the residual stress can be formulated by a multiple of the stress-induced change in the indentation strain and Young's modulus. Predicted stresses from the nanoindent morphologies were comparable with the bending stresses measured from curvature radii excepting in high stress regime above 400 MPa. The stress overestimation under high tensile stress might be discussed from an unexpected plastic zone expansion and significant elastic properties in the metallic glass sample.

1 INTRODUCTION

Several nanoindentation techniques have been developed to characterize various surface mechanical properties such as hardness, Young's modulus (Oliver and Pharr 1992), plastic flow properties (Kim, J.-Y. et al. 2006, Kim, S.H. et al. 2006), fracture toughness (Lee et al. 2006) and residual stress (LaFontaine et al. 1991, Tsui et al. 1996, Zagrebelny & Carter 1997, Suresh & Giannakopoulos 1998, Swadener et al. 2001, Lee & Kwon 2003, Xiao et al. 2014) from raw load-depth data. As a specific issue, the residual stress was roughly probed by observing a shape variation in the raw nanoindentation curve obtained from sample surface subjected to mechanical stress (Zagrebelny & Carter 1997); compressive stress disturbs a penetration of an indenter into sample surface and results in shallower indentation at a given indentation load than that of tensile surface stress. Resulting hardness increases under the compressive stress but decreases under the tensile stress.

However, Tsui et al. (1996) arrived at a conclusion through a series of empirical observation that the hardness is nearly insensitive to the residual stress in elastic regime. Based on this result, few theoretical models for the indentation loading curve appeared. Firstly, Suresh and Giannakopoulos (1998) explained a load decrease at a given indentation depth from an indented surface subjected to tensile residual stress as an addition of imaginary uniaxial stress having the opposite sign and the same amount of the residual

stress to the contact pressure. Lee and Kwon (2003, 2004) tried to extract a deviatoric stress component along the indentation axis from residual stress tensor. This deviatoric stress plays as the imaginary stress in the Suresh and Giannakopoulos model (1998) and causes a decrease in the indentation load at a given indentation depth when the indented surface is subjected to tensile stresses. According to Carlsson and Larsson's previous studies (2001a, b), the theoretical models were, however, applicable to rigid plastic materials which show the stress-independent hardness behavior similar with the empirical data reported by Tsui et al. (1996).

In this study, the indentation strain concept suggested by Milman et al. (1993) was applied to analyze 3D morphologies of nanoindents formed on surfaces subjected to residual stresses. A stress-induced change in the indentation strain was related to contact stress fields by considering elastic unloading or reloading pathway and finally predicted stresses from the indentation stress analysis were directly compared with empirically measured bending stresses.

2 EXPERIMENTAL PROCEDURES

As-cast Zr-based metallic glass having the Vitreloy I composition was machined into a stripe sample of $7 \times 40 \times 0.5$ mm. Both surfaces of the sample were ground with emery papers and finally polished mirror-likely for subsequent nanoindentations. Then it was bent with a 4-point bending jig in Figure 1 and the uniaxial tensile stress applied was estimated from curvature radius measured with a reflective laser profiler (LJ-G015 model, Keyence, Japan); the bending surface stress is given by E·t/(2R), where E = Young's modulus; t = thickness of the stripe sample; and R = curvature radius.

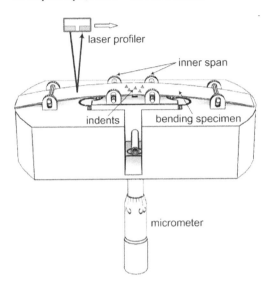

Figure 1.   A 4-point bending jig for stressing stripe samples.

Nanoindentations were done with a MZT-512 nanoindenter (Mityutoyo, Japan) coupled with a Berkovich indenter inside of the inner bending span where the applied stress is uniform. The indentation load increased with the speed of 50 mN/s and hold during 1 s at the peak indentation load of 500 mN. Then the load was relaxed with the speed of -50 mN/s and finally remnant nanoindents were obtained on the bent stripe surface. More than five indentations were repeated at a bent state or an applied stress level.

The indented surface was directly observed with an atomic force microscope (XE-100 AFM, Park Systems, Korea) without removing the bending jig. The indented surface with an area of 30 μm × 30 μm was scanned after locating the remnant nanoindent at the center of the AFM image. Scanning speed was controlled less than 0.2 Hz and whole AFM image acquired was converted into a digital image with a pixel of 512 × 512. Then 3D morphology analysis was done for the remnant nanoindent with a Matlab® image processor.

3   RESULTS AND DISCUSSION

3.1   *3D morphologies of the remnant nanoindents on stressed surfaces*

Curvature radii corresponding to three bending levels were 59.68±1.83, 50.04±1.10 and 29.03±0.48 mm.

By considering E and t of the metallic glass sample as 95 GPa and 0.5 mm, respectively, the stresses estimated from the curvature radii were 398.25±11.78, 474.77±10.29 and 818.23±13.14 MPa. In addition, a fixation of the stripe sample in the bending jig also caused an unexpected curvature in the stripe sample; the curvature radius was 2877.65±98.32 mm and corresponding stress was 8.26±0.29 MPa. A morphological change in the remnant nanoindents obtained from four different stress levels were plotted in Figure 2. Although a morphological change was not clearly identified from the impressions, a significant decay in the material pile-ups was found with an increase of the tensile stresses. This behavior is consistent with the theoretical models (Suresh & Giannakopoulos 1998, Lee & Kwon 2003) assuming a stress-insensitive hardness in elastic regime which is empirically observed by Tsui et al. (1996). The material pile-ups which are displayed as contours around a nanoindent in Figure 2 become weaker with tensile residual stress in order to meet both assumptions of the theoretical models; the indentation hardness is insensitive to the surface stress and the peak indentation depth increases with the increase of tensile residual stress.

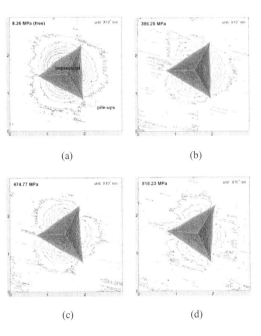

Figure 2.   Variation of pile-up morphology around indents formed under several tensile stresses.

Different from the negligible dependency of 2D nanoindent shape on surface stress, the impression volume or indentation plasticity showed a linear proportional dependency on the uniaxial tensile stress according to an authors' previous study (Lee et al.

2010). This phenomenon is exactly consistent with the stress-dependency of the elastic recovery ratio along the indentation axis which is reported by Xu and Li (2005). Instead of these empirical parameters, a direct relationship between the residual stress and the nanoindentation stress/strain was investigated in this study.

3.2 *Variation of the indentation strain along the loading and unloading paths*

A nanoindentation cycle involves three contact states of 0, 1 and 2 in Figure 3. While the loading path from the State 0 to the State 1 shows elastic/plastic characteristics, the unloading path from the State 1 at the peak indentation load to the State 2 corresponding to remnant nanoindent has reversible or elastic behavior. According to a Milman et al.'s work (1993), two indentation strains are defined: the peak contact strain, ϵ_c at the State 1 and the nanoindent strain, ϵ_i at the State 2. ϵ_c is defined by $-\ln(S_1/S_0)$, where S_1 = contact area beneath a sharp indenter; and S_0 = projected area corresponding to S_1. Negative sign in the above equation means compressive strain and it extends conventional 1D strain expressed by the ratio of elongated length with it original length to 2D areal strain. Similarly, ϵ_i is expressed by $-\ln(S_2/S_0)$, where S_2 = deformed area inside of the remnant nanoindent; and S_0 = projected area corresponding to S_2. Max and Balke (1997) have reported that radial recovery in a remnant nanoindent is negligible comparing to its depth-directional recovery. Thus, the projected area can be given by a common value S_0 regardless of the loaded or unloaded state if radial recovery in the remnant nanoindent is neglected.

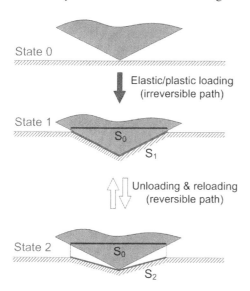

Figure 3. Contact morphologies related to three states consisting of a nanoindentation cycle.

To measure the contact and nanoindent areas from a 3D nanoindent morphology that is observed, a determination of the contact boundary is of importance. Three peak points, D, E and F determined from pile-up lobes were connected with three corners, A, B and C of a Berkovich impression and finally formed a polygonal contact area according to the Saha & Nix's method (2001); S_2 was approximated by adding six triangular patches in Figure 4. The triangle AOF can be selected as one of the six patches and the point O is the nanoindent center. In addition the in-plane area closed with six points, A, B, C, D, E and F was integrated for determining S_0.

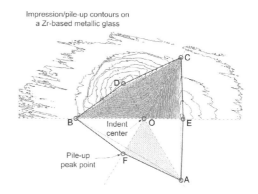

Figure 4. Extraction of a polygonal contact boundary from a remnant nanoindent with pile-up lobes.

From nanoindents formed on the unbent stripe, ϵ_i corresponding to the stress-free state was estimated and compared with ε_i^{res}s measured from stressed states; ε_i^{res} showed a linear proportional relationship with the uniaxial tensile stress, σ_{res}. It means that the impression area may be proportional to the tensile residual stress if the projected contact area is invariant regardless of residual stress. In other words, a depth-directional recovery inside a nanoindent decreases with an increase of the tensile residual stress. This phenomenon is consistent with the reverse proportional relationship between a recovery ratio of the indentation depth and the tensile stress reported in a previous study (Xu & Li 2005).

3.3 *A quantitative prediction of residual stress*

In order to describe an elastic interaction between indentation stress and residual stress, the reversible unloading or reloading pathway between the States 1 and 2 in Figure 3 is considered. When the elastic reloading of the pre-existing nanoindent takes place, the indentation strain along the indentation axis builds up from the remnant nanoindent strain, ϵ_i to its peak value, ϵ_c at the fully loaded state. A change of the indentation strain along the reloading pathway is expressed with the elastic stress components related to uniaxial compression under an indenter and thus whole stresses and strains have a negative sign (see Equation 1). In addition, different from the loading pathway, whole stress and strain components related to the reloading pathway are in the elastic regime.

$$\varepsilon_c - \varepsilon_i = [\sigma_{33} - \nu(\sigma_{11} + \sigma_{22})]/E, \tag{1}$$

where σ_{33} = stress component along the indentation axis; σ_{11} and σ_{22} = stress components parallel to the indented surface; and ν = Poisson's ratio of the indented sample. Here σ_{33} is equivalent to the contact hardness, H.

If the indentation reloading procedure occurs in a sample surface subjected to a pre-existing stress, the change in the contact strain in Equation 1 can be modified into Equation 2 by implicating the residual stress term into the right side of Equation 1. σ_{res} can be arithmetically added to σ_{11} because whole indentation reloading and stress application occur in the elastic regime. Note that the tensile stress application is expressed as a subtraction form in Equation 2 to consider whole stress and strain components in Equation 1 having a negative sign. Here, ε_i is changed into ε_i^{res} which represents the nanoindent strain under the pre-existing stress, while ε_c is assumed to be invariant because the indentation strain at the fully loaded state corresponds to the contact hardness which has not any dependency on the residual stress according to a Tsui et al.'s study (1996).

$$\varepsilon_c - \varepsilon_i^{res} = [\sigma_{33} - \nu(\sigma_{11} + \sigma_{22} - \sigma_{res})]/E. \tag{2}$$

By subtracting Equation 1 from Equation 2, σ_{res} can be summarized by Equation 3. It means the residual stress can be predicted from a stress-induced change in the nanoindent strain and elastic properties of the indented sample.

$$\sigma_{res} = E(\varepsilon_i - \varepsilon_i^{res}))/\nu. \tag{3}$$

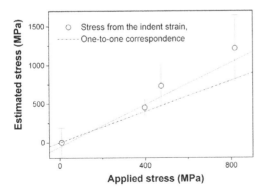

Figure 5. A linear proportional relationship between both estimated and applied stresses.

Residual stress could be calculated by using Equation 3 with information of a change in ε_i^{res} due to the surface stress and elastic properties of the sample; ν of the Zr-based metallic glass is given by 0.36. Especially, a change of the nanoindent strain corresponding to the nearly stress-free state was calculated from a difference between each nanoindent strain and averaged nanoindent strain. Applied stresses measured from the curvature radii of the artificially bent stripes were compared with the predicted stresses from the remnant nanoindent strains (see Figure 5); a clear linear proportional behavior was identified excepting tensile residual stress regime higher than 400 MPa.

The stress overestimation at the high stress level can be explained from few viewpoints. According to a Bolshakov et al.'s study (1996), a rapid expansion of plastic zone occurs under tensile residual stress higher than 99 % of the yield strength. If unexpected expansion of plastic zone occurs under high tensile residual stress, $\varepsilon_c - \varepsilon_i^{res}$ in Equation 2 can increase drastically and this in turn can cause a stress overestimation. In addition, negligible pile-up under high tensile residual stresses makes it difficult to determine boundary of the plastic zone and also brings about effect of the surface roughness. This causes a significant standard deviation under high stress regime. Carlsson and Larsson (2001a, b) have reported that the concept of stress-independent contact pressure is applicable to rigid plastic materials. However, strong elastic properties in a metallic glass sample also can cause a stress deviation from Equation 3.

Notwithstanding some unexpected data, it is valuable to derive a theoretical model for predicting residual stress from the nanoindent strain based on elastic stress and strain interactions. Furthermore this model can be applied general surface stress states and soft metallic materials.

4 SUMMARY

Conventional indentation approaches for measuring residual stress rely on a shape change in nanoindentation load-depth curve due to an application of surface residual stress. However, indirect stress approximations from hardness distribution have been frequently used for welds and joints. Thus, an investigation for predicting residual stress and mechanical properties from remnant indents is needed. In this study, a residual stress model predicting quantitative stress values from remnant nanoindents was developed by combining the nanoindent strain from 3D impression morphologies and the elastic contact stress/strain concept and major conclusions are summarized as below:

1 Remnant nanoindents formed on artificially bent or uniaxially stressed metallic glass stripes showed a linear decrease in pile-ups with an increase of tensile stress.
2 3D nanoindents were approximated as polygonal pyramids, and the contact area inside of the pyramid and its projected area were used for estimating the remnant nanoindent strain. The strain measured showed a linear proportional dependency on the tensile residual stress.
3 An elastic relationship between the contact stress and strain beneath an indenter at the peak indentation load is generally expressed for an indented surface subjected to residual stresses; the indentation stress components parallel to the sample surface is changed into a sum of the indentation stress and residual stress parallel to the surface.
4 If the contact strain under the peak indentation load is independent of the residual stress, the residual stress can be expressed by a multiple of the remnant nanoindent strain and Young's modulus of the indented sample.
5 The predicted stress has a clear proportionality with the applied stress excepting in high stress regime above 400 MPa. This may be attributed to an unexpected expansion of the plastic zone and prevalent elastic properties in the metallic glass samples.

ACKNOWLEDGMENT

This research was partly supported by the Converging Research Center Program funded by the Ministry of Science, ICT and Future Planning (NRF-2014M3C1A8048818) and the Korea Research Council of Fundamental Science and Technology (KRCF) through Creative Agenda Project.

REFERENCES

[1] Bolshakov, A., Oliver, W.C. & Pharr, G.M. 1996. Influences of stress on the measurement of mechanical properties using nanoindentation: Part II. finite element simulations. *Journal of Materials Research* 11: 760–768.
[2] Carlsson, S., & Larsson, P.-L., 2001a. On the determination of residual stress and strain fields by sharp indentation testing. Part I: theoretical and numerical analysis. *Acta Materialia* 49 (12), 2179–2191.
[3] Carlsson, S., & Larsson, P.-L., 2001b. On the determination of residual stress and strain fields by sharp indentation testing. Part II: experimental investigation. *Acta Materialia* 49 (12), 2193–2203.
[4] Kim, J.-Y., Lee, K.-W., Lee, J.-S., & Kwon, D. 2006. Determination of tensile properties by instrumented indentation technique: Representative stress and strain approach. *Surface and Coatings Technology* 201(7): 4278–4283.
[5] Kim, S.H., Lee, B.W., Choi, Y., & Kwon, D. 2006. Quantitative determination of contact depth during spherical indentation of metallic materials-A FEM study. *Materials Science and Engineering* A 415: 59–65.
[6] LaFontaine, W.R., Paszkiet, C.A., Korhonen, M.A. & Li, C.-Y. 1991. Residual stress measurements of thin aluminum metallizations by continuous indentation and X-ray stress measurement techniques. *Journal of Materials Research* 6: 2084–2090.
[7] Lee, J.-S., Jang, J.-i., Lee, B.-W., Choi, Y., Lee, S.G., & Kwon, D. 2006. An instrumented indentation technique for estimating fracture toughness of ductile materials: A critical indentation energy model based on continuum damage mechanics. *Acta Materialia* 54(4): 1101–1109.
[8] Lee, Y.-H. & Kwon, D. 2003. Measurement of residual-stress effect by nanoindentation on elastically strained (100) W. *Scripta Materialia* 49(5): 459–465.
[9] Lee, Y.-H. & Kwon, D. 2004. Estimation of biaxial surface stress by instrumented indentation with sharp indenters. *Acta Materialia* 52(6): 1555–1563.
[10] Lee, Y.-H., Yu, H.-Y., Baek, U.B. & Nahm, S.H. 2010. Analysis of residual stress through a recovery factor of remnant indents formed on artificially stressed metallic glass surfaces. *Korean Journal of Metals and Materials* 48: 203–209.
[11] Marx, V. & Balke, H. 1997. A critical investigation of the unloading behavior of sharp indentation. *Acta Materialia* 45: 3791–3800.
[12] Milman, Y.V., Galanov, B.A. & Chugunova, S.I. 1993. Plasticity characteristic obtained through hardness measurement. *Acta Metallurgica et Materialia* 41: 2523–2532.
[13] Oliver, W.C. & Pharr, G.M. 1992. An improved technique for determining hardness and elastic modulus using load and displacement sensing indentation experiments. *Journal of Materials Research* 7(6): 1564–1583.
[14] Saha, R. & Nix, W.D. 2001. Soft films on hard substrates-nanoindentation of tungsten films on sapphire substrates. *Materials Science and Engineering A* 319–321: 898–901.
[15] Suresh, S. & Giannakopoulos, A.E. 1998. A new method for estimating residual stresses by instrumented sharp indentation. *Acta Materialia* 46(16): 5755–5767.
[16] Swadener, J.G., Taljat, B., Pharr, G.M., 2001. Measurement of residual stress by load and depth sensing indentation with spherical indenters. *Journal of Materials Research* 16: 2091–2102.
[17] Tsui, T.Y., Oliver, W.C. & Pharr, G.M. 1996. Influences of stress on the measurement of mechanical properties using nanoindentation: Part I. Experimental studies in an aluminum alloy. *Journal of Materials Research* 11: 752–759.

[18] Xiao L., Ye D. & Chen C. 2014. A further study on representative models for calculating the residual stress based on the instrumented indentation technique. *Computational Materials Science* 82: 476–482.

[19] Xu, Z.-H. & Li, X. 2005. Influence of equi-biaxial residual stress on unloading behaviour of nanoindentation. *Acta Materialia* 53: 1913–1919.

[20] Zagrebelny, A.V. & Carter, C.B. 1997. Indentation of strained silicate-glass films on alumina substrates. *Scripta Materialia* 37: 1869–1875.

Testing and Measurement: Techniques and Applications – Chan (Ed.)
© 2015 Taylor & Francis Group, London, ISBN: 978-1-138-02812-8

Low-cost machinery fault simulator

K. Shin
Department of Mechanical and Automotive Engineering, Andong National University, Andong, Kyungbuk, Korea

S.H. Lee & S.H. Song
Department of Mechanical Design Engineering, Andong National University, Andong, Kyungbuk, Korea

ABSTRACT: The condition of the most industrial machineries based on rotary motor is monitored by measuring the vibration of the machinery. Also, in terms of reliability and cost, condition-based maintenance has become an important maintenance program and many studies related to condition-based maintenance such as diagnostics and prognostics are carried out. For exact diagnostic and prognostics, the machineries are tested using a Machinery Fault Simulator (MFS). In this paper, we present a low cost MFS for teaching and research and validate the proposed MFS through experiments.

1 INTRODUCTION

Machineries consist of many parts, so that a small fault in one part can create a great expense for maintenance due to the failure of the entire system. Time-based or condition-based maintenance is used to monitor the conditions. But condition-based maintenance has advantages over time-based maintenance in terms of cost (Andrew et al. 2006, Tandon et al 2007). The technologies such as condition-monitoring, diagnostics, and prognostics are necessary for condition-based maintenance. In order to generate an exact diagnostics, the theoretical knowledge and practical skills for the measurement and diagnostics are required. Therefore, machine fault simulators (MFSs) were introduced to study a new theory of diagnostics and for training (Peter et al 2007, Sawalhi & Randall 2008). Since the MFS is a good example for an integrated mechatronics system, it is used to teach theories of vibration, as well as principles of mechatronic systems. However, the commercialized MFS is too expensive to equip several systems for education, and the costs of sensors and data acquisition devices are too high even if the system is built in a laboratory. Therefore, we present a low cost MFS and verify its feasibility as an educational and research tool.

2 DESIGN OF MFS

2.1 Test bed

A test bed, sensors, data acquisition system, and analysis program are main components of MFS.

Misalignment between shafts, bearing faults, unbalanced shafts, and gear faults result vibration and this shows the condition of machinery. The test bed was designed to cover all causes of vibration such as alignment, unbalance, and bearing fault except gear faults. Figure 1 shows the total MFS system where test bed is composed of a motor, coupling, shaft, and shaft support. In order to reduce the cost, we used the standardized parts and reduced the mechanically machined parts. An AC induction motor was used and an inverter was employed to adjust the rotating speed. The cost of test bed was about below $1500.

2.2 Sensor

In general, the properties of vibration exhibits various conditions and fast response to some event of machineries (Linfeng et al. 2013, Luisa et al. 2012). Therefore, vibration-based condition monitoring is popular and we chose the accelerator. We selected the micro-electromechanical systems (MEMS)-based accelerometer made for small digital devices of which sensitivity is 1000 mV/g and its bandwidth is 800 Hz. And theoretical maximum applicable rotating speed is 4800rpm. Since the selected sensor has three degrees of freedom, it is possible to measure vertical, horizontal, and axial vibration. Because the sensor is delivered in a dual flat no lead plastic package, we designed a printed circuit board and developed a module-type sensor. The total cost to make the sensor module was below $10.

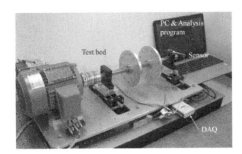

Figure 1. Total MFS system.

2.3 Hardware for data acquisition

Since the DAQ board for noise and vibration measurement requires simultaneous sampling for multiple signals, the price is high. Therefore, we used the multiplexed DAQ and evaluated the feasibility of a low-cost DAQ for multiplexed sampling instead of simultaneous sampling.

Considering the cost and the easiness of programming, we selected DAQ (USB6008, National instruments, USA) which uses a universal serial bus. In particular, a bundle for students that includes the tested DAQ and LabVIEW software can be purchased at low price. Therefore, we could use four differential analog inputs with maximum sampling rate of 2.5 kS/s for the best resolution and sampling rate with the selected DAQ. A multiplex architecture has limits, so we tested the selected DAQ to evaluate its performance with respect to the simultaneous sampler.

We set the maximum number of signals for evaluating simultaneous sampling to two. We measured vibration of rotating shaft and obtained the correlation coefficient. It is expected that periodic signals will show high correlation coefficients in multiplexed data acquisition. However, even in this case, the correlation coefficient changed with respect to the rotating speed of the motor. Figure 2 shows the relationship between rotating speed and the correlation coefficient. Here, the sampling rate was fixed at 5 kHz and the number of datasets acquired for 3 seconds was 50. As the rotating speed increased, a higher correlation coefficient was obtained.

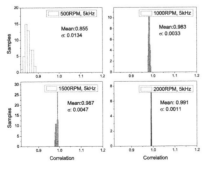

Figure 2. Correlation with respect to rotating speed (5 kHz sampling).

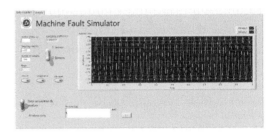

Figure 3. Software for data acquisition and analysis.

Ultimately, we obtained a mean correlation coefficient of 0.991. For 500 rpm, the correlation coefficient was low, and the standard deviation was larger than in the other case. It was expected that these results were caused by the low responsiveness of the sensor module at low frequency. When we examined the power spectrum density of this signal, we came to the same conclusion. The peak at fundamental frequency was less than about 100 times that of the other cases. Furthermore, when we increased the sample number, the standard deviation decreased to a value similar to other cases, but the mean value did not change.

2.4 Software for data acquisition and analysis

The developed program implemented by LabVIEW is composed of two parts, data acquisition and data analysis. We can set configurations such as the sampling rate and number of samples and channels and carry out statistical analysis in each part. Figure 3 shows the front panel of the data analysis part.

3 EXPERIMENT

3.1 Sensor test

We attached two sensors (commercialized and manufactured) on the same lateral line of the cantilever instead of same point and measured the impulse response using a simultaneous sampling data acquisition (DAQ) board. Figure 4 shows the power spectral density of the obtained signals and the frequencies of the peaks of each sensor, though the magnitude of peaks were different due to low sensitivity, matched.

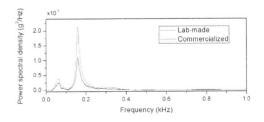

Figure 4. Comparison of two sensors in frequency domain.

3.2 Condition monitoring

The developed sensor module was attached on the bearing support far from the motor. Because it can measure three-directional acceleration, the horizontal, vertical, and axial accelerations were measured. The rotating speed of motor was 1,500 rpm and one dataset was acquired for 3 seconds with a sampling rate of 3 kHz. A total of 10 datasets were obtained. Figure 5 is the result of bearing with an outer race defect. The ball pass frequency of the outer race (BPFO) of the bearing is calculated by equation (1), and the definitions and values of the parameters are given in Table 1:

$$f_{BPFO} = \frac{NB}{2}S\left(1 - \frac{BD}{PD}\cos(PHI)\right) \qquad (1)$$

The BPFO of the used bearing was 77 Hz, which is 3.08 times the rotating speed in revolutions per second. In particular, there are large peaks in the axial direction at 700 Hz and 1,000 Hz, which are close to the harmonics (x9, x13) of BPFO. However, the peaks of the defective bearing were not as clear as we expected. So, we did same test with high-performance sensors (603C01, PCB, USA) and DAQ (USB-4431, National Instruments, USA), but there was no large difference. This unclear result might have been caused by an imprecise defect on the bearing.

Table 1. Definitions and values of parameters for bearing frequency.

Bearing number	Number of balls	Ball dia.	Pitch dia.	Contact angle	Rev/s
	NB	BD	PD	PHI	S
6204	8	0.312	1.358	0	25

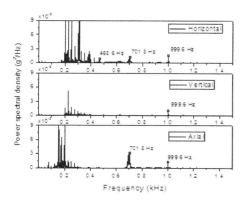

Figure 5. Power spectral density: bearing defect in outer race.

3.3 Correlation analysis

Correlation analysis was carried out to ascertain the possibility of simultaneous sampling of the proposed system. We placed two sensors at different places. The results of the correlation analysis are given in Fig. 6. The results obtained by the proposed system were inferior to those obtained by the high-performance one in terms of standard deviation. Since we obtained a high correlation coefficient in other test, this result was unexpected. The difference in the experimental conditions between the previous experiments and this one was just the number of sensor. Therefore, the probable cause of the decrease in correlation coefficient in relation to the high-performance system is the difference between the two sensor modules. As a result, the correlation coefficients obtained by the proposed system was less than the results obtained by the high performance system. However, since the trends of correlation coefficient variation for each case were the same, the relative qualitative analysis, though the absolute quantitative analysis might be impossible, is possible with the proposed system.

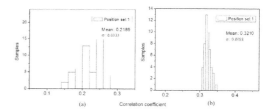

Figure 6. Comparison of the obtain results for a test bed rotating at 2,000 rpm with a 3 kHz sampling rate with proposed sensing system and high performance system.

4 CONCLUSION

We presented a low cost MFS which can be built in a laboratory. In particular, the total system, including the test bed, sensor, DAQ, and analysis program can be fabricated at low expense. The basic tests were carried out and showed the feasibility of condition monitoring. Moreover, in certain conditions, analysis that requires the simultaneous sampling such as the relative correlation analysis, could be performed by the designed MFS. The DAQ system and analysis program can be implemented by the user, so that the proposed MFS would be helpfulfor people in the field of mechatronics.

ACKNOWLEDGMENT

This work was supported by the Daejoo Machinery—Andong National University Cooperative Research Center and Daejoo Machinery Co. Ltd., Daegu, Republic of Korea.

REFERENCES

[1] Andrew K.S et al. 2006 A review on machinery diagnostics and prognostics implementing condition-based maintenance *Mechanical systems and signal processing* 20 1483–1510.

[2] Linfeng Deng & Rongzhen Zhao 2013 A vibration analysis method based on hybrid techniques and its application to rotating machinery *Measurement* 46 3671–3682.

[3] Luisa F.Villa, et al.2012 Statistical fault diagnosis based on vibration analysis for gear test-bench under non-stationary conditions of speed and load *Mechanical Systems and Signal Processing* 29 436–446.

[4] N. Sawalhi & R.B. Randall 2008 Simulating gear and bearing interactions in the presence of faults Part I. The combined gear bearing dynamic model and the simulation of localized bearing faults *Mechanical Systems and Signal Processing* 22 1924–1951.

[5] N. Tandon, et al. 2007 A comparison of some condition monitoring techniques for the detection of defect in induction motor ball bearings *Mechanical systems and signal processing* 21 244–256.

[6] Peter W. Tse, et al. 2004 Machine fault diagnosis through an effective exact wavelet analysis *Journal of sound and vibration* 277 1005–1024.

The sensitivity method to estimate critical speeds for rotors with multilayer laminated damper

W.Chen
Energy China, China Energy Equipment Beijing Technology Centre, Beijing, China

C.Su
Henan Mechanical and Electrical Engineering College, Xinxiang, China

ABSTRACT: The provided method, which uses the transfer coefficient method to compute critical speeds relevant to every scattered stiffness point in the preset stiffness domain, tends to verify the critical speeds of a rotor through analyzing the sensitivity of critical speeds changing with support stiffness. The first step is to simplify the rotor system and partition it into many computed stations, and then to compute the physical characteristics' value of them according to the requirements of transfer coefficient method. The second step is to preset the domain of support stiffness and to scatter stiffness points inside by dichotomy method, then to circularly compute the critical speeds and modes corresponding to every scattered stiffness point until the two neighboring stiffness points with the phenomenon of their modes changing extremely are finally found The third step is to draw the law curve how the critical speeds change with support stiffness and find the sub-domain corresponding to the phenomenon of the critical speeds which change extremely. The last step is to collect or study other cared dynamic characteristics data: modes and responses. It is convenient to modularize this method and helpful to estimate dangerous speed in engineering design or experiment.

1 INTRODUCTION

The multilayer laminated damper (MLD) is a kind of elastic support [1]- [2] and vibration absorber for rotor systems used in light-type gas turbine. It is interested in engineering design because of the characteristics of simple structure, easily assembled, great damper performance, outstanding stability and reliability. But the problem is that its stiffness is difficult to be acquired exactly either by simulating or by testing for its special configuration and the test cost. So, engineers have to face the predicament that they tend to use the uncertain stiffness value from experience or assessment [8] to compute critical speeds and dynamic responses of the rotor system. On the other hand, in practices, fixed stiffness in cool state will be always changing somewhat in a range in working state because of temperature, changed assembly conditions or other random factors[9]- [10].

Therefore, to enhance the quality of theoretical design under the precondition of low cost, a method for calculating critical speeds of rotors with multilayer laminated damper by using uncertain stiffness value is provided in this paper, which is based on transfer coefficient method [11~12] and named the sensitivity method.

2 METHOD

Figure 1 is the chart of calculation procedure, and will be introduced as being low. The first step is to simplify rotor configuration in order to execute the calculation numerically and conveniently. The simplification rules mainly include: ①all parts taking part in rotation will be taken as calculation objects; ②do not concern assembly relationships among all parts and take them as one body; ③ignore the impact on rotor from rotation torque; ④if an assembly consists of inner part and outer part, the physical parameters of outer part should be used in calculation; ⑤mainly care about sectional vibrations; ⑥put the function point of radial bearing on the center of shaft.

The second step is to divide calculation stations. The method instructed in this paper is based on the transfer coefficient method to simulate dynamic performance. So, station division should be done according to the simplified rotor structure and requirements of transfer coefficient method.

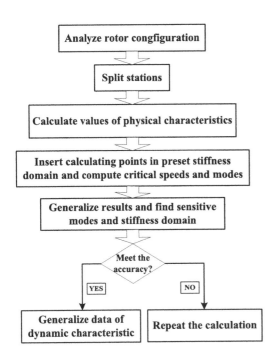

Figure 1.   Procedures of computation.

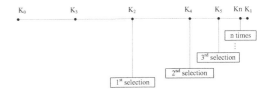

Figure 2.   Chart of selecting stiffness point.

The third step is to calculate physical characteristic value for every station. The correlative rules and requirements could be reference literature [13].

Then, the experiment presets support stiffness domain, scatters stiffness points in it by dichotomy method, and calculates corresponding critical speeds and modes and then tabulates the results to show the relationship between every stiffness point and its dynamic data while at the same time, draws the line of critical speeds vs. support stiffness. Among these, the process of selecting stiffness points in presetting stiffness domain could be expressed as figure 2. As in the chart, the preset stiffness domain is $[K_0 K_1]$ which is selected by engineers. In the first run, the experiment inserts the stiffness point K_2 by the dichotomy method, then respectively calculates data of critical speeds and modes according to the three stiffness points K_0, K_1, K_2. If there are similar modes between K_0 and K_0, we take $[K_2, K_1]$ as the stiffness domain for the second run, otherwise, we take $[K_0, K_2]$ as the stiffness domain for the second run if there are similar modes between K_1 and K_2 and the experiment repeats the selecting process like the above.

Based on the data, we find the sensitive stiffness value domain and the related critical speeds. The former is so sensitive to structure performance that critical speeds could change acutely and it should be avoided in design work.

3   SAMPLE

In this part, the rotor system of a 100kw gas turbine will be the sample to show how to search the sensitive critical speeds and stiffness value by mentioning method.

3.1   Simplification for the rotor

Simplify physical rotor system in order to execute calculation process later. Figure 3 shows a mathematic model for the rotor. Simplify compressor and turbine into mass disks. The K means combined structure stiffness, the C means combined structure damper, No. 1 means left bearing, No. 2 means right bearing. Especially, the combined structure includes rat-cage support, radial bearing and multilayer laminated damper.

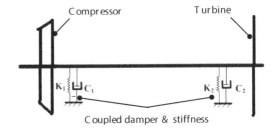

Figure 3.   Simplified configuration of the rotor.

3.2   Station division

According to the requirement of transfer coefficient method, we divide the simplified rotor into some stations and compute correlative physical parameter's value for latter procedure. In this case, 26 stations are divided and part of their physical parameter values are shown in table 1 for not too long paper. For the rotor, key general structure parameters include: total mass is 7.98kg, total length is 302mm, unbalance mass in compressor is 10g·mm, unbalance mass in turbine is 20g·mm.

3.3 Calculation and results

By experience, we take the preset stiffness domain in 4.0e6~1.25e8N/m, and set the damper ratio as 0.1. The critical speeds and modes are calculated and results are listed in table 2. Figure 5 shows the line of critical speeds vs. support stiffness.

From table 2, modes of the first and the second rank change differently in the stiffness domain 1.0e7~1.2e7N/m. From figure 5, in the same domain, the 2nd stiff mode and critical speed show great step change and unsteady. That is to say the stiffness

domain is sensitive to critical speed and mode. To search more accurate location of sensitive stiffness, we could take 1.0e7~1.2e7N/m as the new preset stiffness domain and repeat the calculation process.

We repeat the front steps and list results in table 3. From the mode charts, we can find that they change when the stiffness value is 1.1097e7N/m. This value has met engineering expect in accuracy, so the calculation could be finished finally.

In this case, the stiffness value 1.1097e7N/m is affirmed as the sensitive stiffness.

Table 1. Parameters of computed units.

Station	Name	Length (mm)	Outer diameter (mm)	Inner diameter (mm)	Density (kg/m³)	Mass (kg)
1	/	18	33.62	15	7.831 e3	0.104
2	/	24	29.35	15	7.831 e3	0.094
3	/	20	35.77	15	7.831 e3	0.13
4	compressor	25.5	85.08	15	7.831 e3	1.1
5	compressor	24.5	127.422	15	4.72 e3	1.455
6	/	11	32.81	15	7.831 e3	0.058
7	/	9	42.33	16.5	7.831 e3	0.084
8	bearing	8	39.76	16.5	7.831 e3	0.064
9	fixed	0	/	/	/	/
......	......	......	......	......	......	......
26	......	......	......	......	......	......

Table 2. Results of the first computation.

No.	Stiffness N/m	Damper ratio	1st	Mode	2nd	Mode
1	4.0e6	0.1	5093		11956	
2	7.0e6	0.1	6717		15632	
3	9.0e6	0.1	7600		17592	
4	1.0e7	0.1	8003		18475	
5	1.2e7	0.1	8749		20093	
6	1.5e7	0.1	9752		22228	
...	...	...	...	...	...	...
n	1.25e8	...	...	...	...	...

Table 3. Results of the second computation.

No.	Stiffness N/m	Damper ratio	Critical speed /rpm			
			1st	Mode	2nd	Mode
1	1.1e7	0.1	8385	/	19307	\
2	1.1097e7	0.1	8421	\	19385	⌒
...	...	...	...	...	...	...
m	1.2e7	0.1	9752	\	22228	⌒

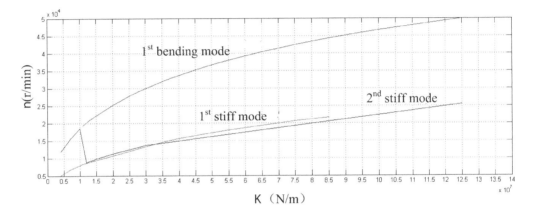

Figure 5. Laws of critical speeds changed with supporting stiffness.

4 CONCLUSION

The sensitivity method could estimate the critical speeds for rotors with multilayer laminated damper effectively. The results are reference for rotor design, test and diagnosis. More work should be done to verify that it is meaningful for those rotors with unknown support stiffness.

REFERENCES

[1] Bugra H, Ertas B, Luo Huageng. Nonlinear dynamic characterization of oil-free wire mesh dampers[J]. Journal of Engineering for Gas Turbines and Power, 2008, 138: 032503–1–8.

[2] Ertas B, Luo Huageng, Hallman D. Dynamic characteristics of shape memory alloy metal mesh dampers[R]. AIAA-2009–2521, 2009.

[3] CHEN Wei, DU Farong, DING Shuiting, etl. Design of rotor system for micro turbojet engine[J]. Journal of Aerospace and Power, 2009, 24(5):1171–1176. (in Chinese).

[4] LIU Shaoquan, ZHANG Yanchun, DU Zhaogang, etl. Prediction of the influence tempreture field on the critical speeds of a rod-fastened rotor[J]. Gas Turbine Technology, 2011, 24(2):20–23. (in Chinese).

[5] ZHU Xiangzhe, YUAN Huiqun, HE Wei. Effect of steady thermal field on critical speeds of a rotor system[J]. Journal of Vibration and Shock, 2007, 26(12):113–116. (in Chinese).

[6] LI Yuxi, WANG Sanmin. Application of the Improved Whole Transfer Matrix Method to the Calculation of Complex Rotor Critical Speed[J]. Journal of Aerospace and Power, 2005, 20(3):413–417. (in Chinese).

[7] CHAI Shan, GANG Xianyue, YAO Fusheng, etl. A whole transfer matrix method for calculating the critical speed of multi-rotor system[J]. J. Shanghai University of Science and Technology, 2002, 24(1):8–12. (in Chinese).

[8] Kobayashi M, Saito S, Yamauchi S. Nonlinear Steady-State Rotor Dynamic Analysis U sing Transfer Coefficient Method[R], DE- Vol. 35, Rotating Machinery and Vehicle Dynamics, ASME 1991.

Testing and Measurement: Techniques and Applications – Chan (Ed.)
© *2015 Taylor & Francis Group, London, ISBN: 978-1-138-02812-8*

Analytically modeling nonlinear dc-gain for op-amps

F.C. Chen & J.Y. Lin

Department of Electrical Engineering, National Chiao-Tung University, Taiwan

ABSTRACT: Op-amp dc-gain is nonlinear, and this can cause distortions in various applications. However, so far no efficient tool is available for analyzing dc-gain nonlinearity; the dc-gain curve can only be generated from a designed op-amp. If the dc-gain nonlinearity is too serious, one has to redesign op-amps. Time-consuming iterations involving transistor-level designs and simulations may be needed before an acceptable result is achieved. In this brief, an analytical op-amp dc-gain curve model is proposed. After entering circuit parameters and process parameters into this model, it can generate op-amp nonlinear dc-gain curves to facilitate subsequent distortion analyses, requiring no transistor- level designs and simulations. This dc-gain model applies to general two-stage op-amps, and can accommodate various 1st stage configurations, including cascode and folded cascode ones. Our analyses show that the smaller V_{DD} is, the more precise the analytical model becomes. Simulations demonstrate that, when $V_{DD} \leq 1.8V$, the error between the curve from the model and the curve generated from SPICE simulation is less than 2%.

1 INTRODUCTION

Operational amplifiers are important devices in analog circuits. In practice, the op-amp dc-gain is nonlinear [1], and this can cause distortions in various applications. For example, the distortion resulted from nonlinear dc-gain can significantly reduce SNDR of sigma-delta modulators (SDMs) [2, 3]. Recently, as supply voltage and device dimension continue to scale down, the effects of dc-gain nonlinearity become more and more severe than ever [4]. However, so far no efficient tool is available for analyzing dc-gain nonlinearity in op-amp designs.

Traditionally, designers who want to evaluate the distortions caused by op-amp dc-gain nonlinearity need to have op-amps designed first. Then, various approaches can be employed to compute distortions caused by the nonlinearity, such as the op-amp dc-gain approach [2]–[7], the input/output approach [8, 9], and the transconductance approach [10]. If the distortion is greater than a tolerance, one has to modify or redesign the op-amp and go through the aforementioned process again. Iterations continue until an acceptable design is achieved. This is a very time-consuming process. In order to significantly speed up the design process, a *nonlinear op-amp dc-gain curve model* was proposed in [11], which produces dc-gain curves directly from op-amp design parameters, eliminating the need of transistor-level op-amp designs and simulations. Unfortunately, this model was trial-and-error in nature and was obtained by finding an equation which fits to several representative nonlinear dc-gain curves, lacking theoretical justifications.

In brief, we propose an *analytical* nonlinear op-amp dc-gain curve model. To the best of our knowledge, no similar results have appeared in literature before. The proposed model applies to the general two-stage op-amps; in particular, it accommodates various 1st stage amplifier configurations, including cascode and folded cascode ones. This model is obtained by decomposing dc-gain into *output resistance* and *generalized transconductance*. After the problem is formulated in Section II, the dc-gain model is derived in Section III. Then, in Section IV, the proposed analytical model is verified against transistor-level circuits. The conclusion follows in Section V.

2 PROBLEM FORMULATION

The purpose of this brief is to propose an *analytical* model of nonlinear dc-gain curves for two-stage op-amps. This model will be of the form

$$A_V\left(V_O\right) = A_0 \ \mathrm{R}\left(\left(\frac{W}{L}\right)_6, \left(\frac{W}{L}\right)_7, V_{DD}, V_{SS}, V_{SG7}, V_{tp7}, V_O\right). \quad (1)$$

After $\left(\dfrac{W}{L}\right)_6$, $\left(\dfrac{W}{L}\right)_7$, V_{DD}, V_{SS}, V_{SG7}, and V_{tp7} are determined, the dc-gain curve is determined and can be plotted against V_O.

3 OP-AMP NONLINEAR DC-GAIN MODEL

In this section, we derive the analytical nonlinear dc-gain curve model from (1). We decompose (1) into two parts: The first part is the *generalized transconductance* $G_m = g_{m1} \times (r_{o2} \| r_{o4}) \times g_{m6}$, consisting of the dc-gain of *1st stage* and the transconductance of second stage. The second part is the *output resistance* $R_{out} = (r_{o6} r_{o7})$ of the second stage. Therefore, (1) is rewritten as

$$A_V = G_m \times R_{out}. \tag{2}$$

3.1 *Output resistance of 2nd stage:* R_{out}

The drain currents can be derived as

$$I_{D6} = I_{D7} = \frac{1}{2} \mu_n c_{ox} \left(\frac{W}{L}\right)_6 \left(V_{GS6} - V_{tn6}\right)^2 \left(1 + \lambda_6 V_{DS6}\right)$$

$$= \frac{1}{2} \mu_p c_{ox} \left(\frac{W}{L}\right)_7 \left(V_{SG7} - |V_{tp7}|\right)^2 \left(1 + \lambda_7 V_{SD7}\right) \tag{3}$$

where W and L are the width and length of the device respectively; μ_n and μ_p are the mobility; c_{ox} is the oxide capacitance per unit channel area; V_{tn6} and V_{tp7} are the threshold voltage; λ_6 and λ_7 are the channel length modulation parameters. The transistors operate in the saturation region.

In (3), $V_{DS6} = V_O - V_{SS}$ and $V_{SD7} = V_{DD} - V_O$. Then, solving for V_{GS6} in (4), one obtains

$$V_{GS6} = \sqrt{\frac{\mu_p \left(\frac{W}{L}\right)_7}{\mu_n \left(\frac{W}{L}\right)_6} \frac{1 + \lambda_7 (V_{DD} - V_O)}{1 + \lambda_6 (V_O - V_{SS})}} \left(V_{SG7} - |V_{tp7}|\right)^2 + V_{tn6}. \tag{4}$$

The meaning of (4) is that, the V_{GS6} of gain stage M_6 can be traced by sweeping V_O. Note that V_{SG7} in (4) can be considered to be constant, since M_5 and M_7 are parallel to M_8.

With I_{D6} and I_{D7} described in (3) and V_{GS6} derived in (4), the output resistance R_{out} can be represented as a function of design parameters, as is shown in (7). Since all of these parameters, except V_O, are constants, the R_{out} expressed in (7) is denoted as

$$R_{out} = \frac{1}{\lambda_6 \left(\frac{1}{2} \mu_n c_{ox} \left(\frac{W}{L}\right)_6\right) \left(\sqrt{\frac{\mu_p \left(\frac{W}{L}\right)_7}{\mu_n \left(\frac{W}{L}\right)_6} \frac{1 + \lambda_7 (V_{DD} - V_O)}{1 + \lambda_6 (V_O - V_{SS})}} \left(V_{SG7} - |V_{tp7}|\right)^2\right)^2 + \lambda_7 \left(\frac{1}{2} \mu_p c_{ox} \left(\frac{W}{L}\right)_7\right) \left(V_{SG7} - |V_{tp7}|\right)^2}$$

$$R_{out} = R_{out}(V_O)\Big|_{\left(\frac{W}{L}\right)_6, \left(\frac{W}{L}\right)_7, V_{DD}, V_{SS}, V_{SG7}, V_{tp7}} = R_{out}(V_O). \tag{5}$$

3.2 *Generalized transconductance:* G_m

With the R_{out} of (3) derived in Part A, here we discuss the G_m of (3).

$$G_m = [g_{m1} \times (r_{o2} \| r_{o4})] \times g_{m6}. \tag{6}$$

In summary, analyses and simulations in this subsection show that variations in g_{m6} and $[g_{m1} \times (r_{o2} \| r_{o4})]$ are generally small, though the former decreases when V_{DD} decreases and the latter is random in nature; moreover, for both g_{m6} and $[g_{m1} \times (r_{o2} \| r_{o4})]$, largest variation typically occurs when V_O is close to the boundaries of its range. Therefore, *it is concluded that variations in G_m are generally small, but variations tend to be larger when V_{DD} is larger, and the largest variation in G_m typically occurs at the boundaries of the range of V_O.*

Table 1. DC-Gain variations of 1st stage.

First stage configuration	$V_{DD}=3.3V$, $V_{SS}=0V$ V_O: 0.3 to 3 V	$V_{DD}=2.5V$, $V_{SS}=0V$ V_O: 0.3 to 2.2 V	$V_{DD}=1.8V$, $V_{SS}=0V$ V_O: 0.3 to 1.5 V
Two-stage	$|A_{min}|=47.14$ $|A_{max}|=48.38$ ΔA_{max}: **1.29%**	$|A_{min}|=42.55$ $|A_{max}|=44.1$ ΔA_{max}: **1.79%**	$|A_{min}|=34.43$ $|A_{max}|=36.85$ ΔA_{max}: **3.39%**
Cascode	$|A_{min}|=4454.2$ $|A_{max}|=4822.5$ ΔA_{max}: **3.97%**	$|A_{min}|=1753.1$ $|A_{max}|=1784.3$ ΔA_{max}: **0.88%**	$|A_{min}|=1435.1$ $|A_{max}|=1530.2$ ΔA_{max}: **3.21%**
Folded cascode	$|A_{min}|=3448.5$ $|A_{max}|=3564.8$ ΔA_{max}: **1.66%**	$|A_{min}|=2209.2$ $|A_{max}|=2247$ ΔA_{max}: **0.85%**	$|A_{min}|=272.16$ $|A_{max}|=293.68$ ΔA_{max}: **3.80%**

3.3 *Proposed analytical nonlinear dc-gain model*

Since $G_m(V_O)$ typically exhibits small variations w.r.t. V_O as is discussed in Part B, the ratio $G_m(V_O)/G_0$ in (13) has a small deviation from 1. Furthermore, the effect of this deviation can be significantly reduced by the shaping provided from $R_{out}(V_O)/R_0$ (to be explained in Section IV). So, $G_m(V_O)/G_0$ in (13) is assumed to be one, and therefore the proposed analytical nonlinear dc-gain model becomes

Model: $\qquad A_V(V_O) = A_0 = \dfrac{R_{out}(V_O)}{R_0}. \tag{7}$

4 SIMULATION RESULTS

The curves obtained from analytical dc-gain model (7) will be compared with those from SPICE

240

simulations on transistor-level circuits. However, in order to quantitatively analyze the error, the (6) is used to represent a real circuit, so that a relative error can be defined by subtracting (7) from (6), and then being divided by A_0, as

$$E_r = \left| 1 - \frac{G_m(V_O)}{G_0} \right| \times \frac{R_{out}(V_O)}{R_0}. \qquad (8)$$

The characteristics of $|1 - G_m(V_O)/G_0|$ and $R_{out}(V_O)/R_0$ are *conceptually* illustrated in Fig. 1, and are summarized here:

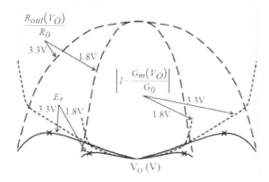

Figure 1. $|1 - G_m(V_O)/G_0|$ and $R_{out}(V_O)/R_0$ have different characteristics in different supply voltages.

1 Curves of $|1 - G_m(V_O)/G_0|$ for 1.8V and 3.3V are plotted in Fig. 1. It can be reasoned from Part B that the curve for larger V_{DD} is higher, and for both curves the highest parts are around boundaries of the range of V_O.
2 Curves of $R_{out}(V_O)/R_0$ for 1.8V and 3.3V are plotted in Fig. 1. R_{out} consists of two parallel resistors, so it is natural for the curves to be tall at middle but fall off at the sides.
3 Curves of E_r for 1.8V and 3.3V are plotted in Fig. 1. E_r is the product of the two previous curves. The largest E_r for 3.3V (crossed point) is larger than the largest E_r for 1.8V (crossed point). For two reasons: First, $|1 - G_m(V_O)/G_0|$ for 3.3V is not only higher, but wider range allows it to go much higher. Second, $R_{out}(V_O)/R_0$ for 3.3V falls off much slower.

The corresponding real circuit simulations of Fig. 1 are given in Fig. 2(a) and 2(b) for 3.3V and 1.8V, respectively. The E_r and $|1 - G_m(V_O)/G_0|$ are actually much smaller than they appear in Fig. 2, due to different scales applied to them. Fig. 1 and 2 are meant to explain why the dc-gain model (8) becomes more accurate when V_{DD} decreases.

In summary, the analytical op-amp dc-gain model (8) is relatively accurate because the deviation $|1 - G_m(V_O)/G_0|$ is generally small and, more importantly, the effect of $|1 - G_m(V_O)/G_0|$ is further reduced by the shaping provided from $R_{out}(V_O)/R_0$. Since errors in Table II for 1.8V cases are all less than 2%, it can be concluded that E_r less than 2% when $V_{DD} \leq 1.8$V.

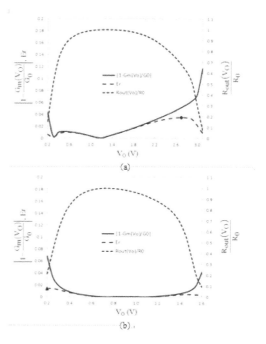

Figure 2. Comparisons between the relative error from different supply voltages. (a) V_{DD}=3.3V, V_{SS}=0V. (b) V_{DD}=1.8V, V_{SS}=0V.

Table 2. The maximum relative error of the proposed model.

	V_{DD}=3.3V, V_{SS}=0V (0.2V to 3.1V)	V_{DD}=2.5V, V_{SS}=0V (0.2V to 2.3V)	V_{DD}=1.8V, V_{SS}=0V (0.2V to 1.6V)
First stage configuration			
Normal	Error(max) =3.52%	Error(max) =2.49%	Error(max) =1.43%
Cascode	Error(max) =6.59%	Error(max) =3.16%	Error(max) =1.43%
Folded cascode	Error(max) =5.18%	Error(max) =3.18%	Error(max) =1.58%

5 CONCLUSION

In this brief, the nonlinear dc-gain model is proposed. We applied dc-gain proportional to output resistance and two-stage op-amp's gain and V_{OS} separation characteristic to generate the nonlinear dc-gain model. We proposed Fig. 3 configuration has one of the output terminal transistor V_{GS}, this V_{GS} is constant. Therefore, the output resistance can be traced by output voltage. On the other hand, we analyzed the error between our model and transistor-level simulation. In the same process parameters, supply voltage decrease to 1.8V, the error decreases obviously because G_m variation is small and the output resistance is more curved in low supply voltage.

This concept is different from traditional design method. Since it does not need to complete the entire circuit, we only determined some parameters $\left(\dfrac{W}{L}\right)_6$, $\left(\dfrac{W}{L}\right)_7$, V_{DD}, V_{SS}, V_{SG7}, V_{tp7}, and A_0, directly produce dc-gain curve. It is not only to eliminate time-consuming, but also to try different op-amp designs rapidly. Besides, the model can be applied to two-stage fully differential op-amp without common-mode feedback configuration by integrating the dc-gain curve.

REFERENCES

[1] B. Razavi, *Design of Analog CMOS Integrated Circuits*. New York: McGraw-Hill, 2002.

[2] V. F. Dias, G. Palmisano, and F. Maloberti, "Harmonic Distortion in SC Sigma-Delta Modulators," *IEEE Trans. Circuits and Systems I, Fundamental Theory and Applications*, vol. 41, no. 4, pp. 326–329, Apr. 1994.

[3] Y. Geerts and W. M. C. Sansen, *Design of Multi-Bit Delta-Sigma A/D Converters*. Norwell, MA: Kluwer, 2002.

[4] A. A. Hamoui, T. Alhajj, and M. Taherzadeh-Sani, "Behavioral Modeling of Opamp Gain and Dynamic Effects for Power Optimization of Delta-Sigma Modulators and Pipelined ADCs," in *Proc. IEEE. Low Power Electronics and Design*, Oct. 2006, pp. 330–333.

[5] F. Medeiro, B. Perez-Verdu, A. Rodriguez-Vazquez, and J. L. Huertas, "Modeling OpAmp-Induced Harmonic Distortion for Switched-Capacitor $\Sigma\Delta$ Modulator Design," in *Proc. IEEE Int. Symp. Circuits and Systems*, Jun. 1994, vol. 5, pp. 445–448.

[6] H. Zare-Hoseini and I. Kale, "On the Effects of Finite and Nonlinear DC-Gain of the Amplifiers in Switched-Capacitor $\Delta\Sigma$ Modulators," in *Proc. IEEE Int. Symp. Circuits Systems*, May 2005, vol. 3, pp. 2547–2550.

[7] J. J. Xu, Y. Liu, S. Y. Zhang, and L. Ding, "A 16Bit $\Sigma\Delta$ Modulator Design with Suppressing Odd-order Harmonic Distortions," in *Proc. IEEE Int. Computer, Mechatronics, Control and Electronic Engineering*, Aug. 2010, vol. 6, pp. 76–78.

[8] K. Abdelfattah and B. Razavi, "Modeling Op Amp Nonlinearity in Switched-Capacitor Sigma-Delta Modulators," in *Proc. IEEE. Custom Integrated Circuits Conference*, Sep. 2006, pp. 197–200.

[9] M. Yavari and A. Rodriguez-Vazquez, "Accurate and Simple Modeling of Amplifier DC-gain Nonlinearity in Switched-Capacitor Circuits," in *Proc. IEEE. Circuit Theory and Design*, Aug. 2007, pp. 144–147.

[10] A. Banerjee, S. Chatterjee, A. Patra, and S. Mukhopadhyay, "An Efficient Approach to Model Distortion in Weakly Nonlinear Gm-C Filters," in *Proc. IEEE Int. Symp. Circuits and Systems*, May 2008, pp. 1312–1315.

[11] F. C. Chen and C. L. Hsieh, "Modeling Harmonic Distortions Caused by Nonlinear Op-Amp DC-gain for Switched-Capacitor Sigma-Delta Modulators," *IEEE Trans. Circuits and Systems II, Express Briefs*, vol. 56, no. 9, pp. 694–698, Sep. 2009.

[12] Y. H. Cheng, M. C. Jeng, Z. H. Liu, J. H. Huang, M. Chan, K. Chen, P. K. Ko, and C. M. Hu, "A Physical and Scalable *I-V* Model in BSIM3v3 for Analog/Digital Circuit Simulation," *IEEE Trans. Electron Devices*, vol. 44, no. 2, pp. 277–287, Feb. 1997.

[13] A. S. Sedra and K. C. Smith, *Microelectronic Circuits*, 5th ed. New York: Oxford, 2004.

Testing and Measurement: Techniques and Applications – Chan (Ed.)
© 2015 Taylor & Francis Group, London, ISBN: 978-1-138-02812-8

An application of the frequency dependent load as a circuit test

N.V. Kinsht & N.N. Petrun'ko
Electrophysics and Electropower Laboratory, Institute of Automation and Control Processes FEB RAS, Vladivostok, Russia

N.V. Silin
Far East Federal University, Vladivostok, Russia

ABSTRACT: The various formulations of the problem of technical diagnosis are possible when coming to the creation and use of analog circuits. One approach allows deviation parameters sufficiently large set of parameters of the elements. A way of testing experiments organization with analog circuit is considered. It means that the different test modes of the target circuit are the connection passive elements with frequency depended parameters and the source frequency vary. The theory of the approach is considered. An example is given.

1   INTRODUCTION

The various formulations of the problem of technical diagnosis are possible when coming to the creation and use of analog circuits. They differ in the source and kind of influence and procedures for the use of diagnostic results. Sometimes in the problem, it is assumed that it is possible for the significant deviation parameters of 1 - 2 elements (Chen Y.M. et al., 2008, Pecht M. et al., 2013). Another approach allows deviation parameters sufficiently large set of parameters of the elements. More generally, to determine the circuit technical condition it is sufficient to determine the parameters of all the elements. The results of this diagnostic can be used to improve the technology of analog circuits or predict their behavior depending on the time of exposure or destabilizing factors.

The theory of circuit diagnosis is a symbiosis of classical circuit theory and formalization approaches inherent in the theory of technical diagnosis. It took considerable time and a lot of theoretical and applied problems are formulated and solved. Operational properties of electrical circuits are uniquely specified by its topology and set of value element parameters which define the technical condition of the circuit of the whole.

Based on the achievements of the classical circuit theory, one of the most natural initial approaches to the diagnostics of electric circuits is a representation of a diagnosed circuit in the form of a passive multipole network. The parameters of this multipole network are subject to definition. The pioneer in the statement and systematic studying of such task is Berkowitz (1962). In its statement possibility of diagnostics of the passive multipole network is

considered, some of the conclusions are considered available both for giving of entrance signals, and for measurement of reactions, and some – partially available on which measurements of reactions are allowed only. Considerable interest in diagnostics of a passive multipole at various assumptions was shown by the Russian researches (Demirchjan K.S. & Butyrin P.A., 1988). Problems of circuit diagnostics are for a long time as a subject of researches Kinsht N.V. et al. (1983), Kinsht N.V. & Petrun'ko N.N. (2013, 2014). Until now, it is possible to formulate new mathematical models and approaches to the electrical circuit diagnosis. So, the power sources (current sources and e.m.f.) are commonly used as a testing influence on the circuit in realization test diagnosis.

Different techniques and methods of influence on the electrical circuit can be used in test diagnostics. The load variation as a test action is one of them. Below the frequency variation approach is considered as a way of the organization n testing experiments with analog circuit.

2   BASIC FREQUENCY MODEL

Let be an active complex circuit A with 2n independent nodes. Parameters of a subset N of its frequency independent elements are incidental to (internal) nodes with numbers $(1,...,n)$ that are to be determined as shown at Fig. 1. To implement the diagnostic procedure, it is connected to a complex addition circuit F which has 2n independent nodes (numbered 1, ..., 2n). And (external) nodes with numbers $((n+1), ..., 2n)$ are available to connect to the circuit test inputs, and to measure reactions (voltages).

It is possible to vary the frequency sources and the k-th test experiment is to measure both the open circuit voltages and load voltages when frequency $\omega=\omega_k$ ($k=1,\ldots,n$). Thus, it is possible to make n independent experiments.

Measurements of node voltages ($V_{(n+1)},\ldots V_{2n}$) can be considered as an indirect measurement of node voltages ($V_1,\ldots,V_n$). It is clear that concerning to the nodes ($(n+1),\ldots,2n$) it is easy to define a matrix of input and transmission parameters, but elements the diagnosed parameters in the analytical expressions of the matrix will appear in the form of elements of rational functions. Let $\mathbf{Y}^N$ be assumed the frequency independent (resistive) nodal admittance matrix of diagnosed circuit with size $n \times n$, and matrix

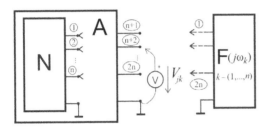

Figure 1. Basic model of the frequency dependent load.

$$\mathbf{Y} = \begin{bmatrix} \mathbf{Y}_{11} & \mathbf{Y}_{12} \\ \mathbf{Y}_{21} & \mathbf{Y}_{22} \end{bmatrix} \quad (1)$$

is the known nodal admittance matrix of complex circuit with size $2n \times 2n$ and matrix

$$\mathbf{Y}^F(j\omega_k) = \begin{bmatrix} \mathbf{Y}_{11}^F(j\omega_k) & \mathbf{Y}_{12}^F(j\omega_k) \\ \mathbf{Y}_{21}^F(j\omega_k) & \mathbf{Y}_{22}^F(j\omega_k) \end{bmatrix} \quad (2)$$

is the known nodal admittance matrix of load circuit $\mathbf{F}$ with size $2n \times 2n$.

Complete system of equations of the open circuit mode for $\omega=\omega_k$ obtained in the form:

(YN+Y11) V'0k+ Y12V"0k= Jk (3)

Y21 V'0k + Y22 V"0k = 0, (4)

We denote:
$\mathbf{V'}_{0k} = col(V_{01},\ldots,V_{0n})$, $\mathbf{V''}_{0k} = col(V_0{}_{(n+1)},\ldots V_{02n})$ and they are nodal open circuit voltages;
$\mathbf{J} = col(J_1,\ldots,J_n)$, – nodal current sources;
for simplicity, it is assumed no sources are connected to external nodes.

After a transformation to Equation 4 $\mathbf{V'}_{0k}$ presents the form:

$$\mathbf{V}_{0k}' = -(\mathbf{Y}_{21})^{-1}\mathbf{Y}_{22}\mathbf{V}_{0k}'' \quad (5)$$

and after substitution to (3) we obtain the Equation:

$$\mathbf{Y}^N\mathbf{V}_{0k}' = \mathbf{J}_k - \mathbf{Y}_{11}\mathbf{V}_{0k}' - \mathbf{Y}_{12}\mathbf{V}_{0k}'' . \quad (6)$$

Let write Equations 7, 8 like Equations 5, 6 for k-th frequency experiment:

$$\mathbf{V}_k' = -(\mathbf{Y}_{21} + \mathbf{Y}_{21}^F(j\omega_k))^{-1}(\mathbf{Y}_{22} + \mathbf{Y}_{22}^F(j\omega_k))\mathbf{V}_k'' \quad (7)$$

$$\mathbf{Y}^N\mathbf{V}_k' = \mathbf{J}_k - (\mathbf{Y}_{11} + \mathbf{Y}_{11}^F(j\omega_k))\mathbf{V}_k' - \\ (\mathbf{Y}_{12} + \mathbf{Y}_{12}^F(j\omega_k))\mathbf{V}_k'' \quad (8)$$

where $\mathbf{V}_k'' = col(V_{k(n+1)},\ldots V_{k2n})$ are measured voltages and $\mathbf{V}_k' = col(V_{k1},\ldots,V_{kn})$, $\mathrm{Jk} = col(J_{k1},\ldots,J_{kn})$.

Let subtract Equations 5, 6 from the Equations 7, 8 and denote:

$$\Delta\mathbf{J}_k = \mathbf{Y}_{11}(i\omega_k)\mathbf{V}_{0k}' - (\mathbf{Y}_{11}(i\omega_k) + \mathbf{Y}_{11}^F(i\omega_k))\mathbf{V}_k' + \\ +\mathbf{Y}_{12}(i\omega_k)\mathbf{V}_k'' - (\mathbf{Y}_{12}(i\omega_k) + \mathbf{Y}_{12}^F(i\omega_k))\mathbf{V}_k'' \quad (9)$$

$$\Delta\mathbf{V}_k = [(\mathbf{V}_1' - \mathbf{V}_{01}'),\ldots,(\mathbf{V}_k' - \mathbf{V}_{0k}'),\ldots,(\mathbf{V}_n' - \mathbf{V}_{0n}')] \quad (10)$$

Let construct square matrices:

$$\Delta\mathbf{V} = [\Delta\mathbf{V}_1,\ldots,\Delta\mathbf{V}_k,\ldots,\Delta\mathbf{V}_n]; \quad (11)$$

$$\Delta\mathbf{J} = [\Delta\mathbf{J}_1,\ldots,\Delta\mathbf{J}_k,\ldots,\Delta\mathbf{J}_n]. \quad (12)$$

Note that elements of the matrices Equations 9, 10 are presented as the complex amplitudes, but they correspond to different frequencies. Combining these data into a single matrix it is possible for an only formal mathematical point of view.

It is ensured by the fact that the matrix $\mathbf{Y}^N$ does not depend on the frequency.

Finally, we obtain the solving for the required nodal admittance matrix of the circuit N:

$$\tilde{\mathbf{Y}}^N = \Delta\mathbf{J}\Delta\mathbf{V}^{-1}. \quad (13)$$

3 EXAMPLE 1

Let's analyze example, with partly inaccessible nodes, as shown in Fig. 2. The elements subset N is signed by dashed line rectangle, it is incident to nodes 1, 2, 3. The passive element parameters of the subset N are to be determined. Numerical values of the circuit element parameters are noted in the dimentionless form on the figure; parameters of the subset N elements are shown in parentheses, they serve only to check of the diagnosis modeling result. Admittances of LC elements are indicated for the dimentionless angular frequency $\omega = 1$. Other mode parameters (voltages and currents) are in the dimentionless form also.

For diagnostics it is made measurements of the node voltages V_4, V_5, V_6, at 3 various frequencies: $\omega_1 = 0.1$, $\omega_2 = 1$, $\omega_3 = 10$.

We will prepare input data according to Equations 1, 2. The fragments of the matrix Y are of the form:

$$\mathbf{Y}_{11} = \begin{bmatrix} y_1 & 0 & 0 \\ 0 & y_2 & 0 \\ 0 & 0 & y_3 \end{bmatrix} = \begin{bmatrix} 9 & 0 & 0 \\ 0 & 10 & 0 \\ 0 & 0 & 8 \end{bmatrix}$$

$$\mathbf{Y}_{12} = \begin{bmatrix} 0 & 0 & -y_1 \\ -y_2 & 0 & 0 \\ 0 & -y_3 & 0 \end{bmatrix} = \begin{bmatrix} 0 & 0 & -9 \\ -10 & 0 & 0 \\ 0 & -8 & 0 \end{bmatrix};$$

$$\mathbf{Y}_{21} = \begin{bmatrix} 0 & -y_2 & 0 \\ 0 & 0 & -y_3 \\ -y_1 & 0 & 0 \end{bmatrix} = \begin{bmatrix} 0 & -10 & 0 \\ 0 & 0 & -8 \\ -9 & 0 & 0 \end{bmatrix};$$

$$\mathbf{Y}_{22} = \begin{bmatrix} y_2 & 0 & 0 \\ 0 & (y_3+y_5) & -y_5 \\ 0 & -y_5 & (y_1+y_4+y_5) \end{bmatrix} = \begin{bmatrix} 10 & 0 & 0 \\ 0 & 9 & -1 \\ 0 & -1 & 16 \end{bmatrix}.$$

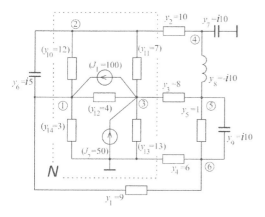

Figure 2. Circuit for example 1.

$$\mathbf{Y}_{11}^F(i\omega)\big|_{\omega=1} = \begin{bmatrix} 5i & -5i & 0 \\ -5i & 5i & 0 \\ 0 & 0 & 0 \end{bmatrix};$$

$$\mathbf{Y}_{22}^F = \begin{bmatrix} (i\omega C_7 - i/\omega L_8) & i/\omega L_8 & 0 \\ i/\omega L_8 & (i\omega C_8 - i/\omega L_8) & -i\omega C_9 \\ 0 & -i\omega C_9 & i\omega C_9 \end{bmatrix};$$

$$\mathbf{Y}_{22}^F(i\omega)\big|_{\omega=1} = \begin{bmatrix} 0 & 10i & 0 \\ 10i & 0 & -10i \\ 0 & -10i & 10i \end{bmatrix}.$$

Open circuit voltage $\mathbf{V}_0$ doesn't depend on frequency: $\mathbf{V}_0 = col[4.48, 0.86, 3.88]$.

The calculated results are shown in Tables 1–4:

Table 1. Measured voltages V"k.

$\omega=0.1$	$\omega=1.0$	$\omega=10$
2.38 - 0.05i	2.53 - 1.71i	0.05 - 0.32i
2.36 - 0.18i	0.72 - 0.83i	1.40 - 0.16i
3.76 - 0.17i	2.49 - 2.06i	1.38 - 0.28i

Table 2. Calculated values ΔJ by Equation (9).

$\omega=0.1$	$\omega=1.0$	$\omega=10$
2.292-1.758i	-5.777-15.88i	-44.82-16.05i
-12.99+0.99i	-5.22 + 4.567i	-19.02+13.50i
11.37-0.590i	2.250 – 1.630i	8.80 – 0.770i

Table 3. Calculated values ΔV by Equation 10.

$\omega=0.1$	$\omega=1.0$	$\omega=10$
-0.38 - 0.130i	-1.09- 1.600i	-3.17- 0.70i
-0.80 - 0.01i	-1.12 – 0.99	-1.25 +0.32
0.08- 0.11	-0.420-0.630i	-0.56 - 0.06i

Comparison of the results of solving the problem with the initial data indicates correctness techniques. Analytic expression for Y_n and digital control values can be used:

$$\mathbf{Y}_n = \begin{bmatrix} (y_{10}+y_{12}+y_{14}) & -y_{10} & -y_{12} \\ -y_{10} & (y_{10}+y_{11}) & -y_5 \\ -y_{12} & -y_5 & (y_{11}+y_{12}+y_{13}) \end{bmatrix} = \begin{bmatrix} 19 & -12 & -4 \\ -12 & 19 & -7 \\ -4 & -7 & 24 \end{bmatrix}.$$

Table 4. Calculated elements of $\tilde{\mathbf{Y}}^N$ by Equation 13.

19.215–0.1574i	-2.30+0.125i	-4.047+0.088i
-12.231+0.1604i	19.136–0.0853i	-6.8993–0.161i
-4.0244 - 0.1411i	-6.964+0.109i	24.155+0.332i

Next, we can discard the imaginary terms like noise and average off-diagonal terms. The finish result showed in Table 4 can be considered satisfactory and confirms the validity of the theoretical model.

4 ADVANCE OF THE FREQUENCY MODEL

The possibility of establishment of an open circuit mode and open circuit voltage measurement were assumed into consideration above the main model. Such mode with circuit disconnect is impossible in many practical tasks. Therefore, it is necessary to develop a new approach to a problem.

Let's $\mathbf{J}'_k$ be according to Equation 8

$$\mathbf{J}'_k = -(\mathbf{Y}_{11} + \mathbf{Y}^F_{11}(j\omega_k))\mathbf{V}'_k - (\mathbf{Y}_{12} + \mathbf{Y}^F_{12}(j\omega_k))\mathbf{V}''_k \ (14)$$

and for pair frequencies ω_i and ω_k (8) will be written in form:

$$\mathbf{Y}^N \mathbf{V}'i = \mathbf{J}'_i + \mathbf{J}(j\omega_i), \tag{15}$$

$$\mathbf{Y}^N \mathbf{V}'i = \mathbf{J}'_k + \mathbf{J}(j\omega_k). \tag{16}$$

We pay attention to Equations 15, 16 based on the different frequencies, in spite of that we carry out formal mathematical transformations. If for the i-th and k-th experiments we can form to equate complex amplitude $\mathbf{J}(j\omega_i)$ with complex amplitude $\mathbf{J}(j\omega_k)$, we subtract the Equation 16 from the Equation 15:

$$\mathbf{Y}^N(\mathbf{V}'_i - \mathbf{V}'_k) = (\mathbf{J}'_i - \mathbf{J}'_k) \tag{17}$$

and denote:

$$\mathbf{Y}^N \Delta\mathbf{V}_{ik} = \Delta\mathbf{J}_{ik} \tag{18}$$

Now we construct square matrices:

$$\Delta\mathbf{V} = [\Delta\mathbf{V}_{12}, ..., \Delta\mathbf{V}_{ik}, ..., \Delta\mathbf{V}_n], \tag{19}$$

$$\Delta\mathbf{J} = [\Delta\mathbf{J}_{12}, ..., \Delta\mathbf{J}_i, ..., \Delta\mathbf{J}_n], i \neq k \quad i, k \in (1, ..., n) \ (20)$$

Finally, we obtain the solving for the required nodal admittance matrix of the circuit N similar to Equation 13. If we make test experiments on the s different frequencies, there is an opportunity to make $s(s-1)/2$ combinations of the (i, k) pairs. It is obvious that $s(s-1)/2 > n$.

The question of what of these combinations are the most informative isn't investigated. To use excess measurements it is applied in the method of the least squares. So, similarly Equations 19, 20 it is possible to make rectangular matrixes $\Delta\mathbf{V}_X$, $\Delta\mathbf{J}_X$:

$$\Delta\mathbf{V}_X = [\Delta\mathbf{V}_1, ..., \Delta\mathbf{V}_r, ..., \Delta\mathbf{V}_m];$$

$$\Delta\mathbf{J}_X = [\Delta\mathbf{J}_1, ..., \Delta\mathbf{J}_r, ..., \Delta\mathbf{J}_m], \quad r = 1, ..., m \leq s(s-1)/2.$$

Now the estimated value of the unknown parameter matrices is obtained as:

$$\tilde{\mathbf{Y}}^N = \Delta\mathbf{J}_X \Delta\mathbf{V}^T_X (\Delta\mathbf{V}_X \Delta\mathbf{V}^T_X)^{-1}.$$

5   EXAMPLE 2

We use conditions of an example 1, however the open circuit mode is forbidden. Let's prepare input data according to Equations 1, 2. In the context of this task

it is not enough to make measurements at three frequencies. We use 4 frequencies: ω=0.1, 0.5, 1.0, 10. Measured voltages are showed in the Table 5.

Table 5.   Measured voltages V"k.

ω=0.1	ω=0.5	ω=1.0	ω=10
2.38 - 0.05i	2.59 - 0.39i	2.53 - 1.71i	0.05 - 0.32i
2.36 - 0.18i	1.98 - 0.84i	0.72 - 0.83i	1.40 - 0.16i
3.76 - 0.17i	3.66 - 0.99	2.49 - 2.06i	1.38 - 0.28i

Table 6.   Calculated values ΔJik by Equation 14.

$\omega_{i,k}$=0.1, 0.5	$\omega_{i,k}$=0.1, 1.0	$\omega_{i,k}$=0.1, 10
-1.5050-7.170i	7.9950-14.185i	-46.085-4.875i
2.4550+ .4800i	7.255+3.4050i	30.955+13.105i
-2.4000-2.880i	-8.640-0.800i	-2.240+ 0.160i

Table 7.   Calculated values ΔVik for Equation 17.

$\omega_{i,k}$=0.1, 0.5	$\omega_{i,k}$=0.1, 1.0	$\omega_{i,k}$=0.1, 10
-0.0500- 0.6100i	-0.7100-1.4700i	-2.810-.5800i
0.070-0.2600i	-0.280-0.9600i	0.4300 +0.330i
-0.0800- 0.3000i	-0.5600-0.5500i	-0.6800+0.00

Calculated elements according to Equation 18 are:

$$\tilde{\mathbf{Y}}^N = \begin{bmatrix} 19.8932-0.0580i & -11.577-0.8811i & -6.735+0.0857i \\ -12.7318+0.3819i & 19.076+0.901i & -5.084-1.3027i \\ -3.9011+1.1436i & 4.8011+0.9767i & 22.952-4.5811i \end{bmatrix}$$

According to Equation 18 after calculations, we obtain $\mathbf{Y}^N$ and finally calculation solves the problem and shows the correspondence of the digital control values in general:

$$\mathbf{Y}^N = \begin{bmatrix} 19.9 & -12.0 & -5.3 \\ -12.0 & 19.1 & -4.9 \\ -5.3 & -4.9 & 3.0 \end{bmatrix} \cong \left\{ \begin{bmatrix} 19 & -12 & -4 \\ -12 & 19 & -7 \\ -4 & -7 & 24 \end{bmatrix} \right\}$$

6   CONCLUSION

The paper provides an analytical description of data processing procedures of test experiments using frequency depend loads as the test influences for resistor parameters diagnosing.

ACKNOWLEDGMENT

The investigation was partly financially supported by the Russian Foundation Basic Researches, grant # 13–08–00924

REFERENCES

[1] Bercowitz R.S. 1962 Conditions for network-element-value solvability. – *IRE Trans. On Circuit Theory, vol.9, March, p. 24–29.*

[2] Chen Y.M. & Wu H.C. & Chou M.W. & Lee K.Y. 2008. Online failure prediction of electrolytic capacitors for LC filter of switching-mode power converters, *IEEE Trans. Ind. Electron., vol. 55, no. 1, pp. 400–406.*

[3] Demirchjan K.S. & Butyrin P.A., 1988. Modeling and Computer Calculation of the Circuits. *High School Publishers, Moskow (in Russian).*

[4] Kinsht N.V. & Gerasimova G.N. & Katz M.A. 1983. Diagnostics of Electrical Circuits. *Energoatomizdat Publishers, Moskow (in Russian).*

[5] Kinsht N.V. & Petrun'ko N.N., Electrical Circuits and System Diagnostics. *Dalnauka Publishers, Vladivostok, (2013) (in Russian).*

[6] Kinsht N.V. & Petrun'ko N.N. 2013 Some opportunities of circuit test diagnosis *Applied Mechanics and Materials. Vols. 373–375. pp.927–930.*

[7] Kinsht N. V. & Petrun'ko N.N., 2013. Certain problems of the circuit test diagnosis. *Universal Journal of Electrical and Electronic Engineering Vol. 1, No 3, Horizon Research Publishing Corporation pp. 76 – 80.*

[8] Kinsht N. V. & Petrun'ko N.N. 2014. Load modes variation in the circuit test diagnosis. *Sci. and Eng. Publ. Comp., Int. Journal of Automation and Power Eng., vol. 3, iss. 1, pp 23–27.*

[9] Pecht M. & Vasan A.S.S. & Long B., 2013 . Diagnostics and Prognostics Method for Analog Electronic Circuits", *IEEE Trans. Industrial Electronics, Vol. 60.*

Testing and Measurement: Techniques and Applications – Chan (Ed.)
© 2015 Taylor & Francis Group, London, ISBN: 978-1-138-02812-8

Velocity profile measurement technique for scour using ADV

S. Das, R. Das & A. Mazumdar
School of Water Resources Engineering, Jadavpur University, Kolkata, India

ABSTRACT: This paper presents a technique for turbulence measurement for clear water scour tests having different shaped piers under different pier widths, densimetric Froude number, inflow depth and pier Reynolds number on a sand bed. Many researchers have conducted various studies to determine the velocity profiles during scour. The contours and spatial distributions of the time-averaged velocities, turbulence intensities, turbulent kinetic energy and Reynolds stresses at different azimuthal planes are determined. Velocity vector plots of the flow field at azimuthal planes are used to show further flow features. The vorticity and circulation of the horseshoe vortex are determined by using forward difference technique of computational hydrodynamics and Stokes theorem, respectively. The flow and turbulence characteristics of the horseshoe vortex are discussed from the standpoint of similarity with velocity and turbulence intensity characteristic scales. These detailed measurements are obtained with a non-intrusive instrument, the Acoustic Doppler Velocimeter (ADV), which measures three-dimensional instantaneous velocities. This instrument is very useful for the measurements related to scour reduction. This instrument can also be used as a tool during the analysis for enhancement of sediment transport as an alternative to dredging mechanism in future which is to be explored. It is therefore felt that more studies are still required to enhancement of sediment transport effectively for dredging purpose and also to measure the turbulence using ADV.

1 INTRODUCTION

On a river or waterway, a new bridge is required beside an existing bridge to meet the demand of transportation for different kinds of vehicles including buses, trains, etc. Sometimes along the flow, a pier of the new bridge is positioned eccentrically downward with respect to the pier of the old bridge. It is like a two-pier arrangement with an in-line front pier and an eccentric rear pier. The horseshoe vortex develops upstream where the eccentric rear pier comes in contact with the wake vortex formed downstream of the in-line front pier. The combined strength of both vortices enhance scour hole of the rear pier thereby increasing the rate of sediment transport, moving the sediment away from the upstream axis of symmetry as the flow approaches in a downward direction. Eccentrically arranged piers can cause sediment to move towards the flume wall flowing downstream. The horseshoe vortex in the middle of two piers can enlarge the scour hole of the downstream pier while shifting the sediments away from the downstream plane of symmetry. Therefore, one needs a set of measured horseshoe vortex data for the downstream pier like turbulent kinetic energy, vorticity, and circulation etc to shed light on this scope. The suitability and effectiveness of this kind of natural occurrence with particular reference to enhancement of sediment transport may be explored as an alternative to dredging mechanism in future.

A large number of studies have been carried out to predict scour depth in the base of piers by Raudkivi and Ettema (1983), Barbhuiya and Dey (2004), Khwairakpam et al. (2012), Das et al. (2014a,b,c). These studies have been performed primarily by means of laboratory-flume experiments and include the use of non-dimensional equations, which can lead to semi-empirical equations for predicting the maximum scour depth.

For a better understanding of the sediment transport dynamics at clear water equilibrium scour condition around piers, many researchers like Dey et al. (1995), Istiarto and Graf (2001), Dey and Raikar (2007), Das et al. (2012, 2013a,b,c, 2014c) have focused on the flow field, vortex strength and turbulence field around piers. Most of these studies have been confined to single piers. Das et al. (2013a,b,c) addressed equilibrium scour geometry around single piers observed under different pier width, densimetric Froude number, inflow depth and pier Reynolds number by conducting several clear water scour tests in a tilting flume. For the same number of tests, an experimental investigation has also been done to find out the flow fields, time-averaged velocity vectors, and strength of horseshoe vortex as well as the wake vortex within the equilibrium scour hole measured by an ADV. Das et al. (2014d) also carried out ADV measurements around a triangular pier to obtain flow patterns and characteristics of the horseshoe vortex system.

2 FLOW PATTERN AT A CIRCULAR PIER

The *flow pattern past a circular pier*, protruding vertically from a horizontal plane boundary, in uniform open channel flow is complex in detail, and the complexity increases with the development of a scour hole at the base of the circular pier. Studies of flow patterns for such a case have seen reported by many previous researchers. The component features of the flow pattern are the down flow in front of the cylinder, cast-off vortices and wake, boundary layer, horseshoe vortex and bow wave.

The approach flow velocity goes to zero at the upstream face of the pier, in the vertical plane of symmetry, and since the approach flow velocity decreases from the free surface downward to zero at the bed, the stagnation pressure, $0.5\rho U^2$, also decreases. This downward pressure gradient drives the down flow. The down flow, in the vertical plane of symmetry, has at any elevation a velocity distribution, with zero in contact with the pier and again some distance upstream of it. The v_{max} value according to the experimental data at any elevation is 0.05 to 0.02 circular pier diameters upstream of it, being closer to the circular pier lower down and increases towards the bed. Just above the bed, if no scour hole is present, v_{max} is approximately $0.4U$. The maximum of v_{max}/U occurs when the depth of the scour is about two times the pier diameter or more. This maximum is located at about one pier diameter below the approach flow bed and is about 80% of the mean approach flow velocity U.

The so-called horseshoe vortex develops as the result of the flow separation at the upstream rim of the scour hole; it is a *lee eddy* similar to the eddy or ground roller downstream of a dune crest. The horseshoe vortex extends downstream, a few pier widths past the sides of the pier, before losing its identity and becoming part of general turbulence. The horseshoe vortex also pushes the maximum downflow velocity within the scour hole closer to the pier. A comment on the horseshoe vortex is inserted here because of the frequent reference in literature to it as the scouring agent. Uniform flow at velocity U has vorticity $-U$ per unit stream wise length which is advected at approximately $0.5U$. The rate of advection is $-0.5U^2$ and if the vorticity filaments were caught on the pier the vorticity $-0.5U^2t$ would increase there monotonically.

Generation of vorticity occurs at the boundaries but only where there is a pressure gradient. In approaching the pier the flow meets an adverse pressure gradient. This leads to generation of vorticity of the opposite sense to that in the boundary layer. The circulation decreases and goes to zero over depth on the centre line of a cylindrical pier protruding from a plane bed (as does the velocity). The vortex filament divides along this line, as vorticity is generated on the

cylinder and base plate by the pressure gradients, and passes the pier. The parts on the circular pier separate and form *Karman vortex street* (cast-off vortices). There is no accumulation of vorticity upstream of the pier. The concentration of vorticity develops with the scour hole as alee or separation eddy.

The stagnation pressure apart from down flow also causes a sideways acceleration of flow past the circular pier. The flow separates at the sides of the pier and the surfaces of separation enclose a wake downstream of the pier. The discontinuity in the velocity profile from flow to wake leads to the development of concentrated vortices in the interface, the cast-off vortices, which translate with the flow. Near the bed these vortices interact with the horseshoe vortex causing the trailing parts of it to oscillate at the vortex shedding frequency n of approximately $nb/U \approx 0.2$. The cast-off vortices with their vertical low pressure centres lift sediment from the bed like miniature tornados.

3 VELOCITY MEASUREMENT USING ADV

3.1 *ADV (make: Sontek, USA)*

An ADV (or acoustic Doppler velocimeter) is a single-point current meter that accurately measures the three components of water velocity in both high and extremely low flow conditions. Velocities are measured in a sampling volume located a distance away from the probe head (Horizon ADV, 2005). The probe head is made up of a single transmitter located in the center of the probe head and either two or three receivers mounted on arms. The transmitter generates a narrow beam of sound that is projected through the water. Reflections from particles or "scatterers" (such as suspended sediment, biological matter, or bubbles) in the water are reflected and sampled by the highly sensitivity receivers.

The intersection of the receiver axes designates the location of the sampling volume. A 16 MHz MicroADV is optimal for use in the laboratory. High frequency sampling (up to 50 Hz) combined with a tiny sampling volume makes this system perfect for measuring low flow conditions. This system excels in applications such as the measurement of turbulence, orbital velocities in a wave field, and precise flow field studies. The sampling volume is typically 5 cm from the probe. Ping is a single estimate of water velocity. A sample refers to the collection of several pings to produce a mean estimate of the water velocity. This mean value is output for storage and analysis. In addition to velocity, signal amplitude and correlation coefficient are output with each sample. Sampling rate is the rate of output for mean velocity data. This rate is user-programmable in the range from 0.1 to 50 Hz, depending on the acoustic frequency of the ADV.

The volume of water in which the ADV makes velocity measurements is known as sampling volume. Data accuracy increases if it is increased. The center of the sampling volume is nominally located a fixed distance from the probe tip: 5 cm for 16-MHz ADV probes. The exact distance from the probe tip to the center of the sampling volume is encoded in the probe configuration file. The sampling volume is cylindrical shaped, with the axis along the axis of the acoustic transmitter. For the 16-MHz ADV probes, the cylinder is about 6 mm in diameter and 9 mm long. The resolution is 0.01 cm/s and the accuracy level is 1% of the measured velocity. Water salinity in parts per thousand and water temperature is used for sound speed calculations. Sound speed is used to convert the Doppler shift to velocity.

A cylindrical polar coordinate system is used to represent the flow and turbulence fields for the four experiments. A three-beam 5 cm down-looking ADV (16 MHz MicroADV Lab Model) are used to measure the instantaneous three-dimensional velocity components. A sampling rate of 50 Hz and cylindrical sampling volume 0.09 cm³ having 1–5 mm sampling height are set for the measurements. Sampling heights 5 and 1–2 mm are used for measurement of velocity components above and within the interfacial sub-layer (when velocity component decreases rapidly), respectively.

A sampling duration of 120–300 s may be considered ensuring a statistically time-independent averaged velocity. The sampling durations will be relatively long near the bed. The ADV readings are taken along several vertical lines at different azimuthal planes. The lowest horizontal resolution of the ADV measurements is normally 1 cm. The output data from the ADV are filtered using software WinADV32. During the filtering of ADV data, the minimum SNR and correlation parameter were maintained at 15 and 70%, respectively.

3.2 VectrinoPlus (make: Nortek AS, Norway)

It is a high-resolution acoustic velocimeter used to measure turbulence and 3D water velocity in a wide variety of applications from the laboratory to the ocean. The basic measurement technology is coherent Doppler processing, which is characterized by accurate data and no appreciable zero offset. The VectrinoPlus uses the Doppler effect to measure current velocity by transmitting a short pulse of sound, listening to its echo and measuring the change in pitch or frequency of the echo (NorTek AS, 1996). The signal-to-noise ratio (SNR) is defined as

$$SNR = 20 \log_{10} \left(Amplitude_{Signal} / Amplitude_{Noise} \right) \quad (1)$$

Some noise will always be present, so Amplitude signal should read Amplitude signal + noise.

However, for SNR values in the magnitude applicable to typical VectrinoPlus situations, the difference is negligible. Sampling rate sets the output rate for the velocity, amplitude, correlation and pressure data. The VectrinoPlus allows data collection rates up to a sampling rate of 200 Hz. Sampling volume height is cylindric with a user selectable height of 3–15 mm. Sampling diameter is 6 mm. Distance from the probe is 5 cm. If the transmit length is increased then SNR ratio will be increased. Speed of sound depends upon salinity of water. The salinity is 0 for fresh water and typically 35 ppt for the ocean.

A four-beam 5 cm down-looking acoustic Doppler type VectrinoPlus probe is used to measure the instantaneous velocity components. It acted with an acoustic frequency of 10 MHz. The ADV operates on a pulse-to-pulse coherent Doppler shift to provide instantaneous three-dimensional velocity components at a sampling rate of 100 Hz. An adjustable cylindrical sampling volume of 6 mm diameter having 5–1 mm sampling height is usually set for the measurements. Sampling heights of 5 and 2 mm are usually used for the measurement of discharge above and within the interfacial sub-layer (when velocity component decreases rapidly), respectively. The lowest horizontal resolution during the ADV measurement is 1 cm.

The measurements at the zone closer than 4.5 mm to the boundary are beyond the ability of the ADV. Because of the interference due to echoes from the bed, the received signal is disturbed, which results in inaccurate velocity measurement. In the near-bed flow zone, the data captured by Vectrino sometimes contained spikes due to the interference between incident and reflected beams. Therefore, the data were filtered by a spike removal algorithm (Blanckaert and Lemmin, 2006). The output data from the ADV are filtered, using software WinADV32, to obtain the time mean and root mean square values from the entire velocity record. Throughout the experiments, the SNR was maintained 17 or above. In general, signal correlations between transmitted and received pair of pulses for Vectrino are larger than 70%, which was the recommended cut-off value. The ADV has little flow-field interference if the measurement samples are positioned about 5 cm away from the probes. Having used the stationary analysis, the sampling durations may be varied from 120 to 180 s in order to achieve a statistically time independent average velocity. Here also, the sampling durations will be relatively long near the bed. Das et al. (2013a,b) adopted the smaller time duration equal to 120 s.

4 CONCLUSIONS

Extreme search on scouring mechanism at eccentric pier groups in published literature has revealed that almost no experimental work has been done in this regard. Only a few experiments were carried out with special emphasis on clear water scour mechanism and turbulence effect around two piers arranged eccentrically. The increased rate of sediment transport and strength of the vortex are maximum for the square pier and minimum for the circular pier. Therefore, the square pier is the better alternative for enhancing the scour and thereby improving the dredging mechanism. It may be also important if the longitudinal spacing increases after a certain limit then the rate of sediment transport will gradually decrease. It indicates that if the piers are extremely separated then the reference pier will not be influenced by the rear pier. The circulation of the horseshoe vortex at the positive and negative azimuthal planes of the same magnitude may not always be similar due to the strong mixing of turbulence, separation of the streamline and bluff body effect. Further studies are still required for flow visualizations and pressure measurements may be carried out by conducting the same laboratory experiments. The enhancement of sediment transport mechanism and its flow dynamics may be observed using three or more number of piers of different shapes as well as different eccentricity and longitudinal spacing. The same may be observed at different the flow and bed conditions.

REFERENCES

[1] Barbhuiya, A. K. & Dey, S. 2004. Local scour at abutments: A review. *Sadhana, Academy Proceedings in Engineering Sciences*, 29(5) : 449–476.

[2] Blanckaert, K. & Lemmin, U. 2006. Means of noise reduction in acoustic turbulence measurements. *Journal of Hydraulic Research*, 44 (1) : 3–17.

[3] Das, S., Ghosh, S. & Mazumdar, A. 2012. Flow field past a Triangular Pier due to Sediment Transportation at Clear Water Equilibrium Scour Hole. *International Journal of Emerging Trends in Engineering and Development*, 7(2) : 380–388.

[4] Das, S., Das, R. & Mazumdar, A. .2013a. Circulation characteristics of horseshoe vortex in the scour region around circular piers. *Water Science and Engineering*, 6(1) : 59–77.

[5] Das, S., Midya, R., Das R. & Mazumdar, A. 2013b. A Study of Wake Vortex in the Scour Region around a Circular Pier. *International Journal of Fluid Mechanics Research*, 40(1) : 42–59.

[6] Das, S., Das, R. & Mazumdar, A. 2013c. Comparison of Characteristics of Horseshoe Vortex at Circular and Square Piers. *Research Journal of Applied Sciences, Engineering and Technology*, 5(17) : 4373–4387.

[7] Das, S., Das, R. & Mazumdar, A. 2014a. Vorticity and Circulation of Horseshoe Vortex in Equilibrium Scour Holes at Different Piers. *Journal of The Institution of Engineers (India): Series A, Springer*, 95(2) : 109–115.

[8] Das, S., Ghosh, R., Das, R. & Mazumdar, A. 2014b. Clear Water Scour Geometry around Circular Piers. Ecology, Environment and Conservation, 20(2) : 479–492.

[9] Das, S., Das, R. & Mazumdar, A. 2014c: Variations of clear water scour geometry at piers of different effective width. *Turkish Journal of Engineering and Environmental Sciences*, 38(1) : 97–111.

[10] Das, R., Khwairakpam, P., Das, S. & Mazumdar, A. 2014d. Clear-Water Local Scour around Eccentric Multiple Piers to shift the Line of Sediment Deposition. *Asian Journal of Water, Environment and Pollution*, 11(3) : 47–54.

[11] Das, S., Ghosh, S. & Mazumdar, A. 2014e. Kinematics of Horseshoe Vortex in a Scour Hole around Two Eccentric Triangular Piers. *International Journal of Fluid Mechanics Research*, 41(4) : 296–317.

[12] Dey, S., Bose, S. K. & Sastry, G. L. N. 1995. Clear Water Scour at Circular Piers: A Model. *Journal of Hydraulic Engineering*, 121(12) : 869–876.

[13] Dey, S. & Raikar, R. V. 2007. Characteristics of horseshoe vortex in developing scour holes at piers. *Journal of Hydraulic Engineering*, 133(4) : 399–413.

[14] Horizon ADV. 2005. Sontek ADV User Guide. SonTek/YSI, Inc., CA, USA.

[15] Istiarto, I. & Graf, W. H. 2001. Experiments on flow around a cylinder in a scoured channel bed. *International Journal of Sediment Research*, 16(4) : 431–444.

[16] Khwairakpam, P., Ray, S.S., Das, S., Das, R. & Mazumdar, A. 2012. Scour hole characteristics around a vertical pier under clear water scour conditions. *ARPN Journal of Engineering and Applied Sciences*, Vol. 7(6) : 649–654.

[17] NorTek AS. 1996. *ADV Operational Manual*, Vollen, Norway: Nortek AS.

[18] Raudkivi, A.J. & Ettema, R. 1983. Clear-Water Scour at Cylindrical Piers. *Journal of Hydraulic Engineering*, 109(3) : 338–350.

Testing and Measurement: Techniques and Applications – Chan (Ed.)
© *2015 Taylor & Francis Group, London, ISBN: 978-1-138-02812-8*

A novel electromagnetic concentrative probe and pulse eddy current nondestructive testing

C.Y. Xiao & C.W. Yang

School of Automation Science and Electrical Engineering, Beihang University, Beijing, China

ABSTRACT: The probe is one of the crucial components for the eddy current testing system, determining the detection of useful signals and the measurement sensitivity. In this paper, a novel electromagnetic concentrative probe is designed through simulations by finite element method, which consists of a special-shaped iron core, excitation coils and a magnetic sensor. Through experiments of pulsed eddy current nondestructive testing, the probe's good performance for detecting deep defect are verified, and its sensitive factors such as excitation frequency, duty ratio and defect position are also studied. The excitation frequency of 5 Hz and duty ratio of 5% are conducive to deep defect detection in the experiments.

1 INTRODUCTION

The probe is a key component in the eddy current nondestructive testing system, determining the detection of useful signals and the high measurement sensitivity in time and space. In 1988, Ueno et al. proposed an 8-shape coil probe used for functional magnetic stimulation[1]. The structure of the probe preferably ensured the eddy current density strongly induced in the conductor below the symmetry axis of the coil and avoid the undesirable zone [2]. Xiao and Zhang presented the analytical solutions of transient pulsed eddy current problem due to elliptical electromagnetic concentrative coils [3]. The electromagnetic concentrative coils are indispensable in the functional magnetic stimulation and have potential applications in nondestructive testing. These probes with air core coils can focus electromagnetic fields and induce high current intensities in the specimen, so it is an obvious solution to detect the defect in the deepest region of specimen [4, 5]. However, producing strong eddy current by these air core probes needs high current excitation or lots of winding turns. Thus the size of the probe becomes big.

In this paper, based on the above research, a novel probe was designed by simulations of finite element method to detect deep defect, to decrease the probe size and to improve the detection sensitivity. By the experiments of pulse eddy current nondestructive testing, the magnetic concentrative performance of the probe was verified. The experimental results also showed the effects of the pulse parameters on the detection depth and the detection signals.

2 THE NOVEL ELECTROMAGNETIC CONCENTRATIVE PROBE AND ITS CONCENTRATIVE PERFORMANCE

2.1 Structure of the novel electromagnetic concentrative probe

The novel probe for nondestructive testing, which can concentrate the eddy current in the specimen consists of three parts: a special C-shape iron core, excitation coils and a magnetic sensor, as shown in Figure 1a.

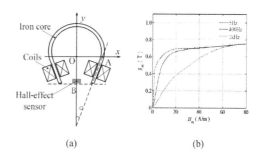

(a) (b)

Figure 1. The electromagnetic concentrative probe. (a) The structure of the probe. (b) B-H curve of iron core made of soft magnetic alloy 1J79.

The special C-shape iron core can improve the excitation magnetic field intensity and focus the current density in a small zone in the specimen, so it contributes to detecting deep defect. The iron core is made of soft magnetic alloy 1J79, which has the characteristics of high permeability and low coercivity [6]. B-H curve of 1J79 at different frequencies is shown in Figure 1b. The Hall-effect magnetic sensor was used to detect the magnetic flux density in order to obtain the high detection sensitivity in space. The magnetic sensor is set over the specimen according to the location and the size of the defect in the specimen. The coils are made of copper windings.

2.2 The electromagnetic concentrative performance analysis of simulations

In order to study the detection performance of the novel probe, the nondestructive eddy current testing model was established and the electromagnetic fields were simulated by finite element method.

The rectangular coordinate system $O\text{-}xyz$ is shown in Figure 2, where coordinate origin is located on the symmetrical line and12 mm away from the bottom surface of the iron core. The dimensions of the iron core, as shown in Figure 1a, are OA=10 mm, OB=12 mm, α=30°, and the rectangular cross-section is 3mm×28mm. The lift-off is 0.3 mm. The three-dimensional nondestructive testing model is established in Ansoft/Maxwell, shown in Figure 2.

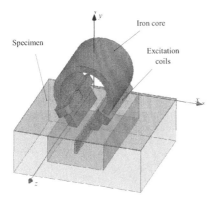

Figure 2. The eddy current testing model of electromagnetic concentrative probe.

The coils' conductivity is 5.8×10^7S/m and relative permeability $\mu_r=1$. The specimen is made of aluminum alloy, its dimensions are 50mm× 20 mm× 20 mm(length×width×thickness), its conductivity σ =3.4×10^7S/m and relative permeability $\mu_r=1$. The through-hole defect with the dimensions of 50mm ×1 mm ×1 mm(length×width×height), is located at the center of the specimen's surface along the z-axis direction. The sinusoidal excitation voltage is 6 V and the frequency is 1 kHz.

The current density distributions on the surface of the specimen by simulations are shown in Figure 3. It can be seen that: i) Due to the addition of C-shape iron core, the concentrative effect of the novel probe is better than the conventional one, and the eddy current distributions are more concentrated in a small zone of

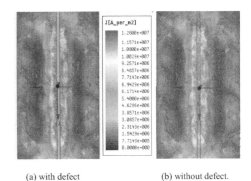

(a) with defect (b) without defect.

Figure 3. The distributions of the electric current density on the specimen surface. (a) with defect. (b)without defect.

the specimen. For this reason, under the same excitation voltage, this probe can produce a stronger excitation magnetic field, which is conducive to detecting the deep defect in the specimen and improving the detection signal sensitivity. ii) The flow of eddy current along the x-axis direction is impeded because of the existing of defect, shown in Figure 3a, so the magnetic flux intensity over the surface of the specimen when there is no defect in the specimen is stronger than that when there exists a defect. Thus, compared with the case of no defect shown in Figure 3b, the whole magnetic flux intensity in the case of existing defect is bigger.

These different signals along the y-axis direction can be detected by a magnetic sensor. Measuring the component along the y direction is because that it is convenient for a Hall-effect sensor to be fixed. The sensor was set at the center of the bottom surface of the probe, which is 0.3mm away from the surface of the specimen.

3 EXPERIMENTS

3.1 Experimental system

In order to verify the detection performance of the probe and to study the effects of pulse parameters on detection signal, the experimental system of pulse eddy current testing was designed. This system consists of the pulse excitation source (signal generator and bipolar DC power supply), the electromagnetic concentrative probe, specimen, and digital oscilloscope. The probe in the experiments is made according to the three-dimensional model established in simulations, as shown in Figure4. Considering the excitation voltage and coils' impedance, the diameter of 200-turnenameled windings were selected as 0.31mm, and the maximum current is allowed 5A. The coils' resistance is 1.73Ω, and its inductance is 0.53mH. The excitation voltage is 4 V. Allegro's Hall-effect sensor of A1321-type was used as a magnetic sensor. The Hall-effect sensor provides a voltage output that is proportional to the applied magnetic flux intensity from −440 Gs to +440 Gs and a sensitivity of 5.0mV/G. The sensitive area of magnetic sensor is less than 0.01mm², which has a high spatial resolution.

Figure 4.   The photograph of the concentrative probe.

Referring to the civil aviation standards [7], the specimen made of aluminum alloy LY12R is produced as shown in Figure 5. Two different sizes of defect are fabricated on the specimen, assuming defect with sizes of 10 mm × 3 mm × 4 mm (length×width×depth) is donated as A, defect with10 mm × 1 mm × 1 mm (length×width× depth) sizes is donated as B. The thickness of aluminum alloy plate varies from1 mm to 4 mm. When the defect is covered with the aluminum alloy plate, the deep defect can be simulated, whose depth varies from 1 mm to 4 mm.

Figure 5. Aluminum alloy specimen to fabricate the deeper defects.

3.2   Experimental results and analysis

The Hall-effect sensor was set at the center of the bottom surface of the probe, which is 0.3mm away from the surface of specimen. The magnetic flux intensities over the surface of specimen 0.58 mm were measured.

3.2.1   Detection signals at different frequencies

The location and the position of a probe is the same as the above description in Section 2. At different exciting frequencies, we measured the voltages when the specimen respectively has the surface defect and the deep defect under the surface of 3 mm. The experimental results are shown in Table 1.

Table 1.   Measured voltage peak at different frequencies.

Frequency	Voltage (defect under 3mm)	Voltage (no defect)	Voltage difference
Hz	mV	mV	mV
5	120	106	14
50	75	67	8
500	15	13	2

As can be seen from Table1, at the frequency of 5 Hz, 50 Hz and 500 Hz, the peak of the measured voltage are respectively 14 mV, 8 mV and 2 mV when the defect is under the surface of 3 mm. It means that the lower excitation frequency is more effective for detecting deep defect because of skin effect, while the excitation frequency is too small, the voltage difference will be also small because of the less eddy currents. In the experiments, we found that when the excitation frequency is less than 5 Hz, the measured voltage becomes unstable and the voltage difference is too weak to be identified. In the following experiments, we use the excitation frequency of 5 Hz.

3.2.2   Detection signals at different duty ratio pulse excitation voltage

When the duty ratio of pulse excitation voltage varies from 1% to 10%, the experimental results obtained by the magnetic sensor are shown in Figure 6.

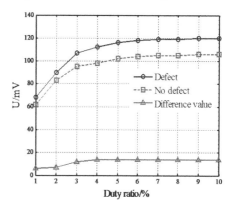

Figure 6.   Measured voltage peak at different duty ratios.

From Figure 6, we can know that the measured voltages increase as the increasing of duty ratio, and the voltage difference between the defect and no defect also increases as the increasing of duty ratio. When the duty ratio is less than 5%, both the voltage and the voltage difference increases rapidly; when the duty ration is more than 5%, the voltage difference almost keep the constant value of 14 mV. So in the following experiments, the duty ratio of pulse excitation voltage is set as 5%.

3.2.3 Detection signals of different size defects

The testing parameters are the same as above sections, and the two kinds of defect are described in Section 3.1. The experimental results of detecting two defects are respectively shown in Table 2.

Table 2. Measured voltage peak when measuring different size defects.

Defect	Voltage (surface defect)	Voltage (no defect)	Voltage difference
	mV	mV	mV
A	145	106	39
B	135	106	29

It can be seen from Table 2 that the voltage difference for defect A is 39 mV and that for defect B is 29 mV, it means that the greater the defect, the smaller the eddy current and the greater the voltage difference peak. Detection signal for no defect are shown in Figure 7.

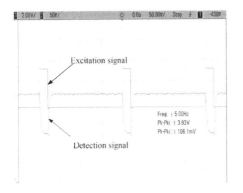

Figure 7. Measured voltage peak when there is no defect in the specimen.

Due to the voltage overshoot at the rising edge and falling edge in excitation coils, the output voltage of Hall-effect sensor has a pulse peak at the rising edge and the falling edge of the excitation signal. Experimental results of pulse eddy current nondestructive testing by the novel probe show that this voltage peak is sensitive to the defect.

3.2.4 Detection signals at different defect depth

Suppose that the distance between the defect's upper surface and the specimen surface is the defect depth d. When the defect depth varies and other parameters are the same as above sections, the experimental data are listed in Table 3.

It can be seen from Table3 that the voltage difference decreases as the increasing of defect depth.

Table 3. Measured voltage peak at different defect depth.

The depth of defect d	Voltage (defect)	Voltage (no defect)	Voltage difference
mm	mV	mV	mV
0	145	106	39
1	136	106	30
2	125	106	19
3	120	106	14
4	110	106	4

The concentrative eddy current probe can detect the defect located under the surface from 0 mm to 4 mm.

4 CONCLUSIONS

A novel probe is proposed to strengthen magnetic field and concentrate eddy current in the specimen based on simulation by finite element method. The concentrative performance is significantly improved, which contributes to detecting the deep defect with high sensitivity compared to the traditional eddy current probes. Experimental results of pulse eddy current nondestructive testing by the novel probe show that the voltage peak is sensitive to the defect. The excitation frequency of 5 Hz and of duty ratio 5% is conducive to deep defect detection in our experiments.

REFFERENCES

[1] Ueno, S.&Tashiro, T&Harada, K.1988.Localized stimulation of neural tissues in the brain by means of a paired configuration of timevarying magnetic fields. *Journal of Applied Physics*, 64(10): 5862–5864.

[2] Kolyshkin, A.A.&Rémi, V.1999. Series solution of an eddycurrent problem for asphere with varying conductivity and permeability profiles. IEEE Transactions on Magnetics, 35(6): 44–45.

[3] Xiao, C.Y.&Zhang, J. 2010. Analytical solutions of transient pulsed eddy current problem due to elliptical electromagnetic concentrative coils.*Chinese Physics B*, Vol.19, No.12: 120302.

[4] Sivaprakasam, P. &Karthikeyen, S&Hariharen, P.2012. A study on non destructiveevaluation of materials defects by eddy current methods.International Conference on Mechanical, Automotive and Materials Engineering .Jan. 7–8, 2012.Dubai.

[5] Li, S.&Huang, S.&Zhao, W.&Yu, P.2007. Development of differential probes in pulsed eddy current testing for noise suppression. *Sensors and Actuators*, V135: pp.675–679.

[6] Wu,H.Z. 2001. *Handbook of materials on electrical machines*.Xian: Shanxi Science Education Press, pp.509–516.

[7] MH/T 3002.5–1997, 1997. *Nondestructive testing in aircraft Eddy current testing*. Beijing: People's Republic of China Civil Aviation Industry Standard.

Testing and Measurement: Techniques and Applications – Chan (Ed.)
© *2015 Taylor & Francis Group, London, ISBN: 978-1-138-02812-8*

Measurement for the parameters of Permanent Magnet Synchronous Motor

D.D. Zhu & X.P. Chen
Soochow University, Suzhou, China

Y. Tan
Nanrui Group Co., Ltd, Nanjing, China

ABSTRACT: Based on the mathematical model of PMSM, we studied the measurement methods for stator winding resistance and inductance of PMSM, specifically, by inputting suitable Space Vector Pulse Width Modulation (SVPWM) wave on the stator winding to calculate the inductance and employing appropriate Pulse-Width Modulation (PWM) waves to evaluate the stator winding resistance. Compared to other measurement methods, we don't have to consider the dead time and also have no force on the position of the rotor. The theory related is simple and the measurement is powerful adaptability. In addition, our measuring results have verified the effectiveness of the methods.

1 INTRODUCTION

In recent years, with the rapid development of power electronic technology, microelectronic technology, new motor control theory and the rare earth permanent magnet materials, the AC servo control technology has made a great progress. Compared with the traditional synchronous motor, the AC Permanent Magnet Synchronous Motor (PMSM) has the advantages such as reliability, simple structure, low consumption and high efficiency [1], so that it gains the extensive attention of the whole academia and industry. Moreover, acquiring the precise parameters is the prerequisite for using the PMSM effectively [2–4]. Thus, the measurement of PMSM's parameters is quite a hot topic nowadays.

The stator's self and mutual inductances both under the influence of wind's and rotor's Magnetic motive force (MMF). As a consequence, the values of d-axis and q-axis inductances are determined by the stator's current and the position of the rotor [5–8]. Literature [9] measures the three-phase inductance of motor by other measuring equipment to avoid the direct measurement of the stator's self and mutual inductances. This method is simple and less calculation. However, it's based on the situation that the motor is in open circuit. Not only is this method poor engineering application, but also the accuracy of parameters depends on the measuring tool. Literature [10] calculates the inductance and resistance parameters by inputting PWM wave and SVPWM wave that the accuracy of resistance and inductance will be influenced by the

dead time despite the measuring theory is uncomplicated. Specially, the rotor must be rotated to position which is a lot of inconvenience in some situations of engineering application, such as in the condition to drive the elevator, in which the PMSM needs to trice elevator hoist when rotating the rotor to the specified position. Otherwise, the lift will slide due to inertia. Based on the two methods above, the method without depending on other measuring tools which eliminates the influence of dead time on the resistance was proposed. Moreover, the measurement of inductance, which not only reduces the effect of dead time on the measurement results, but also has no force on the rotor's position, brings great convenience to the engineering application.

2 PERMANENT MAGNET SYNCHRONOUS MOTOR

PMSM can be divided into sine wave PMSM and the ladder of PMSM according to the induction electromotive force waveform of the stator windings. The sine wave PMSM is chosen as the research objection who serotor adopts permanent magnets instead of electrical excitation and the stator is composed of three-phase windings which are often connected as type Y and short distance distributed that can be simply equivalent as the model in Figure.1. According to the installation positions of the permanent magnet, the PMSM is divided into three categories: convex mounted type, embedded type and buried type [11].

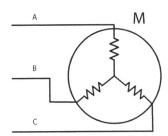

Figure 1. The simple equivalent model of stator windings.

3 THE PRINCIPLE FOR MEASUREMENTS

3.1 *The principle of inductance measurement*

The convex mounted sine wave PMSM, whose inductances of d-axis and q-axis are equal, was adopted in this study. The mathematical model of voltage and flux for PMSM in synchronous rotation d-q axis is shown as equation (1) and (2)[12–13]:

$$u_d = R_S i_d + p L_d i_d - \omega_e L_q i_q \tag{1}$$

$$u_q = R_S i_q + p L_q i_q + \omega_e \psi_f + \omega_e L_d i_d \tag{2}$$

p is differential operator; ω_e is the rotating angular velocity; R_S is the stator's resistance; i_d, i_q are the current of stator in d-axis and q-axis; u_d, u_q are the voltage of stator in d-axis and q-axis; ψ_f is rotor flux which is constant. The angles appear in this paper are all electrical degree.

According to the mathematical model above, if the u_d, u_q motivations are given, there will be i_d, i_q output as feedback. Meanwhile, keeping the motor's rotor static, so $\omega_e = 0$. According to equation (1), the equation (3) can be acquired:

$$u_d = R_S i_d + p L_d i_d \tag{3}$$

Then to derive the equation (4):

$$\frac{L_d}{R_S} \frac{di_d}{dt} + i_d = \frac{u_d}{R_S} \tag{4}$$

On the basis of the step response feature of the inertial system, if u_d signal is given, the current response should be equation (5):

$$i_d = \frac{u_d}{R_s} \left(1 - e^{-\frac{R_s}{L_d} t} \right) \tag{5}$$

Transforms the current response equation (5) to the inductance equation (6):

$$L_d = \frac{-R_s t}{\ln \left(1 - \dfrac{i_d R_s}{u_d} \right)} \tag{6}$$

According to the equation (6), the keys to calculate L_d lie in the acquirements of u_d and i_d, similarly to L_q. The method to measure L_d proposed by literature [10] is to rotate the rotor to $0°$ position and then apply the space voltage vector of $0°$. At this point, $U_d = U_a = \frac{2}{3} U_{dc}, i_d = i_a$. Similarly, the rotor should be rotated to $270°$ position when L_q be measured. This method which has a mandatory requirement of rotor position will be a lot of inconvenient in practical application. The method proposed in this paper which solved the problem brings great convenience for measuring the inductance of the stator.

Having obtained the current position of rotor firstly and applied the relevant space voltage vector on the stator, $U_O = U_d$. According to the principle of SVPWM, space voltage vector is the composition of the basic voltage vector. Taking the first vector, for example, the vector composition diagram is shown in Figure.2[14]:

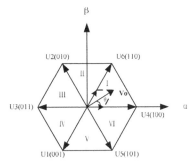

Figure 2. Basic space vector composite image.

According to the volt second balance principle that is shown in equation (7):

$$\begin{cases} |U_d| \cos\theta = \dfrac{T_4}{T_s} |U_4| + \dfrac{T_6}{T_s} |U_6| \cos\dfrac{\pi}{3} \\ |U_d| \sin\theta = \dfrac{T_6}{T_s} |U_6| \sin\dfrac{\pi}{3} \end{cases} \tag{7}$$

$|U_4| = |U_6| = \frac{2}{3} U_{dc}$, so equation (8) can be obtained:

$$|U_d| = \frac{2\sqrt{\left(T_4^2 + T_6^2 + T_4 T_6 \right)}}{3 T_s} U_{dc} \tag{8}$$

T_s is the period of PWM; T_4, T_6 are the working time of U_4 and U_6; U_{dc} is the bus voltage which can be achieved via A/D sampling. When the rotor is in

258

other sectors, the $|U_d|$ can be calculated according to equation (8) as well.

We get the i_a, i_b by A/D sampling and then calculate the i_d through Clark and Park transforms, and record the corresponding sampling moment t. When R_s is known, we calculate the L_d according to equation (6).

Similarly, q-axis inductance can be gained on the basis of U_q, i_q that we should apply the voltage vector vertical to the rotor, so $U_O = U_q$. In the experimental process, the vector action time should be short since the motor's rotor must keep static. The motor in test has the feature that $L_d = L_q$, therefore, only the L_d needs to be measured.

3.2 The principle for resistance measurement

Generating three-phase sine waves have $120°$ a phase difference by inverter circuit to drive the PMSM according to the SVPWM principle. The inverter circuit is composed of six IGBT transistors as it is shown in the Figure.3:

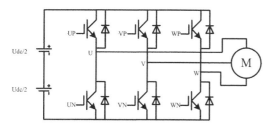

Figure 3. Inverter circuit.

Literature [10] presented a method that used voltage pulse to measure winding's resistance which not completely eliminated the dead time effect on the results. The method that the PWM waves were employed only on one bridge arm was proposed in this paper which can eliminate the influence of dead time.

The method was based on the Ohm's law that $R = U / L$. We inputted the fixed duty ratio of the voltage pulse to equivalent DC voltage due to the fact that the IGBT transistor can't open for a long time continuously. U_{dc} was acquired through A/D sampling and the duty ration $per\%$ was recorded. i_o is the total current that can be Figured out by i_a, $i_b.i_a$, i_b obtained by A/D sampling. In order to reduce the sampling error, and deal the current with filter processing, the resistance can be Figured out in equation (9):

$$R = \frac{U_{dc} \times per\%}{i_o} \qquad (9)$$

The value of $per\%$ defines the accuracy of measurement results. To eliminate the work and use error

of dead time, we open the corresponding under bridge arm firstly and then input the voltage pulse from the upper bridge arm. The equivalent circuit is shown in Figure.4:

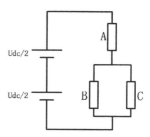

Figure 4. The stator resistance measurement equivalent circuit.

4 THE CIRCUIT DESIGNED FOR CURRENT AND VOLTAGE

The resistance and inductance of PMSM is related to the voltage and current. The circuit of A/D sampling is designed to measure U_{dc} and i_a, i_b. The STM32F407 has 16 A/D channels which are divided into regular channel and injection channel. The sampling precision is up to 12/bit and rate can reach 2.4M. The injection channel and regular channel are used to sample i_a, i_b and U_{dc} respectively.

The PMSM's working current reaches ampere level. As the large current is not suitable for direct measurement, the current transformer is adopted to attenuate the original signal for about 2000 times and reduce the mutual influence in the circuit. The current signal should be transformed to voltage signal as the A/D channel of core chip is fit for voltage. Moreover, the direction of i_a, i_b is not unique, so the voltage bias circuit shown in Figure.5 is designed to transform the current to the range of $0 \sim V_{ref}$ for sampling, V_{ref} is the reference voltage of A/D sample.

The bus voltage of the inverter is more than 300V. In order to make it fit for A/D sampling, the voltage

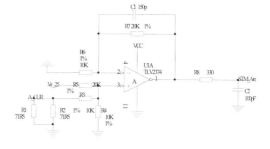

Figure 5. Voltage bias circuit.

isolation and amplifier circuit shown in Figure.6 is designed. The DC signal suffers the voltage divider circuit and then the operational amplifier circuit is composed of TLV2372. The isolation circuit compose by A7840 can reduce the mutual influence in the circuit.

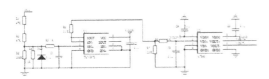

Figure 6.   Voltage isolation and amplifier circuit.

There will be two different signals after optical couple. The voltage differential circuit is shown in Figure.7 which combines the two signals into one signal for A/D sampling.

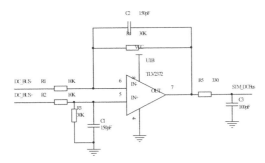

Figure 7.   Voltage differential circuit.

5   SOFTWARE DESIGN

5.1   *Inductance measuring*

The experiment adopts the motor whose $L_d = L_q$, so the software designed mainly aims at the measurement of L_d. The first step is to get the position of the motor's rotor and then apply the space voltage vector corresponding. The key to measure the inductance is to obtain R_s, U_d, i_d according to equation (6). The detail of measuring R_s will be introduced below.

According to equation (8), the U_d associates with the action time of each vector and the bus voltage. The STM32F407ZE, which is applied in the area of PMSM controlling widely, is chosen as the control chip that the speed is up to 144M. The complementary PWM waves can be generated through TIM1 in STM32F407. We read the CCR1, CCR2, CCR3 to calculate the action time of each vector. Attention to depress the influence of dead time is set for avoiding the inverter short.

The i_a, i_b should be filtered to reduce the A/D sampling error. Using the data in the first 2.5ms can

improve the accuracy of the results, as i_d is an index curve that tends to be stable with the time growth. The overall software design flow chart is shown in Figure.8.

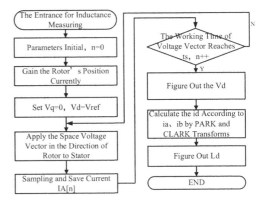

Figure 8.   Direct axis inductance measuring process.

5.2   *Resistance measuring*

The duty ratio of voltage pulse has a great influence on the accuracy of resistant. Accordingly, adjusting the duty cycle of voltage pulse based on the feedback value of i_o in the process of measuring resistance. The reference range is $0.5 I_r \sim 2 I_r$ in the paper, and I_r is the rated current of the motor. Identifying the resistance study is effective only when the feedback current reaches the reference range and record the voltage pulse duty ratio and the bus voltage at the same time. The software design flow of resistance measurement is shown in Figure.9.

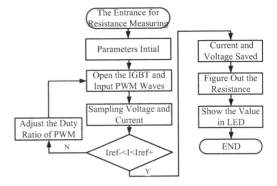

Figure 9.   Stator resistance measuring process.

6   TEST RESULTS

6.1   *Inductance measurement results*

The actual inductance of PMSM is mH class, so the sampling frequency should be set as largely as possible. In the software, the sampling frequency and SVPWM frequency are both4KHz.

According to equation (5),the i_d is an index curve. When the rotor is located at the 0 position, $i_d = i_a$. The i_a measured by the FLUCK oscilloscope hat is shown in Figure.10 consists of the theory value. The channel A is the equivalent voltage after i_a suffered voltage bias circuit. The channel B is one of SVPWM waves.

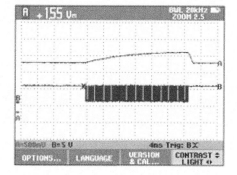

Figure 10. The curve of $\mathbf{i_a}$ in FLUCK.

Sampling the anterior values of i_a, i_b with $250\mu s$ interval and calculating i_d.Data-fitting of i_d is shown in Figure.11 which conforms to the Figure.10.

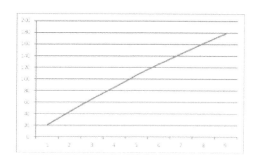

Figure 11. The fitting curve of $\mathbf{i_d}$.

According to the method of measuring inductance provided by literature [9], we rotate the rotor to the $0°$ position, then measure and calculate the reference L_d which is about $46.2mH$ by LCR4263B measuring instrument from Agilent Company. The measuring result is shown in Tab.1.

Inductance is the important parameter in the PI regulator of the motor's current loop. The accuracy of inductance will influence the motor's stability. The measuring result is stable and in a reasonable range.

Table 1. Inductance measurement results.

POSITION	INDUCTANCE (mH)	ERROR(%)
$0°$	48.35	4.65%
$30°$	48.35	4.65%
$60°$	48.14	4.20%
$90°$	47.02	1.77%
$120°$	46.64	0.95%
$150°$	47.54	2.90%
$180°$	47.62	3.07%
$210°$	48.14	4.20%
$240°$	48.33	4.61%
$270°$	48.14	4.20%
$300°$	47.93	3.74%
$330°$	48.20	4.33%

6.2 Resistance measurement results

The experiment adjusts the duty ratio through the software automatically. The reference resistance is 4.2Ω tested by the LCR measuring instrument of Agilent Company. The experimental result is shown in Tab.2.

Table 2. Resistance measurement results.

Ref Value(Ω)	Resistance(Ω)	Error(%)
4.2	4.4	4.76

The resistance is an important parameter in PI regulator of current loop as well. The experiment's result is in a reasonable range and it can meet the actual requirement.

7 CONCLUSION

Accurate or not of motor's measuring parameters impacts the speed and position adjustment of motor significantly. In this paper, we input appropriate PWM wave in order to evaluate the motor's resistance based on the stator's structure. The method proposed doesn't have to consider the dead time compared to other methods. In addition, by inputting the corresponding voltage vector according to the current position of the rotor, motor's d-axis inductance can be Figured out. This method is widely applicable no matter what the position of rotor currently is. This feature will bring a great convenience for some engineering applications. Finally, the experiment results verify the feasibility of the measurement methods.

ACKNOWLEDGMENTS

This study was supported by the Prospective Project on the Integration of Industry, Education and Research of Jiangsu Province No. BY2012110.

REFERENCES

Liu Xianxing, PuYanzhu, Hu Huangwen, Zhu Yuqiu. The permanent magnet synchronous motor control system based on the precise linear resolve space vector decoupling of [J]. Proceeding of the CSEE, 2009, 27(30):55~59.

Rahman M A.High efficiency IPM motor drives for hybrid electric vehicle[C]. Canadian Conference on Electrical and Computer Engineering(IEEE CCECE 2007), Vancouver, 2007.

Park J W, Koo D H, Kim J M,et al. Improvement of control characteristics of interior permanent-magnet synchronous motor for electric vehicle [J]. IEEE Transactions on Industry Application, 2001, 37(6):1750~1560.

Li Kunpeng, Hu Qiansheng, Huang Yunkai. The analysis on the winding's inductance based on circuit model of the permanent magnet of brushless dc [J]. Proceeding of the CSEE, 2004, 24(1): 77~80.

Cheng Shukang, Yu Yanjun, Chai Feng, et al.Inductance parameters of built-in permanent magnet synchronous motor research [J]. Proceeding of the CSEE, 2009, 29(18): 94~99.

Stumberger B, Stumberger G, Dolinar D, et al. Evaluation of saturation and cross-magnetization effects in interior permanent-magnet synchronous motor [J]. IEEE Transaction on Industry Application, 2003, 39(5): 1264~1271.

Han Guangxian, Wang Zongpei, Cheng Zhi, et al.The accurate of hybrid stepping motor nonlinear simulation model [J]. Proceeding of the CSEE, 2002,22(5): 116~120.

Dajaku G, Gerling D. The correct analytical expression for the phase inductance of salient pole machines [C]. 2007 IEEE International Conference on Electric Machines and Drives(IEMDC 2007), Antalya, Turkey, 2007.

Liu Jun, Wu Chunhua, Huang Jiangming, Yu Jingshou. The method measures the parameters of permanent magnet synchronous motor [J]. Power Electronic Technology, 2010, 44(1): 46~48.

Wu Jiabiao. Motor servo drive system and its parameter identification [D]. Zhejiang: Zhejiang university, 2013: 19~28.

Zhou Yunxia. Control system of position servo [D]. Guangdong: Zhongshan University: 9~18.

Chen Rong. Research the system of permanent magnet synchronous motor [D]. Jiangsu:Nanjing university of aeronautics and astronautics, 2004:10~13

Li Bingqiang, Lin Hui.Current vector control technology of surface mounted permanent magnet synchronous motor [J].Proceeding of the CSEE, 2011,31(Added):288~294.

Liu Jun, Chu Xiaogang, Bai Huayu. Direct torque control research based onmodulation strategy of flux linkage reference voltage space vector of permanent magnet synchronous motor [J]. Proceeding of the CSEE, 2005,20(6): 11~15.

Testing and Measurement: Techniques and Applications – Chan (Ed.)
© *2015 Taylor & Francis Group, London, ISBN: 978-1-138-02812-8*

Experiment and numerical study of flow resistance characteristics of automotive three-way catalytic device

Y. Wang, Y.L. Jiang & L. Huang

College of Aerospace Engineering, Nanjing University of Aeronautics and Astronautics, Nanjing, China

ABSTRACT: In this paper, based on the study of flow resistance of traditional straight type three-way catalytic converter and by the method that combined with numerical simulation and experiment, the flow resistance of a new type of three-way catalytic converter structure using metal carrier (by-pass type three way catalytic converter) was investigated. After the comparative analysis with the three way catalytic converter, some conclusions could be obtained. Under the same inlet air flow rate and exhaust pressure, the pressure loss (flow resistance) of the by-pass type three way catalytic converter is smaller than that of the straight type three-way catalytic converter. In addition, the airflow uniformity of the by-pass type three way catalytic converter is much better. In general, the design of the by-pass type three-way catalytic converter contributes to reducing the flow resistance and prolonging the service life.

1 INTRODUCTION

Nowadays with the growth of cars, automobile exhaust pollution has become more and more serious. To reduce the exhaust pollution, three way catalytic device which has the function of transforming harmful gas into CO_2, N_2 and H_2O[1] is installed in car's exhaust system. But with the installment of the three way catalytic device (TWC for short), flow resistance as well as engine residual pressure is potentially increasing. This phenomenon reduces the amount of fresh air breathed in, which causes the reduction of air-fuel ratio, thereby affecting the overall performance of the engine [2].

Therefore it is necessary to take imperative and in-depth research on reducing flow resistance without affecting catalytic efficiency of the TWC. In this paper, numerical simulation of traditional straight type TWC and a new type of by-pass TWC is proposed based on the finite volume method, and then experimental study is set up to make a comparison. The results have some reference value for the design and optimization of TWC design.

2 NUMERICAL SIMULATION AND ANALYSIS

2.1 *Physical composition*

Traditional straight type TWC (Figure 1) is composed of a diffuser inlet, a carrier and a diffuser outlet[3]. The carrier will be blocked after long-term use of the TWC. Therefore a by-pass type is designed and bypass hole as well as diversion pipe are set up to reduce resistance.

As is shown in Figure 2, by-pass type TWC is made up of diffuser inlet, inside diffuser inlet, by-pass channel, by-pass hole, inside diversion pipe, carrier, carrier cavity, board and diffuser outlet [4]. In operating condition, most vehicle exhaust flows through the carrier, leaving a small part flows through the by-pass hole. But when the carrier is clogged, exhaust gas mainly flows from the by-pass hole, making the engine operate normally.

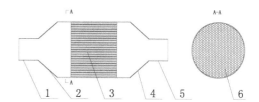

1. inlet, 2.diffuser inlet, 3, 6.carrier, 4.diffuser outlet, 5 outlet

Figure 1. Structure of straight type TWC.

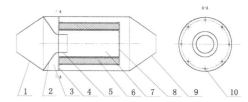

1. diffuser inlet, 2.inside diffuser inlet, 3.by-pass channel, 4,10.by-pass hole, 5.inside diversion pipe, 6.carrier, 7.carrier cavity, 8.board, 9.diffuser outlet

Figure 2. Structure of by-pass type TWC.

2.2 Flow resistance characteristics

Flow resistance of the TWC includes the resistance of diffuser, friction, carrier resistance and Tapered vascular resistance

2.2.1 Resistance of diffuser

Air flows into and outside the device through the diffuser, the local pressure loss of which is:

$$\Delta P = k\frac{\rho(U_1 - U_2)^2}{2} \tag{1}$$

In this formula: k is correction factor, as spread angle is 60°, according to experimental data provided by Gibson; U_1 is inlet velocity, m/s; U_2 is outlet velocity, m/s.

2.2.2 Friction

In this paper, air that flows into the TWC is considered as uncompressible gas. According to fluid dynamics, pressure loss along the way is:

$$\Delta P = \lambda\frac{L}{d}\frac{\rho U^2}{2} \tag{2}$$

In this formula: λ is friction factor; d is diameter of the pipe, m; L is length of the pipe, m; ρ is air density, kg/m³; U is average velocity of air in TWC, m/s.

Meanwhile λ can be calculated by function of Re and relative roughness Δ/d:

$$\lambda = f(\text{Re}, \Delta/d) \tag{3}$$

2.2.3 Carrier resistance

1 Carrier resistance of straight type TWC

Resistance inside the carrier of straight type can be calculated by Formula Darcy:

$$\Delta P' = \lambda\frac{L}{d_h}\frac{\rho u^2}{2} = \frac{\text{Re}\,\lambda}{2}\frac{\mu L u}{d_h^2} \tag{4}$$

In this formula, λ is loss coefficient along the way; ρ is flow density; μ is air viscosity coefficient; L is carrier length; d_h is hydraulic diameter of pipe; u is flow velocity inside the carrier.

2 Carrier resistance of by-pass type TWC

By-pass type has a complex internal structure and factors affecting flow resistance are various. From the flow trend of figure 2, it can be seen that except carrier resistance, flow resistance consists of local resistance and on-way resistance of bend, water hammer pressure resistance, eddy current loss and so on.

Local resistance of bend can be calculated by formula (1), the local loss coefficient can be get as follow:

$$\zeta_3 = k\frac{\theta}{90} \tag{5}$$

In this formula:

$$k = 0.131 + 0.159(\frac{D}{R})^{35} \tag{6}$$

Water hammer pressure loss:

$$\Delta P'' = \rho c U \tag{7}$$

In this formula c is spread velocity of compression wave in pipe, U is flow velocity, ρ is flow density.

From formulas above, the local press loss of carrier could be obtained as follows:

$$\Delta P_3 = f(\text{Re}, U, \zeta, \mu, \rho) \tag{8}$$

2.2.4 Local pressure loss of diffuser

Local pressure loss of diffuse is:

$$\Delta P = \zeta\frac{\rho U_2^2}{2} \tag{9}$$

In this formula ζ is coefficient of local resistance; U1 is air inflow velocity, m/s; U2 is air outflow velocity, m/s.

In formula (9)

$$\zeta = \frac{\lambda}{8\sin\sin(\theta/2)}[1 - (\frac{A_2}{A_1})] + \frac{\theta}{10000}, \lambda \text{ is by-pass}$$

loss coefficient after reducing. A1 and A2 are cross-section areas before and after reducing.

From formula(2)-(4) it can be concluded:

$$\Delta P = f(U^2, \text{Re}, \Delta/d, \rho, \zeta) \tag{10}$$

As is shown above, total pressure loss of TWC is a function of square of average flow velocity, Reynolds number, pipe roughness and air density [5]. In this function average flow velocity changes with inlet velocity linearly. Meanwhile inlet and outlet diffuser angle, length of inside diversion pipe, number and area of by-pass hole, thickness of filter wall and the particulate layer, capacity of carrier all have influence on the flow resistance of the carrier.

2.3 Model and grid forming

CFD software ANSYS ICEM is applied for the modeling of straight type TWC and by-pass type TWC.

Figure 3.   Grid of straight type TWC.

Figure 4.   Grid of by-pass type TWC.

2.4   *Boundary conditions*

k-ε turbulence model is applied with the following assumptions:(1) Flow process in the TWC is considered steady; (2) Air in the TWC is considered as incompressible flow; (3) Wall of TWC is in adiabatic condition [6]; (4) Carrier is set as porous media, the porosity of which is set at 0.3, and viscous resistance coefficient of porous media is set as well.

Inlet velocity is set under 60m/s (including 10m/s, 20m/s, 30m/s, 40m/s, 50m/s), "second-order upwind" format and SIMPLE pressure-velocity coupling algorithm are selected [7][8].

2.5   *Simulation results*

Pressure distribution of traditional straight type TWC and by-pass type TWC is displayed in figure 5 and 6.

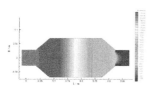

Figure 5.   Pressure distribution of straight type TWC.

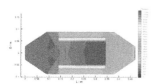

Figure 6.   Pressure distribution of by-pass type TWC.

Pressure distribution in figures above is integrated with flow rate in both straight and by-pass type TWC, and then are compared as follows:

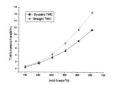

Figure 7.   Comparison of pressure loss from simulation of straight Type TWC and by-pass type TWC.

From figure 7 it can be concluded that the change trends of press loss with inlet flow are similar in straight and by-pass type TWC, they both have exponential relationship. However, under smaller inlet flow the pressure loss of the two kinds of TWC is basically the same. With the increasing of inlet flow and application of by-pass hole, press loss of the two kinds of TWC shows significant difference. Flow resistance of by-pass type is less than that of straight type, and this phenomenon is more obvious under higher inlet flow.

3   EXPERIMENT AND ANALYSIS

3.1   *Experiment bench*

Air flow is applied instead of the exhaust gas. Schematic diagram is shown in figure 9.

The experiment bench is composed of fan inverter, high pressure fan, vortex flowmeter, micro pressure sensor, the TWC, silencer and temperature sensor.

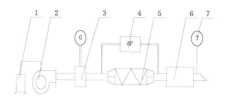

1. Fan inverter; 2.high pressure fan; 3.Vortex flowmeter; 4. Micro pressure sensor; 5. the TWC; 6. Silencer; 7. Temperature sensor

Figure 8.   Experiment design of TWC.

Figure 9.   Picture of the experiment bench.

3.2 *Results and comparison*

Relationship curves of pressure loss and flow in straight and by-pass type TWC are compared as follows.

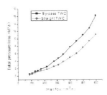

Figure 10. Comparison of pressure loss from experiment data of straight type TWC and by-pass type TWC.

From figure 10, we can see with the increase of inlet flow rate, pressure loss change rate of straight type is higher than that of by-pass type. This means under low flow rate, flow resistance of the two types is nearly the same, but under high flow rate, by-pass type TWC has smaller flow resistance.

During measurement process in experiment, due to some reasons such as the accuracy of instrument, lack of measurement method and undulation of experimental environment, there must exist a certain difference between measured result (i.e., the measured value of the measured or instrument value) and real result. From the results of figure 11 and figure 12, we can conclude that relative error of experiment results and simulation data are decreasing gradually with steady increase of flow rate. By analyzing the figures and calculating relative error, it is found that the test results of the measured data coincides with simulation data, and the accuracy of the simulation results is also demonstrated by the test.

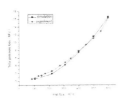

Figure 11. Comparison of experiment and simulation results of straight type TWC.

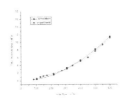

Figure 12. Comparison of experiment and simulation results of by-pass type TWC.

4 CONCLUSIONS

In this paper, based on finite volume method, traditional straight type TWC and a new by-pass type TWC are simulated, and experimental verification is made according to TWC device bench. Following conclusions are obtained by comparing the two types of TWC:

1 Under different flow rate, both in the traditional straight type converter and the new by-pass converter, pressure loss increases exponentially with the increase of flow rate.

2 Under certain inlet flow rate, total pressure loss and carrier pressure loss of by-pass type TWC are less than that of straight TWC; this means the design of by-pass type TWC is beneficial for reducing flow resistance and prolonging the life of catalytic converters.

3 Experiment results and simulation data are compared and analyzed; the relationship between pressure loss and flow rate obtained from experiment and simulation are basically consistent. This demonstrates the possibility of applying numerical simulation for analyzing the flow resistance of TWC.

REFERENCES

[1] Kong Xianghua. Numerical simulation of three-way catalyst [D]. Kunming University of Science, 2005.
[2] Gong Jinke, Zhou Liying, Liang Yu. A study on the effect of pressure loss in three-way catalytic converter on engine performance[J]. Automotive Engineering, 2004, 26(4): 413–416.
[3] Xiu Hengxu. Numerical simulation and optimize design of flow field of three-way catalysts in NGV [D]. Chongqing University, 2004.
[4] Shuai Shijin, Wang Jianxi, Zhuang Renxie. et al. Application of CFD in design and optimize of catalysts in vehicles[J]. Automotive Engineering, 2000, 18(2):129–133.
[5] Liu Biao, Liang Yu Zhou Liying. The effect of monolith on the flow field and pressure loss in catalytic converter [J]. Journal of Hunan University (Natural Sciences), 2004, 31(1): 17–20.
[6] Elemwntary fluid mechanics [M]. Press of BUAA, 1990.
[7] Tsinoglou D N, Koltsakis G C, Missirlis D K, et al. Transient modelling of flow distribution in automotive catalytic converters[J]. Applied Mathematical Modelling, 2004, 28(9): 775–794.
[8] Han Zhanzhong. Examples and analysis of FLUENT-fluid simulation and calculation of engineering [M]. Beijing: Beijing Institute of Technology, 2009.

Testing and Measurement: Techniques and Applications – Chan (Ed.)
© 2015 Taylor & Francis Group, London, ISBN: 978-1-138-02812-8

The design of storage measurement device in projectile's motion parameters testing under nose to tail loading state

Y. Hang, F. Shang, D.R. Kong, J. Wang, J.Q. Zhang & Y.Z. Li
Mechanical Engineering School, Nanjing University of Science & Technology, Nanjing, China

ABSTRACT: Since the force condition of projectile under nose to tail loading state is quite different from projectile under common single launch, nose to tail loaded projectiles have a quite different motion state. In this paper, we put forward a storage measurement method to test projectile's interior ballistic motion parameters on the basis of conducting a force analysis on nose to tail loaded projectile. We also designed a kind of storage measurement device which could measure projectile's base pressure, nose pressure and its motion acceleration based on carrying out the sensor selection, analyzing circuit characteristics and many other works. This device can be used to acquire data about projectile's motion parameters under nose to tail loading state and has certain significance of analyzing the projectile's motion law.

KEYWORDS: Interior ballistic, Projectile, Motion parameters, Force analysis.

1 INTRODUCTION

When it comes to common projectile's interior ballistic motion, projectile is mainly affected by propellant gas pressure which produced in the base of the projectile. If multiple projectiles are loaded nose to tail in a single gun barrel with propellant packed between them, igniting the propellant to launch projectiles sequence, there is only a gap of few milliseconds between two projectiles' launch. For a certain projectile (except the first one), the projectile is not only affected by its propellant gas pressure, but also influenced by the gas pressure from the launch of previous projectile, thus causing its motion characteristic's change. It is quite essential to measure motion parameters such as projectile's base pressure, nose pressure and its motion acceleration to discuss projectile's force condition and motion characteristic.

Storage measurement method [1] is a widely used method to acquire projectile's motion parameters. Under the premise of not destroying artillery's structures, putting miniature data acquisition and storage device into bore to acquire and store information realtime, recovering the device and reading the data after the experiment. At present, storage measurement devices that are used commonly are usually massive and can only acquire a projectile's base pressure and motion acceleration under common launch method [2]. They may not meet the requirement of projectile's motion parameters test under nose to tail loading state.

Based on force analysis, this paper puts forward a design of storage measurement device which can be used to measure small caliber projectile's motion parameters under nose to tail loading state. The device can measure the projectile's base pressure, nose pressure and its motion acceleration at the same time, and has certain significance on the test of projectile's motion parameters testing under nose to tail loading state.

2 FORCE ANALYSIS ON PROJECTILE UNDER NOSE TO TAIL LOADING STATE IN BORE

Take one projectile (except the first one) as the research object, excluding gravity, only take the force paralleled to the direction of projectile's motion. Its force condition is shown in Figure 1. OX is paralleled to the bore axis.

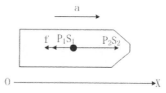

Figure 1. Force analysis of projectile under nose to tail loading state in bore.

According to Newton's second law, the projectile's force equation can be established:

$$P_2 S_2 - P_1 S_1 - f = ma \qquad (1)$$

P_1 is the resistance to the projectile nose caused by the gas pressure from the launch of previous projectile, P_2 is propellant gas pressure produced in this projectile's bottom. S_1, S_2 are the equivalent areas of the projectile's base and nose. f is caused by projectile-barrel friction. m is the mass, a is the instantaneous acceleration of projectile.

Apparently, mechanical model of projectile under nose to tail loading state is distinguished from common projectile in their interior ballistic motion. Projectile under nose to tail loading state is affected by projectile base's propellant gas pressure as well as gas pressure from the launch of previous projectile, which is a resistance for the projectile. Since f is much less than P_1 and P_2, we ignore it in the following part of this paper.

In a word, projectile's base pressure, nose pressure and its motion acceleration should be acquired at the same time to obtain projectile's accurate interior ballistic motion rule under nose to tail loading state.

3 OVERALL DESIGN ON PROJECTILE MOTION PARAMETER STORAGE MEASUREMENT DEVICE

According to the testing requirement about projectile's interior ballistic motion parameters, this system should be used to measure three parameters as mentioned before. Sampling rate per channel is 100 kHz, sampling time is 20ms, pre-sampling time is 10ms, two channels' pressure range are both 0~ 400MPa, acceleration range is 0~20000g. The overall design is shown in Figure 2.

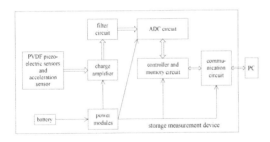

Figure 2.   Overall design of the testing system.

This system is composed of sensors, power modules, charge amplifiers, filter circuit, ADC circuit, controller and memory circuit and communication circuit. The output of the sensors is conditioned to become into a voltage signal which is proportional to the pressure or acceleration by charge amplifiers. Signal with inappropriate band will be filtered

by filter circuit, and then the designed ADC circuit acquires signals and change the analog signal into digital form with the help of controller circuit. At last the system executes data storage to put data into memory unit. After the experiment PC will acquire testing data by communication circuit and process data.

4 DESIGN OF SIMULATED TESTING PROJECTILE

We design a simulated testing projectile to test its motion parameters in the process of interior ballistic motion. Pressure sensors are installed in the projectile's base and nose, acceleration sensors, circuit boards, battery are put inside of the simulated projectile. Packaging simulated projectile's weight, volume and structure the same as actual projectile. Launching this simulated projectile by artillery under nose to tail loading state and the device will acquire projectile's interior ballistic motion parameters and record the testing data. Structure diagram of simulated testing projectile is shown in figure 3.

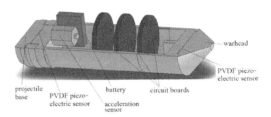

Figure 3.   Structure diagram of simulated testing projectile.

We design simulated testing projectile following actual projectile's structure parameters, the PVDF piezoelectric is stacked on the bottom and top of simulated projectile laminated by projectile base and warhead. Projectile base and warhead are used to pass pressure to the sensors and keep sensors from instantaneous high temperature and propellant gas's damage. All sensors are fixed inside the simulated projectile to test parameters with the motion of simulated projectile.

Polyvinylidene fluoride (PVDF) is a new kind of piezoelectric material [3] with strong piezoelectric properties which can be made into a thin film thinner than 100μm. The material also has a high mechanical strength and toughness. The PVDF piezoelectric sensor has the advantages such as wide frequency response, strong stability and is quite suitable for our testing.

With the limitless of test space, we separate all circuits into three circuit boards and connect them integrally. Sensors are connected to the circuit boards by signal lines. The system is powered by two paralleled polymer lithium batteries to prevent the system from instantaneous power failure.

After the system's installation, filling the inner space of a projectile with potting material [4] to resist insulation, insulate heat and anti-overload impact.

5 DESIGN OF CIRCUIT

The design of system's circuit should be in the principles of low power consumption and simplification. We design charge amplifiers, filter circuit, ADC circuit, controller and memory circuit and communication circuit to make up the circuit system.

5.1 *Charge amplifiers*

The output of the PVDF piezoelectric sensor and acceleration piezoelectric sensor is an electric charge signal. Piezoelectric sensor has extremely high insulation resistance to make sure electric charge will not leak. Common amplifiers shouldn't be used to measure electric charge signal because of their low input resistance. While, charge amplifier with high input resistance and low output resistance is the dedicated preamplifier circuit for piezoelectric sensor. The output voltage of charge amplifiers is proportional to the input electric charge signal. The charge amplifier circuit is shown in Figure 4.

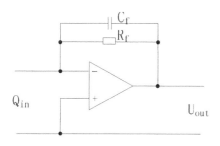

Figure 4. The charge amplifiers circuit.

We use the amplifier OPA2132 to build circuit. This chip has high input impedance and high open-loop gain. The relationship between input and output is:

$$U_{out} = Q_{in} / C_f \qquad (2)$$

A high-value resistor R_f is used to provide direct feedback, reducing the output of amplifier's zero drift, making the charge amplifier stable.

5.2 *Filter circuit*

The system is quite easy to be disturbed by a high-frequency signal. An effective band of pressure in-bore is 5kHz. Set 10kHz at an upper cutoff frequency of the filter circuit. According to the circuit mentioned in 3.1, 100mV zero drift is still existed even R_f is introduced to reduce the output of amplifier's zero drift. Setting 0.01Hz as the lower cutoff frequency of the filter circuit is an effective way to relieve the problem. Take limitless space into consideration, the system uses an order passive filter circuit.

5.3 *ADC circuit*

ADC circuit is designed to change the output voltage of the signal conditioning circuit into digital signal which can be easily identified, transmitted and stored by computers and micro-controllers.

We choose a type of 12-bit plus sign SAR ADC to acquire a signal. The ADC can accept true bipolar analog input signals, it has a high speed serial interface (SPI) that can operate at throughput rates up to 1 MSPS. SPI is a full-duplex synchronous communication bus, occupying only four pins of the chip that simplify PCB layout greatly.

5.4 *Control and storage circuit*

Data acquisition, storage and communication are required to be under the control of microcontroller. The system uses 32 bit STM32F1 Series MCU based on ARM-Cortex-M3 core as microcontroller. This embedded SCM has low power consumption. Its maximum operating frequency can reach 72MHz, 64KB SRAM and 512KB Flash memory, integrated on-chip can replace the external FLASH to store tested data, which will simplify the circuit design.

STM32 has abundant peripheral modules such as SPI, USART, and a variety of optional low power consumption mode. Its internal DMA controller can release the data transmission from the control of the CPU, which improves the speed of data transmission greatly. To sum up, STM32 is very suitable to control the system's operation as well as the data transmission and storage.

5.5 *Communication circuit*

This system communicates and exchange data with the host computer via RS232 serial port. The system can receive control orders from the host computer and send acquired data to the host computer when it is connected to the host computer through serial port.

RS232 serial port uses the positive and negative logic level to represent 1 and 0 signals. The micro controller STM32 uses TTL level. So the MAX232 chip is used to complete bidirectional level conversion.

6 SYSTEM SOFTWARE DESIGN

The duration of projectile's interior ballistic motion is about ten milliseconds. STM32 get working under the control of software after power on. There is still a time gap before the launch of a projectile. It must cause memory overflow and power consumption if the storage measurement device keeps collecting and storing data at a high speed. Leading the system to delay a few minutes under low power consumption mode without data collecting and storing is very essential. So waiting trigger before the start of the acquisition, beginning formal acquisition during the projectile's motion and setting the system into sleep mode after testing work to reduce power consumption are the key points in the design of software.

The software workflow chart is shown in Figure 5:

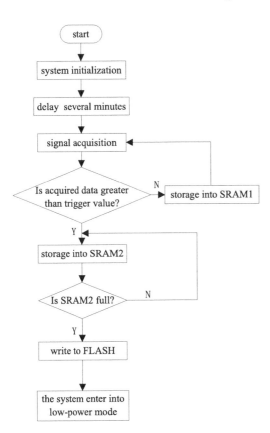

Figure 5. Software workflow chart.

The system starts to initialize after power on and delays several minutes under low-power mode. Since when the system starts a formal test is still unknown after the delayed time, the negative delay sampling is

designed. Firstly, the system compares the acquired digital data with a preset trigger value. If the acquired data is smaller than the preset trigger value, it means that the pressure in bore does not increase, obviously and the projectile does not begin to move. Storing the testing data into SRAM1 is a pre-divided block of SRAM. Covering storage circularly to ensure the acquired and stored data are motion parameters measured 5~10ms before the formal acquisition. When the acquired signal reaches the trigger value, it means that the projectile is under the effect of propellant gas pressure and begin to move. In the meanwhile, storing the testing data successively into SRAM2 is another block of SRAM. When the space of SRAM2 is full, the test finishes. The system writes the SRAM data into on-chip FLASH and then enters the low power consumption mode waiting for recovering and uploading data.

SRAM features in a fast, convenient read-write and data will miss after power off. FLASH takes longer time in writing data, but is not very easy to lose data. So the software adopts twice storage method to make the two kinds of memory work together and complete the data acquisition and storage in short time.

7 CONCLUSION

On the basis of analyzing the motion characteristic of projectile in bore under nose to tail loading state, this paper raises the motion parameters needed to be tested to describe the projectile's motion characteristic: projectile's base pressure, nose pressure and its motion acceleration. This paper puts forward method to test projectile's motion parameters and designs a storage measurement device with the design of the simulated testing projectile, hardware circuit and software workflow. The designed device will solve the problem of projectile's motion parameter testing under nose to tail loading state, and has certain significance on projectile's interior ballistic motion law analysis.

REFERENCES

[1] Jiang, X.H. 2002. Design of a Micro-stored Data Acquisition System Based on Micro Converter. *Chinese Journal of Scientific Instrument* 23(6): 588–591.
[2] You,W.B. 2012. Projectile Bottom Pressure Storage Testing Based on Acceleration Correction. *Journal of Detection & Control* 34(4):6–9.
[3] Wang, G.L. 2004. Design and Implementation of the Three-Point PVdFPiezo-Film Sphygmo-Transducer, *Journal of Transcluction Technology* 4: 688–692.
[4] Gao, H. 2003. Application of Pouring Technology to Electronic Production. *Electronics ProcessTechnology* 24(6): 257–259.

Testing and Measurement: Techniques and Applications – Chan (Ed.)
© 2015 Taylor & Francis Group, London, ISBN: 978-1-138-02812-8

Magnetic flux leakage field within the detector unit optimization based on Comsol

L.J. Yang, W.T. Cui & S.W. Gao

School of Information Science & Engineering, Shenyang University of Technology, Shenyang, Liaoning, P. R. China

ABSTRACT: In order to make the pipeline in the critical state of saturated magnetization and magnetic field, and also be able to detect magnetic flux leakage signals of the defects, the excitation unit inside the pipeline magnetic flux leakage detector is optimized. Studying the magnetic flux leakage detection principle and the permanent magnet excitation intensity of the pipeline can achieve magnetic saturation. Based on Maxwell's equations, we use Comsol Multiphysics to establish a three-dimensional model of the pipeline leakage field, change permanent magnet size, check the leakage magnetic field of magnetic induction intensity of axial and radial component, quantitative analysis of magnetic saturation degree, the size of the corresponding relations between permanent magnet. The results show that: for pipe wall thickness to 8mm, width of 80mm permanent magnet and steel brush can make pipeline near saturation magnetization, at more than 5% of wall thickness defect has to check.

1 INTRODUCTION

The pipeline is one of the most important ways of energy, transport, which is the main artery of the country's energy, the detection of oil and gas pipelines and maintenance is important. Magnetic flux leakage testing method is one of the current oil and natural gas transmission pipeline defect detections, the most effective way is to automatically detect and develop for the purpose of non-destructive testing techniques [1]. The principle is the magnetic flux leakage into the detector inside the pipe, the pipe wall motion detector along the real-time detection, while storing the results and analysis of data, can accurately describe the case of pipeline defects by processing data [2–3]. However, the current domestic and magnetic flux leakage testing equipment generally relatively large degree of magnetization in general are in the pipeline over the saturation level, optimizing and testing equipment volume and improve the ability to detect magnetic flux leakage pipeline.

This paper studies the magnetic saturation degree of pipeline magnetic flux leakage inspection, the influence of change of permanent magnet and steel brush size to optimize magnetic flux leakage detection. Based on the simulation model of pipeline leakage magnetic field, using Comsol Multiphysics software to establish a three-dimensional model of the pipeline leakage field, change permanent magnet size, check the leakage magnetic field of magnetic induction intensity of axial and radial component, analysis of the magnetic saturation degree and the size of the corresponding relation, permanent magnet excitation unit to simulation optimization.

2 MAGNETIC FLUX LEAKAGE DETECTION PRINCIPLE

2.1 Magnetic flux leakage detection principle

Magnetic flux leakage testing is a magnetic phenomenon through the magnetic sensitive element probe to detect the magnetic iron near surface defect, in order to determine a detection technique with defects being detect [4].

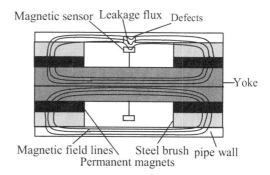

Figure 1. Magnetic flux leakage testing schematics.

As shown in Figure 1, when the magnetizing apparatus magnetized ferromagnetic materials under test, if the material of the material is continuous, uniform, the material of the magnetic induction line will be constrained in the material, magnetic flux is parallel to the surface, almost no magnetic induction line wears out from the surface, the surface of the work-piece is tested without magnetic field. However, when there are cutting lines defects in materials and defects on the surface of the material or organization status change can make the magnetic permeability changes, because the defects of permeability is very small, magnetic resistance is very big, distortion in the magnetic circuit of the magnetic flux, magnetic induction line to change, in addition to the part of the magnetic flux directly through the defects or to bypass by material internal defects, and part of the magnetic flux would spill over the material surface, routing around the defects via air back into the material again, and defect formation leakage magnetic field in the material surface [5].

2.2 *Magnetic magnetization process*

The magnetization process is divided into four stages: the first stage is the reversible domain wall displacement; The second phase is not reversible magnetization; The third stage is the magnetic domain magnetic moment of rotation. The fourth stage is becoming saturated. The four stages can be summarized as two basic ways: one is the domain wall movement; The second is the rotation of the magnetic domain magnetic moment. With the increase of magnetic field intensity H, magnetic induction intensity B is also increasing, reaching the change of the magnetic field strength leveled off after magnetic saturation, and little change in the defect of the leakage magnetic field [6].

Said magnetization process with mathematical formula: because of the ferromagnetic material in the exchange coupling interaction existing between the adjacent electronics, in the absence of additional magnetic field, they can spin magnetic moment in every small area lined up according to the same direction and form spontaneous magnetization small region, called the magnetic domain. In each domain, the magnetic moment is lined up in a direction, in a state of saturation magnetization. Per unit volume of the saturated magnetic moment with saturation magnetization M_s said, using a V_i said the volume of magnetic domain, each domain is $M_s V_i$ magnetic moment. When there is no additional magnetic field, magnetic moment direction of the magnetic domain is disordered, present state of the random distribution, and the magnetic moment cancel each other out, the ferromagnetic material does not display the magnetic, formula is as follows:

$$\sum M_s V_i \cos\theta_i = 0 \qquad (1)$$

When a ferromagnetic material is in an external magnetic field, due to the external magnetic field interacts with the magnetization of the magnetic domains, resulting in changes in the volume of magnetic domains under external magnetic field. By moving domain wall changed V_i, rotating magnetic domain magnetic moment changed θ_i, M_s unchanged magnetized, so:

$$\Delta M_H = M_s \left[\sum V_i \Delta\left(\cos\theta_i\right) + \sum \cos\theta_i \Delta V_i \right] \qquad (2)$$

Therefore, in the ferromagnetic material it will show the macroscopic magnetic properties, and its strength increases with the increase of the applied magnetic field, until all of the magnetic domains are aligned along the magnetic field to achieve the saturation.

3 COMSOL MULTIPHYSICS FINITE ELEMENT SIMULATION

Comsol Multiphysics is based on the finite element method, by solving partial differential equations (single field) or partial differential equations (more than) to achieve true simulation of physical phenomena, using mathematical methods to solve real-world physical phenomena. ACDC module which contains the low-frequency simulation applied to simulate the electrical size (unit size / electromagnetic wavelengths) of less than a tenth of the wavelength of the electromagnetic field. This module can be used for static as well as dynamic and static magnetic field analysis, and other physical modules can be freely coupled. Select the static magnetic field under the low-frequency simulation module ACDC no current module (Magnetic Fields No Current), conducted simulation model.

3.3 *Comsol finite element modeling*

Comsol finite element three-dimensional model into the defective pipe, steel brush, yoke iron and permanent magnet four parts.

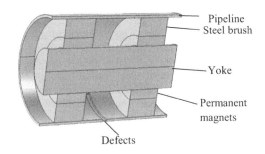

Figure 2.   Three-dimensional solid model of the pipeline map.

Piping and the permanent magnet, the yoke, the brush form a closed loop of steel, wherein the yoke reluctance of the magnetic circuit system can be reduced, increasing the critical part of the magnetic flux density. Steel brush the magnetic field lines coupled into the pipes, steel brush has a certain carrying capacity, the permanent magnet from mechanical damage [7]. According to the principle of magnetic flux leakage testing lines of magnetic force in a magnetic circuit by not only, part of dissipative throughout the space outside the magnetic circuit, the model of the outermost sets the air layer, limits the area of magnetic field lines, improve the simulation accuracy.

3.4 *Define material properties and boundary conditions*

Comsol defined material properties to be accurate, three-dimensional simulation of the pipeline using *X52* pipeline steel, low carbon steel yoke using *St37* steel. The relative permeability of air is 1. Pipeline, yoke iron, steel brush the description of the material is described using *B-H* magnetization curve, using a hysteresis loop instrument can measure the magnetic material of the *B-H* curve, the hysteresis loop, the measured *B-H* curve of input into Comsol material attribute set. The permanent magnet adopts the *NdFedB* permanent magnetic material. It possesses the characters of high remanence, high magnetic energy product and high coercivity.

In the physics Comsol software selection static magnetic field is no current module (mfnc), this module needs you to set the appropriate field boundary condition, the natural boundary conditions:

$$n \cdot \left[\left(\mu_0 \nabla V_m - M_1 \right) - \left(\mu_0 \nabla V_m - M_2 \right) \right] = -n \cdot \left(B_1 - B_2 \right) = 0 \quad (3)$$

M is the unit of magnetization, B_1, B_2 are the magnetic flux density of the two kinds of medium, V_m is magnetic scalar. It also needs to be defined magnetic insulation boundary conditions:

$$n \cdot \left(\mu_0 \nabla V_m - \mu_0 M_0 \right) = n \cdot B = 0 \quad (4)$$

In the surrounding air region boundaries that tangential magnetic field and the boundary, the formula (4) defines the magnetic induction intensity of the normal component of zero boundary conditions, and defines zero magnetic potential at the same time:

$$V_m = 0 \quad (5)$$

Solved using Comsol, must be given a point (at least one point) of the magnetic potential. Equation

(5) makes the magnetic potential of boundary condition into the zero value. As a result, the unique solution can be calculated by Comsol.

Meshing should be proceeded after material property setting in Comsol. According to the division of different objects, user control mesh. After division, steady state is solved.

3.5 *Permanent magnet size effect on the simulation results*

Application of Comsol software to the pipe wall thickness is 8mm, the wall thickness of pipeline engraved 10% injury simulation, changing the size of the permanent magnet contrast, continue to reduce magnetic saturation. In the simulation of permanent magnets and steel brush width decreases were followed by 100mm, 80mm, 60mm, 40mm, 20mm. As shown in Figure 2, Permanent magnets for 100mm pipe three-dimensional simulation model diagram. Pipe length 360mm, diameter of 323.9mm; permanent magnet length of 100mm, thickness of 30mm; steel brush length of 100mm, a thickness of 56.45mm; yoke length of 360mm, a diameter of 67.5mm; Defect length of 10mm, thickness of 20mm. The outermost layer is the air layer, inside the pipe wall thickness of 8mm.

Comsol post-processing of the streamline method to calculate the defect near the distribution of the magnetic field lines, figure 3 for the defective pipe of the three-dimensional entity model of the magnetic induction curve floor plan. No defects on the pipe side, the normal distribution of the magnetic field lines within the duct; defective in that the side part of the magnetic field lines or defects through the bypass pipe; defect part through the lines of magnetic force, back to the pipeline; also part of the magnetic field lines leaked into the air, and then back to the bypass pipe defects.

Figure 3. Three-dimensional model of magnetic induction curve pipeline plan.

One-dimensional drawing group to view defects axial magnetic flux density is more intuitive than the two-dimensional radial flow chart.

Figure 4, 5 for the five groups of permanent magnets of the same size pipe defects radial and axial magnetic flux density map, in the two figures it can be more clear to see defects in the magnetic field of the permanent magnet with decreasing size decreases, for 10% of injuries in the permanent magnet width of 20mm, 1m lift-off value, the magnetic flux is small.

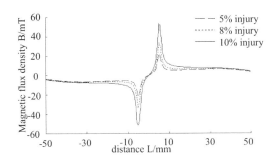

Figure 6. Defect depth impact on magnetic flux leakage signals.

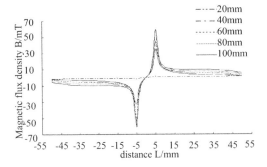

Figure 4. Five groups of radial magnetic flux density curve.

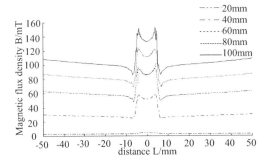

Figure 5. Five groups of axial magnetic flux density curve.

By two figure known as permanent magnet and steel brush size reduced to 40mm, radial magnetic flux density is 35.13mT, reduced to 20mm, the magnetic field strength is less than 20mT, is unsaturated. It can be concluded from the simulation for the pipe wall thickness is 8mm pipe, size of the permanent magnet is greater than 20mm.

3.6 *Defect depth on the simulation results*

The simulation results for the pipeline wall thickness 10% of injury, the permanent magnets and steel brush size 80mm magnetic field is saturated, then compare the depth of the impact of different defects in magnetic flux leakage signal contrast as shown.

By comparison, it shows that the permanent magnet and steel brush size is 80mm, for pipe wall thickness 5% and 8% of the injury, but it also can be checked.

4 CONCLUSION

According to the principle of magnetic flux leakage detection, the definition of magnetic saturation magnetic flux leakage detection, combined with the finite element method, using Comsol Multiphysics simulation software simulation pipeline magnetic flux leakage testing to verify, it can be drawn curve axial component and a radial component of the span distance between peaks and valleys of the axial length of the defect, according to the curve shape that can be judging defect types and geometry size change direction; For pipe wall thickness to 8mm, width is 80mm pipe magnetized permanent magnet and steel brush to near saturation state, for the wall thickness more than 5% of the defects are testing.

ACKNOWLEDGMENT

Foundation item:
   The Twelfth Five Year Plan of Ministry of national science and technology support funded project, China (No.2011BAK06B01–03); The national high technology research and development program funded 863 projects, China (No.2012AA040104); The Ministry of science and technology special major national instrument, China (No.2012YQ090175).

REFERENCES

[1] Liu, H.F, Zhang, P, Zhou, J.J & Yu, L. 2008. Oil and gas pipelines in the forest situation and development trend of corrosion detection technology. Pipeline Technology and Equipment (5): 46–49.
[2] Ke, M.Y, Liao, P. 2010. Real-time Data Mining in Magnetic Flux Leakage Detecting in Boiler Pipeline. IEEE computer society (10): 130–133.

[3] Yang, L.J, Zhang, S.L & Gao, S.W. 2013. Effects of plate thickness of magnetic flux leakage detection. NDT (10): 10–13 .

[4] Tang, Y, Pan, M.C & Luo, F.L, etc. 2011. Based on the three dimensional field measuring pulse magnetic flux leakage detection technology. Journal of instruments and meters 10 (32):2297–2302.

[5] Ahmad, K.R, James, P.R, Natalia, K.N, James, R.H & Sabir, P. 2009. Machine Learning Techniques for the Analysis of Magnetic Flux Leakage Images in Pipeline Inspection. IEEE transaction on magnetic 8(45): 3073–3084.

[6] Hu, Z, Ren, J.L. 2005. Analysis and magnetic memory elements detect ferromagnetic material. NDT 27 (7): 355–359 .

[7] Shan, S.Q, Chen, S.L & Jin, S.J. 2012. Data acquisition system for three-axis high definition pipeline inspection tool. Transducer and Micro-system Technologies (31): 118–121.

Testing and Measurement: Techniques and Applications – Chan (Ed.)
© 2015 Taylor & Francis Group, London, ISBN: 978-1-138-02812-8

Measurement of underwater shockwave pressure characteristics by explosive energy and electric pulse power energy

H. Sakamoto, Y. Watabe, S. Honda & Y. Ohbuchi
Graduate School of Science and Technology, Kumamoto University, Japan

ABSTRACT: The high speed fracture phenomena of glass container was clarified for low cost glass recycling method development. The underwater shock wave initiation and propagation were visualized by the high speed image converter and high speed color video camera and the pressure transitions was measured by using the tourmaline sensor method. The shock wave pressures produced by the explosive and the electric pulse energy were measured and the optimum crushing conditions were discussed.

1 INTRODUCTION

Many containers made of the glass are widely used. This paper discussed about the raw material recycling of glass container crushing technique by the explosive and the largest electric pulse energy. The small glass fragment crushed is called "Cullet". Here, the "Cullet" generation process is visualized by high speed camera methods. Furthermore, the pressure variation under underwater shock wave which is caused by these energies was measured.

2 MESUREMENT METHOD

2.1 Shock wave propagation and fracture behavior

The experiment was carried out by using the underwater shock wave experimental unit in the Institute of Pulsed Power Science. The outline of the bottle fracture experiment by using an underwater shock wave of explosive energy is shown in Figure 1.

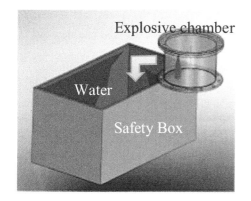

Figure 1. Experimental unit.

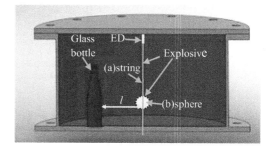

Figure 2. Explosive unit.

The experimental unit is composed of an explosive chamber and a safety box which had been filled with water. The explosive box is set in a specimen and an explosive (Figure 2).

The explosive used was PETN (explosive rte:6308m/s) and an electric detonator was used as an igniter. In order to visualize the underwater shock wave propagation and bottle fracture behaviors caused by explosive energy, the shock wave propagation was taken by high-speed camera (HTV-1) with Xenon flash (H1/20/50 type). The optical observation system is shown in Figure 3. The bottle fracture behavior was observed by a high speed video camera (Ph54) and high-power mercury lamp. The layout of measurement system and the specification of high speed video camera (Ph54) are shown in Figure 4 and Table 1, respectively.

2.2 Underwater shock wave pressure change

In the measurement of the explosive underwater shock wave pressure, the pressure sensor and the explosive were set in the pressure container which was filled with water shown in Figure 5. The explosive used was

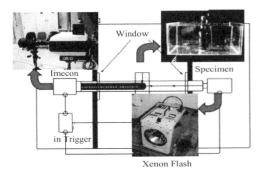

Figure 3. Optical observation system of shock wave propagation.

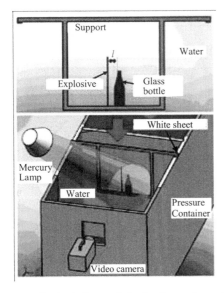

Figure 4. The bottle fracture behavior observation system.

Table 1. The specification of high speed video camera.

Model-type	Ph05
Product Co.	Nobby Tech.
Pixel	256×256
Pps	1/36036
Exposure time	25.75μs

the same as propagation observation's one. A pressure sensor and an explosive are set in the support frame. The support frame is stored in the pressure container.

In order to measure underwater shock wave pressure, one tourmaline sensor was used. When the tourmaline receives pressure, the voltage is generated by the piezoelectric effect. By measuring this voltage

with an oscilloscope, the underwater shock wave pressure can be obtained.

The frame of which a metallic wire and the pressure sensor set are stored in the small tank filled with the water shown in Fig.6. After that the electrode is connected and the high power current discharged to the metal wire. The underwater shock wave pressure and the actually discharged current are measured. The pressure sensor used the same sensor as the experiment of the explosive.

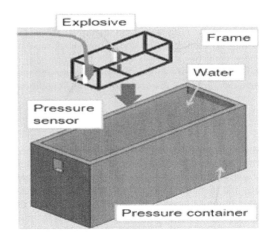

Figure 5. Experimental equipment (explosive).

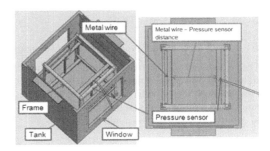

Figure 6. Experimental equipment (electric pulse).

2.3 *Experimental condition*

First, in the case of explosive, the shape and quantity of explosive were shown in Table 2 and the distance between the pressure sensor and the explosive was changed as shown in Table 3.

Next, in the case of high current pulse, the aluminum wires were used and each length is the same. The wire's length and the distance from the pressure sensor to the wire and the diameter of metal wire are shown in Table 3 and Table 4, respectively.

Table 2. Experimental conditions (explosive).

Shape	Explosive length	Explosive Quantity	Blasting depth
String type	200 [mm]	2.0 [g]	1000 [mm]

Table 3. Explosive conditions (explosive).

Distance between explosive and sensor l[mm]						
350	500	650	800	950	1100	1250

Table 4. Experimental condition (electric pulse).

Diameter of aluminum wire [mm]		
0.5	0.8	1.0

Distance metal wire – sensor [mm]		
250	350	450

3 RESULTS AND DISCUSSION

3.1 *The shock wave propagation behaviors*

The glass bottle was charged the underwater shock wave by explosive shown in Figure 2. Here, the behaviors of an underwater shock wave of explosive energy were observed by the high-speed photography method. The photograph was taken by using high speed camera (HTV-1) and two xenon flashlights (H1/20/50type) are shown in Figure 3.

Figure 7(a)-(d) and Figure 8(a)-(d) show the behaviors of shock wave propagation in the case of string type and sphere type explosive, respectively. The appearance of the shock-wave propagation can be clearly observed from these figures.

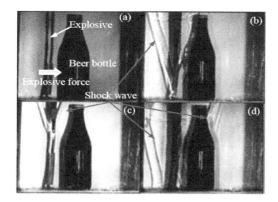

Figure 7. Shock wave propagation (string type explosive).

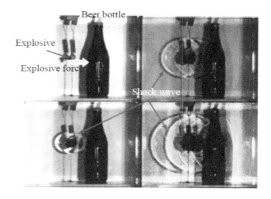

Figure 8. Shock wave propagation (sphere type explosive).

3.2 *The fracture behaviors*

Figure 9 shows the fracture process in the case of ball type explosive. It was confirmed that the small crack was generated on the surface of the bottle after the shock wave passed the bottle, and the crack spread to the whole surface of the bottle radially shown in Fig.9 (a) and (b). Next, the explosive gas reaches to the bottle. A negative pressure occurs when the explosive gas reaches the bottle, and the bottle is fractured shown in Fig.9 (c)

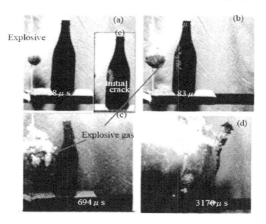

Figure 9. Fracture process of glass bottle by underwater shock wave (sphere type explosive).

3.3 *Results of pressure measurement*

The result of the underwater shock wave by an explosive is shown in Fig. 10. Figure 10 (a) and (b) shows the result at the 350–800mm distance and 950–1400mm distance between the explosive and the sensor, respectively. The result of the underwater shock wave by the electric pulse is shown in Fig. 11. Figure 11 (a) and (b) shows the result of the 0.5 mm and 0.8 mm diameter in metal wire, respectively. In the case of the 1.0mm diameter in metal wire, since an intense noise occurred when the electric current discharged, it was unable to be measured. By comparing the pressure by explosive and electric pulse at the 350mm distance, the pressure 15~18 MPa was obtained.

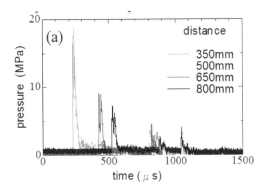

Figure 10(a). Change of shock wave pressure explosive-sensor distance : 350–800 (explosive).

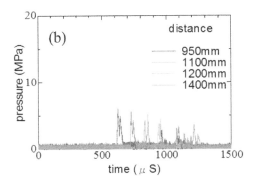

Figure 10(b). Change of shock wave pressure explosive-sensor distance : 950–1400 (explosive).

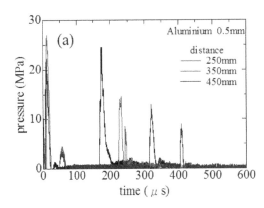

Figure 11(a). Change of shock wave pressure (electric pulse).

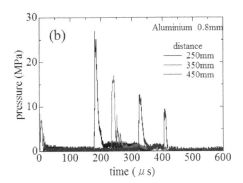

Figure 11(b). Change of shock wave pressure (electric pulse).

The detonation velocity and the effective shock wave impulse were calculated from these results. Each result is shown in Fig.12 and Fig.13, respectively. The effective shock wave impulse was calculated from time to receive the pressure of 3.5MPa or more. The pressure 3.5 MPa is the allowable crushing stress of glass bottle. Each result of the detonation velocity is about 1500 m/s. However the effective shock wave impulse turned out that the value of the electric pulse is approximately a half compared with the explosive.

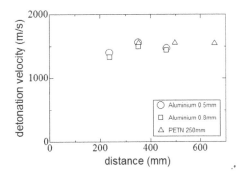

Figure 12. Detonation velocity.

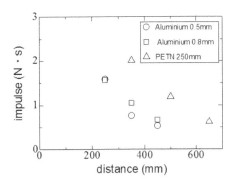

Figure 13. Effective shock wave impulse.

280

4 CONCLUSIONS

The shock wave propagation, glass fracture behaviors by shockwave and shockwave pressure change produced by the explosive and the electric pulse energy were observed and measured. By this behavior of the shockwave propagation and pressure change, the optimum crushing conditions ("Cullet" generation system) for glass container recycling was discussed. The results obtained are as follows.

1 The appearance of the shock wave propagation wave and glass containers fracture behaviors were clarified by taking the high-speed photograph and high-speed video system.
2 The small crack was initiated on the surface of the bottle after the shock wave passed the bottle, and the crack spread to the whole surface of the bottle radially.
3 The difference of the shock wave impulse characteristic between the explosive and the electric pulse is considered as one of the major factors which cannot make small cullet by using the electric pulse.

REFERENCES

[1] Kobayashi A.S.,1973. Experimental Techniques in Fracture Mechanics.
[2] Kawabe S. & Sakamoto H.,2008. The study about "Cullet" generation technique and application for recycling system by using underwater shockwave, Materials Science Forum 566, 231–236.
[3] Sakamoto H. & Kawabe S.,2010. High-speed fracture phenomena of glass bottle by underwater shock wave, Materials Science Forum 654–656, 2543–2547.
[4] Sakka S.,1988. The Dictionary of Glass, Asaka Press.

Application and protection on low-voltage power line carrier communication technology

J. Fan, X. Chen, X.W. Chu, Y. Zhou & Y.X. Yi
Jiangsu Electric Power Research Institute & Sate Grid Key Laboratory of Electric Energy Measurement, Nanjing, China

ABSTRACT: Easy-installation and cheap-price of Power Line result in the wide application of Low-Voltage Power Line Carrier Communication Technology, but the interference of noise is great, for example harmonic signal, so the efficiency of power line carrier communication is affected and the device of power line carrier communication is damaged.

1 INTRODUCTION

Power line communication release information transferred by using the power line available now. The power line is installed conveniently and preserved simply. Besides, the price is cheaper than others, so the power line communication is widely applied to low-voltage reading device. But as a special communicating medium, the impedance of low-voltage power line, attenuation and interference of noise change in large extent in the process of transmission, which can result in the difficult communication and damage the device of the power line communication with carrier even.

This passage analyses the problem which is found in the course of using power line communication and puts forward solutions from the aspect of principle, so the device of power line communication is protected and the power line communication runs reliably and stably.

2 POWER LINE COMMUNICATION WITH CARRIER

The technology of power line communication with carrier has two kinds, one is the narrow-band communication with the carrier, the other one is a broadband communication with the carrier, low-voltage power line communication adopts the narrow-band communication with the carrier. The frequency of carrying less than 500 kHz is defined as Low-voltage narrow-band communication with the carrier, its' rate ranges from several hundred pips to several kbps. The technology of modulation with carrier includes some kinds, such as Frequency Shift Keying and Phase Shift Keying. Besides, the technology of handling data information with high speed can be used to improve the efficiency of communication, such as the modulation with several carriers. Orthogonal Frequency Division Multiplexing is one of the modulations with several carriers.

3 THE ANALYSIS OF BREAKDOWN IN THE COURSE OF APPLYING

Nowadays, the low-voltage power line communication is applied, but the device in the circuit is damaged during the applying. We found some similarities among the problems: the users have old or Electromagnetic Interference charger of electric-bike; its' voltage supplied is higher than 260 V; it has a long and independent line; it does not have another Electromagnetic Releasing device in parallel with the line; the temperature of the surroundings is high. When the same phenomenon appears, the collecting device absorb and use up the harmonic wave interfered, so the collecting device becomes hot and cause damage to the device of the carrier's circuit. The collecting device can't work or communicate normally at the end.

The quality of the electric-bike's charger is bad and it does not have restrained circuit, which as shown in fig 1. It can produce a high harmonic wave when charging, which is shown as the yellow curve in Fig 2. The harmonic wave produced can cause damage of the collecting device. Besides, the Electro Magnetic Compatibility interference produced can make the TCC082C, a kind of chip, abnormal. Then the serial port for communication is occupied all the time and the indicator lamp for communication is lighting. At the same time, the collecting device can't read the information of meter.

Figure 1. The circuit board of the charge lacking of restrained circuit.

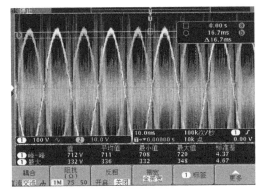

Figure 2. The harmonic wave curve produced when the bad charger is charging.

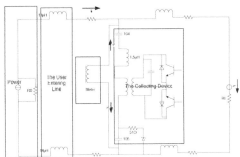

Figure 3. The circuit in the laboratory with the surroundings of scene.

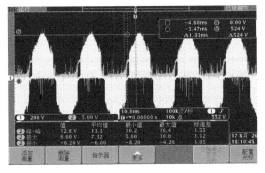

Figure 4. The harmonic wave curve produced in the laboratory.

A circuit with the surroundings of the scene is imitated in the laboratory. Fig 3 shows the circuit in the laboratory. T1 is on behalf of a normal voltage regulator, T2 is on behalf of a special voltage transformer; B1, B2 is on behalf of the bad charger which is running at full capacity; C is on behalf of the collecting device which is placed in a small space. The harmonic wave of collecting device is displayed on the screen of an oscilloscope, fig 4 shows the highest harmonic wave curve after adjusting the voltage. The experiment proves that the harmonic wave can directly cause the damage of collecting device. Besides, the Electro Magnetic Compatibility interference produced can make the TCC082C, a kind of chip, abnormal. Figure 5 shows the structure diagram of carrier module hardware.

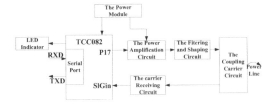

Figure 5. The structure diagram of carrier module hardware.

4 THE PROTECTING MEASURES

After analyzing the reason of a breakdown, three solutions are raised. The first one is protected by combining PTC with TVS, the second one is protected by a filter with a carrier, the last is protected by

semiconductor ignitron. The specific introduction and comparison is as below:

Protecting means by combining PTC with TVS: the schematic diagram is shown as fig 6, a PTC is added in front of the P6KE22CA and TVS is placed near the PTC, when the range of voltage becomes very high, the TVS will be broken down and become hot quickly, therefore the PTC will become hot and be turned into the state of protection. The circuit is reliably protected by the combination of two devices. After the interference of harmonic wave turn small, the PTC will become normal and the communication with carrier turn regular also.

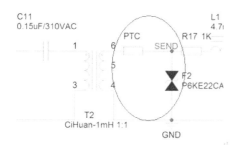

Figure 6. The improved schematic diagram of PLC by using PTC and TVS.

Protecting means by a filter with carrier: fig 7 shows the schematic diagram, the circuit which contains L1 & C1 is linked to the null line, C2 is placed between null line and live line at the side of the user. Besides, there is another circuit containing L2 & C3. The frequency of these two circuits should be set at the frequency of the carrier. If the frequency is high, the C2 expresses as low-impedance and has the effect of filtering. This means showing the effect of protection is good, but thick line is needed to connect L1, C1 and the power line.

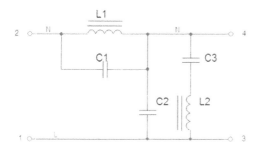

Figure 7. The improved schematic diagram of PLC by filtering circuit.

Protecting means by semiconductor ignitron: semiconductor ignitron is a module of PNPN and is also called as solid ignitron. It can be seen as normal silicon, which is controlled by voltage. If the voltage is higher than avalanching voltage, the semiconductor will discharge to the module at a moment. If keeping on increasing, the semiconductor turn connecting due to the negative impedance. After the current turn less than maintaining current, the module is resetting and turns to the state of high impedance.

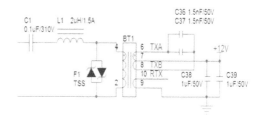

Figure 8. The improved schematic diagram of PLC by using TSS.

5 CONCLUSIONS

The quality of the electric-bike's charger is bad and it does not have restrained circuit. It can produce a high harmonic wave when charged. The harmonic wave produced can cause the damage of collecting device directly. Besides, the Electro Magnetic Compatibility interference produced can make the TCC082C, a kind of chip, abnormal.

REFERENCES

[1] FAN Jian-xue, SHENG Xin-fu. Research on Low-Voltage Power Line Carrier Communication Technology [J]. Electrical Measurement & Instrumentation, 2005,42(2):36–38.
[2] LV Zhong-yu, MENG Li, LI Lu. The solution to the interference of Low-Voltage Power Line Carrier Communication [J]. Electrical Measurement & Instrumentation,2008,(3):36–40.
[3] QIN Wen-hua, TIAN Hai-feng, LIU Xin. Reading the Data by Using the Technology of Low-Voltage Power Line Carrier Communication [J]. Network & Communication, 2005,(6): 58–61.

Testing and Measurement: Techniques and Applications – Chan (Ed.)
© 2015 Taylor & Francis Group, London, ISBN: 978-1-138-02812-8

Optimizing configuration of supply chain with survival assessment model

Felix T. S. Chan

Department of Industrial and Systems Engineering, The Hong Kong Polytechnic University, Hung Hum, Hong Kong

B. Niu

Department of Industrial and Systems Engineering, The Hong Kong Polytechnic University, Hung Hum, Hong Kong
College of Management, Shenzhen University, Shenzhen, China

A. Nayak, R. Raj & M. K. Tiwari

Department of Industrial Engineering and Management, Indian Institute of Technology, Kharagpur, West Bengal, India

ABSTRACT: This paper adopts Biased Random Key Genetic Algorithm (BRKGA) to optimize the configuration of a robust supply chain for decision making using survival analysis. To perform proportional hazard class of survival analysis, Cox-PH model is developed to compare the significance of structural and logistics flexibility. Illustrative simulation models are provided to demonstrate different aspects of the proposed methodology.

KEYWORDS: Optimization, supply chain configuration, Cox-PH model, survival analysis, BRKGA.

1 INTRODUCTION

Flexibility refers to the ability to configure (and re-configure) assets and operations to react dynamically to emerging customer trends at each node of the supply chain. "The preserving of flexibility when faced with uncertainty" is not a lost cause, as according to Jones and Ostroy [1] who established the significance of flexibility when responding to uncertain situations as the work presents flexibility as a neglected approach to deal with uncertain situation, but it gives different positions to take when it comes to decision making. There are different dimensions of flexibility in supply chain like developing new distribution channels, finding new source points, opportunity specific outsourcing, contracts with different firms, machine flexibility when it comes to manufacturing level flexibility, and volume flexibility through the resources [2]. In this work, robustness is defined as the maximum cost a firm experiences due to demand uncertainty. Robustness are highly correlated to flexibility [3]. This paper combines the flexibility to develop effective mitigation strategy (robustness). To respond to changes in the market, different dimensions of flexibility are used to make the supply chain robust by process flexibility (structural and logistics) and strategic inventory through investment flexibility. Biased Random Key Genetic Algorithm (BRKGA) heuristic [4,5] encodes a solution of the combinatorial optimization problem as a vector of random maintaining the elitism of solutions. In this paper, chromosomes are a vector of real

values such that the constraints are satisfied in any chromosome. Thus, during the encoding of the digits, it is assured that feasible solutions are produced. This paper is focused on the genetic algorithm family of evolutionary algorithm and attempts to implement a new variant BRKGA. When the problem is complex, the number of variables is very high, even in a small sample. In such cases, traditional optimization tools can not work and evolutionary algorithms are applied and it is observed that BRKGA provides efficient and better solutions than GA. BRKGA maintains elitism and thus converges faster than GA which is required when problem sizes increase. Cox-PH is a survival analysis technique which is a statistical method to find out how a particular variable affects the system and if it is statistically significant or not affecting the performance of the system [6-8]. Cox-PH model does not assume any distribution. Cox-PH model enables us to control many variables simultaneously, which may be confusing if ANOVA or MANOVA are used. The impact of a variable on how it affects the hazard rate is determined by the coefficient for that variable in the hazard function. Both the sign and the magnitude of the coefficient are important in understanding the effect of the variable in accelerating the system towards the event. The model is based on hazard function and the coefficients of the variables are obtained using the maximum likelihood function. For example, if the failure of a firm to meet the demand is the event for study, then different explanatory variables like investment, raw materials, forecasting, production capability and logistic and strategic decision like resource allocation affect the event. These variables are interdependent on each other, that is, investment depends on forecasting, raw materials and production capability.

2 BUILDING BRKGA MODEL AND COX-PH MODEL

2.1 Building the BRKGA model

The optimal values of the different variables are obtained using the evolutionary algorithm- Biased Random Key Genetic Algorithm. BRKGA maintains elitism. BRKGA is a variant of GA and it outperforms GA. In this paper, an attempt is made to show the implementation and effectiveness of BRKGA which could be employed in the supply chain. To show the implementation of BRKGA and its difference from GA, a comparative study is done.

2.1.1 Representation scheme

In this paper, a real number chromosome structure is used to represent a solution. The length of the chromosome is 21 digits. The strings are made such that the digits follow the resource constraints. Each digit represents a variable. The chromosome structure is shown in Fig. 1.

Figure 1. Chromosome structure.

2.1.2 Fitness function

The fitness of a chromosome is obtained by evaluating the chromosome with Eq. 1 The first term calculates revenue while it is succeeded by investment cost, inventory holding cost, cost in stage with structural flexibility, cost in stage with logistics flexibility, backordering cost, overstock cost and understock cost.

$$TR = \sum_{i=1}^{n} O_{it} * sp_i - [\sum_{i=1}^{n}(x_{ip} * p_i) + \sum_{j=1}^{d} y_{fp} * p_j' + \sum_{it=1}^{n}\sum_{t=1}^{T} a_{it} * h_{it} + \sum_{it=1}^{n}\sum_{k=1,k\neq i}^{n} y_{ijt}^{k} SC *_i$$
$$+ \sum_{i=1}^{n}\sum_{k=1,k\neq i}^{n} y_{ikt}^{k} lc *_i + \sum_{it=1}^{n}\sum_{t=1}^{T} b_{it} * S_{it} + \sum_{it=1}^{n}\sum_{t=1}^{T} OS_{it} * osc_{it} + \sum_{it=1}^{n}\sum_{t=1}^{T} US_{it} * usc_{it}] \qquad (1)$$

2.1.3 GA operators
2.1.3.1 Mutation
Mutation is used in genetic algorithms to avoid entrapment in local minima. In BRKGA, mutation is implemented by introducing mutants in the population. In BRKGA, in mutation, completely random sample is obtained. The random sample is obtained

using the same method as used in initialization. A number of mutant individuals are randomly generated and placed in the new population in the partition.

2.1.3.2 Cross-over

Crossover in BRKGA uses elitist strategy. In BRKGA, crossover is done between two chromosomes, one from the elite group and the other from the non-elite group of the population. In this crossover, two parents give rise to a single offspring while in GA two offspring are generated by crossover operation. In crossover, the real value crossover function is used in this paper. For crossovers, real value crossover is done with respect to a digit randomly selected in a chromosome in the elite group.

2.2 Building the Cox-PH model

Cox-PH model is used in the work presented in the paper to find how the event of interest is explained by these factors interacting amongst themselves. The Cox PH model is a semi parametric model [6] which makes no assumptions about the form of $h(t)$ (nonparametric part of model) and assumes parametric form for the effect of the predictors on the hazard. The hazard function, also called the conditional failure rate function, is defined in Eq. 2 as

$$h(t) = \lim_{\Delta t \to 0} \Pr(t \leq T \leq t + \Delta t \mid T \geq t) = \frac{f(t)}{S(t)} \qquad (2)$$

where

$h(t)$ = hazard function
$f(t)$ = probability density function of T
$S(t)$ = Pr (T>t), the survival function of T
Notations
$\mu' = [\mu_1, \mu_2, \mu_3, ...\mu_n]$: set of coefficients of the different variables (covariates) in the model
$z' = [z_1(t), z_2(t), z_3(t), ...z_n(t)]$: set of decision variables for the model such that
$z_i(t)$: Value of variable i in time t
t': Time at which the prediction is made
$S(t \mid z(t'))$: The survival probability of the system at time t given that the system has survived till time t'.
$S_o(t)$: Base survival function
$h_o(t)$: Base hazard function
Cox-PH model is used to describe the relationship between system hazard function and covariates as shown in Eq. 3.

$$h(t \mid z(t)) = h_0(t)^{\mu' z(t)} = h_0(t)^{\sum_{i=1}^{n} \mu_a z_a(t)} \qquad (3)$$

Hazard rate is shown by Eq. 4.

$$h(t) = \frac{-S'(t)}{S(t)} \qquad (4)$$

By solving Eq. 4 and simplifying it, Eq. 5 is obtained.

$$S(t \mid z(t)) = \exp(\int_0^t h_o(t) \exp(\sum_{i=1}^n \mu_a z_a(t)))) \qquad (5)$$

The conditional probability of the survival of the system $\Pr(T > t \mid t' z(t'))$ that is the system survives for a time T given that it has survived till time t' is given by Eq. 6 as

$$\frac{S(t \mid z(t))}{S(t' \mid z(t'))} = S(t \mid z(t')) \qquad (6)$$

Using Eq. 6, Eq. 5 can be simplified to Eq. 7 as

$$S(t \mid z(t')) = \left(\frac{S_o(t)}{S_o(t')}\right)^{\exp[\mu' z(t)]} \qquad (7)$$

3 ANALYSIS OF THE SIMULATION MODEL

As the time increases, the number of events increases, and thus the survival function dips. When there is a failure event, the survival function showing the Kaplan Meier estimation dips. After the factors are fitted as covariates, the coefficients of these factors are shown in Table 1. It is observed that since there are some statistically insignificant factors, a model refitting has been done. The variables with higher p- values are considered as statistically insignificant. Thus, a refit is done based on statistically significant factors.

The simulation model has been developed and run in matlab. The Cox-PH simulation model has been run in the windows 7 platform using R Software package. The R-Square value obtained for the Cox-PH model fitting obtained is 0.474 which can be considered as a good fit. The standard error for the fitting is 0.032. Time taken for the above simulation to run was observed to be 0.32 and 0.26 seconds respectively for two Cox-PH fitting. The *p-value* for logrank test, wald test and likelihood ratio test is approximately 0. Since, the fact that the lower the *p-value*, the more statistically significant a model is, thus, it can be considered that the model proposed is a good model which is statistically significant. Logrank test is a hypothesis test to address the variable and their effect in survival model. Wald test is used to find the relationship between the data items. The Log likelihood test is used to compare two models, null model and alternate model and estimates whether the data fits which model better. For addressing a real life situation with many more factors and interactions, first a pre-screening may be used to avoid over-fitting and under-fitting of the Cox-PH model to bring the number of factors to a level so that it best represents the supply chain under study.

4 RESULT INTERPRETATION

The Cox-PH model finds the coefficients corresponding to different variables. These exponential coefficients are interpretable as the multiplicative effects of the variable. Thus, for example, holding other covariates (factors in the model) constant, a change of one unit of variable 3 from the optimal values as obtained

Table 1. Coefficients of the factors after Cox-PH model fitting.

factors	coef	Exp (coef)	se(coef)	z	Pr(>|z|)	Statistically Significant at:
y_{310}	0.159938	1.173438	0.046841	3.414	0.000639	0
y_{120}	-0.17548	0.839058	0.035374	-4.961	7.03E-07	0
y_{220}	-0.31057	0.733031	0.036851	-8.428	0	0
y_{320}	0.044825	1.045845	0.013322	3.365	0.000766	0
y_{11l}^2	-0.27162	0.762141	0.142544	-1.906	0.056709	0.05
y_{11l}^3	-0.29089	0.747597	0.135338	-2.149	0.031606	0.01
y_{21l}^1	0.28507	1.329796	0.14058	2.028	0.042577	0.01
y_{21l}^3	0.32864	1.389902	0.142415	2.308	0.02102	0.01
y_{31l}^1	-0.31295	0.731287	0.136943	-2.285	0.022298	0.01
y_{31l}^2	0.24614	1.278814	0.134112	1.835	0.066457	0.05
TI	0.283526	1.327804	0.134764	2.104	0.035389	0.01
FS	0.01312	1.013206	0.003626	3.618	0.000296	0

from BRKGA increases the chances of occurrence of an event by the factor $e^{0.1845} = 1.202$ where an event means revenue of the firm in one production cycle is $< 15 \times 10^4$ units. Structural flexibility and logistics flexibility provide opportunities for meeting the changes in the demand. Significance of structural and logistics flexibility must be compared. For the comparison two terms shall be defined- structural cushion and logistics cushion. Structural cushion is defined as the ability of the supply chain to correspond to changes in product demand during the stage with structural flexibility. Logistics cushion is defined as the ability of the supply chain to correspond to changes in product demand during the stage with logistics flexibility. Comparison between the cushion provided by the two flexibilities can be done by the paired t-test. The difference between the cushion provided by the flexibilities is assumed to be normal. The results show that the difference follows the normal distribution with Mean Square value of 0.0422. It is observed that the effect of the structural flexibility is statistically significantly higher than that of the logistic flexibility since the test statistics value is 12.584 as compared to the statistic value of 1.648 at 95% confidence index. Thus an inference can be drawn that structural flexibility is more efficient in meeting to the demand variation than the logistics flexibility under uncertain demand conditions. It is observed that the structural flexibility is more significant in realizing the flexibility of the supply chain. It is also observed from the coefficients of the variables in the Cox-PH model that structural flexibility affects the performance of the supply chain more significantly than the logistics flexibility. Statistically the variables pertaining to the structural flexible are more significant. This result of comparing the two flexibilities is also a significant contribution of the paper.

5 CONCLUSION

In this paper, the focus has been on developing a robust supply chain capable of performing well under uncertain demand conditions. Investment decision and adopting flexibilities in their supply chain have been widely used by the managers to conduct decision making to reduce unexpected profit losses and thus to reduce the incorrect decision making. A supply chain may have inherent errors due to inaccuracy in the forecasting ability of the demand due to different factors responsible for the demand variation. In this paper, a systematic model and approach is developed to find the optimal supply chain configuration and resource allocation at stages with structural and logistics flexibility. This paper brings into light few of the novel approaches to solve uncertain demand conditions.

ACKNOWLEDGEMENT

The work described in this paper was supported by grants from the Research Grants Council of the Hong Kong Special Administrative Region, China (Project No. PolyU 510311); the NSFC (Project No. 71471158); and The Hong Kong Scholars Program Mainland–Hong Kong Joint Postdoctoral Fellows Program (Project No.: G-YZ24). The authors also would like to thank the Hong Kong Polytechnic University Research Committee for financial and technical support.

REFERENCES

[1] R.A. Jones, J.M. Ostroy, Flexibility and uncertainty. Review of Economic Studies 51 (1984) 13–32.
[2] A.M. Sánchez, M.P. Pérez, Supply chain flexibility and firm performance: A conceptual model and empirical study in the automotive industry, International Journal of Operations & Production Management 25:7 (2005) 681–700.
[3] J.A. Van Mieghem, Investment strategies for flexible resources. Management Science 44:8 (1998) 1071–1078.
[4] J.F. Goncalves, M.G.C. Resende, Biased random-key genetic algorithms for combinatorial optimization, Journal of Heuristics 17:15 (2011) 487–525.
[5] J.F. Goncalves, J.R. Almeida, A hybrid genetic algorithm for assembly line balancing, Journal of Heuristics 8:6 (2002) 629–642.
[6] D.R. Cox, Regression Models and Life-Tables, Journal of the Royal Statistical Society 34:2 (1972) 187–220.
[7] S.C. Zhou, C. Sievenpiper, Failure event prediction using the Cox proportional hazard model driven by frequent failure signatures, IIE Transactions 39:3 (2007) 303–331.
[8] Y. Yuan, S. Zhou, C. Sievenpiper, K. Mannar, Event log modeling and analysis for system failure prediction, IIE Transactions, 43 (2010) 647–666.

Design of a coal quality detection management system for power plants based on UML

Z. Liu

Hebei Engineering Research Center of Simulation & Optimized Control for Power Generation (North China Electric Power University), Baoding, China

ABSTRACT: The coal quality detection is an integral part of the fuel management, including coal sampling, coal sample preparation, and coal detection three parts. Based on the idea of object-oriented analysis and design, the requirements analysis model for a Coal Quality Detection Management System (CQDMS) is established by using the Unified Modeling Language. Three-level electronic coding based on bar code technology is designed and the statistic model and the dynamic model of the system are visually modeled using UML. Finally, the CQDMS is developed by the VC.NET technology.

1 INTRODUCTION

The coal quality detection is an integral part of the fuel management, and the coal quality and the economic benefit of power plants depend on coal detection results. On the other hand, the coal quality results can provide a significant guarantee for safe operation and environmental protection in power plants [1]. However, some shortcomings exist in most coal quality detection management processes, such as non-fully networked, non-standard operation, heavy workload, low data reliability, and so on. Thus, it is necessary to design a Coal Quality Detection Management System (CQDMS) combined with detection and management to realize the standardization, institutionalization, modularization.

The key to develop CQDMS is the process of analysis, design, and modeling, and it is important to improve the quality, reliability, and reusability of the software and to shorten the software development cycle. Moreover, the Unified Modeling Language (UML), a set of object-oriented modeling notations [2], is very rich in expression ability, and can provide various types of diagrams, such as the use case diagram, class diagram, interaction diagram, and activity diagram. Thus, UML is widely used in modeling and development of softwares.

The aim of the study is to design a CQDMS for power plants by modeling with UML. Based on the requirement analysis, a structure of use cases is framed, followed by the establishment of static and dynamic models. The static model describes the implement of use cases and the relationships between different classes, whereas the dynamic model describes the interactions between different objects.

2 REQUIREMENT ANALYSIS

After a study on coal quality detection laboratories in power plants, it is determined that the main application of CQDMS is to manage the coal quality detection process, and the management involves a series of activities. In the coal sampling process, CQDMS generates a sampling note according to the coal batch message. Then, an outside code and an inside code are generated for each sample bag based on the sampling note. When these sample bags are sent to the sample preparation room, CQDMS checks the codes and classifies these bags for sample preparation. When the sample preparation is completed, operators package sample bags and stick the sample preparation bar codes on bags. After the sample bags are sent to the detection lab, first, CQDMS identifies the bar codes on the received sample bags and allocates a laboratory's code for each sample, which is the unique identification code for a sample in the following detection and data processing. In the laboratory, all detection instruments are connected with CQDMS via a network, and all the measurement data from each detection instrument are collected and updated in the database automatically in real-time. When a coal quality parameter of a detected sample needs to be calculated, CQDMS will perform a series of statistics on the selected detection data according to the national standards and the results of the parameter are stored in a specified database and displayed in a list. Once all the detection results of a sample are calculated, an original report is generated. After the report is audited, CQDMS automatically calculates the parameters as received basis and dry basis of the sample. Subsequently, all coal quality parameters of the sample are audited, followed by a coal quality report that is generated automatically based on the qualified results.

Based on the above analysis, the main operators of CQDMS are sampling operators, sample preparation operators, laboratorians, auditors, and an administrator. The use case diagram is shown in Figure 1 and it explains the behavior and related participants of CQDMS [2].

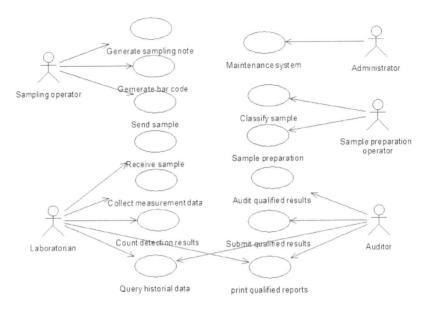

Figure 1. The use case diagram for CQDMS.

3 SYSTEM DESIGN

3.1 *Three-level electronic coding based on bar code technology*

Based on the data of sampling, sample preparation, and detection, the CQDMS establishes a three-level coding to manage the three processes. In the sampling, after the sampling note is scanned, the system automatically generates random sample codes related to the coal batch number and establishes the data clues between coal batches and coal sampling information. In the sample preparation, the system automatically identifies sample codes and classifies sample bags for the same batch according to the batch number. When sample preparation is completed, the system automatically generates random sample preparation codes according to the sample preparation position and sets up the data clues between the coal sampling and the sample preparation information. In the detection, the system identifies sample preparation codes, automatically generates detection codes, and establishes the data clues between the sample preparation and the coal quality information. PDF417 code is chosen as the code because it has excellent characteristics, such as large information capacity and coding range, strong fault tolerance, high reliability, low cost and so on. The flow chart of coding is shown in Figure 2.

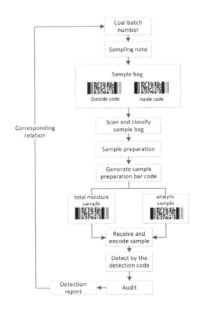

Figure 2. The flow chart of coding.

3.2 System design and modeling based on UML

3.2.1 Static structure modeling

The purpose of static modeling with UML is to abstract the optimum classes from related use cases or scenarios, to analyze the relationship between these classes, and to determine the attributes and operations of classes [3]. Class diagrams play a central role in describing software structure, and provide a static description of system components [4]. In the UML specification, the system design classes are divided into boundary classes, entity classes, and control classes according to their responsibilities [5,6]. Boundary classes are in charge of the communication between the system and the external environment, and they are usually used for system interfaces which are able to provide interfaces between the system and user (or the system and other systems) [7]. Control classes are in charge of the coordination of other classes, and they receive little information from other classes, but they send messages to other classes in the form of commissioned responsibility. Entity classes are mainly descriptions of things that actually exist, including necessary information to be stored and related behavior. The main entity class diagram for CQDMS is shown in Figure 3.

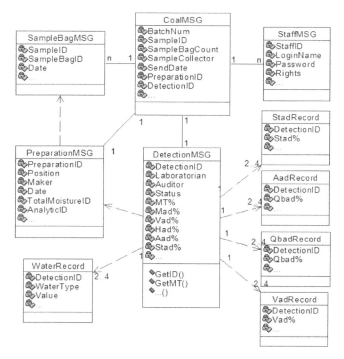

Figure 3.   The entity class diagram for CQDMS.

3.2.2 Dynamic behavior modeling

The statistic model of CQDMS just describes the structure to implement use cases, whereas a systematic dynamic model also needs to be established to describe dynamic characteristics and the behavior of use cases in the object-oriented analysis method. The dynamic model mainly expresses interactions between different objects in CQDMS, which is established by using UML based on interaction diagram, activity diagram, and statechart diagram [8]. The interaction diagrams describe the interrelationships among objects and the message-passing between objects, which are used to model the interaction process of multiple objects in the system, including the sequence diagram and the collaboration diagram. The sequence diagram focuses on the exact sequence of a message [9], whereas the collaboration diagram focuses on describing the relationship between objects.

Figure 4 shows the sequence diagram for 'Calculate detection results'. The sequence diagram involves five objects interacted with each other - Laboratorian, LabHome, StatisticsWin, StadRecord, and DetectionMSG. The 'Calculate detection results' includes 12 processes. First, a laboratorian enters the LabHome, and then he can enter the StatisticsWin by clicking the 'Count' button. In the StatisticsWin, he can check all of the results of one coal quality parameter by a detection code or by a detection date. When the records need to be counted, the system will automatically calculate the detection results, store the qualified results in a database and display them in the list on the Statistics. If no faults are found in the results, he can submit the detection results to the DetectionMSG and exit the StatisticsWin by clicking the 'OK' button. Finally, the detection results are shown in the list of the LabHome.

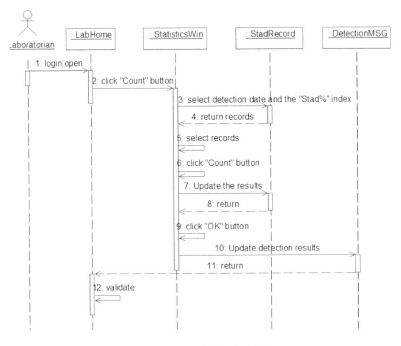

Figure 4.   The sequence diagram of 'calculate detection results' for CQDMS.

The state diagram is a supplementary description of the class diagram, which shows all of the possible states of a kind of object, as well as the state transition when some events occur. In addition, the state diagram shows the state sequence that an object experiences, the events that causes the state transition and the action that accompanies with the state transition. Figure 5 is a state diagram of coal quality detection results for a sample. It includes five states - new, to audit, qualified, unqualified, and submitted, and four events - detection, re-detection, audition, and submission.

The activity diagram is a model view used to describe the behavior of the CQDMS, which can be used to describe the action and the results without considering the events that causes the state change. The activity diagram for a receive sample process is shown in Figure 6.

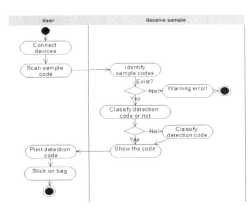

Figure 6.   The activity diagram for a receive sample process.

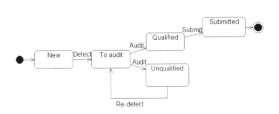

Figure 5.   The state diagram of coal quality detection results for a sample.

294

| Lab Code | 20131217001 | | Date | 2013/12/18 | | Auditor | | ∨ | User Name | lz |

Data Display

Total moisture Mt(%)	10.85	Air-dry volatile matter Vad(%)	29.88	Audit
Air-dry moisture Mad(%)	4.52	Dry,ash-free basis volatile matter Vdaf(%)	38.08	
Air-dry ash Aad(%)	17.02	Air-dry bomb calorific value Qb,ad(J/g)	25846.97	Count
Air-dry Hydrogen Had(%)	4.47	Air-dry gross calorific value Qgr,ad(J/g)	25769.81	Submit
Air-dry total sulfur Stad(%)	0.82	As-received basis net calorific value Qnet,ar(MJ/Kg)	22.95	Print Report
Dry basis total sulfur Std(%)	0.86			History

Data Sheet

Lab Code	Date	Auditor	Mt(%)	Mad(%)	Aad(%)	St , ad(%)	Vad(%)	Qb , ad(J/g)	Had(%)
20131217001	2013-12-18		10.85	4.52	17.02	0.82	29.88	25846.97	4.47
20131217002	2013-12-18		0.00	0.00	0.00	0.00	0.00	0.00	0.00
20131217003	2013-12-18		0.00	0.00	0.00	0.00	0.00	0.00	0.00
20131218001	2013-12-19		0.00	0.00	0.00	0.00	0.00	0.00	0.00

Figure 7. The detection interface of CQDMS.

4 IMPLEMENTATION

According to the requirements of the users, combined with the analysis of the system model structure, a CQDMS is developed based on Client/Server (C/S) mode. A coal quality detection system is implemented by using the C++ language and SQL 2008 database technology in the framework MFC program of VC.NET. The CQDMS has been applied successfully in the feed coal management in a coal-fired plant in Tianjin, and it runs steadily and reliably. Figure 7 shows the detection interface of CQDMS.

5 CONCLUSION

Based on the functional requirements of CQDMS, requirements analysis and design model of the system are performed, and the use case diagram, class diagrams, interaction diagrams, and the deployment diagram are established by using UML. The efficiency of data processing is improved, and the man-made errors and frauds in data acquisition are avoided. Thus, the cost of fuel management and resource consumption decreases.

REFERENCES

[1] Liu, Z., Dong, Z., Han, P. 2014. Developing a coal quality detection management system for power plants. *Journal of Software* 9(6): 1634–1644.

[2] Gong, D.C., Wang, Y.T. 2011. UML presentation of a conceptual green design control system to react to environmental requirements. The International Journal of Advanced Manufacturing Technology 52(5-8): 463–476.

[3] Gomaa, H. 2001. Designing concurrent, distributed, and real-time applications with UML. Proceedings of the 23rd International Conference on Software Engineering, IEEE Computer Society: 737–738.

[4] Karasneh, B., Chaudron, M.R.V. 2013. Extracting UML models from images. Computer Science and Information Technology (CSIT), 2013 5th International Conference on, IEEE: 169–178.

[5] Feng, H.H. 2012. Object-Oriented Requirements Analysis and Modeling with UML. Tsinghua University Press: Beijing.

[6] Liu, X.S. 2014. Primer of Modeling with UML. China Machine Press: Beijing.

[7] Gonzalo, G., Llorens, J., Fraga, A. 2014. Metamodeling generalization and other directed relationships in UML. Information and Software Technology 56(7): 718–726.

[8] Henric, A., Herzog, E., Johansson, G., Johansson, O. 2010. Experience from introducing unified modeling language/systems modeling language at Saab Aerosystems. Systems Engineering 13(4): 369–380.

[9] Misbhauddin, M., Alshayeb, M. 2010. Extending the UML Metamodel for Sequence Diagram to Enhance Model Traceability. Software Engineering Advances (ICSEA), 2010 Fifth International Conference on, IEEE: 129–134.

Testing and Measurement: Techniques and Applications – Chan (Ed.)
© *2015 Taylor & Francis Group, London, ISBN: 978-1-138-02812-8*

Study on method of obtaining the low-frequency characteristic of shock wave pressure measurement system based on quasi-static calibration

F. Yang, D.R. Kong & J.Q. Zhang
Nanjing University of Science and Technology, Nanjing, China

L. Kong & J.J. Su
Modern Chemistry Institute, Xi'an, China

ABSTRACT: When measuring shock wave pressure in explosive environments, the shock wave pressure sensor should be reconstructed by heat insulation and vibration isolation, so it would cause errors if the sensitivity of the bare sensor by manufacturers is directly used to calculate shock wave pressure. To obtain the transmission characteristics of shock wave pressure measurement system under actual working conditions, it is needed to research on the calibration method of shock wave pressure measurement system. In this paper, based on drop-hammer, hydraulic pressure calibration device, according to the principle of quasi-static calibration, the shock wave pressure measurement system was calibrated. Through the experiment of quasi-static calibration, typical experimental data were obtained. By using the appropriate data processing method, the low-frequency transmission characteristics of shock wave pressure measurement system was obtained.

KEYWORDS: Quasi-static calibration, pressure sensor, comparison calibration, low-frequency transmission characteristics.

1 INTRODUCTION

The peak pressure, overpressure duration and specific impulse of shock wave are important indexes to assess the damage power of explosion. So accurately measuring and recording the shock wave pressure are of great significance to obtain the above indexes. For the common method of shock wave pressure measurement-electrometric method, the mainly used pressure sensors are piezoresistive pressure sensors and piezoelectric pressure sensors. Piezoresistive sensors have great low-frequency characteristics, but they are vulnerable to the effects of temperature and light. Dynamic characteristics of piezoelectric sensor are good, and this type of sensor is often used to measure the dynamic signal. Under damage conditions, due to the specificity of exploding field, such as the dramatic changes in temperature and vibration, there are spurious responses of shock wave sensors. To suppress spurious response, it is needed for the sensor to be reconstructed by heat insulation and vibration isolation. Because of the changes on equivalent mass and equivalent stiffness, the sensitivity of the reconstructed sensors installed on the test platform is different from that of bare sensors. Therefore, calibration method of shock pressure measurement system needs to be studied.

Commonly used calibration methods contain static calibration and dynamic calibration [1]. Static calibration is that standard pressure which generates from standard calibration equipment, acts on calibrated sensor, and a linear relation between standard pressure and calibrated voltage can be obtained, so the static sensitivity of calibrated sensor can be calculated. Because the insulation resistance of the pressure sensor and the input impedance of amplifier cannot be infinite [2], the output electric charge will reduce in the process of static measurement, so it is better not to carry out static calibration on piezoelectric pressure sensor. Dynamic calibration is commonly executed by shock tube. Because the duration of step signal is finite (usually 4-8ms), the obtained low-frequency characteristics are unreliable, and only the characteristics above 1000Hz can be obtained. Based on a drop-hammer, hydraulic pressure calibration device, this article discusses the quasi-static calibration method of shock wave pressure measurement system.

2 PRACTICE OF QUASI-STATIC CALIBRATION METHOD

2.1 *Principle of quasi-static calibration*

Quasi-static is a method between static calibration and dynamic calibration [3]. Drop-hammer, hydraulic pressure calibration device generates half sine signals with different peak pressure and pulse width, which realizes comparison calibration between standard sensor and calibrated sensor.

The peak pressure and pulse width of half sine signal relate to the height of the drop- hammer, the area of the piston, the volume of initial cavity, the extending height of piston rod and the mass of the drop - hammer. By adjusting the above physical quantities, half sine signal with needed peak pressure and pulse width can be obtained, and the range of its pulse width is 1-12ms. By analyzing amplitude spectral density of half sine signal, it can be concluded that the smaller the impulse width, the wider the frequency range of the signal energy mainly distributing.

2.2 *The composition of quasi-static calibration experiment*

Quasi-static calibration experiment system consists of a standard pressure monitoring system and a calibrated pressure measurement system, shown in Figure 1. The Standard pressure monitoring system consists of a standard sensor, resistance strain gauge, a PXI data acquisition system and signal wires. 8530B-1000 piezo resistive sensor of Endevco Company is chosen as a standard sensor after being tracked by the method of directly comparing by the national metrology institutes and getting its sensitivity and the measurement range is 0-6.9MPa. The chosen resistance strain gauge is SDY2107B. The calibrated pressure measurement system consists of calibrated sensor, ICP signal conditioner, PXI data acquisition system and signal wires. ICP Piezoelectric sensor 211M0160 of Kistler Company is chosen as the calibrated sensor and its measurement range is 0-6.9MPa. The model of ICP signal conditioner is PCB482C05, which provides the constant current source and signal conditioning. The series of NI PXI-1042Q is chosen as a data acquisition system. According to shock wave pressure sensor, calibration range of the experimental system is 0-6.9MPa. The drop-hammer free falls and impact piston and the half sine pressure signal acts on the standard sensor and the sensor calibration at the same time.

Figure 1. The composition of quasi-static calibration experiment system.

Drop-hammer hydraulic pressure calibration device whose composition is shown in Figure 2 realizes comparison calibration between standard sensor and calibrated sensor. The composition of drop-hammer hydraulic pressure calibration device is shown in Figure 2, and the device is mainly composed

of drop-hammer, piston and cylinder. The standard sensor and the calibrated sensor symmetrically arranged in the cylinder to ensure that the two sensors feel the same pressure. The cavity fills with pressure transmission medium, and air inside should be tried to be exhausted.

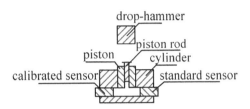

Figure 2. The composition of drop-hammer hydraulic pressure calibration device.

2.3 *Practice of quasi-static calibration experiment*

By analyzing amplitude spectral density of half sine signal with different pulse width, it can be known that frequency characteristics of the system within 0-1000Hz can be obtained by signal with 1ms pulse width. Through experiment, five groups of typical signals in standard channel and voltage signals in calibrated channel whose pressure range is 0-6.9MPa and pulse width is 1ms (±10%) are obtained, shown in Table.1.

Table 1. Five groups of typical signals with 1ms(±10%) pulse width.

Peak pressure in standard channel(MPa)	Pulse width in standard channel(ms)	Voltage in calibrated channel(V)
1.2302	1.0720	0.8959
2.8896	0.9380	2.2740
3.7487	0.9160	2.9965
4.3563	0.9060	3.5642
5.1371	1.0480	4.3213

3 OBTAINMENT OF QUASI-STATIC SENSITIVITY

Though tracked by the method of directly comparing by the national metrology institutes, the static sensitivity of standard sensor can be obtained as 0.0318V/MPa. And its low-frequency characteristic is proved to be smooth and is able to meet the requirements. So this sensitivity can be used to transfer quantity.

The nominal sensitivity of calibrated sensor is 0.814V/MPa. Divide signals into several groups due to their different pulse width. For every group, take pressure values in a standard channel as the X axis and voltages in the calibrated channel as Y axis; a straight line can be drawn by linear least squares fitting. And the slope of the line is the quasi-static sensitivity. Take the Obtainment of quasi-static sensitivity of signal with 1ms pulse width, for example, the voltage-pressure curve is shown in Figure 3. By the same token, quasi-static sensitivity of signal with other pulse width can be obtained. From calculation, it can be known that quasi-static sensitivity of signal with 1ms pulse width is 0.8174V/MPa, and its linearity is 2.17%. Obviously, the calculated quasi-static sensitivity is consistent with the nominal sensitivity.

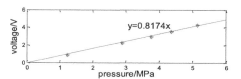

Figure 3.   Curve of voltage-pressure.

4   METHOD OF OBTAINING THE LOW-FREQUENCY TRANSMISSION CHARACTERISTICS OF SHOCK WAVE PRESSURE MEASUREMENT SYSTEM

4.1   *The basic principle of obtaining the low-frequency transmission characteristic*

It is not exact that the dynamic characteristic of the system is reflected only by dynamic sensitivity [4]. The system model can fully characterize the system dynamic characteristic, and frequency characteristic is a commonly used non-parameter model.

Pretreatment of signal contains removing the trend terms, and converting the output voltage signal in a standard channel to standard pressure signal $x(t)$ by sensitivity of standard sensors. Then add windows to signal for the purpose of reducing the energy leakage. Design low pass filter, in order to reduce the high-frequency interference and realize anti-aliasing. By Fast Fourier Transform and chirp-z transform, the amplitude spectral density $X(\omega) = FFT[x(t)]$ in a standard channel whose frequency resolution is 1Hz. By the same token, we can get the amplitude spectral density $Y(\omega) = FFT[y(t)]$ of voltage signal $y(t)$ in calibrated channel. Based on the equation of frequency characteristic-$H(\omega) = Y(\omega)/X(\omega)$, the frequency characteristic of shock wave pressure measurement system within 0-1000Hz with frequency resolution of 1Hz can be obtained.

4.2   *Obtainment of low-frequency characteristic of typical sensor*

Taking the obtained signal whose peak pressure is 4.3563MPa and pulse width is 0.9060ms for example, we can get the frequency characteristic of shock wave pressure measurement system. The sensitivity of the standard pressure monitoring system is transmitted by the standard calibration equipment.

Because the resistance strain gauge can realize the automatic balance, which ensures the mean value of signal in standard channel is zero before pressure arrives. So only the trend term in calibrated channel should be removed. The negative drift before pressure arrives should be calculated and subtracted.

The purpose of this paper is to obtain low-frequency characteristic within 0-1000Hz, so frequency component above 1000Hz can be filtered. A FIR digital filter with linear phase is designed. Choose Hanning window to achieve the truncation of infinite impulse response of ideal low-pass filter, in order to suppress the Gibbs phenomenon. Signals after pretreatment and being filtered are shown in Figure 4.

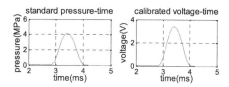

Figure 4.   Signals after pretreatment and being filtered.

In order to reduce the spectrum leakage, the pressure signal in standard channel and the voltage signal in calibrated channel 1 are added to Hanning window. The sidelobe of Hanning window is small, so the energy is concentrated in the mainlobe, which can effectively inhibit the spectral leakage phenomenon.

The original frequency resolution is only 61Hz. In order to reduce the barrier effect and improve the frequency resolution, using Chirp-z transform, the frequency resolution within 0-1000Hz can be increased to 1Hz. Chirp-z transform is a kind of method of spectrum zoom in z-plane [5], which applies to the frequency bands where no serious line interference occurs. The amplitude spectral density with Hanning window and spectrum zoom is shown in Figure 5.

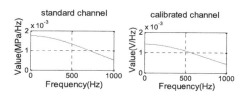

Figure 5.   The amplitude spectral density with Hanning window and spectrum zoom.

Finally, the obtained low-frequency characteristic of shock wave pressure measurement system is shown in Figure 6. It can be seen from the figure that, the frequency characteristics of the system within 0-1000Hz are stable, and the measured sensitivity is consistent with the sensitivity provided by manufacturers, which indicates that the method of obtaining low-frequency characteristic of shock wave pressure measurement system can realize the calibration of sensitivity whose magnitude is traceable.

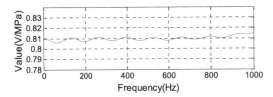

Figure 6. Low-frequency characteristic of shock wave pressure measurement system.

5 CONCLUSION

Based on a drop-hammer, hydraulic pressure calibration device, by calibrating the shock wave pressure measurement system, it can be proved that the obtained sensitivity is consistent with the sensitivity provided by manufacturer, which proves the feasibility of the method on obtaining the sensitivity of the shock wave pressure measurement system by quasi-static calibration as stated.

The transmission characteristic within 0-1000Hz shows substantially flat by obtaining the low-frequency transmission characteristic of the shock wave pressure measuring system. The stability of the low-frequency transmission characteristic is proved.

When the sampling frequency is high and the amount of data is small, using chirp-z transform for spectrum zoom can effectively improve the frequency resolution.

REFERENCES

[1] Deren Kong, Yongxin Li, Mingwu Zhu. Quasi-static Calibration Measure of Pressure sensor for Shock Wave[J]. Chinese Journal of Scientific Instrument, 2002,S1:16–17.
[2] Yushan Zhang, Hutang Qi. Quasi-static Calibration Method of Piezoelectric Transducers[J].Metrology & Measurement Technique, 2009,01:33–34.
[3] Deren KONG, Mingwu ZHU, Yongxin LI, PU Xiong zhu. Quasi-static absolute calibration on pressure-measuring sensors[J]. Journal of Transducer Technology, 2001,12:32–34.
[4] Ting Zhang. Some problems concerning dynamic calibration and evaluation of dynamic behaviours for pressure transducers[J]. Acta Metrologica Sinica, 1981,02:146–152.
[5] Kangpan, Chenghao Ding, Weihua Li. Spectrum analysis comparison between ZFFT and chirp-z transform[J]. Journal of Vibration and Shock, 2006,06:9-12+174.

Artificial intelligence and application

Classification fusion of global and local G-CS-LBP features for accurate face recognition

S. Nikan & M. Ahmadi

Department of Electrical and Computer Engineering, University of Windsor, Windsor, Ontario, Canada

ABSTRACT: A face recognition algorithm based on the combination of discriminative feature extractors and fusion of local and global classification results has been proposed in this paper. Gabor and Centrally Symmetric Local Binary Pattern (G-CS-LBP) are combined to extract distinctive features insensitive to appearance variations. Decision fusion of classification results on local histograms and global decision on the concatenated coefficients, enhances the recognition accuracy. Moreover, regulating the effect of global decision by assigning a variable weight affects the identification accuracy. Experimental results on the challenging FERET, FRGC and AR databases, under illumination, expression and occlusion, show improved performance of the proposed technique compared to the state of the art algorithms.

1 INTRODUCTION

Human face recognition includes a wide range of applications in law enforcement, social networking, border crossing monitoring and access control. There are many face recognition technologies available. However, intra-subject variations caused by lighting condition, facial expression and partial occlusion on the face, reduce the effectiveness of the recognition algorithms significantly. Therefore, the main focus is on increasing the robustness of identification techniques against appearance changes. Applying the effective preprocessing approaches to enhance the image and reduce the effect of aforementioned appearance variations increases the accuracy of face recognition. The orientated local histogram equalization proposed in (Lee et al. 2012), compensates illumination and explores edge orientations. An illumination insensitive representation was proposed in (Nikan&Ahmadi 2014) based on the image gradient. However, execution time is very crucial in real-time applications. Thus, excluding the preprocessing stage and employing discriminative image descriptors which are invariant against degrading conditions, reduces the computational complexity. Gabor wavelet (Yanget al. 2013)and different configuration of local binary pattern (LBP)(Ahonenet al. 2006) are some examples of feature extraction techniques which describe the local special image invariant against intra subject variations. Based on the human perception, in recognizing people, a fusion of complementary information leads to more accurate recognition

performance. In (Zhanget al. 2005)the concatenation of LBP histograms of Gabor images was used as the feature vector. The proposed approach in (Mehta et al. 2014)image characteristics at global and local levels by fusion of LBP histograms of directional images at four partitioning levels. In this paper, we proposed a face recognition algorithm which is the combination of two feature extractors and decision fusion of local and global application of classification technique. Figure 1 shows the proposed algorithm, where the Gabor filter bank extracts the image characteristics at different scales and orientations and centrally symmetric LBP (CS-LBP) texture descriptor, proposed in (Heikkila et al. 2009),is applied on the Gabor filtered images. Through a block-based strategy, we subdivided the G-CS-LBP images into small sub blocks. Concatenation of the histograms of the sub blocks at different scales and orientations are fed into the local extreme learning machine (ELM) (Huang et al. 2006) classifiers. A decision fusion technique combines the local decisions on the class label of the probe image. In order to increase the classification confidence, we also contribute the global decision on the whole face image in final identification result. In the global-based classification, we combine all extracted G-CS-LBP coefficients and apply principal component analysis (PCA) (Turk&Pentland 1991) to reduce the size of feature space. ELM is employed to find the class label of the whole face image. Decision fusion scheme combines the local and global class labels and finds the class label with the maximum vote.

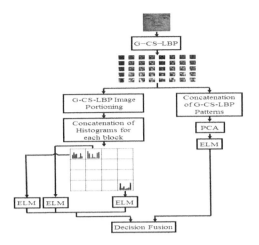

Figure 1. Block diagram of the proposed face recognition.

The rest of paper is organized as follows. In sections 1, 2 and 3, feature extraction, classification and fusion techniques are described in details. The experimental results are evaluatedin section 5. The paper is concluded in section6.

2 FEATURE EXTRACTION

The proposed G-CS-LBP feature extraction technique in this paper is the combination of Gabor filter bank and CS-LBP as a local texture descriptor.

2.1 Gabor filter bank

Gabor filter is able to describe signal aspects in different spatial scales and orientations similar to the response of the human vision cells which are sensitiveto scale and orientation. Gabor filter is a powerful feature extraction technique extensively used to capture the image characteristics at different spatial frequencies which are directionally selective. The explored spatially local features are insensitive to illumination and facial expression. The Gabor filtered images are obtained by convolving the image by the following filters at different scales and orientations(Lei et al. 2011):

$$\psi_{s,o}(x, y) = \frac{q_{o,s}^2}{\sigma^2} \cdot e^{-\left(\frac{z^2 q_{s,o}^2}{2\sigma^2}\right)} \cdot \left[e^{\left(jzq_{s,o}\right)} - e^{\left(-\frac{\sigma^2}{2}\right)} \right] \quad (1)$$

Where's and o are the scale and orientation and $q_{s,o} = q_s \exp(j\theta_o) = [\pi / 2(\sqrt{2})^s] \exp(j\pi o / 8)$ (in this paper we defined 5 scales and 8 orientations). $z = (x, y)$, and $\sigma = 2\pi$ (Lei et al. 2011). Fig 2 shows the result of Gabor filtering.

2.2 CS-LBP

Local binary pattern is a local texture analysis which is invariant to illumination variation due to compensating the intensity differences between neighbor pixels and its execution is simple and fast. However, we can reduce the size of extracted features and increase the stability against flat texture areas, by employing centrally symmetric LBP (Heikkila et al. 2009). Rather than comparing a center pixel with its neighbors in a circular neighborhood in LBP, in CS-LBP neighbors are compared with their center-symmetric pairs. Therefore, as shown in Figure2, for P neighbors around a pixel, the following equation gives the binary pattern.

$$CS_{pattern}(x, y) = \sum_{k=0}^{\left(P/2\right)-1} F\left(g_k - g_{k+\left(P/2\right)}\right) 2^k$$

$$where \quad F(u) = \begin{cases} 1 & u \geq th. \\ 0 & otherwise. \end{cases} \quad (2)$$

g_k is the gray value of kth neighbor of the pixel at (x, y) position. Thus, in this feature extraction approach the number of binary patterns is reduced to half of LBP patterns. We have $2^{(P/2)}$ decimal values for image pixels, which reduces the length of histogram bins and is computationally cost effective. Also, by comparing the intensity differences with a small threshold (th), rather than 0, in LBP, the robustness against flat areas is increased (Heikkila et al. 2009). The threshold value in this paper is equal to 0.1, obtained through the exhaustive search.

2.3 PCA dimensionality reduction

One of the most popular linear dimensionality reduction approaches is principal component analysis (PCA), which projects the high dimension data space to a smaller size subspace based on the eigenvectors of the covariance matrix corresponding to the large eigenvalues. We calculate the average image of the gallery set of n samples, as the mean value μ, which is subtracted from gallery samples to obtain the covariance matrix as follows (Turk & Pentland 1991):

$$C = \frac{1}{n} \sum_{k=1}^{n} (S_k - \mu) \cdot (S_k - \mu)^T. \quad (3)$$

Where S_k is the kth gallery sample. PCA projection-matrix, V, is constructed using the eigenvectors, v, of

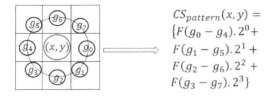

$$CS_{pattern}(x,y) = \{F(g_0 - g_4).2^0 + F(g_1 - g_5).2^1 + F(g_2 - g_6).2^2 + F(g_3 - g_7).2^3\}$$

Figure 2.   CS-LBP pattern for a pixel at (x,y) position.

The covariance matrix which are associated with m largest eigenvalues, $\propto$, and is employed to transform gallery samples to the lower dimension subspace as follows(Turk &Pentland 1991):

$$\propto .v = C.v, \qquad \widehat{S_k} = V^T.(S_k - \mu). \qquad (4)$$

Where, $V = [v_1, v_2, ..., v_m]$ and S_k is the kth projected gallery sample. Similar projection technique is applied on the probe samples to reduce the dimensionality of the feature space.

3   CLASSIFICATION APPROACH

The proposed classifier in this paper is extreme learning machine (ELM) which is applied locally on the local histograms of concatenated block histograms for different G-CS-LBP images. Also, ELM finds the globalclass label of the combination of image characteristics at different scale sand orientation. ELM is a leaning algorithm for single hidden layer feed-forward neural network, as shown in Figure3, which is extremely faster than the conventional gradient decent techniques (Huang et al. 2006). Despite the iterative tuning of the weights and biases in gradient decent to minimize the cost function, ELM assign random values to hidden layer biases and input weights. If the activation function,$f(.)$, is infinitely differentiable, by assigning random values to hidden layer matrix, the output weights can be calculated analytically by using Moore Penrose inverse of hidden layer matrix as follows:

$$\sum_{k=1}^{M} \bar{\rho}_k f(\bar{w}_k.\bar{a}_l + b_k) = \bar{t}_l. \qquad (5)$$

$$H\rho = T \xrightarrow{\substack{Moore-Penrose \\ Inverse}} \hat{\rho} = H^\dagger T \qquad (6)$$

Where $\bar{a}_l$ and $\bar{t}_l$ are input and target vectors and $\bar{w}_k$, $\bar{\rho}_k$ are weight connections from M hidden nodes to the input and output nodes, respectively, and b_k is the bias of kth hidden node. $l = 1, 2, ..., L$ (L is number of training samples). $T = [t_{l1},..., t_{lh}]$ and H are the target and hidden layer matrices, respectively.

4   DECISION FUSION

Local class labels of the image subblocks and the result of global based classification are fed into the majority voting scheme and the class label with maximum vote is selected as the final decision on the identity of the probe sample. Every subblockclass label has one vote. However, the number of votes for the global decision varies to regulate the effect of global decision regarding the effect of degrading condition. By varying the weight, we find the value which leads to the highest recognition accuracy.

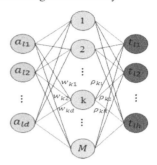

Figure 3.   Extreme learning machine configuration.

5   SIMULATION RESULTS

Experimental simulations of the proposed algorithm have been evaluated on three challenging face databases. FERET, FRGC 2.4 and AR. Image size is 128×128 pixels, and 8×8 subblock gives the best accuracy through the exhaustive search.

5.1   FERET

In order to evaluate the performance of the proposed face recognition techniqueon a large number of low quality images with illumination and expression variation and one sample per subject problem, gallery subset of FERET database (Phillips et al. 2000)with 1196 subjects is used as the gallery. Fb, Fc, DupI and DupII are employed as four probe sets with 1195, 194, 722 and 234 images, respectively. Table 1 compares the performance of different algorithms.

Table 1.   Recognition accuracy (%) for FERET.

Subset	Fb	Fc	DI	DII
Method				
SRC(Wright et al. 2011)	86.9	77.3	51.6	33.3
LGBPHS(Zhang et al. 2005)	94	97	64	53
LBP-w(Ahonen&Hadid 2006)	97	79	66	64
MTP(Khan et al. 2013)	98	83	71	67
ProposedMethod	94.9	96.9	71.5	68

5.2 FRGC 2.4

Experiment 4 of the second version of face recognition grand challenge (FRGC) (Phillips et al. 2005)is a large scale database containing images in uncontrolled illumination, expression conditions and blur and aging effect. It includes 466 individuals with 16028 and 8014images in the target and query sets, respectively. In this paper we took one sample per subject from target and query sets in the gallery and probe sets, respectively. Table 2 shows recognition accuracy of the proposed algorithm compared to reported results of the state of the art algorithms.

Table 2.   Recognition accuracy (%) for FRGC 2.4.

Method	Accuracy (%)
LBP (Ahonen&Hadid 2006)	11.99
OLHE (Lee et al. 2012)	20.53
LGBFR (Nikan&Ahmadi 2014)	32.62
Proposed Method	33.26

5.3 Occluded images of AR

AR database consists of 2600 images of 100 subjects with 26 images per subject taken in two sessions (13 images per session)(Martinez &Benavente 1998). The images are affected by illumination variation, facial expression, scarf and sunglasses occlusion and appearance changes in two weeks. In this paper, in order to evaluate the effect of occlusion, 8 images per subject with facial expressions from sessions 1 and 2 are used in the gallery set and 6 occluded images per subject, with scarf and sunglasses, in both sessions are employed in two probe sets. Table 3 illustrates the performance of proposed technique in comparison with the state of the art approaches.

Table 3.   Recognition accuracy (%) for AR.

Method	Scarf	Sunglass
SRC (Wright et al. 2011)	59.5	87
RPCA (Luan et al. 2014)	89.5	90.5
GRRC-L2 (Yang et al. 2013)	79	93
Proposed Method	85.5	98.5

5.4 Global weight variation

In this subsection we calculated the accuracy percentage in the above experiments versus the variation in the weight which is assigned to the global decision in majority voting fusion. Figure4 illustrates the recognition accuracy for FERER, FRGC and AR databases.

As is shown in this figure, very small values of weight reduce the accuracy. The weights equal to 9, 15 and 20 give the maximum accuracy for experiments on FERER, FRGC and AR, respectively. Thereafter, the accuracy decreases and remains constant.

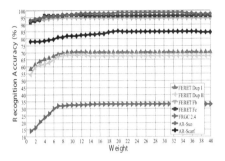

Figure 4.   Recognition accuracy (%) versus global weight value.

6 CONCLUSION

In this paper, a combination of Gabor filter bank and centrally symmetric local binary pattern as G-CS-LBP extracts the local characteristics of image at different scales and orientation. Application of CS-LBP reduces the size of feature space which increases the execution speed. Fusion of the local class labels on image subblocks and global decision on the whole patterns at different direction and spatial frequencies leads to improved recognition accuracy compared to popular methods in this area, for large number of images in FRGC, FERET and occluded images in AR database under illumination and facial expression variation. Increasing weights on the votes of global decision in the majority voting fusion, has improved the performance of the proposed algorithm because of the complementary performance of the local and global based results under different conditions.

REFERENCES

[1] Ahonen, T., Hadid, A. &Pietikainen, M.2006. Face description with local binary patterns: application to face recognition.*IEEE Trans. Pattern Anal. Mach. Intell.* 28 (12): 2037–2041.
[2] Heikkila, M., Pietikainen, M. &Schmid, C. 2009. Description of interest regions with local binary patterns. Pattern Recognition 42 (3): 425–436.
[3] Huang, G.B.,Zhu, Q.Y.&Siew,C.K. 2006. Extreme learning machine: theory and applications. Neurocomputing 70 (1-3): 489–501.
[4] Khan, A., Bashar, F., Ahmed, F. &Kabir, M.H. 2013 Median ternary pattern (MTP) for face recognition. Proc. Int. Conf. Informatics, Electronics and Vision (ICIEV): 1–5.

[5] Lee, P.H., Wu, S.W. & Hung, Y.P. 2012. Illumination compensation using oriented local histogram equalization and its application to face recognition. IEEE Transactions on Image Processing 21 (9): 4280–4289.

[6] Lei, Z., Liao, S.,Pietikäinen, M.&Li, S. 2011.Face recognition by exploring information jointly in space, scale and orientation.IEEE Trans.Image Process.20 (1): 247–256.

[7] Luan, X., Fang, B. & Liu, L. et al. 2014. Extracting sparse error of robust PCA for face recognition in the presence of varying illumination and occlusion. Pattern Recognition47(2): 495–508.

[8] Martinez A. &Benavente, R. 1998. The AR face database, CVC. Tech. Rep. 24.

[9] Mehta,R., Yuan,J.&Egiazarian, K.2014. Face recognition using scale-adaptive directional and textural features.Pattern Recognitionhttp://ieeexplore.ieee.org/xpl/RecentIssue.jsp?punumber=8347 (5): 1846–1858.

[10] Nikan, S. & Ahmadi, M. 2014. A Local Gradient-based illumination invariant face recognition using LPQ and multi-resolution LBP fusion. IET Image Processing: in press.

[11] Phillips, P.J., Moon, H., Rizvi, S.A. &Rauss, P.J. 2000. The FERETevaluation methodology for face-recognition algorithms.IEEE Trans. Pattern Anal. Mach. Intell. 22 (10): 1090–1104.

[12] Phillips, P.J., Flynn, P.J. & Scruggs, W.T. et al. 2005. Overview of theface recognition grand challenge. Proc. IEEE Comput. Soc. Conf. Comput. Vis. Pattern Recognit.: 947–954.

[13] Turk, M.&Pentland, A. 1991. Eigenfaces for recognition.J. Cogn. Neurosci.3 (1): 71–86.

[14] Wright, J., Yang, A., Ganesh, A., Sastry, S. & Ma Y. 2009. Robust face recognition via sparse representation. IEEE Trans. Pattern Anal. Mach. Intell. 31 (2): 210–227.

[15] Yang, M., Zhang, L., Shiu, S.C.K. & Zhang, D. 2013. Gabor feature based robust representation and classification for face recognition with Gabor occlusion dictionary. Pattern Recognition 46 (7): 1865–1878.

[16] Zhang,W., Shan,S., Gao,W., Chen,X. & Zhang,H. 2005. Local gabor binary pattern histogram sequence (lgbphs): a novel non-statistical model for face representation and recognition. ICCV: 786–791.

Testing and Measurement: Techniques and Applications – Chan (Ed.)
© 2015 Taylor & Francis Group, London, ISBN: 978-1-138-02812-8

Gas identification with pairwise comparison in an artificial olfactory system

M. Hassan & A. Bermak
Hong Kong University of Science and Technology, Hong Kong

A. Amira
University of the West of Scotland, Paisley, Scotland

ABSTRACT: An artificial olfactory system, referred to an electronic nose, is a multi-sensor platform used for gas classification. Lack of selectivity and low repeatability of the gas sensors are the major challenges in all gas identification problems. Pattern recognition algorithms are combined with a sensor array to address these challenges. The implementation of these algorithms is another challenge for the hardware friendly system. In this paper, we introduce a hardware friendly algorithm for gas identification. In this algorithm, we use sensitivity difference of any two sensors in the array as an input feature and a subset of the features is extracted by evaluating the capability of each pair of sensor to split the gases into two branches. The learning process of the pairs of sensors continues at every split point on the way until all individual gases are identified. The learned pairs of sensors at each split point are used for the identification of a new test response pattern and plurality voting is used for the distribution of the gases in cases of contention among the pairs. In order to assess the performance of our approach, a 4x4 tin-oxide gas sensor array is used to acquire the data of three gases in a laboratory. Accuracy rate of 100% is achieved with our algorithm on this experimental data set.

1 INTRODUCTION

An artificial olfactory system, referred to an electronic nose, attempts to mimic the human olfactory system to sense and discriminate between different odors. Usually, a number of sensors with varying characteristics are used in an electronic nose to discriminate more odors and obtain a unique pattern corresponding to each odor. The development of the electronic nose is mainly driven by the requirement to classify chemical substances and detect leakage of combustible and explosive gases. Since its development, it has been tested in many real life applications(Boilot et al., 2003), (Falasconi et al., 2012), (Holmberg et al., 1995), (McEntegart et al., 2000).

Sensors in an electronic nose usually suffer from poor selectivity, non-linearity and small inter-class sensors response variability. Hence, robust pattern recognition algorithms are needed for discrimination among odors. A variety of pattern recognition algorithms have been used for odors discrimination (Shi et al., 2008), (Bhattacharyya et al., 2008), (Boilot et al., 2003), (Distante et al., 2003). These algorithms perform considerably well but their implementation is not hardware friendly.

In this paper, we introduce hardware friendly algorithm for the gas identification. In this algorithm, we use sensitivity difference of each sensor pair as a feature and a tree, containing root node and internal decision nodes, and based approach is used for the realization of our algorithm. All possible pairs of sensors are explored at every decision node of the tree to find the best split of gases data during the learning stage. Based on the difference between selected sensors' sensitivities at each decision node, the positive data values are routed to one branch, while the negative values are routed to another branch. The distribution of the data samples continues until individual gases are identified. In the presence of multiple sensors pairs for supporting similar distributions at a decision node, plurality voting is used for the classification decision when applying on test vectors. If more than one distribution of gases is possible, then the distribution accompanied with the maximum sensors pairs is adopted. Experimental data acquired with 4x4 tin-oxide gas sensor array, corresponding to exposure of three gases, is used to evaluate the performance of our algorithm. We use similar approach with an array of seven commercial gas sensors in (Hassan and Bermak, 2014)with a single sensor pair at every decision node but this approach may exhibit over fitting.

The paper is organized as follows. Section 2 explains our proposed gas classification algorithm. Section 3 describes the experimental setup. Section 4 evaluates the performance of our algorithm and compares its results with existing approaches. Finally, Section 5 concludes this paper.

2 CLASSIFIER BASED ON PAIRWISE COMPARISON

Our objective is to extract suitable sensor pairs from all available pairs that can classify gases, with maximum performance. In order to implement our algorithm, we build a tree(Murthy et al., 1994) which contains a root node, internal nodes and a leaf nodes. It starts with the root node and terminates at leaf nodes. A leaf is labeled with an associated class code as an output. We use the difference between two sensors' sensitivities to split the data into only two branches at every decision node. An evaluation function is required to analyze the suitability of each sensor pair to distribute gases data in two branches. In our algorithm, we use split index as an evaluation function. At each decision node, all sensor pairs are explored by evaluating split index and data is distributed in branches based on the selected sensor pairs that offer best split index. The distribution of gases continues until all individual gases are identified.

Data impurity is evaluated at each node before the evaluation function is applied in order to find whether data samples belong to one gas or multiple gases. Data impurity can be measured in different ways, such as by the entropy function and misclassification rate (Alpaydın, 2010). We use the misclassification rate as the impurity measure. Let us assume that N_d is the total samples arriving at node d and N_d^i samples belong to gas i; then the probability of gas i at node d is given in Equation 1. Misclassification rate is measured in Equation 2 to find impurity in data. If the misclassification rate is zero at a node, then this node is declared a leaf node, which implies that all the data samples belong to one gas and there is no need to further split the data.

$$p_d^i = \frac{N_d^i}{N_d} \tag{1}$$

$$I_d = 1 - \overset{max}{i}[p_d^i] \tag{2}$$

If the data at node d belongs to different gases, then these gases are split into two branches by exploring the difference between the sensitivities of each sensor pair j, and gas samples with positive difference values are routed to one branch, while the gas samples with negative difference values are routed to another branch. The probability of gas i, taking branch k, is given in the following equation

$$p_{dk}^i = \frac{N_{dk}^i}{N_d^i} \tag{3}$$

where N_{dk}^i represents the number of samples of gas i along any branch k, and p_{dk}^i describes the probability

of gas i along the k branch. The probability of gases present at the node d and taking branch k is defined in Equation 4 in order to know whether all the data samples at node d are routed to one branch or distributed in two branches.

$$p_{dk}^t = \frac{\sum_{i=1}^{C_d} p_{dk}^i}{C_d} \tag{4}$$

where C_d is the number of gases whose data is available at node d. When no data sample takes a branch k then p_{dk}^t is zero, and if all the samples at node d take branch k then p_{dk}^t is 1.

Split index S_d^j is computed in Equation 5 for each sensors pair j in order to evaluate its fitness for available gases distribution at node d in two branches.

$$S_d^j = \frac{\sum_{i=1}^{C_d} |p_{dx}^i - p_{dy}^i|}{C_d} - [|p_{dx}^t - p_{dy}^t|]^+ \tag{5}$$

where p_{dx}^i and p_{dy}^i represent probability of gas i samples at node d taking branch x and y respectively. p_{dx}^t and p_{dy}^t describe the probability of data samples at node d taking branch x any y respectively. The last term of the Equation 5 is a rectifier function and is defined as follows:

$$[z]^+ = \begin{cases} z \ if \ z = 1 \\ 0 \ if \ z \neq 1 \end{cases}$$

It implies that if all of the samples of all gases take only a single branch, then the value of the rectifier function is 1 and this situation results in a 0 split index. On the other hand, if gases are distributed in two branches, then the rectifier function is 0 and results in a non-zero split index value. If the split index at node d is one, then it implies that some gases with all their samples are routed to one branch and the remaining gases with all their samples are routed to the other branch. The sensor pair corresponding to the maximum split value is selected for classification of test vectors. If there are more than one sensorpairs offering the best split value for the same gas distribution, then plurality voting is considered for the classification decision of the test vectors. There might be several sensors pairs with the same best split index value, but supporting multiple gases distribution. In this case, the distribution with the maximum sensors pairs will be adopted first. If there is classification error with this distribution in the learning phase, then other distributions with their supporting sensors pairs are examined in order to achieve maximum classification performance. Here, we present an example to illustrate this idea. Suppose there are three gases, C1, C2 and C3, at the root node; then there might be various sensors pairs supporting possible gases distribution cases, as shown in Figure 1.

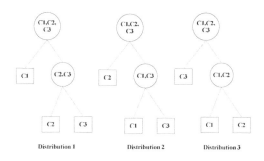

Figure 1. A simplified example showing possible gas distributions with three gases.

Some sensors pairs may split gas C1 data into one branch and gases C2 and C3 data into the second branch, as shown in distribution 1 of Figure 1 and some may support distributions 2 and 3. In this situation, the gas distribution with the highest accompanying sensors pairs is selected. In the case of classification error, the procedure starts over by selecting the gas distribution with the second highest accompanying sensors pairs and the process continues until maximum classification is achieved.

3 EXPERIMENTAL SETUP

In this paper, we use in-house fabricated gas sensor array to acquire the data of three gases: carbon mono oxide (CO), ethanol (C_2H_6O) and hydrogen (H_2).

3.1 Gas sensor array

To identify and discriminate gases, we use a 4 x 4 tin oxide gas sensor array which is fabricated by using an in-house 5μmCMOS process(Guo et al., 2007). Different post treatment schemes are implemented across each row and column to obtain diverse response. Three different catalysts Pt, Pd, and Au, are deposited on columns 2, 3 and 4 respectively. Ion implants of B, P and H are carried out along rows 2, 3 and 4 respectively. There is no catalyst in the first column and no ion implantation in the first row.

3.2 Data acquisition

Experimental setup of data acquisition is shown in Figure 2. The setup consists of mass flow controllers (MFCs), cylinders of the target gases, cylinder of dry air, 4x4 gas sensor array embedded in a gas chamber, and an interface computer for data processing. Dry air is used to achieve baseline response (the response

without any gas) of the sensors. MFCs are used to adjust the flow rate of each gas. Gases are injected into the gas chamber, and their concentration is controlled by mixing the gases and air at different rates. A data acquisition card is installed in the interface computer to capture the response of the gas sensor array. After acquiring the response of the sensor array, pattern recognition algorithms are implemented in the computer for classification of target gases.

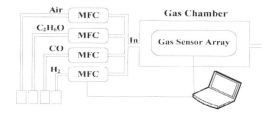

Figure 2. Experimental setup to characterize gas sensor array with three target gases.

4 PERFORMANCE EVALUATION

In order to evaluate the performance of our proposed algorithm, data is collected at different gas concentrations. The gas concentration is controlled by mixing gases with air at different flow rates. The tin oxide gas sensor's response is usually slow and takes a long time to reach steady state. Hence, a sensor's baseline response is achieved by exposing it to dry air for 750 seconds, and its gas response is achieved by exposing it to the gas for another 250 seconds. Sensors response is acquired at every second during exposure to target gas and dry air. Different gas concentrations from 20ppm to 200ppm with 20ppm incremental steps are used in the experiments. After acquiring the sensors' responses with the gas sensor array, the data is analyzed in the computer for gases classification.

With our proposed algorithm, different sensor pairs appeared to be the candidates for the different gases distributions with our training data sets. Majority of candidate pairs that support the same type of distribution are taken as a trained combinations or pairs and the test vectors are tested on these trained combinations. With our training data, the trained sensors pairs at the root node split CO gas samples into one branch and the C_2H_6O and H_2 gases samples into other branch as shown in Figure 3. The trained sensors pairs at the internal decision node distribute C_2H_6O gas samples into one branch and H2 gas samples into other branch depending upon the polarity of the difference between sensor pair. There are five sensors pairs which support distribution at the root node and three sensors pairs which support distributions at the internal decision node. In order to compare the performance of our algorithm, we also implement some statistical pattern recognition algorithms namely KNN, MLP, and SVM. We also transform the data from the original dimensional space to reduced dimensional space by using PCA and LDA, and use KNN classifier for identification. The performance of all the algorithms is summarized in Table 1. It is evident from the results that our proposed algorithm exhibits higher performance compared to other algorithms despite its simple implementation.

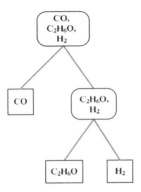

Figure 3.   Distribution of gases with our algorithm.

Table 1.   Performance comparison of gas identification algorithms.

Classification Method	Classification Accuracy (%)
KNN	95.2
MLP	93.3
SVM (Lin.)	87.8
SVM (RBF)	92.2
PCA with KNN	80
LDA with KNN	96.7
Our Work	100

5   CONCLUSION

Gas identification using an array of gas sensors has many potential applications in numerous fields. The gas sensors usually suffer from non-linearity and low repeatability. Therefore, it is essential to integrate pattern recognition algorithms in order to discriminate gases. Existing gas classification algorithms perform comparably well but are not suitable for hardware friendly implementation. In this paper, we introduced

an implementation friendly algorithm which also offer high performance as compared to commonly used pattern recognition approaches in artificial olfactory system.

ACKNOWLEDGEMENT

The presented research work is a part of an ongoing research project funded by Qatar National Research Fund (QNRF) under the National Priority Research Program (NPRP) No. 5-080-2-028.

REFERENCES

[1] Alpaydın, E. 2010. *Introduction to Machine Learning Second Edition*. MIT Press.
[2] Bhattacharyya, N., Bandyopadhyay, R., Bhuyan, M., Tudu, B., Ghosh, D., and Jana, a. 2008. Electronic Nose for Black Tea Classification and Correlation of Measurements With "Tea Taster" Marks. *IEEE Trans. Instrum. Meas.* 57, 1313–1321.
[3] Boilot, P., Hines, E., Gongora, M., and Folland, R. (2003). Electronic noses inter-comparison, data fusion and sensor selection in discrimination of standard fruit solutions. *Sensors Actuators B Chem.88*, 80–88.
[4] Distante, C., Ancona, N., and Siciliano, P. 2003. Support vector machines for olfactory signals recognition. *Sensors Actuators B Chem.88*, 30–39.
[5] Falasconi, M., Concina, I., Gobbi, E., Sberveglieri, V., Pulvirenti, a., and Sberveglieri, G. 2012. Electronic Nose for Microbiological Quality Control of Food Products. *Int. J. Electrochem.* 1–12.
[6] Guo, B., Bermak, A., Member, S., Chan, P.C.H., and Yan, G. 2007. An Integrated Surface Micromachined Convex Microhotplate Structure for Tin Oxide Gas Sensor Array. *IEEE Sens. J.7*, 1720–1726.
[7] Hassan, M., and Bermak, A. 2014. Gas Classification Using Binary Decision Tree Classifier. *in International Symposium on Circuits and Systems*. 2579–2582.
[8] Holmberg, M., Winquist, F., Lundström, I., Gardner, J.W., and Hines, E.L. 1995. Identification of paper quality using a hybrid electronic nose. *Sensors Actuators B Chem.27*, 246–249.
[9] McEntegart, C., Penrose, W., Strathmann, S., and Stetter, J. 2000. Detection and discrimination of coliform bacteria with gas sensor arrays. *Sensors Actuators B Chem.70*, 170–176.
[10] Murthy, S.K., Kasif, S., and Salzberg, S. 1994. A System for Induction of Oblique Decision Trees. *J. of Artificial Intelligence Research*. 2, 1–32.
[11] Shi, M., Bermak, A., Member, S., Chandrasekaran, S., and Member, S. 2008. A Committee Machine Gas Identification System Based on Dynamically Reconfigurable FPGA. *IEEE Sens. J.8*, 403–414.

Three-dimensional acoustic tomography for hot spot detection in grain bulk

H. Yan, L. Zhang & L. J. Liu
School of Information Science and Engineering, Shenyang University of Technology, Shenyang, China

ABSTRACT: A hot spot in a grain bulk is a localized high temperature zone and normally spoilage beginning in this location. In this paper, three-dimensional (3D) acoustic temperature tomography for detecting a hot spot in grain bulk is investigated. The influence of sound transceiver layouts on 3D temperature field reconstruction accuracy is investigated. 16 sound transceivers are arranged around the measurement space in two different layouts. 1000 single-hot temperature fields are reconstructed by using simulation data respectively resulting from these two transceiver layouts. Better transceiver layout is chosen based on reconstruction accuracy.

1 INTRODUCTION

Temperature field reconstruction based on acoustic tomography [1-2] uses the dependence of sound speed in materials on temperature along the sound propagation path. It has many advantages such as non-contact, fast response, wide temperature range and suitable for larger space. It has already been applied to temperature measurement of industrial furnaces [3-4], whereas monitoring the temperature distribution of stored grain by acoustic tomography [5] is a new application research being explored.

A hot spot in a grain bulk is a localized high temperature zone and normally spoilage beginning in this location. If a hot spot cannot be found in time, a great loss of grain will occur. Compared with contact method, non-contact method is a better one for hot spot detection in grain bulk, since grain is the poor conductor of heat [6].

In stored grain, sound is propagated principally through the gas in the narrow passageways between the grain kernels. Propagation through the solid grain matrix appears almost nonexistent because of the friction between the kernels. Therefore, the temperature distribution of stored grain can be monitored via acoustic tomography [5].

Reconstruction of 2D temperature fields by acoustic tomography is often reported. But in many cases, for example, hot spot detection in grain bulk, reconstruction of 3D temperature field is needed. In this paper, the influence of sound transceiver layouts on the reconstruction accuracy of 3D temperature field is investigated for hot spot detection in grain bulk.

2 PRINCIPLE OF MEASUREMENT

Temperature measurement by acoustic method is based on the principle that the sound velocity in a medium is a function of the medium temperature. The sound velocity c in a gaseous medium at an absolute temperature T is given by $T = 1/(Bc2)$, where B is a constant decided by gas composition [4-5].

In grain bulk, sound is propagated principally through the gas in the passageways between the grain kernels. The relationship between the sound velocity c in free space and the measured sound velocity c_m in grain bulk with the same gas and temperature can be expressed as $c_m = c/\lambda$, where λ is the speed conversion coefficient of grain. Using parameter λ, the measured sound velocity in stored grain can be converted into the sound velocity of free space with the same gaseous medium and temperature. λ is related to the frequency of the sound, the combined thermodynamic parameter of the gas, and the acoustic propagation properties of the grain. In practice, λ can be determined by calibration method [5].

To reconstruct the temperature distribution in a space by acoustic tomography, several acoustic transceivers should be installed on its periphery. On the basis of transceiver positions and the travel-time measurements along each effective sound wave path, the temperature field can be reconstructed by using suitable reconstruction algorithms [5].

3 RECONSTRUCTION ALGORITHM

Assuming the distribution of temperature is $T(x, y, z)$, the distribution of reciprocal of sound velocity is $f(x, y, z)$, we have

$$T(x, y, z) = 1 / [B \cdot f(x, y, z)]^2 \tag{1}$$

$$t_k = \int_{p_k} f(x, y, z) dp_k \quad (k = 1, 2, \dots K) \tag{2}$$

Where K is the number of the effective sound travel paths, and t_k is the sound travel-time of the kth sound path.

Divide the measurement space into M cells. Let the center coordinates of mth cell be (x_m, y_m, z_m). Expand $f(x, y, z)$ over a finite set of basis functions as follows

$$f(x, y, z) = \sum_{m=1}^{M} \varepsilon_m e^{-\beta \sqrt{(x-x_m)^2 + (y-y_m)^2 + (z-z_m)^2}} \tag{3}$$

Where β is the shape parameter of exponential basis functions, and it is related to the size of the space and the layout of the sound sources/receivers.

Substituting equation (3) into (2), we have

$$t_k = \sum_{m=1}^{M} \varepsilon_m \int_{p_k} e^{-\beta \sqrt{(x-x_m)^2 + (y-y_m)^2 + (z-z_m)^2}} dp_k = \sum_{m=1}^{M} \varepsilon_m a_{km}$$
$$a_{km} = \int_{p_k} e^{-\beta \sqrt{(x-x_m)^2 + (y-y_m)^2 + (z-z_m)^2}} dp_k \tag{4}$$

(4) can be further rewritten as follows.

$$t = A\varepsilon,$$
$$A = (a_{km})_{k=1,\dots,K, m=1,\dots,M}, t = (t_1, \dots, t_K)^T, \varepsilon = (\varepsilon_1, \dots, \varepsilon_M) \tag{5}$$

$$\varepsilon = \sum_{i=1}^{p} \left(\frac{\sigma_i^2}{\sigma_i^2 + \mu} \right) \frac{u_i^T t}{\sigma_i} v_i \tag{6}$$

Where u_i and v_i are the left and right singular value vectors of matrix A; $\sigma_1 \geq \sigma_2 \geq \dots \geq \sigma_p > 0$ are the singular values of matrix A; p is the number of non-zero singular values, μ is non-negative regularization parameter which controls the weight of measured data and experience in solution. Usually, the regularization parameter is chosen by experience.

Since the location of sound transceivers and cell centers are known, matrix A and its singular value decomposition can be determined. Then vector ε can be calculated by using measured or simulated sound travel-time vector t and equation (6), and the temperature distribution can be obtained by using equation (3) and (1).

4 TRANSCEIVER LAYOUTS

In this paper, the space to be measured is 1m×1m×1m, and 16 acoustic transceivers are amounted on its periphery in two different layouts. Figure 1 shows layout 1 and its 92 effective sound wave paths. Figure 2 shows layout 2 and its 112 effective sound wave paths.

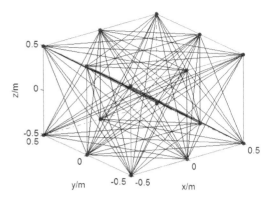

Figure 1. Layout 1 and its 92 effective sound wave paths.

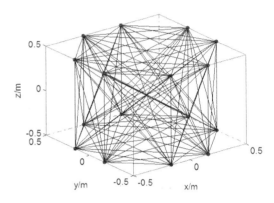

Figure 2. Layout 2 and its 112 effective sound wave paths.

5 TEMPERATURE FIELD RECONSTRUCTION

5.1 Temperature field models

In this paper, the space to be monitored is evenly divided into 10×10×10=1000 cells. In order to investigate the ability of 3D acoustic tomography for hot spot detection in grain bulk, 1000 one-hot temperature fields are reconstructed from simulated travel-times. The jth temperature field to be reconstructed can be expressed as

$$T(x, y, z) = 293 + 60ei^{-6[(x-xj)^2 + (y-yj)^2 + (z-zj)^2]} \tag{7}$$

This model describes a temperature field with a single hot spot. The center of the hot spot is (xj, yj, zj), and the temperature of the hot spot is 343K. The hot spot is positioned at the midpoint of the jth cell, $j=1$, 2,..., 1000.

5.2 Estimation of reconstruction error

In order to evaluate the quality of reconstruction temperature field, hot position error d and hot temperature error E_{max}, root mean squared error E_{rms} are used.

$$d = \sqrt{(rx_o - x_o)^2 + (ry_o - y_o)^2} \qquad .(7)$$

$$E_{max} = \frac{T_{rh} - T_h}{T_h} \times 100\% \qquad (8)$$

where (x_o, y_o) and (rx_o, ry_o) are the center coordinates of true hot spots and reconstructed hot spot respectively; T_h and T_{rh} are the temperature value of true hot spot and reconstructed hot spot respectively.

$$E_{rms} = \frac{\sqrt{\frac{1}{n}\sum_{j=1}^{n}\left[T(j) - \hat{T}(j)\right]^2}}{T_{mean}} \times 100\% \quad j = 1, ..., n \qquad (9)$$

Where $\hat{T}(j)$ is the reconstructed temperature of jth pixel, $T(j)$ is the temperature of jth pixel in model temperature field, and T_{mean} is the mean temperature of the model temperature field.

5.3 Reconstruction results

In order to test the ability of 3D acoustic tomography in detecting a hot spot in grain bulk, 1000 single-hot temperature fields are reconstructed by using simulation data respectively resulting from transceiver layout 1 and layout 2. In the reconstruction, the shape parameter of basis function β is set at 1e-4, and the regularization parameter μ is set at 0 since the simulation travel-time data are noisy-free.

To any one of the 1000 single-hot temperature fields, the hot position error d is zero, no matter the travel-time data comes from transmitter layout 1 or layout 2. The hot temperature error E_{max} and root-mean-squared error E reconstructed from the 1000 temperature fields are given in Figure 3 ~Figure 6, as a function of hot spot position.

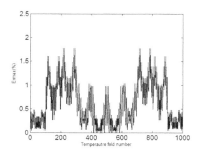

Figure 3. The hot temperature error resulting from transceiver layout 1.

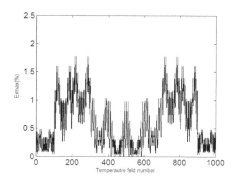

Figure 4. The hot temperature error resulting from transceiver layout 2.

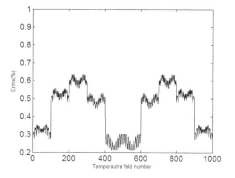

Figure 5. The root-mean-squared error resulting from transceiver layout 1.

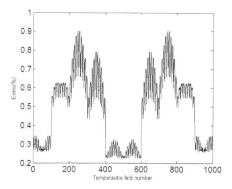

Figure 6. The root-mean-squared error resulting from transceiver layout 2.

The reconstruction result of a 3D temperature field is scalar volume data, which comprises four 3D arrays. The X, Y, and Z arrays specify the coordinates of the scalar values (i.e., the reconstructed temperature values) in the array T. Volume visualization is the creation of graphical representations of data sets that are defined on 3D grids. In general, scalar data is best viewed with iso-surfaces, slice planes, and contour slices.

In this paper, slice planes are used. As an example, Figure 7 shows the model field with a hot spot positioned at (0.05,-0.05, -0.25). The reconstruction fields from transducer layout 1 and layout 2 are given in Figure 8 and 9 respectively.

6   CONCLUSIONS

Following conclusions can be found from the reconstruction results given in section 5.

1   3D acoustic tomography is capable of reconstructing single-hot spot fields and good at finding a hot position.
2   Transceiver layouts have an influence on 3D temperature field reconstruction accuracy. Transceiver layout 1 is better than layout 2 in terms of reconstruction accuracy for the 1m×1m×1m measurement space surrounded by 16 sound transceivers.

In this paper, simulation travel-time data used are noise-free and the regularization parameter is set to zero. Future work will be focused on the selection of the regularization parameter for real travel-time data since measurement noisy is avoidless.

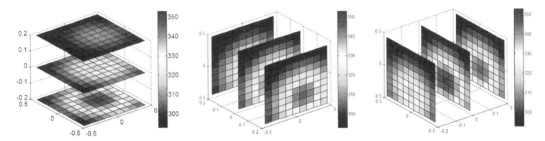

Figure 7.   The model field with a hot spot positioned at (0.05,-0.05,-0.25).

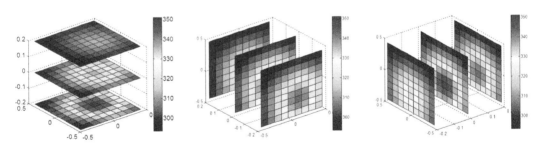

Figure 8.   The reconstruction field from transducer layout 1.

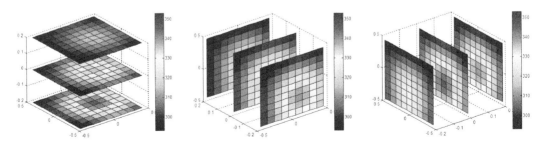

Figure 9.   The reconstruction field from transducer layout 2.

ACKNOWLEDGMENT

The authors thank the National Natural Science Foundation of China (No. 60772054, 61372154), the Program for Liaoning Excellent Talents in University (No. LR2013005) for supporting this research.

REFERENCES

[1] Barth M and Armin R 2011 Acoustic tomographic imaging of temperature and flow fields in air. Meas. Sci. Technol. 22 035102.

[2] Holstein P, Raabe A, P. Hostein, A. Raabe, Müller R, Barth M, Mackenzie D and Starke E 2004 Acoustic tomography on the basis of travel-time measurement. Meas. Sci. Technol. 15 1420- 28.

[3] Bramanti M, Salerno A E, Tonazzini A, Pasini S, and Gray A 1996 An acoustic pyrometer system for tomographic thermal imaging in power plant boilers. IEEE T. Instrum. Meas. 45 159–67.

[4] Srinivasan K, Sundararajan T, Narayanan S, Jothi T J S, Rohit and Rohit Sarma C S L V 2013 Acoustic pyrometry in flames. Measurement 46 315–23.

[5] Yan H, Chen G, Zhou Y and Liu L J 2012 Primary study of temperature distribution measurement in stored grain based on acoustic tomography. Exp. Therm. Fluid Sci. 42 55–63.

[6] Manickavasagan A, Jaya D S, White N D G, et al 2006 Thermal imaging of a stored grain silo to detect a hot spot. Appl. Eng. Agric. 22, 891–897.

Testing and Measurement: Techniques and Applications – Chan (Ed.)
© 2015 Taylor & Francis Group, London, ISBN: 978-1-138-02812-8

Overload prediction and its performance evaluation of the underwater robot platform

J.W. Lee, Y.H. Choi & J.H. Suh
Korea Institute of Robot and Convergence, Korea

ABSTRACT: Underwater robot platform has been applied in many harsh environments such as in deep sea or artificial water tank. An operator that operates the robot platform will come across many difficulties while the robot is underwater. In order to compensate for this, we proposed a procedure and method that can provide high-level information from low-level data. And as an example, we implemented the prediction of external overload to enhance operational efficiency and work continuity and evaluated its performance. Finally, it is proved that the prediction accuracy of the external overload is about 96%.

1 INTRODUCTION

As the work area extends into the shallow and deep sea, the robot platform technology has been evolving fast in recent years. In addition to the ocean, the robot platform is also applied to manmade waterways and the water tanks.

These underwater environments are very difficult for humans to work, so the robot technology has been applied to improve the work efficiency.

In the past 30 years, underwater robots have been used to lay submarine cable and pipeline. These robots can be divided into the skid-type and the crawler-type (or track-type) according to the driving method. The skid-type robot is normally selected for the seabed soil below 30kPa and the crawler-type for the hard seabed ground.

The need for underwater robot platform in order to maintain the artificial water tank and the waterway is growing, and the underwater robot platform is used with means for replacing manpower, cutting the cost and increasing work efficiency.

The underwater robot platform technology is rapidly evolving as its application prospects increase. One of the many difficult things in operation is the lack of information about robot platform's status and surrounding environment.

In recent years, the operator obtains information by mounting many sensors on the robot platform, but it does not provide the high level information which is required for operator.

It is especially difficult to determine for the external load due to the work load. That is the direct cause of the fatigue of the underwater robot platform which will affect the work efficiency, i.e. work amount and work time.

In this paper, we targeted the track driving based underwater robot platform for cleaning the artificial water tank in the industry and described the decision rule to predict the external overload of the robot platform. Then we showed the prediction results on the basis of the data gathered in the field operation. Finally we described the conclusions and future work.

2 RELATED WORK

In this chapter we survey previous work for the detection and decision about system status. Dexter and Pakanenm[1] suggested FDD(Fault Detection and Diagnostic) methodology, so they have constructed the hierarchical decision-making rule set, and broken down the real-time diagnostics test into three steps consisting of identification, fault detection and fault isolation. Jeffrey and Steven[2] described specifications of functional goals by applying FDD methodology to HVAC(Heating, Ventilating, and Air Conditioning) systems which are the subject of decisions and layered by configuring a set of rules using the data of the actual HVAC systems and implemented them to execute the decision-making framework. Yang[3] proposed the condition-based failure prediction methodology for preventive maintenance of equipment and techniques and it was implemented to run the scheme to estimate the predictive value condition by treatment with a Kalman filter for the sensor data. Feng and Derong[4] proposed a method of neural network based classification in the working state for oil pipeline system. Then they classified the results derived for neural network in an individual state of system operation and separated by more than

86% accuracy based on a threshold value of the pressure and leakage in the pipeline.

Previous research has been performed in a number of techniques and applications for the status detection of system that has the system architecture by defining the conditions and decisions criteria from the lower level of the sensor data to run hierarchical rules commonly. And low level data characteristic of the robot platform which moves in the environment is dependent on the movement position and time, so it is necessary to update the threshold of conditions and decisions for determining the state.

In this paper, we proposed a novel procedure and method for estimating the external overload to robot platform when the sensor data is changed when the robot moves in the environment.

3 THE TRACK-BASED UNDERWATER ROBOT PLATFORM AND CONTROL STRUCTURE

The underwater robot platform has a track-based electrical driving unit, rotating screws and suction equipment for cleaning operations in the artificial industrial water tank.

The external load of the left and right track drive is shown directly on the robot platform. The configuration of the driving unit is shown in Table 1.

Table 1. Configuration of the driving unit.

Part	Specification	In/Out	Sensing
Track(Crawler)	Metal Support + Lubber Band	-	-
Motor + Reducer	Each Driving Part 750W and Screw Part 500W + 1/70 Reducer	48V 25A (85% Efficiency)	Encoder Pulse
Motor Driver	ELMO General Purpose (Cello)	Max. 60V, 30A	Internal Parameters
Underwater Umbilical Cable	Power Line 2/2.5mmSQ x 4	Safety Current 10/15A x 4	-

Each motor is connected to the reduction gear for driving the tracks, and the motor driver operating the motor control supplies electrical power to the motor and receives encoder pulse signal. The main controller transmits a control target velocity or a target current to the motor driver, and receives data of control value and operating status of motor drive. In addition, the main controller collects data about the actual current consumption of each motor by adding the external current sensor on motor power line. This relationship described is shown in the following Figure 1.

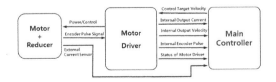

Figure 1. Relationship between motor, motor driver and motor controller.

Detailed specification of the parameter collected for each driving unit is the same as in Table 2, and the parameter list contained values like control speed, tracking current, tracking velocity, encoder pulse and RPM. It also includes status flags such as the fault signal for the motor driver, operation error signal for the motor, current over-limit signal, over-voltage signal, under-voltage signal and over-temperature signal. Among these values, the parameters representing the generated overload is the current limit warning signal and under-voltage signal, and the other values are not related to the overload occurs.

Table 2. Parameter list of motor driver and sensor.

	Parameter List	Description	Unit	Range
Motor Driver	ELMO_ControlSpeed	Input control velocity of each driving motor	mm/s	-1000~1000
	ELMO_CurrentActual	Output tracking current of each driving motor	A	0~40
	ELMO_VelocityActual	Output tracking velocity of each driving motor	mm/s	-1000~1000
	ELMO_Encoder	Output encoder pulse of each driving motor	Pulse	-1000000000 ~1000000000
	ELMO_RPM	Output RPM of each driving motor	Revolutions per minute	4000
	ELMO_Fault	Output fault flag of motor driver	Flag	0 or 1
	ELMO_MotorFailure	Output operation error flag of motor	Flag	0 or 1
	ELMO_Error_CurrentLimit	Output current over-limit flag	Flag	0 or 1
	ELMO_Error_UnderVoltage	Output under-voltage flag	Flag	0 or 1
	ELMO_Error_OverVoltage	Output over-voltage flag	Flag	0 or 1
	ELMO_Error_Temperature	Output over-temperature flag in motor driver	Flag	0 or 1
Current Sensor	ExternalCurrent	Output induced current of external sensor	A	0~40

4 PROCEDURES AND METHODS FOR ESTIMATING THE OVERLOAD TO THE UNDERWATER ROBOT PLATFORM

While operating the underwater robot platform to clean the artificial water tank in the field of industry, one of the most important things is that the operator should recognize the degree of the external overload and the status of the platform. An external overload of the robot platform may reduce the work efficiency. Continuous external load applied to the platform may cause problems, such as a direct damage or modification, and it takes a lot of money and time to repair the problem.

If it is possible to predict the occurrence of overload, we will be able to improve the control algorithms and that will allow the robot to work for a longer time. Thus, we can expect high stability, reliability and efficiency of the robot.

The first step for executing the overload prediction is to select the parameters within the target system related to the state of the load estimation.

The robot system has a variety of parameters. There are robot control input variables such as a target velocity and a target current of each axis and control output variables such as tracking velocity, current, encoder pulse and RPM. And there are robot status variables such as a motion sensor, water leak detection, temperature and trip signal of circuit protector. The select of valid parameters which can describe the available target state can be chosen through an automated analysis method, for example factor analysis and linear discriminant analysis or selected by the domain knowledge manually.

The following Table 3 shows the coefficient of the Linear Discriminant Analysis[5]. The result is a dataset consisting of two classes. Class A is a dataset consisting of data of the section used to perform the prediction, and in this result, effective parameters for the classification of the class must have a larger coefficient value like as bottom screw's one.

Table 3. Coefficients of two classes in linear discriminant analysis.

Left Motor	Coefficient (A)	Right Motor	Coefficient (A)	Front Motor	Coefficient (A)	Bottom Motor	Coefficient (A)
EC	0.719	EC	7.792	EC	-0.405	EC	12.705
XC	12.569	XC	0.096	XC	22.738	XC	-45.616
TV	0.450	TV	0.428	TV	10.762	TV	-12.992
EV	4.175	EV	7.455	EV	-13.061	EV	31.045
RM	-5.383	RM	-9.738	RM	2.311	RM	-17.817

Left Motor	Coefficient (B)	Right Motor	Coefficient (B)	Front Motor	Coefficient (B)	Bottom Motor	Coefficient (B)
EC	-0.069	EC	-0.929	EC	10.652	EC	-1.368
XC	4.123	XC	1.207	XC	-19.451	XC	17.675
TV	0.580	TV	-0.381	TV	-0.035	TV	-6.380
EV	0.659	EV	-0.214	EV	1.271	EV	4.108
RM	-1.009	RM	1.713	RM	1.053	RM	1.927

EC(Motor driver output current), XC(External induced current), TV(Control velocity), EV(Motor driver tracking velocity), RM(Motor driver tracking RPM)

Another method is applied to a Factor Analysis[6] to extract the valid parameters; each factor value represents one characteristic of the class, as shown in Table 4 below.

Table 4. Coefficients of two classes in factor analysis.

A	Left Motor					Right Motor				
	EC	XC	TV	EV	RM	EC	XC	TV	EV	RM
F1	0.656	0.526	-0.805	-0.252	-0.463	0.715	0.658	-0.792	-0.381	-0.556
F2	0.546	0.728	0.212	0.493	0.482	0.425	0.599	0.158	0.411	0.407
F3	-0.066	-0.106	0.048	-0.035	0.005	0.012	-0.082	0.066	0.139	0.167

A	Front Motor					Bottom Motor				
	EC	XC	TV	EV	RM	EC	XC	TV	EV	RM
F1	0.476	0.552	-0.587	0.217	-0.308	0.170	0.326	-0.523	-0.171	-0.221
F2	0.194	0.432	0.135	0.250	0.220	-0.406	-0.219	0.402	0.171	0.339
F3	0.370	0.218	0.128	0.059	0.253	0.702	0.750	-0.012	0.647	0.526

B	Left Motor					Right Motor				
	EC	XC	TV	EV	RM	EC	XC	TV	EV	RM
F4	0.606	0.838	-0.815	-0.305	-0.612	0.590	0.811	-0.821	-0.487	-0.728
F5	0.524	0.146	0.163	0.450	0.327	0.509	0.153	0.162	0.387	0.301
F6	-0.300	0.200	-0.102	-0.029	-0.099	-0.347	0.159	-0.107	-0.016	-0.078

B	Front Motor					Bottom Motor				
	EC	XC	TV	EV	RM	EC	XC	TV	EV	RM
F4	0.343	0.744	-0.351	0.057	-0.074	0.296	0.683	-0.520	0.262	-0.043
F5	0.356	0.037	0.447	0.428	0.474	0.504	0.184	0.305	0.365	0.437
F6	-0.554	0.244	0.508	0.308	0.428	-0.533	0.206	0.405	0.044	0.271

EC(Motor driver output current), XC(External induced current), TV(Control velocity), EV(Motor driver tracking velocity), RM(Motor driver tracking RPM)

Factors obtained through factor analysis are subject to various parameters of axis motor drivers. Electrical current related parameters are the most common. In particular, factor F3 and F6 are the reflection of the common features of the bottom screw in each class, so we can know that the operation of the bottom screw has a significant effect on the overload that is to be determined.

In a real situation, the external load affects the robot platform. The robot can't be controlled to follow the velocity and current, so the velocity and current's difference is generated between the motor driver's output current and the input control values.

Another feature is the difference between the output current value of the motor driver and the induced current value measured by the external sensor. This causes an imbalance in the motor driver's power output(Source) and the motor's power consumption (Sink). The cause of this phenomenon is thought to be due to the external load.

Compared to the increase of the output of the electrical current of the motor driver, as shown in Figure 2 below, the induced current of the external current sensor is not able to follow them. Even though the current supply of the motor driver is raised to the current limit, the difference of electrical current continues and cuts off the output after about 2 seconds. To reflect this characteristic in the overload prediction, this feature was applied to the creation of estimation equation.

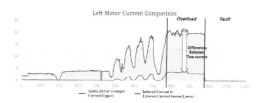

Figure 2. Comparison of motor driver's output current and induced current of external current sensor.

After the selection of the parameters used to determine the target state of the overload prediction, the condition or classification equation is generated using these parameters. These values such as the current or velocity are not simple parameters that can distinguish the state by a specific reference value, so they were classified through linear classification algorithm.

The training dataset for classification algorithm consists of motor driver's parameters which are normalized output current, induced current, control velocity and RPM value. Results of the large linear classification algorithm[7] based on Cross-validation is the same as in Table 5. The row expresses the classification result and the column represents actual state that the overload that has occurred. In the classification result, the number of prediction failure is

44,547, which did not predict the occurrence of the actual overload. Therefore, it is necessary to generate a new dataset consisting of the calculated values of the complex parameters.

Table 5. State classification results of linear classification algorithm.

Classification	Real State	
	False (1,072,415)	True (44,834)
False (1,107,726)	1,063,179	44,547
True (9,523)	9,236	287

#total sample = 1,117,249

In order to derive the elements of the new data set, several prediction discriminant equations are generated based on characteristics such as differences in two different types of current and velocity, and they were run to obtain the results which are shown in Table 6 below.

The prediction discriminant equations in the Table 6, equations from No.1 to No.4 by utilizing the gap in the amount of current in different measuring methods were to be classified through the simple or amplified difference values. Equations from No. 5 to No.7 by utilizing the gap in the amount of velocity between the control input and the tracking output of motor driver were to be classified the dataset.

Table 6. Generates prediction discriminant equations.

No	Prediction Discriminant Equation	Description
1	$ELMOCurrent > a \& (ELMOCurrent - ExternalCurrent) > b$	Uses the simple difference of amount of two currents larger than a reference value
2	$\left(norm(ELMOCurrent)^b - norm(ExternalCurrent)^c\right) > d$	Uses the amplified difference of amount of two currents
3	$\left((1 - norm(ELMOCurrent))^e - (1 + norm(ExternalCurrent))^f/c\right) > d$	Same as the above equation
4	$\frac{(1+norm(ELMOCurrent))^e}{b} - \frac{(1+norm(ExternalCurrent))^f}{d} > e$	Same as the above equation
5	$TargetVelocity > a \& RPM < b$	If the target velocity is high but the actual running velocity is low
6	$TargetVelocity > a \& (TargetVelocity - ELMOVelocity) > b$	Uses the simple difference of amount of two velocities larger than a reference value
7	$\left(norm(TargetVelocity)^a - norm(ELMOVelocity)^b\right) > c$	Uses the amplified difference of amount of two velocities

The results of evaluating the accuracy of measuring the prediction discriminant equations can be obtained when the associated actual overload occurs within 5 seconds after the prediction as shown in Table 7 below.

Each equation was executed to obtain the accuracy of prediction classified. The degree of correctness of the prediction equation from No.1 to No.4 using the gap of the amount of two currents is about 70 ~ 85%, and the success rate of equations from No.5 to No.7 using the difference of between the control velocity

and the tracking velocity output is up to 64%. And the overall determination of the prediction obtained by the discriminant equations has a 70% accuracy which was based on the boundary value when the degree of prediction precision is the highest through automated repeat loop.

Table 7. The accuracy of prediction discriminant equation.

No	Prediction Discriminant Equation	#Prediction	Accuracy
1	$ELMOCurrent > 19 \&$ $(ELMOCurrent - ExternalCurrent) > 17$	2,823	70.70%
2	$\left(norm(ELMOCurrent)^6 - norm(ExternalCurrent)^2\right)$ > 0.05	755	85.17%
3	$\left((1 - norm(ELMOCurrent))^2\right.$ $\left. - (1 + norm(ExternalCurrent))^6/64\right) > 0.05$	804	80.35%
4	$\left(\frac{(1+norm(ELMOCurrent))^5}{32}\right.$ $\left. - \frac{(1+norm(ExternalCurrent))^2}{4}\right) > 0.05$	755	84.90%
5	$TargetVelocity > 501 \& RPM < 1000$	1,325	58.57%
6	$\left(norm(TargetVelocity)^5 - norm(ELMOVelocity)^2\right)$ < 0.04	1,417	64.43%
*	All Prediction Equation	3,037	70.83%

After the combination of these results, a new dataset is generated by reconfiguring the calculation expression in the prediction discriminant equation.

Finally, the result of applying a linear classification algorithm in the new dataset was shown in Table 8.

Table 8. Classification results of linear classification algorithm at the new data set.

Classification	Real State	
	False (1,072,415)	True (44,834)
False (1,073,409)	1,071,866	1,543
True (43,840)	549	43,291

#total sample = 1,117,249

The row of the table is the overload prediction result by the classification algorithm, and it was predicted for the overload when it is 'True'. The column is the number of actual overload that occurred, and it is the cumulative sum of the results of overload-related signals from 5 seconds after predicted time.

The results of Table 8, the actual overload related signal occurred 44,834 times and the number of estimated value by the classification algorithm occurred 43,840 times. Therefore, the overload prediction is at least about 97% accurate.

The number of generation of the actual overload is 43,291 times, and the accuracy of prediction is more than 98%.

The number of the false alarm, which means there is an overload but it was predicted, is 1,543 samples. The prediction failure is up to 4% of the actual overload.

5 CONCLUSION

A track-based underwater robot platform is widely applied in the oceanic environment such as the shallow and deep sea, and in artificial environments such as man-made waterways and industrial water tanks. The underwater robot platform selects a track-type driving method for moving on the seabed and industrial facilities for heavy work.

We proposed a procedure and method to generate high level information for recognizing the workload status of robot platform from a lower level of the data which the operator is required to operate the robot platform in the underwater environment.

To evaluate the overload on the robot platform, the load-related valid parameters were chosen within various system parameters, and prediction discriminant equations were generated by combining these parameters. The prediction performance was evaluated by applying a linear classification algorithm. The overload prediction results by the prediction equation have over 96% accuracy.

The overload prediction related to the robot platform working underwater or in artificial water tanks will be common to the entire track-based underwater robot platform.

ACKNOWLEDGEMENT

This work was supported by grant No. 10043928 from the Industrial Source Technology Development Programs of the MOTIE(Ministry of Trade, Industry & Energy), Korea. And this research was also a part of the project titled 'Establishment of the underwater glider operational network for the real-time ocean observation' funded by the Ministry of Oceans and Fisheries(MOF) and Korea Institute of Marine Science & Technology Promotion(KIMST), Korea.

REFERENCES

[1] Dexter, Arthur & Pakanenm, Jouko. 2001. Demonstrating automated fault detection and diagnosis methods in real building, *VTT Building Technology.*

[2] Jeffrey, Schein & Steven, T. 2008. A hierarchical rule-based fault detection and diagnostic method for HVAC systems, *American Society of Heating, Refrigerating and Air-Conditioning Engineers, HVAC&R Research, Vol. 12, No.1 , 1 January 2006.*

[3] S. K. Yang. 2003. A condition-based failure-prediction and processing-scheme for preventive maintenance, *IEEE Transactions on Reliability, Vol. 52, No. 3, September 2003.*

[4] Feng, Jian & Derong, Liu. 2006. Detection algorithm and application based on work status simulator, *6th World Congress on Intelligent Control and Automation, 21–23 June 2006.*

[5] http://en.wikipedia.org/wiki/Linear_discriminant_analysis (last access: November 2014).

[6] http://en.wikipedia.org/wiki/Factor_analysis (last access: November 2014).

[7] http://www.csie.ntu.edu.tw/~cjlin/liblinear/ (last access: November 2014).

Testing and Measurement: Techniques and Applications – Chan (Ed.)
© *2015 Taylor & Francis Group, London, ISBN: 978-1-138-02812-8*

Gaussian muti-scale fast registration algorithm and its application in machine vision

X.H. Cao, J.H. Yang & Y. Dai
School of Automation, Northwestern Polytechnical University, Xi'an, China

ABSTRACT: To meet the real-time requirement of image processing in machine vision applications, a novel Gaussian Muti-scale Fast Registration (GMFR) algorithm is proposed. GMFR avoids the problem of losing details of images in sub-sampling process by means of Gaussian filtering based on muti-scale images. The advantage of image Cross-Correlation (CC) and Cross-Power Spectrum (CPS) are integrated by the presented method to achieve both accuracy and efficiency of image registration. The performance of GMFR is analyzed quantificationally based on the pre-defined performance function, which shows that the advantage of GMFR algorithm increases exponentially as the image gets larger. A machine vision environment is built for surface defects detection of lead frames and verifies the accuracy, robustness and high computing efficiency of GMFR in noisy image registration.

1 INTRODUCTION

Image registration refers to the process of overlaying two or more images of the same scene taken at different times, from different viewpoints, under different light conditions and/or different sensors (Zitova and Flusser, 2003). It aligns the sensed image to the reference image (Goshtasby, 2012). When machine vision (Davies, 2004, Wang et al., 2013) is used to achieve intelligent defect detection of industry products (Yen and Sie, 2012), image registration is the crucial procedure to match the detecting image with the template. However, the efficiency, robustness, anti-noise capacity of image registration algorithm restricts the application of machine vision. Therefore, a novel image registration algorithm which has these characteristics is needed.

In this paper, the widely used image Cross-Correlation (CC) (Kaneko et al., 2003) algorithm in spatial domain and image cross-power spectrum (CPS) algorithm in frequency domain are introduced. Based on CC and CPS, muti-scale image analysis is introduced. Then Gauss Muti-scale Fast Registration (GMFR) is proposed, a performance function is defined to analyze the efficiency of the presented registration algorithm. Finally, GMFR is verified in a machine vision environment which is built up for acquiring lead frame images.

2 IMAGE REGISTRATION

Let $g(x, y)$ ($x=0,1,2,...,M-1$; $y=0,1,2,...,N-1$) denote the reference image (a digital grayscale image with $M \times N$ pixels) (Gonzalez et al., 2004) and $f(x, y)$

denote the sensed image $f(x, y)=g(x-x_0, y-y_0)$. (x_0, y_0) should be obtained in order to make $f(x, y)$ align to $g(x, y)$.

2.1 CC and CPS methods

The CC function of $f(x, y)$ and $g(x, y)$ is given by

$$R_{fg}(x, y) = \frac{1}{MN} \sum_{m=0}^{M-1} \sum_{n=0}^{N-1} f^*(m,n)g(x+m,y+n) \quad (1)$$

where $f^*(x, y)$ is the complex conjugate of $f(x, y)$ (Madisetti, 2010). The location of peak point of $R_{fg}(x, y)$ means the most exact matched position from the sensed to the reference image. The size of $R_{fg}(x, y)$ sets to $M \times N$. Because of the incompleteness of the CC, the peak point will not appear in the position (x_0, y_0) but in (x_0', y_0') which is near to the central zone of $R_{fg}(x, y)$. Registration parameter (x_0, y_0) is obtained by central transformation

$$x_0 = x_0' - N / 2, y_0 = y_0' - M / 2 \quad (2)$$

In frequency domain, let $F(u, v)$ and $G(u, v)$ denote the 2-D discrete Fourier transform of $f(x, y)$ and $g(x, y)$ respectively, and it is given by

$$F(u,v) = \frac{1}{MN} \sum_{x=0}^{M-1} \sum_{y=0}^{M-1} f(x,y)\exp[-j2\pi(\frac{ux}{M}+\frac{vy}{N})] \quad (3)$$

$$G(u,v) = F(u,v)\exp[-j2\pi(\frac{ux_0}{M}+\frac{vy_0}{N})] \quad (4)$$

$G(u, v)$ is based on Fourier Shift Theorem. The normalized CPS of two images is given by

$$\frac{F(u,v)G(u,v)^*}{\left|F(u,v)G(u,v)^*\right|} = \exp[\,j2\pi(ux_0 + vy_0)] \qquad (5)$$

where $G(u, v)^*$ is the complex conjugate of $G(u, v)$ (Gonzalez et al., 2004). It computes the CPS of the sensed and reference images and looks for the location of the peak (x_0', y_0') in its inverse Fourier transform. The registration parameters (x_0, y_0) need central transformation

$$x_0 = x_0' - (N/2+1), y_0 = y_0' - (M/2+1) \qquad (6)$$

The CPS method shows strong robustness against the correlated and frequency dependent noise and non-uniform, time varying illumination disturbances. The computational time savings are more significant if using FFT. However, FT shows limitation in dealing with local image where CC methods can deal. The GMFR is proposed by combining the advantages of these two methods and using muti-scale analysis.

2.2 Muti-scale image analysis using Gaussian filtering

Muti-scale image analysis is realized by sub-sampling. For a 2-D image, sub-sampling refers to interlaced extraction in orthogonal directions of the image. After one time sub-sampling, the size of image is reduced to one-quarter (see Fig. 1). $\lambda = 0, 1, 2...$ is a scale parameter and the 2-D grayscale image could be expressed by $f(x, y; \lambda)$. $\lambda = 0$ is corresponding to the original image. The greater λ is the lower resolution the image has. Although low-scale image is fuzzy, it still shows the tendency of image shifting. The computational time will decrease apparently when using low-scale image in registration. If an image is sub-sampled directly, lots of image details will lose and the SNR of image will drop sharply which may result in registration failure.

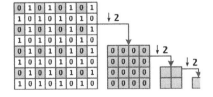

Figure 1.   Expression of muti-scale analysis.

In order to save details in low-scaled image, Gaussian filtering is used in muti-scale analysis. Gaussian filter is given by

$$h = [\,\frac{c}{2}\ \ \frac{b}{2}\ \ a\ \ \frac{b}{2}\ \ \frac{c}{2}\,]^T \times [\,\frac{c}{2}\ \ \frac{b}{2}\ \ a\ \ \frac{b}{2}\ \ \frac{c}{2}\,] \qquad (7)$$

where $a+b+c = 1$ and $a+c = b$ to ensure the normalization of an image and the central point contributing more than the surrounding points (J A Hne, 2004).

Gaussian filter saves the details in low-scale image, and also suppresses image noise. Figure 2 is the results of image registration in Figure 5 using CPS methods when $\lambda = 3$.

It is obvious that Figure 2b without filtering in low-scale image has many noise waves and it is hard to find the true result. Therefore, using Gaussian filter can save details in low scale images, it also reduce the noise of the images and improve the anti-noise capacity in registration.

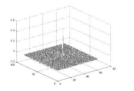 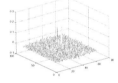

(a) With filtering   (b) Without filtering

Figure 2.   Comparison of registration result between sub-sampling and guassian sub-sampling image.

2.3 GMFR algorithm

The Gaussian filtering muti-scale expression of reference and sensed image are denoted by $f(x, y; \lambda)$, $g(x, y; \lambda)$ respectively, for $\lambda = 0, 1, 2,..., s$. $s = max\ \lambda$ refers to the lowest scale of the image in muti-scale analysis. The parameter s cannot be set to small value; it should depend on the size of original image and ensure that the lowest scale image have enough details of original image. (x_{0s}, y_{0s}) is registration result of s-scale image using CPS. After that, computing the local CC in spatial domain in (s-1)-scale images in the range of $(2x_{0s} - \Delta x : 2x_{0s} + \Delta x, 2y_{0s} - \Delta y : 2y_{0s} + \Delta y)$ In muti-scale analysis the size of single direction of (s-1)-scale images is two times of s-scale, so $min\ \Delta x = 2, min\ \Delta y = 2$. Only 5×5 local area of image CC is computed in (s-1)-scale and given by

$$R_{fg}(x, y; \lambda) = \frac{1}{5 \times 5} \sum_{m=2x_{0s}-2}^{2x_{0s}+2} \sum_{n=2y_{0s}-2}^{2y_{0s}+2} f^*(m, n; \lambda) g(x+m, y+n; \lambda) \qquad (8)$$

Repeat the procedure of (8) until iterate to $\lambda = 0$, the original scale of the image. Then the final registration result (x_0, y_0) is obtained.

3 THE PERFORMANCE OF GMFR ALGORITHM

To simplified the expression, the size of original image is modified to $M \times M$. In computational complexity analyzing, without regarding to add operation, only multiplication is taken into account. $OCC(M)$ denotes multiplication times in one CC of two same size images, and is given by

$$O_{CC}(M) = M^4 \qquad (9)$$

To complete a DFT of a grayscale image, M^4 times multiplication are needed. If using FFT, multiplication times could be reduced to $(Mlog_2M)^2$. $O_{CPS}(M)$ denotes multiplication times in one CPS of two same size images, and is given by

$$O_{CPS}(M) = 2 \times [(M\log_2 M)^2 + 4M^2 + 4(M\log_2 M)^2] \qquad (10)$$
$$= 2 \times [5(M\log_2 M)^2 + 4M^2]$$

The first item refers to multiplication of FFT; the second item denotes complex multiplication in frequency domain; the last item refers to complex multiplication of iFFT;

To complete the proposed GMFR algorithm, Let $s = 3$, the required number of multiplication times is denoted by

$$O_{GMFR}(M) = 2[5(\frac{M}{2^3}\log_2 \frac{M}{2^3})^2 + 4(\frac{M}{2^3})^2] + \frac{525}{16}M^2 \qquad (11)$$

Let $M = 2^n$, define a performance function P

$$P_{CC/GMFR}(n) = \frac{O_{CC}(M)}{O_{GMFR}(M)} = \frac{2^{2n}}{\dfrac{10(n-3)^2}{2^6} + \dfrac{527}{16}} \qquad (12)$$

Two P function curves are shown in Figure 3. The advantage increases rapidly as a function of n. For instance, when $n = 10$ (1024 pixels per image), the GMFR has nearly a 2.6×10^4 to 1 advantage over CC. Compared to CPS, the advantage of GMFR is obvious since the upper curve increases rapidly than the lower one. The performance of GMFR is more significant if the images, which are to be registered, are large.

4 MACHINE VISION ENVIRONMENT AND EXPERIMENT

A vision environment is built for acquiring lead frame images in the using of its defect detection. To maximize verification of GMFR performance, a low-cost

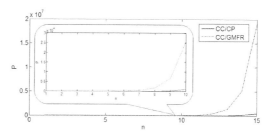

Figure 3.   Algorithms Performance function P curve.

area-CCD camera and white light source are used. The machine vision hardware environment is shown in Figure 4.

Figure 4.   Image acquisition system of machine vision.

Figure 5a-5b are acquired from the machine vision in Figure 4, where Figure 5a refers to reference image and Figure 5b refers to sensed and   noised image and expression of muti-scale. Here, let $s = max \; \lambda = 3$. Figure 5c are registered results of Figure 5b using GMFR and Figure 5d-5e are the registration results of each scale image respectively.

The shaded area of Table 1 is the final result of GMFR. GMFR algorithm has the same accuracy with CC algorithm. When sensed images are noised, GMFR could also maintain the correct result and accuracy. Using central transformation (6), the registration parameter (x_0, y_0) is: (-60,-39).

Table 2 gives a comparison of computational time between GMFR, CPS and CC under the same experiment environment. The advantage of GMFR in Table 2 is not as obvious as that in Figure 3. That is because the computational time includes image preprocessing, camera driving, etc. However, the efficiency of GMFR is improved obviously.

5 CONCLUSION

This paper proposes a novel GMFR algorithm to solve the problem of low efficiency of image registration algorithm in order to fulfill the real-time requirements in industry machine vision. The GMFR dramatically

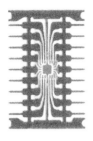 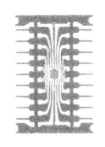 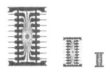 

(a) Template image   (b) Image to be registered & muti-scale expression (noised)   (c) Registration result

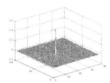   

(d) λ = 3 CPS result   (e) λ = 2 local CC result   (f) λ = 1 local CC result   (g) λ = 0 local CC result

Figure 5.   Registration results of lead frame image using GMFR algorithm (noised) (size:768*608).

Table 1.   Results of GMFR algorithm.

Scale	CPS result	Local CC result	result
λ = 3	(30,43)		(30,43)
λ = 2		(4,3)	(61,86)
λ = 1		(3,4)	(122,173)
λ = 0		(3,2)	(244,345)

Table 2.   Computational time of three algorithms.

Algorithm	GMFR	CC	CPS
Time(s)	0.586639	8.55176s	0.847500

improves the efficiency and anti-noise capacity of registration algorithm. A performance function is defined to analyze the computational complexity. The experiment results of lead frame image demonstrate that the GMFR has good robustness when dealing with noised image, and the efficiency is improved obviously. The proposed method could make machine vision more applicable and extensive in industry.

REFERENCES

[1] DAVIES, E. R. (2004) Machine vision: theory, algorithms, practicalities, Elsevier.
[2] GONZALEZ, R. C., WOODS, R. E. & EDDINS, S. L. (2004) Digital image processing using MATLAB. Upper Saddle River, NJ Jensen: Prentice Hall,.
[3] GOSHTASBY, A. A. (2012) Image registration methods. Image Registration., Springer.
[4] J A HNE, B. (2004) Practical handbook on image processing for scientific and technical applications.
[5] KANEKO, S., SATOH, Y. & IGARASHI, S. (2003) Using selective correlation coefficient for robust image registration. Pattern Recognition, 36, 1165–1173.
[6] MADISETTI, V. (2010) Digital signal processing fundamentals, CRC press.
[7] WANG, C., JIANG, B. C., LIN, J. & CHU, C. (2013) Machine Vision-Based Defect Detection in IC Images Using the Partial Information Correlation Coefficient. Semiconductor Manufacturing, IEEE Transactions on, 26, 378–384.
[8] YEN, H. & SIE, Y. (2012) Machine vision system for surface defect inspection of printed silicon solar cells..
[9] ZITOVA, B. & FLUSSER, J. (2003) Image registration methods: a survey. Image and vision computing, 21, 977–1000.

Testing and Measurement: Techniques and Applications – Chan (Ed.)

A three-step model for Tsunami warning and evacuation framework

C. Srisuwan & P. Rattanamanee
Department of Civil Engineering, Faculty of Engineering, Prince of Songkla University, Hat Yai, Thailand

ABSTRACT: Evacuation seems to be the best and the only option for the human to deal with tsunami hazards. In this research work, a set of predictive models is developed and tested for use in a comprehensive framework for preparing for the evacuation. The models are set to provide an estimation of the tsunami arrival time, confirmation of the event, and prediction of the possible impact of the waves that are washing ashore. The models are integrated as a single unit which functions in priority of urgency. Validation of previous tsunami data shows that the framework would be able to fulfill the need in tsunami warning and evacuation plan with timely and accurate information.

1 INTRODUCTION

Tsunamis have been regarded as one of the most devastating natural disasters in the human history. Early occurrences of tsunamis were dated back to thousands of years, while recent events including the 2004 Indian Ocean and the 2011 Tohoku tsunamis were reported to cause over 200,000 and 16,000 deaths, respectively (Satake et al. 2006; Fuse and Yokota 2012). Tsunamis may be triggered by earthquakes, landslides, or volcanic eruptions, but earthquake-induced tsunamis are the most common and often most powerful threatening phenomena (Di Risio and Beltrami 2014).

Tsunamis are unpreventable, leaving evacuation as the best action for the human being to avoid possible catastrophe. Modeling of tsunami waves serves to simulate the event and provides essential information for warning and planning for the evacuation. Many sophisticated tsunami models are available which may be classified into arrival time models (Titov et al. 2005; Zhang et al. 2009), wave transformation models (Kowalik et al. 2005; Kirby et al. 2013), and wave run-up models (Tinti et al. 1994; Madsen and Fuhrman 2008). Real-time observation of tsunamis by use of in-situ or remote technology can also be utilized to better assist the tsunami warning and evacuation (e.g. Merrifield et al. 2005; Kulikov 2006).

In the present article, three types of models are developed and integrated for use in a new conceptual framework for avoiding casualties and damages due to tsunamis. The framework will be comprehensive, covering the tsunami generating until washing ashore, yet straightforward enough that a layperson could understand the computation processes. To ensure a timely and appropriate response, the modulus function in three hierarchical stages to permit: an early warning for the tsunami potential, a confirmation or cancellation of the warning, and a prediction of the possible impact. In the sections that follow, the tsunami wave theory is first outlined and the proposed models are described. The models are tested against available data and implementation of the new framework is discussed.

2 THEORETICAL DESCRIPTION OF TSUNAMIS

Tsunamis are surface gravity waves that are almost always considered shallow water waves based on the water depth (h) to wavelength (L) ratio. Linear wave theory may be adopted to describe tsunami waves with the governing equations that follow (e.g. Dean and Dalrymple 1991) where ∇^2 is the Laplacian operator imposed on the inviscid, irrotational velocity potential ϕ, together with the conservation of energy that yields in which ∇ is the gradient operator, and P is the fluid pressure. Both Equations (1) and (2) are applied over the entire water column $-h \geq z \geq \eta$, where η is the surface displacement assumed to feature the periodic form of where H_o and ω are the height and the angular frequency of the wave; k indicates the wave number ($2\pi/L$) in the wave propagation direction x.

$$\nabla^2 \phi = 0 \tag{1}$$

$$\frac{P}{\rho} = -\frac{\partial \phi}{\partial t} - gz - \frac{1}{2}\nabla\phi \cdot \nabla\phi \tag{2}$$

$$\eta(t) = \frac{H_o}{2}\exp(-i[kx + \omega t]) \tag{3}$$

With appropriate assumptions and boundary conditions, the above set of governing equations can be linearized to allow the one-dimensional shallow water wave equation where c is the wave celerity or the wave phase speed.

$$\frac{\partial^2 \eta}{\partial t^2} - c^2 \left(\frac{\partial^2 \eta}{\partial x^2} \right) = 0 \tag{4}$$

Nonlinear wave theory is sometimes argued to be a better approach to describe tsunami waves (e.g. Titov et al. 2005; Kirby et al. 2013), particularly the solitary wave solution of a permanent wave form of infinite wave length, analogous to the presence of linear shallow water waves. Such a wave, for example, may be governed by the well-known Boussinesq wave equation (Daily and Stephan Jr 1952) in which $o^{\varepsilon+}$ implies higher-order derivative terms, with the theoretical wave profile η given following

$$\frac{\partial^2 \eta}{\partial t^2} - c^2 \left(\frac{\partial^2 \eta}{\partial x^2} \right) - \frac{3c^2}{h} \left(\frac{\partial \eta}{\partial x}^2 + \eta \frac{\partial^2 \eta}{\partial x^2} \right) + o^{\varepsilon+} = 0 \tag{5}$$

$$\eta(t) = H_o \operatorname{sech}^2 \left[\frac{ct}{h} \sqrt{\frac{3}{4} \frac{H_o}{H_o + h}} \right] \tag{6}$$

The wave solutions outlined here can be solved for important wave parameters required in the description of tsunami wave motion. These may include wave celerity, water particle velocity, wave energy dissipation, and some others that are associated with wave transformation processes.

3 HIERARCHICAL MODELS FOR TSUNAMI WAVES

An efficient tsunami evacuation plan requires timely and accurate information of the propagating waves. The work presented here aims at introducing a new framework that employs three hierarchical models as shown in Figure 1. The models will be set to function upon the urgency of the evacuation preparation at each stage.

The first model in the system will be for the prediction of the tsunami arrival time, which sends an alert to the users immediately. The verification model will then be executed to confirm or cancel the warning of the tsunami waves. In the case of confirmation, the third model will perform to predict the possible impact of the waves in the area where the user is located. Primary concepts and computation techniques in these models are described below.

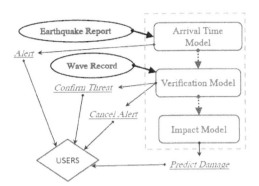

Figure 1: Overview of the models and their functions for use as a tsunami warning and evacuation framework.

3.1 Arrival time model

For life safety, the ultimate goal for the task here is to provide the most immediate alert for the time remaining before the waves start to wash ashore. The time estimation, therefore, should be relied on a simple calculation based on the wave traveling distance and its phase speed.

Typically, seismic monitoring networks can detect an earthquake and its epicenter in an order of few to ten minutes (e.g. Bondár et al. 2004), allowing a reasonable estimation of the tsunami propagating distance to a particular location. A few options for the estimation of the wave phase speed are outlined and evaluated here among the choices of wave theory and configuration of the wave propagation domain.

If the linear wave theory is used (i.e. Equations (1) and (2)), the wave phase speed can be determined using the dispersion relation of the wave to follow where c is the phase speed or wave celerity. For a long wave with $kh < 1/4$, the hyperbolic tangent of kh will asymptotically approach the value of kh itself and the celerity thus reduces to $c = \sqrt{gh}$, which is the shallow water linear wave celerity. Due to its appearance, tsunamis may also be described by the solitary wave theory (i.e. Equations (3) and (4)), with the wave celerity given as (Rayleigh1976) in which the major difference to the linear wave solution is that the wave height H_o is required to be known.

$$c^2 = \frac{g}{k} \tanh(kh) \tag{7}$$

$$c^2 = gh \left(1 + \frac{H_o}{h} \right) \tag{8}$$

Besides the choice of wave theory, determination of the water depth h is also crucial to the wave celerity and three options may be outlined here. The first and the most ideal one would be to utilize actual bathymetry data which, however, are not always available and also make the wave propagation path become another dependency. Another option is to assume a single depth for the entire domain, such as an average depth in an oceanic region. This method may be claimed to be sufficient for the task (e.g. Zhang et al. 2009), but a third method is introduced here for describing a more realistic bathymetry of the domain near the shoreline.

In the new method suggested, the depth profile from a location of the epicenter will be generated from the shoreline with a constant local slope until the depth approaches the mean depth of the ocean. Figure 2 shows tsunami arrival times computed using this method compared to that of the constant depth assumption. The differences between the results depend on the nearshore slope and, more primarily, on how close to the shore the tsunami has occurred. It is found that if the traveling distance is closer than 100 km, such a distance could be up to 10 times, but then vanishes totally for any distance greater than 10,000 km.

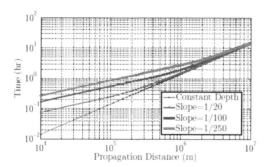

Figure 2: Comparison between arrival times estimated using the constant depth assumption with and without inclusion of nearshore slope.

Figure 3 shows a set of tsunami arrival times estimated using the linear and the solitary wave theories in comparison to the actual arrival times from observation. The estimation is performed for the 2004 Indian Ocean Tsunami based on available data and suggested parameters in Merrifield et al. (2005), Kowalik et al. (2005), and Zhang et al. (2009), summarized in Table 1. It turns out the results from the two wave theories that are not showing any visible difference at all because the wave height to water depth ratio (H_0/h) is very small, $\approx O(10^{-4})$. The estimation results agree very well with the observed arrival times with the duration differences all lower than 5% except for the two furthest locations where the errors are around 15%.

3.2 Verification model

Once a tsunami threat is alerted, the actual occurrence of the waves should be confirmed and announced again to the users. This necessity arises under the facts that not all earthquake generates tsunamis, and also that the waves could be so small that an evacuation is clearly not needed. A modeling technique for executing this task is outlined here as part of the tsunami warning and evacuation framework.

3.3 Verification Model

Once a tsunami threat is alerted, the actual occurrence of the waves should be confirmed and announced again to the users. This necessity arises under the facts that not all earthquakes generates tsunamis, and also that the waves could be so small that an evacuation is clearly not needed. A modeling technique for executing this task is outlined here as part of the tsunami warning and evacuation framework.

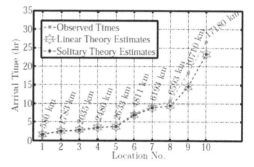

Figure 3: Arrival times computed using linear and solitary wave solutions compared to observed values.

Table 1: Locations where arrival times of the 2004 Indian Ocean Tsunami are estimated as illustrated in Figure 3.

	Location	Coordinates	Distance (km)
1	Phuket, Thailand	7.88°N, 98.40°E	580
2	Colombo, Srilanka	5.93°N, 79.97°E	1,783
3	Chennai, India	13.04°N, 80.17°E	2,035
4	Male, Maldive	4.18°N, 73.52°E	2,480
5	Diego Garcia, British Indian	7.28°S, 72.40°E	2,633
6	Salalah, Oman	16.93°N, 54.00°E	4,811
7	Lamu, Kenya	2.27°S, 40.90°E	6,193
8	Portland, New Zealand	38.33°S, 141.60°E	6,593
9	Jackson Bay, New Zealand	43.98°S, 168.62°E	10,710
10	Arica, Chile	18.22°S, 70.21°E	17,180

The proposed model will be set to work based on real-time records of ocean surface displacement, such as those obtained from existing wind-wave gauges (e.g. Cecioni et al. 2014; Di Risio and Beltrami 2014). Considering any waves from any directions, this surface displacement may be described in the form of Fourier series in which the subscript n indicates each frequency component of the waves; A_n and B_n are Fourier coefficients related to the magnitudes of the waves. In a time series of the displacement, three contributions may include short waves, tides, and tsunamis, if exist, with different frequency bands in the orders of seconds, minutes to a few hours, and quarter to half a day, respectively. These differences in the frequency bands are illustrated in Figure 4 where a time series is simulated based on parameters from the 2004 Indian Ocean Tsunami (Merrifield et al. 2005).

$$f(t) = \sum_{n=0}^{N} A_n \cos(\omega_n t) + B_n \sin(\omega_n t) \tag{9}$$

It should be insightful to note that that the distinction in Figure 4 appears obvious only when the wave components are separated; a wave sensor can never see or provide a notice of it without a special algorithm. For detection and quantification of tsunamis, the technique suggested here involves a transformation of the wave time series into wave energy spectra in a frequency domain. Each type of the waves can then be analyzed according to their frequency bands with a threshold level of energy density that indicates the existence of tsunamis.

analysis of the above relations. Figure 5 shows a set of wave spectra derived over the time for a time series combining the short waves and the tsunami waves simulated in Figure 4; the tide was neglected since practically it could be filtered out from a time series easily and very accurately (e.g. Pawlowicz et al. 2002).

$$A_n = \sqrt{(E(\omega_n)\Delta\omega} * \cos(\Psi) \tag{10}$$

$$B_n = \sqrt{(E(\omega_n)\Delta\omega} * \sin(\Psi) \tag{11}$$

In Figure 5, the analyzing result shows that the level of wave energy in the tsunami frequency band does not rise before the third hour when the first and the largest tsunami wave starts to present in the time series. As the sequential tsunami waves occur, some significant levels of the energy still appear in the frequency band. In the late hours, the attenuation of the tsunami wave amplitudes may cause some reduction in the energy densities, but the existence of such tsunami waves is still evident compared to the resulting spectra without them.

Since the technique presented here works in a quasi real-time sense, i.e. a sectioning of the time series is required, the spectral analysis may be completed repeatedly for two purposes. One is to provide a quick confirmation of the tsunamis as soon as the first wave arrives at the measurement station, justified upon the threshold of the energy intensity. Meanwhile, the other purpose is to quantify the magnitude of the waves which is then passed through the next model for the prediction of the possible impact of the phenomenon.

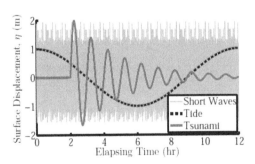

Figure 4: Example of time servies simulation representing wind waves, tide, and tsunamis.

In relation to a surface wave energy spectrum E, represented as a function of wave frequency, the Fourier coefficients An and Bn in Equation (9) may be described following where Ψ is a random phase function. The determination of the wave spectra may simply be achieved by use of the fast Fourier transform technique which numerically applies a Fourier

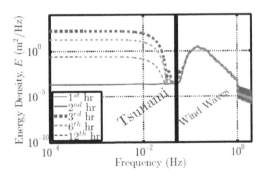

Figure 5: Surface wave energy spectra derived according to the time series simulated in Figure 4.

3.4 Impact model

In a tsunami evacuation plan, a misperception is often found when the waves are analyzed according to their size and speed rather than the impact that they could make. This fact leads to the introduction of the impact model here for the prediction of possible initial damage once the waves wash ashore.

The model is developed to focus on the spatial extent that the waves could reach, represented by wave run-up elevation. This factor is intended for guiding people to evacuate from any risky are-as-rather than to make a decision based on possible damage in such areas.

The domain setup employs a sloping beach connected to a constant offshore depth (h_o) as adopted in the arrival time model. Assuming a linear monochromatic wave, this setup allows the solution for the surface displacement in the offshore region ($h = h_o$) that follows (Keller and Keller 1964)

$$\eta_o(x,t) = \frac{H_o}{2}\exp\left[-i(\omega t + k_o x)\right] + i\left(\frac{H_r}{2}\right)$$
$$\exp\left[-i(\omega t - k_o x + \varepsilon)\right] \tag{12}$$

in which H_o and H_r are the incident and reflected wave heights; z is a random phase difference; and k_o is the non-dimensional wave number equal to $\omega/\sqrt{gh_o}$. Note that x is directed shoreward from the shoreline. In the nearshore zone with a sloping bottom, the surface displacement may be solved based on the linear long wave equation (Equation (4)), which yields (Lamb 1932) where R is the wave amplitude at the shoreline taken also as the run-up elevation; $J_0(\sigma)$ is the zeroth-order Bessel function of the first kind as a function of the transformation parameter

$$\eta_s(x,t) = iRJ_0(\sigma)\exp\left[-i\omega t\right] \tag{13}$$

$$\sigma = 2\omega\sqrt{\frac{x}{gS}} \tag{14}$$

with S representing the bottom slope. In order to estimate the wave run-up based on the offshore wave height, Equations (12) and (13) are matched at the toe of the sloping profile ($x = x_o$) under the hydrostatic and kinematic conditions that respectively follow

$$\eta_o(x_o, t) = \eta_s(x_o, t) \tag{15}$$

and

$$\frac{\partial}{\partial x}\left[\eta_o(x_o, t)\right] = \frac{\partial}{\partial x}\left[\eta_s(x_o, t)\right] \tag{16}$$

which lead to the relation

$$RJ_0(\sigma)\exp\left[-i\omega t\right] = \frac{-iH_o}{2}\left[\frac{2J_0(\sigma)}{J_0(\sigma_o) - iJ_1(\sigma_o)}\right]$$
$$\exp\left[-i(\omega t + k_o x_o)\right] \tag{17}$$

where $J_0(\sigma)$ is the first-order Bessel function of the first kind and σ_o is the factor in Equation (14) evaluated at $x = x_o$. For convenience, this relation may be rearranged to read which is simply the wave run-up to the offshore wave height ratio. The Bessel functions involved may be eliminated using an asymptotic approximation method, but they are kept here since a modern computer is able to evaluate them promptly.

$$\frac{R}{H_o} = \left|\frac{1}{J_0(\sigma_o) - iJ_1(\sigma_o)}\right| \tag{18}$$

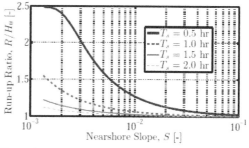

Figure 6: Wave run-up to wave height ratio estimated by use of Equation (18) for the prediction of tsunami impact.

Figure 6 shows the wave run-up to wave height ratio computed using Equation (18), and its variation of the nearshore slopes and tsunami wave periods. The wave run-up appears to increase monotonically, but not linearly, for milder slopes and shorter wave periods. In a geometric sense, these sensitivities may be interpreted by the relative extent of sloping shelf and the wavelength (L). For example, a very long period, wave would feature an extended wave length to which any cross-shore profile would appear and react as a vertical section which leads to a reduction of the wave runup.

4 IMPLEMENTATION GUIDELINES

The three models introduced above are proposed for uses in a tsunami warning and evacuation framework. The framework is intended to be partially self-directed, promoting awareness of individual users and supporting them during the event. The information provided is optimized in priority of urgency and should be sent out to personal handheld devices.

Here, the models are reviewed with suggestions for their implementations.

The arrival time model is set to receive an earthquake detection and estimates the wave traveling time between the epicenter and the user's location. The distance between these two positions should be obtained easily via GPS. The depth of the ocean of the tsunami origin and the slope of the local shoreline is also required, which practically should be prepared in advance as database. These parameters can then be called based on the epicenter and the GPS locations. This model should be configured to work independently and the alert should be sent to the user most rapidly.

The verification model will be activated by the tsunami alert. The most important input here is the ocean surface time series, which should be obtained from the wave measurement station closest to the tsunami epicenter. Overlapping sections of the time series, e.g. of duration up to 1 hour, are repeatedly processed for wave energy spectra. The existence of tsunamis is then identified based on energy in the corresponding frequency band and a confirmation of the event must readily be sent out to the user. Cancellation of the warning is also permitted when such energy is not present, but this action should be proceeded with a very serious caution.

The analysis of the ocean surface time series is also aimed for estimating magnitudes and periods of the tsunamis. Note that this should be done separately from the tsunami detection as explained previously in the model description. These wave parameters are utilized by the impact model to predict possible damage to the coastal zone using the wave run-up elevation as a nominal indicator. It is suggested that the numeric result be interpreted to descriptive word for the warning, e.g. lowest to highest threat. This type of information should be helpful not only for people seeking evacuation but also for emergency managers for search and rescue purposes.

5 CONCLUSIONS

The warning and evacuation plan for a tsunami disaster requires timely and accurate information. The framework that is introduced in this study provides such information using three types of models developed according to priority in responding to the event. In the model tests, arrival times of the waves, which seem to be the most critical factor, were predicted by a linear wave model of a simplified domain with typical errors of less than 5%. It is proposed that the user be first alerted with this estimated duration.

Surface wave data collected at a nearby measurement station are set to be analyzed in another model

to confirm an existence of the waves. The difference in the frequency bands between tsunamis and other waves is the key in the identification. This technique was applied on wave energy spectra derived using available data from the 2014 Indian Ocean Tsunami where the tsunami waves could be clearly distinguished.

Possible impact due to the approaching waves is also set to be predicted and provided in the framework. The prediction is achieved in the other model which utilizes the linear long wave solution to solve for wave run-up elevation that is treated as a representative indicator for the impact. A sensitivity test on the model shows that the threat becomes higher for a shorter tsunami wave period and a milder nearshore slope. This variation agrees with both theory and observation of tsunami wave run-up.

Some important notes on the implementation of the framework are outlined in this article, but another step forward toward such an important task would be appropriate still. Acquisition and preparation of the required inputs are the two most challenging topics to be discussed. Validation of the models and evaluation of the framework based on other tsunami events would also be necessary before their active use for actual tsunami warning and evacuation.

REFERENCES

Bondár, I., S. C. Myers, E. R. Engdahl, and E. A. Bergman (2004). Epicentre accuracy based on seismic network criteria. *Geophysical Journal International 156*(3), 483–496.

Cecioni, C., G. Bellotti, A. Romano, A. Abdolali, P. Sammarco, and L. Franco (2014). Tsunami early warning system based on real-time measurements of hydro-acoustic waves. *Procedia Engineering 70*, 311–320.

Daily, J. W. and S. C. Stephan Jr (1952). The solitary wave: its celerity, profile, internal velocities and amplitude attenuation in a horizontal smooth channel. *Coastal Engineering Proceedings 1*(3), 13–30.

Dean, R. and R. Dalrymple (1991). *Water wave mechanics for scientists and engineers*, Volume 2. World Scientific, Singapore.

Di Risio, M. and G. Beltrami (2014). Algorithms for automatic, real-time tsunami detection in wind-wave measurements: Using strategies and practical aspects. *Procedia Engineering 70*, 545–554.

Fuse, A. and H. Yokota (2012). Lessons learned from the japan earthquake and tsunami, 2011. *Journal of Nippon Medical School 79*(4), 312–5.

Keller, J. B. and H. B. Keller (1964). Water wave run-up on a beach. Technical report, ONR Research Rep. Contract NONR-3828(00). Dept. of the Navy, Washington, D.C. 40 pp.

Kirby, J. T., F. Shi, B. Tehranirad, J. C. Harris, and S. T. Grilli (2013). Dispersive tsunami waves in the ocean: Model equations and sensitivity to dispersion and coriolis effects. *Ocean Modelling 62*, 39–55.

Kowalik, Z., W. Knight, T. Logan, and P. Whitmore (2005). Numerical modeling of the global tsunami: Indonesian tsunami of 26 december 2004. *Science of Tsunami Hazards 23*(1), 40–56.

Kulikov, E. (2006). Dispersion of the Sumatra tsunami waves in the Indian Ocean detected by satellite altimetry. *Russian Journal of Earth Sciences 8*.

Lamb, H. (1932). *Hydrodynamics*. Cambridge University, Cambridge.

Madsen, P. A. and D. R. Fuhrman (2008). Run-up of tsunamis and long waves in terms of surf-similarity. *Coastal Engineering 55*(3), 209–223.

Merrifield, M., Y. Firing, T. Aarup, W. Agricole, G. Brundrit, D. Chang-Seng, R. Farre, B. Kilonsky, W. Knight, L. Kong, et al. (2005). Tide gauge observations of the Indian Ocean tsunami, December 26, 2004. *Geophysical Research Letters 32*(9).

Pawlowicz, R., B. Beardsley, and S. Lentz (2002). Classical tidal harmonic analysis including error estimates in mat-lab using t tide. *Computers & Geosciences 28*(8), 929–937.

Rayleigh, L. (1876). On waves. *The London, Edinburgh, and Dublin Philosophical Magazine and Journal of Science 1*(5), 257–279. Satake, K., T. T. Aung, Y. Sawai, Y. Okamura, K. S. Win, W. Swe, C. Swe, T. L. Swe, S. T. Tun, M. M. Soe, et al. (2006). Tsunami heights and damage along the Myanmar coast from the December 2004 Sumatra-Andaman earthquake. *Earth Planets and Space 58*(2), 243–252.

Tinti, S., I. Gavagni, and A. Piatanesi (1994). A finite-element numerical approach for modeling tsunamis. *Annals of Geophysics 37*(5), 193–212.

Titov, V., F. Gonzlez, E. Bernard, M. Eble, H. Mofjeld, J. Newman, and A. Venturato (2005). Real-time tsunami forecasting: Challenges and solutions. In *Developing Tsunami-Resilient Communities*, pp. 41–58. Springer Netherlands.

Zhang, D. H., T. L. Yip, and C. O. Ng (2009). Predicting tsunami arrivals: Estimates and policy implications. *Marine Policy 33*(4), 643–650.

Testing and Measurement: Techniques and Applications – Chan (Ed.)
© *2015 Taylor & Francis Group, London, ISBN: 978-1-138-02812-8*

The wear recognition on guide surface based on the feature of radar graph

Y.H. Zhou, W.M. Zeng & Q. Xie
School of Mechanical Engineering, Xiangtan University, China

ABSTRACT: In order to solve the wear recognition problem of the machine tool guide surface, a new machine tool guide surface recognition method based on the radar-graph barycenter feature is presented in this paper. Firstly, the gray mean value, skewness, projection variance, flat degrees and kurtosis features of the guide surface image data are defined as primary characteristics. Secondly, data Visualization technology based on radar graph is used. The visual barycenter graphical feature is demonstrated based on the radar plot of multi-dimensional data. Thirdly, a classifier based on the support vector machine technology is used, the radar-graph barycenter feature and wear original feature are put into the classifier separately for classification and comparative analysis of classification and experiment results. The calculation and experimental results show that the method based on the radar-graph barycenter feature can detect the guide surface effectively.

1 INTRODUCTION

Guide surface wear is the main reason for the decline of machine accuracy in the process of working. Machine vision detection technology is used to identify whether a guide surface has been worn or not and the degree of wear, to analyze the reasons of guide surface wear, and then adopt corresponding measures to alleviate wear or repair guideway, it is of significance in ensuring the quality of product processing [1,2,3].

From the perspective of machine vision detection, guide surface wear can be regarded as surface defects [4]. Some data features, extracted from the two-dimensional image of guideway, are used to characterize the various types of guide surface wear condition. These data features mainly include gray, geometry, projection, texture, image sequences and other physical features.

The conditions of guide surface wear are complex, and the types of defect images are various, one kind of data feature of image defect is difficult to comprehensively reflect them. So the type of data features of image defects should be chose reasonably to reduce the calculated amount of data feature extraction, it is the key step to realize the application of machine vision detection project on the guide surface wear successfully. The visualization, which can transform features data into graphs is an effective method to recognize the wear conditions on guide surface. The graph is much better to explore the internal relationship between data than other forms of visualization representation, researchers can truly observe their simulations and process results of practical problems with data visualization [5,6]. Therefore, this paper proposes a method of wearable recognition on the guide surface based on the feature of radar graph.

2 FEATURE SELECTION AND DATA EXTRACTION

Wear defects can be divided into three categories based on guide usage, corresponding to the normal (almost no wear), mild wear and severe wear (affect the results). A total of 390 sample images of linear guide surface has been achieved by means of the MV-VS1200 machine vision image processing platform in this study, numbered 1-390 unified composition image data sample set of guide surface wear, including normal 112, mild wear 186, severe wear 92, as is shown in Fig.1. The basic parameters are as follows: Pixel size 256x256, field of view of 1.5×1.1mm.

(a) normal (b) mild wear (c) severe wear

Figure 1. Image of guide surface sample.

For any of the guide surface image data, it can be expressed in the form of a matrix in machine vision, That is the defect image A can be expressed as gray value matrix form as Eq. 1. Where $a_{i,j}$ is the grey value of pixel point (i, j).

$$A = \begin{bmatrix} a_{1,1} & a_{1,2} & \cdots & a_{1,256} \\ a_{2,1} & a_{2,2} & \cdots & a_{2,256} \\ \vdots & \vdots & a_{i,j} & \vdots \\ a_{256,1} & a_{256,2} & \cdots & a_{256,256} \end{bmatrix}, (i,j=1,2,3,...,256) \quad (1)$$

Guide surface gray characteristics can be obtained by the gray histogram, which is a probability statistics of the rail surface gray distribution. $S(x_i)$ is defined as a number, which is the amount of pixels whose characteristic value of x_i in image A. The total amount of pixels is defined as N,

$N = \sum_i S(x_i)$. Then normalization processing is

made for $S(x_i)$, The feature data can be calculated as follows:

$$h(x_i) = \frac{S(x_i)}{N} = \frac{S(x_i)}{\sum_i S(x_i)} \quad (2)$$

$$v^2 = \sum_{x_i=0}^{n-1} (x_i - \overline{x}_i)^2 h(x_i) \quad (3)$$

The gray histogram of image A is $H(A) = [h(x_1), h(x_2), ..., h(x_n)]$, n is the amount of number whose features can be calculated. Then the gray mean value, skewness, kurtosis features are gotten in turn as follows:

$$\overline{x}_i = \sum_{x_i=0}^{n} h(x_i).x_i \quad (4)$$

$$H_s = \frac{1}{v^2} \sum_{x_i=0}^{n-1} (x_i - \overline{x}_i)^3 h(x_i), \quad (5)$$

$$H_k = \frac{1}{v^4} \sum_{x_i=0}^{n-1} (x_i - \overline{x}_i)^4 h(x_i) - 3 \quad (6)$$

The gray value of target point in this image is calculated numerically accumulate along the 90° direction, a set of image projection data can be obtained, the variance of data reflect the discrete degree of gray values.

$$F_{v^2} = \sum_j (\rho_j - \overline{\rho}_j)^2 b(j), (i,j=1,2,...,256) \quad (7)$$

Where $a_{i,j}$ is the gray value of pixel point (i,j), ρ_j is the gray value that pixel point j projecting onto the axis, $b(j)$ is the total gray values of pixel point j on the projection axis.

3 BARYCENTRE FEATURE OF RADAR-GRAPH

Radar graph, which is also called Bratu or Spider graph, is a modeling tool that reflects the qualitative problem with quantitative indicators. Radar can be able to rent the multi-dimensional data into a two-dimensional plane, it is easy to study the relationship between feature data points for samples and thus classified.

The steps of five features data represented by radar graph are as follows: Firstly, make a circle, the circumference is divided into five equal portions, drawing the corresponding 5 radius; Secondly, these five radius, in turn, are defined as the axis of data features and labeled with the appropriate calibration (0-1); Thirdly, five feature values of the sample image data were normalized processing, and marked on the corresponding coordinate axis respectively, and then connected them into a 5 five-sided polygon. Fig.2 is the radar-graph feature of three sample image (06,13,25) which are typical wear defects of guide surface.

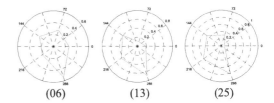

(06)　　　　(13)　　　　(25)

Figure 2. The feature of radar graph.

On the radar graph, a triangle can be composed by the two adjacent features data in irregular polygon and the center, each of which has a center of gravity. Its barycentre amplitude (distance from the center of graph) and angle features can be calculated as follows:

$$G_{ij} = \sqrt{\left(\frac{r_{ij}}{3}\sin\omega_{ij}\right)^2 + \left(\left(r_{ij}\cos\omega_{ij} - \frac{r_{ij+1}}{2}\right)/3 + \frac{r_{ij+1}}{2}\right)^2} \quad (8)$$

$$\theta_{ij} = \arcsin\frac{\frac{r_{ij}}{3}\sin\omega_{ij}}{G_{ij}} + \frac{2\pi(j-1)}{d}$$

Where i is the sequence of sample graph, $j = 1,...,5$. In the i graphics, r_{ij} is the distance between j barycentre feature and center, w_{ij} is the angle between feature and the 0° axis, d is the dimension of the data. The barycentre feature of radar graph are shown in Fig.3.

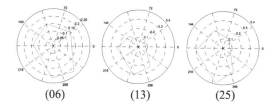

(06)	(13)	(25)

Figure 3.   The feature of radar graph.

4   EXPERIMENTAL VERIFICATION

Support vector machine (SVM) has been widely used as a mature image classification method, it is a machine learning method of small sample based on statistical learning theory, it can classify small sample effectively.

Training sample set is set to $x_i = (T_i, z_i), i = 1, 2, ..., 390$, x_i is the sample, $T_i = (G_{ij}, \theta_{ij})$ is the feature vectors that extracted above. $z_i \in \{1, 2, 3\}$ is the category label, the number of training samples is 390. Selection of "one to one" combination structure three categories of two types by SVM: normal and mild wear, mild wear and severe wear, normal and severe wear. Then the RBF kernel function and the optimal classification function are calculated as follows:

$$K(x, x_i) = \exp\left(-\frac{\|x - x_i\|^2}{2\sigma^2}\right) \tag{9}$$

$$f(x) = \text{sgn}\left(\sum_{i=1}^{l} \alpha_i^* y_i K(x, x_i) + b^*\right) \tag{10}$$

Where sgn( ) is the sign function, σ is the feature variance of sample data, $\alpha_i *$ is the lagrange multiplier corresponding to each sample, $b*$ is the classification threshold. Actually the non-support-vector samples corresponding $\alpha_i *$ is 0, sum only support vectors, its value determines the category of x_i.

The experiment was conducted on Matlab R2009b, with computer environment for Intel Core i5 3337U 1.8 GHz CPU and 2GB RAM. Three groups of images have been randomly selected each 20 as the training set, the training model to create two categories: the classification model of wear feature data and radar-graph barycentre feature data. Then in the 390 sample, 40 images are randomly selected for classification recognition tests, the results are shown in Tables 1 and 2.

Table 1.   Classification results based on wear feature.

Wear defects	Correct	False	Total	Accuracy
normal	18	2	20	90%
mild	12	3	15	80%
severe	2	3	5	40%

Table 2.   Classification results based on barycentre feature.

Wear defects	Correct	False	Total	Accuracy
normal	19	1	20	95%
mild	14	1	15	90%
severe	5	0	5	100%

5   CONCLUSIONS

In this work, wear features of the guide surface image is extracted, and be visualized with radar graph, Support vector machine technology is used for the classification study of the radar-graph barycentre feature. The following points can be concluded from this work:

1 Using the gray value, skewness, kurtosis, projection variance, area perimeter ratio data and others can effectively express the wear condition of the guide surface at different angles;
2 The features data, which are visualized with radar graph, can directly express the coupling relationship between the image feature and wear original feature of guide surface.
3 Results show that: the method based on the radar-graph barycentre feature can detect the guide surface more effectively than wear original feature.

ACKNOWLEDGMENTS

This study is supported by National Natural Science Foundation (Grant No51375419), and the Natural Science Foundation of Hunan Province and Xiangtan city, China (Grant No12JJ8010). Aid program for Science and Technology Innovative Research Team in Higher Educational Institutions of Hunan Province (Grant No2012-318).

REFERENCES

[1] S. Dutta, S.K. Pal and S. Mukhopadhyay, et al: Application of digital image processing in tool condition monitoring: A review[J]. CIRP Journal of Manufacturing Science and Technology, 2013, 6(3): 212–232.

[2] Hassan M H, Diab S L. Visual inspection of products with geometrical quality characteristics of known tolerances[J]. Ain Shams Engineering Journal, 2010, 1(1): 79–84.

[3] Tang Kai-yong. [J]. The abrasion of the slideway of the machine too l and its influence on the precision of themachine tool. Mechanical Research & Application, 2006,10(5):12~14.

[4] WANG Jin-jia, LI Jing, et al. Distinguishing Visual Feature Extraction Method Using Quadratic Map and Genetic Algorithm [J]. Journal of System Simulation, 2009,8(16):80~83.

[5] M. Chen, D. Ebert, H. Hagen, et al. Data Information and Knowledge in Visualization .IEEE Computer Graphics and Applications,2009,29 (1): 12~19.

[6] LIU Wen-yuan, LI Fang, HONG Wen-xue. Research on classifier of multi-dimensional data based on radar chart mapping[J].Computer Engineering and Applications, 2007,43(22):161–164.

Testing and Measurement: Techniques and Applications – Chan (Ed.)
© 2015 Taylor & Francis Group, London, ISBN: 978-1-138-02812-8

Moving object tracking in the low-rank representation

X.F. Kong, F.Y. Xu, H. Wang, G.H. Gu & Q. Chen
School of Electronic and Optical Engineering, Nanjing University of Science & Technology, Nanjing, China

ABSTRACT: Object tracking is one of the most important research directions in the image processing. However, most existing tracking algorithms only consider two adjacent frames at a time and ignore the continuity of the whole sequence. When the object is occluded, it is difficult to locate the object in the right trajectory after the occlusion. In this paper, an object tracking algorithm in the low-rank representation has been proposed to cope with the occlusion. We decompose the observation sequence into a low-rank matrix, which is composed of the moving object, and a sparse support matrix, which is composed of the occlusion. We use the variation between two adjacent frames to locate the moving object in the sequence, and the sparse support matrix to cope with the occlusion. Several experiments on real video sequences indicate that the object tracking algorithm in the low-rank representation works effectively under conditions of the occlusion.

KEYWORDS: Object tracking, Low rank, Occlusion.

1 INTRODUCTION

Since the wide use of image processing, moving object tracking has become an important direction in the field of computer vision [1].

The proposed visual tracking algorithm in this paper is developed based on the point tracking. The key point of point tracking is to establish a correspondence between the candidate target and the reference target [2].

However, due to the occlusion, feature points in moving object may disappear, enter and leave the view field. In this situation, the trajectories of the moving object are always incomplete in typical point tracking methods.

In this paper, a robust moving object tracking algorithm in the low-rank representation [3] has been proposed. The main advantage is that it decomposes the sequence observation matrix into a low-rank matrix, which is composed of the moving object, along with a sparse support matrix, which is composed of the occlusion. We use the variation between two adjacent frames to locate the moving object, and the sparse support matrix to cope with the occlusion, so that it can keep the trajectories of the moving object complete even in situations involving the occlusion.

2 RELATED WORK

In this section, we first discuss the most relevant methods to our work.

KLT [6] is one of the most widely used tracking methods for feature points. The KLT tracking method makes use of the spatial intensity gradient of the images to iteratively track feature points between frames. The main shortage of KLT tracking is that it only considers two adjacent frames at a time, and ignores the feature information of previous frames, so KLT usually fails when occlusions happen.

In order to avoid the loss of the feature points, we introduce the low-rank representation [3] into our method. The low-rank representation has proven that a matrix can be decomposed into a low rank matrix and a sparse matrix. In our algorithm, the low rank matrix is composed of the moving object and the sparse matrix represents the occlusion.

3 MOVING OBJECT TRACKING IN THE LOW-RANK REPRESENTATION

In this section, we focus on the problem of moving object tracking in the low-rank representation. First, we give the definitions of the matrices and variances we use in this paper. Then, we establish the models we use to track the moving object.

3.1 *Definitions*

In this paper, we use following notations. Given a video sequence, we choose the moving object in the first frame manually, and use a rectangle window with height h and width w to frame it. Then, we denote the same size in other frames, and use $I_j \in R^m$ to denote the j th frame in the video sequence, which is written as a column vector consisting of $m = h \times w$ pixels. The i th pixel in the j th frame is denoted as ij.

Denote $W_{m \times n} = [I_1, I_2, ..., I_n] \in R^{m \times n}$ to represent all n frames of the sequence, in which each column is a vector of one frame.

Denote $\tau = [d_{xx}, d_{xy}, d_{yx}, d_{yy}, dx, dy] \in R^{1 \times 6}$ to represent the deformation between two adjacent frames.

Denote $T_{m \times n} \in R^{m \times n}$, we call it the moving object matrix. It is a matrix with the same size of $W_{m \times n}$, and denotes the moving object in the video sequence.

Denote $S \in \{0,1\}^{m \times n}$, we call it the occlusion support matrix [3]. It is a binary matrix denoting the occlusion in the video sequence. Here we give the definition of S as follow:

$$S_{ij} = \begin{cases} 0, & \text{if } ij \text{ is in the moving object} \\ 1, & \text{if } ij \text{ is the occlusion} \end{cases} \quad (1)$$

We use $P_s(X)$ to represent the orthogonal projection of a matrix X onto the linear space of matrices supported by S, and $P_{s^\perp}(X)$ to be its complementary projection.

$$P_s(X)(i,j) = \begin{cases} 0, & \text{if } S_{ij} = 0 \\ X_{ij}, & \text{if } S_{ij} = 1 \end{cases} \quad (2)$$

$$P_s(X) + P_{s^\perp}(X) = X \quad (3)$$

3.2 Model establishments

Our goal is to estimate the moving object matrix T as well as the occlusion support matrix S, when given the sequence observation matrix W.

Based on the low-rank representation [3], a matrix can be decomposed into a matrix which has low rank K, along with a matrix which is sparse. In our algorithm, we decompose the sequence matrix W into a low-rank matrix T, along with a binary sparse matrix S with 0 and 1, which represents the occlusion. ε represents the Gaussian noise [4].

$$W = T + S + \varepsilon \quad (4)$$

Since the object is moving frame by frame, we need to compensate for the motion. Here, we use the 2D parametric transforms [5] to model the translation, rotation, and planar deformation of the moving object.

$$W \circ \tau = T + S + \varepsilon \quad (5)$$

3.2.1 Target model

Except for the occlusion, the intensity of the moving object should be unchanged over the sequence. Thus, moving object images in the tracking sequence are linearly correlated with each other, which can form a low-rank matrix T based on the low-rank representation. Besides the low-rank property, we don't make any additional assumptions on the moving object. Thus, we only impose the following constraint on T:

$$rank(T) \leq K \quad (6)$$

where K constrains the complexity of the target model.

3.2.2 Occlusion model

We denote that the occlusion on object in the video sequence composes the occlusion support matrix S. The occlusion is defined as any objects that move differently from the moving object or block the moving object in the view. It represents intensity changes that cannot be fitted into the low-rank matrix T of the moving object. Thus, we call them outliers in the low-rank representation. Usually, we have a prior knowledge that the occlusion should be sparse with relatively small size. The binary 0 and 1 in the occlusion support matrix S can be naturally modeled by a Markov Random Field [6]. Consider a graph $g = (v, \varsigma)$, where v is the set of vertices denoting all $m \times n$ pixels in the video sequence and ς is the set of edges connecting spatially or temporally neighboring pixels. Then, the energy of S is given by the Ising model [7]:

$$E(S) = \sum_{ij} u_{ij}(S_{ij}) + \sum_{ijkl,} \upsilon_{ij,kl} |S_{ij} - S_{kl}| \quad (7)$$

where u_{ij} denotes the unary potential of S_{ij} being 0 or 1, and the parameter $\upsilon_{ij,kl} > 0$ controls the strength of dependency between S_{ij} and S_{kl}. To prefer $S_{ij} = 0$ that indicates the sparse occlusion, we define the unary potential u_{ij} as

$$u_{ij}(S_{ij}) = \begin{cases} 0, & \text{if } S_{ij} = 0 \\ \upsilon_{ij}, & \text{if } S_{ij} = 1 \end{cases} \quad (8)$$

where the parameter $\upsilon_{ij} > 0$ penalizes $S_{ij} = 1$. For simplicity, we set $\upsilon_{ij} = \beta$ and $\upsilon_{ij,kl} = \gamma$, where $\beta > 0$ and $\gamma > 0$ are positive constants.

3.3 Optimization establishment

We propose to minimize the following energy to estimate T, S, and τ_n through combing the two models mentioned above:

$$\min_{T,S_{ij}\in\{0,1\}}\frac{1}{2}\sum_{ij:S_{ij}=0}\left([W_{1:n-1},D\circ\tau]-[T_{1:n-1},T_n]\right)^2$$
$$+\beta\sum_{ij}S_{ij}+\gamma\sum_{(ij,kl)\in\varepsilon}|S_{ij}-S_{kl}| \qquad (9)$$
$$s.t. \quad rank([T_{1:n-1},T_n])\leq K$$

Formulation (9) shows that the moving object should form a low-rank matrix and fit the observed sequence in the least squares sense except for the occlusion region that is sparse and contiguous. $D\circ\tau$ denotes the new coming frame, and $W_{1:n-1}$ denotes the 1th to $n-1$ th frame in the matrix.

To make the energy minimization easier to solve, we can use the nuclear norm to substitute the rank operator on T. It has been proven that the nuclear norm is an effective convex surrogate of the rank operator. Besides, it can avoid overfitting. Rewritten (9), we obtain the final form of the energy function:

$$\min_{T,S_{ij}\in\{0,1\}}\frac{1}{2}\left\|P_{S^\perp}\left([W_{1:n-1},D\circ\tau]-[T_{1:n-1},T_n]\right)\right\|_F^2$$
$$+\alpha\|T\|_*+\beta\|S\|_1+\gamma\|Avec(S)\|_1 \qquad (10)$$

where A is the node-edge incidence matrix of g, and $\alpha>0$ is a parameter associated with K.

4 ALGORITHM OF OPTIMIZATION EQUATION

Formulation (10) is nonconvex and it includes both discrete and continuous variables. It is extremely difficult to find out the solutions of τ_n, T, and S. Hence, we can adopt an alternating algorithm that separates the formulation over τ_n, T, and S into three steps.

4.1 Estimation of the deformation matrix τ

τ can be estimated by given $\hat{T}$ and $\hat{S}$ through iteratively minimizing formulation (10):

$$\hat{\tau}=\arg\min_{\tau}\|P_{\hat{S}^\perp}([W_{1:n-1},D\circ\tau]-\hat{T})\|_F^2 \qquad (11)$$

Here, we use the incremental refinement [8] to solve this parametric motion estimation problem. At each iteration, we update $\hat{\tau}$ by a small increment $\Delta\tau$ and linearize $D\circ\tau$ as $D\circ\hat{\tau}+J_{\hat{\tau}}\Delta\tau$, where $J_{\hat{\tau}}$ denotes the Jacobian matrix. Thus, τ can be updated in the following way:

$$\hat{\tau}\leftarrow\hat{\tau}+\arg\min_{\Delta\tau}\|P_{\hat{S}^\perp}$$
$$([W_{1:n-1},D\circ\hat{\tau}]-\hat{T}+J_{\hat{\tau}}\Delta\tau)\|_F^2 \qquad (12)$$

The minimization over $\Delta\tau$ in formulation (12) is a weighted least squares problem which has a close-formed solution [9].

4.2 Estimation of the low-rank matrix T

Given an estimate of the occlusion support matrix $\hat{S}$ and an estimate of the deformation matrix $\hat{\tau}$, the minimization in (10) over T turns out to be the matrix completion problem [10]:

$$\min_{T}\frac{1}{2}\left\|P_{\hat{S}^\perp}\left([W_{1:n-1},D\circ\hat{\tau}]-[T_{1:n-1},T_n]\right)\right\|_F^2+\alpha\|T\|_* \qquad (13)$$

The optimal T in formulation (13) can be computed efficiently through the algorithm of SOFT-IMPUTE [11], which makes use of the lemma proposed in Ref.[12]. The solution is given as:

$$\hat{T}\leftarrow\Theta_\alpha(P_{\hat{S}^\perp}([W_{1:n-1},D\circ\hat{\tau}])+P_{\hat{S}}(T)) \qquad (14)$$

4.3 Estimation of the occlusion support matrix S

The energy can be rewritten as follows:

$$\min_{T,S_{ij}\in\{0,1\}}\frac{1}{2}\left\|P_{S^\perp}\left([W_{1:n-1},D\circ\tau]-[T_{1:n-1},T_n]\right)\right\|_F^2+\beta\|S\|_1+\gamma\|Avec(S)\|_1$$
$$=\frac{1}{2}\sum_{ij}\left([W_{1:n-1},D\circ\hat{\tau}]-\hat{T}\right)^2(1-S_{ij})+\beta\sum_{ij}S_{ij}+\gamma\|Avec(S)\|_1 \qquad (15)$$
$$=\sum_{ij}(\beta-\frac{1}{2}\left([W_{1:n-1},D\circ\hat{\tau}]-\hat{T}\right)^2)S_{ij}+\gamma\|Avec(S)\|_1+C$$

$$C=\frac{1}{2}\sum_{ij}\left([W_{1:n-1},D\circ\hat{\tau}]-\hat{T}\right)^2 \qquad (16)$$

where C is a constant when $\hat{\tau}$ and $\hat{T}$ are fixed. The above energy in formulation (15) is in the standard form of the first-order MRFs with binary labels [13], which can be solved exactly using graph cuts [14].

4.4 Summary of the tracking algorithm

All steps of the algorithm we proposed are summarized below.

Algorithm: Moving Object Tracking in Low-Rank Representation.

1 Locate the moving object in the first frame manually. The size of the window is $h\times w$.
2 Use KLT tracking method to track the moving object when the number of frame n is less than 15. When n is larger than 15, our algorithm begins, go to step 3.

3 Initialize $W = [I_1, I_2, ..., I_n]$, $\hat{\tau} = 0$, $\hat{T} = T$, $\hat{S} = S$. We first give a rough estimate to the rank of the target model $K = \sqrt{n}$ in formulation (6). We initialize α to be the second largest singular value of W, initialize a relatively large β, and $\eta_1 = \eta_2 = 0.5$.

4 Use formulation (12) to estimate τ, until convergence.

5 Use formulation (13) to estimate T, until convergence.

6 If $rank(\hat{T}) \leq K$, we reduce α by a factor $\eta_1 < 1$ and repeat SOFT-IMPUTE until $rank(\hat{T}) > K$

7 Use the variance of $W_{ij} - \hat{T}_{ij}$ to estimate $\hat{\sigma}^2$.

8 Reduce β by a factor $\eta_2 = 0.5$ after each iteration until β reaches $4.5\hat{\sigma}^2$.

9 Use formulation (15) to estimate S, until convergence.

10 Output $\hat{\tau}, \hat{T}, \hat{S}$.

5 SIMULATIONS AND EXPERIMENTS

In this section, we present several experiments to compare our moving object tracking algorithm with the KLT tracking method.

The sequence we use called "dudek". It has 300 frames with a person sitting in front of the desk. From the 100th frame to the 110th frame, the person used his right hands to cover his face. After the 110th frame, the occlusion disappears. Figure 1 shows the result of our algorithm. At the beginning of the sequence, we locate the face by a red bounding box in the size of 110×130. The yellow crosses denote truth data on the face and the red crosses are the tracking results. Figure 2 shows the result of the KLT tracking algorithm.

Figure 1. The result of our algorithm. From left to right, top to bottom are the 1th, 68th, 105th, 114th, 116th and 217th frame.

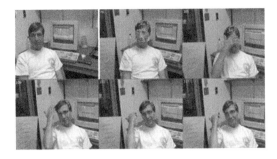

Figure 2. The result of KLT algorithm. From left to right, top to bottom are the 1th, 76th, 105th, 109th, 111th and 112th frame.

From the results we can see that the KLT tracking algorithm can track the face well in the simple scenario. However, when the face has been occluded for a while, KLT tracking algorithm cannot track it in the right trajectory. Compared with KLT tracking algorithm, our algorithm proposed in this paper can keep on tracking after the occlusion.

6 CONLUSIONS

In this paper, we have proposed a robust object tracking algorithm based on the low-rank representation. We decompose the sequence matrix into a low-rank matrix with the moving object, along with a sparse matrix with the occlusion. We use the variation between two adjacent frames to locate the moving object in the video sequence, and the sparse matrix to cope with the occlusion.

In the future research, we are looking forward to focusing on the feature points tracking instead of the tracking of the whole object regions. Besides, the method based on the low-rank representation can be developed into a two-dimensional principal component analysis (2DPCA), we will do more researches on this in the future.

REFERENCES

[1] Wax N, 1995, Signal to Noise Improvement and the Statistics of Tracking Populations [J]. Journal of Applied Physics, 26(5): 586–595.

[2] Hou Z, Han CZ, 2006, A Survey of Visual Tracking [J]. Institute of Automation, 32(4): 603–617.

[3] Zhou X, Yang C, Yu W. 2013, Moving object detection by detecting contiguous outliers in the low-rank representation [J]. Pattern Analysis and Machine Intelligence, IEEE Transactions on, 35(3): 597–610.

[4] N. Oliver, B. Rosario, and A. Pentland, 2000. "A Bayesian Computer Vision System for Modeling

Human Interactions," IEEE Trans. Pattern Analysis and Machine Intelligence, vol. 22, no. 8, pp. 831–843, Aug.

[5] R. Szeliski, 2010, Algorithms and Applications. Computer Vision: Springer.

[6] S. Geman and D. Geman, 1984, "Stochastic Relaxation, Gibbs Distributions, and the Bayesian Restoration of Images," IEEE Trans. Pattern Analysis and Machine Intelligence, vol. 6, no. 6, pp. 721–741, Nov.

[7] S. Li, 2009, Markov Random Field Modeling in Image Analysis. SpringerVerlag,.

[8] R. Szeliski, 2010,Computer Vision: Algorithms and Applications. Springer.

[9] Y. Peng, A. Ganesh, J. Wright, W. Xu, and Y. Ma, 2010. "RASL: Robust Alignment by Sparse and Low-Rank Decomposition for Linearly Correlated Images," Proc. IEEE Conf. Computer Vision and Pattern Recognition.

[10] R. Mazumder, T. Hastie, and R. Tibshirani, 2010. "Spectral Regularization Algorithms for Learning Large Incomplete Matrices," J. Machine Learning Research, vol. 11, pp. 2287–2322.

[11] B. Recht, M. Fazel, and P. Parrilo, 2010. "Guaranteed Minimum-Rank Solutions of Linear Matrix Equations via Nuclear Norm Minimization,"SIAM Rev. vol. 52, no. 3, pp. 471–501.

[12] J. Cai, E. Cande `s, and Z. Shen, 2010. "A Singular Value Thresholding Algorithm for Matrix Completion," SIAM J. Optimization, vol. 20, pp. 1956–1982.

[13] Y. Boykov, O. Veksler, and R. Zabih, 2001. "Fast Approximate Energy Minimization via Graph Cuts," IEEE Trans. Pattern Analysis and Machine Intelligence, vol. 23, no. 11, pp. 1222–1239, Nov.

[14] V. Kolmogorov and R. Zabih, 2004. "What Energy Functions Can Be Minimized via Graph Cuts?" IEEE Trans. Pattern Analysis and Machine Intelligence, vol. 26, no. 2, pp. 147–159, Feb.

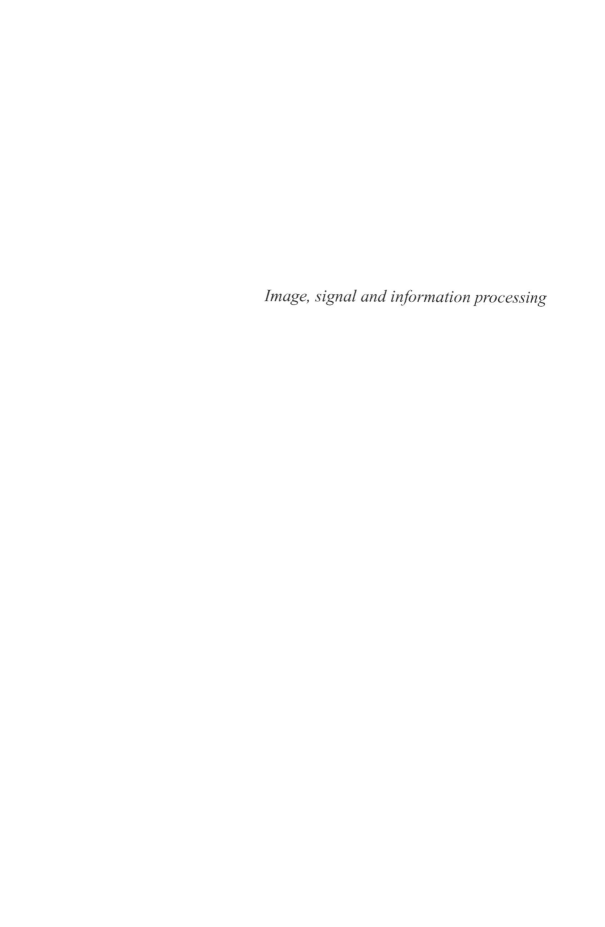

Image, signal and information processing

Testing and Measurement: Techniques and Applications – Chan (Ed.)
© *2015 Taylor & Francis Group, London, ISBN: 978-1-138-02812-8*

A study of image processing based hole expansion test

S. H. Oh, S.H. Yang & Y.S. Kim
School of Mechanical Engineering, Kyungpook National University, Daegu, Republic of Korea

ABSTRACT: In the automotive industry, reducing the CO_2 emission and automobile safety during a crash is one of the areas where receiving the most interest is the production of vehicles. Achieving both objectives simultaneously can be done by choosing materials such as dual-phase steel and ferrite bainite steel. These steels are used in the automobile chassis and body parts and are formed by hole flanging to meet strength and design requirements. The formability of sheet material is determined empirically through the hole expansion test, and the judgment is subjective as it relies on the eye and experience of the tester. This manual judgment involves many errors and large deviations. This paper presents an automatic crack recognition system that finds cracks based on an image processing technique to overcome the current method's dependence on human senses.

1 INTRODUCTION

As the automotive industry moves to address environmental concerns, an important priority has been reducing CO_2 emissions while maintaining passenger safety against collisions. Reducing the weight of automobiles is a primary method of lowering fuel consumption. The two basic approaches involve automobile design and materials selection, which are closely related to each other.

Instead of nonferrous light materials such as aluminum and magnesium, high-strength steel sheets have been widely used to meet the above two goals.

An automobile chassis contains a sub-frame and low arm; it is a skeletal frame upon which various mechanical parts such as the engine, tires, axle assemblies, brakes, and steering are bolted. The chassis is considered to be the most significant component for maintaining the driving stability of an automobile. With the increase in demand for lightweight automobiles, the design and material of chassis parts have been considered. Milliken (2002) provides comprehensive technical notes on chassis design.

At present, high-strength steels such as ferrite-bainite base fine blanking steel (FB540 / FB590) and ferrite-martensite base dual-phase advanced high-strength steel (DP590) have been adopted for the chassis components of automobiles. During the manufacture of chassis components, the forming process follows the sequence of draw forming the overall part and then hole punching (or piercing)/ expanding to join it with the engine sub-frame and wheels.

The quality of the punched and hole-expanded area is crucial to the fatigue strength of arms and has a critical effect on automobile safety. The formability of sheet steels during hole expansion is called "stretch-flangeability" and is determined by the Hole Expansion Ratio (HER) in the hole expansion test, which is a useful measure of a material's susceptibility to edge cracking.

There are two main hole expansion tests for determining the hole expansion property. The first uses a conical punch as per the Japan Iron and Steel Federation Standard and ISO 16630 Standard (2009), and the other uses a flat-topped punch employed by Corus Standard (Fang et al., 2003).

In the hole expansion test, a punched sheet with a specified hole size is stretched under tension by the punch penetration until a visible edge crack in the punched hole appears. The occurrence of edge cracking during the test is highly dependent on the material characteristics at the blanked/sheared edge and die clearance. Much research has been performed to clarify the effects of material and process variables such as the hole fabrication method (e.g., conventional punching, laser cutting, water-jet cutting) and die clearances on the HER and fatigue strength of the automobile components.

Fang et al. (2003) investigates the effects of mechanical properties on the hole expansion test using a flat-topped punch and reveals that a high ratio of the yield strength to the ultimate tensile strength provides good hole expansion properties. Kumar et al. (2011) performs hole expansion tests on hot-rolled steel sheets to clarify the various factors that influence the HER and concludes that the ultimate tensile strength and carbon equivalent have the most pronounced effect on the HER. Shimizu et al. (2004) develops a new precipitation-hardened and high-strength steel with a high HER and higher fatigue limit for the automobile suspension and chassis. Konieczny et al. (2007) shows that the conventional

Forming Limit Curve (FLC) fails to predict failure during the sheared edge stretching and that laser-cut edges display superior stretchability for AHSS steels compared with conventional punching. Kadarno et al. (2014) develops a special punching process that uses a taper punch and step die, which thickens the sheared edge and improves the fatigue strength of the punched high-strength steel sheets. Park et al. (2010) performs the conical punch hole expansion test and finite element analysis (FEA) on DP590, FB590, SAPH440 sheets and discusses the effect of hole fabrication methods and clearance on the HER. They reveals that the HER values for DP590 and FB590 sheets shows maximum values of 58.7% and 116.4%, respectively, near a clearance of 13.5%. Hance et al. (2013) performs hole expansion tests for various steel sheets with different hole fabrication methods and finds that water-jet cutting and laser cutting exhibits relatively high hole expansion performance compared to pierced-hole preparation.

Because the crack occurrence in hole expansion tests is currently judged manually based on the human eye and experience, the process is prone to many errors and large deviations. Thus, the results are not robust and objective. Another method for judging the crack occurrence and HER is to carefully monitor the edge expansion with a video extensometer or check for a sudden drop in the punch force on the punch force–penetration depth curve. Dunckelmeyer et al. (2009) uses a video camera to monitor the edge expansion in the hole expansion test and investigate the effects of the clamping force and penetration speed on the punch force–displacement curve to indicate the occurrence of edge failure.

These methods may provide reasonable results for some materials such as very high–strength steel steels and sheet material with little ductility, which show sudden failure occurrence on the edge area. However, in the case of very thin sheets or sheet material with high ductility, the drop in punch force upon crack occurrence is negligible, so this method fails to detect edge crack occurrences. Some researchers have used FEA to accurately simulate a blanking process using both a remeshing method for the large mesh distortion in the sheared zone and a finite element separation method in order to investigate the effect on the hole expansion performance.

Takuda et al. (1999) uses the ductile failure criterion to predict the initiation of fracture for flat-, hemispherical-, and conical-headed punches and shows good results with regard to the fracture initiation sites and critical stroke for fracture occurrence in experiments. Hatanaka et al. (2003) performs numerical simulations for the blanking process based on FEA using the ductile failure criterion, and their results shows good agreement with blanking experiments for various clearances and materials.

Kacem et al. (2013) numerically predicts the limits of the hole expansion process for aluminum sheets using a fracture criterion based on local strain measurements in tension and clarified the effect of process parameters on the occurrence of damage. Chung et al. (2011) simulates hole expansion tests of TWIP940, TRIP590, and 340R steels using a damage model with a triaxiality-dependent fracture criterion and hardening behavior with stiffness deterioration to numerically predict the HER.

In order to address these issues, we developed an automatic edge crack recognition system that finds visible edge cracks using an image processing method via a Charge-Coupled Device (CCD) camera. For image processing, we use the Visual C++ and Open CV libraries (Bradski and Kaebler, 2008), and we use MFC for the interface working in a Windows environment. The developed method improves the accuracy of hole expansion tests and increases the reliability of the tested results.

2 HOLE EXPANSION TEST

2.1 System to evaluate hole expansion ratio

A hole expansion test is a useful tool for estimating strength-flangeabiliy. Figure 1 shows a schematic view of a hole expansion test performed on a universal sheet metal forming simulator.

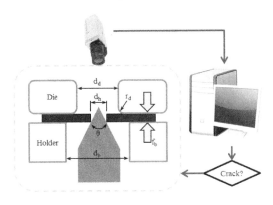

Figure 1. Schematic view of hole expansion test.

In this study, sheet specimens with dimensions of 135 mm × 135 mm are used. To prepare the specimen for the hole expansion test, it is previously punched with a punch having a diameter(d_h) of 10 mm. During the hole expansion test, the specimen is clamped tightly between an upper die and lower blank holder by a holding force of 10 tons. The inner diameter of the upper die(d_d) is 50 mm, and the die corner radius(r_d) is 5 mm. An angle(q) of 60° conical punch is used to expand the hole at a constant velocity of 10 mm/min.

The sheared edge (burr) formed during the punching of the hole is placed upward against the punch during hole expansion to facilitate fracture. The movement of the conical punch is terminated just after a visible edge crack is initiated near the edge of sheared hole. Then, the diameter of the hole is measured. Hole expansion ratio(HER) is determined according to Equation 1:

$$HER(\%) = \frac{D_f - D_i}{D_i} \times 100 \qquad (1)$$

Where D_f and D_i are the hole diameters just after failure and in the initial state, respectively.

To apply an image processing technique to the hole expansion test, the punch surface is coated with TiN having a hardness of 2000–2500 Hv in order to distinguish the conical punch from the edge of the sheared hole in the image. To clearly capture the deformed image of the hole edge for every punch penetration depth, we install a CCD camera with a high-brightness LED at the center of the top of the system. This is shown in Figure 2, which presents the experimental setup of this study.

CCD camera

Tool and conical punch

Main frame

Figure 3. Developed hole expansion system based on image processing method showing main frame, CCD camera, tool, and conical punch.

2.2 Image processing using Canny edge detector

The image processing scheme for the hole expansion test is as follows: image capture, preprocessing to eliminate noise, edge line extraction, clustering, and pattern recognition.

For pattern recognition of the hole edge, we uses the Visual C++ and Open Source Computer Vision (OpenCV) libraries. The occurrence of edge cracks on the sheared hole is judged from the final image extracted in the window. If the system judges that visible edge cracks are initiated near the edge, the test automatically will terminate. We obtained one frame of a captured image every 0.1 s by using the timer function in the C language.

In the captured images, the characteristic features of the hole edge are recognized with the Canny edge detector. This is a commonly used image processing tool for detecting edges in a very robust manner. Canny (1986) develops a computational algorithm to define detection and localization criteria for edge detection, where edges are marked at maxima in the gradient magnitude of a Gaussian-smoothed image.

Figure 3 shows the sequential process for Canny edge detection from a CCD camera image.

Figure 2. Process of Canny edge detection (a) grayscale image, (b) edge detection without Gaussian filtering, (c) edge detection with Gaussian filtering.

As shown in Figure 3(a), the CCD camera image is first converted into a gray-scale image using the *cvCvtcolor* function in openCV because the RGB image has a three-channel value, and it is difficult to compare the brightness. The edge detection taken directly from the RGB image shows a complicated image with a great deal of noise, as shown in Figure 3(b). Thus, classifying the boundaries is very confusing. Here, the innermost circle indicates a tool shape, the middle circle is the inside edge of the expanded hole, and the outermost circle is the outside edge of the expanded hole, which is easily prone to fracture. The remaining symbols such as points and small dashed lines indicate noise that can be ignored.

To get a clear edge image of the area of concern (i.e., the outside edge of the expanded hole), a Gaussian filtering scheme is used to eliminate the various sources of noise caused by reflection and surface conditions, as shown in Figure 3(c).

2.3 Feature extraction

Because our area of concern is the outside edge of the expanded hole, the image is extracted by eliminating all other images. The elimination scheme of non-useful images excluding the outside edge is as follows. First, we assumes that the edge line image of the area of concern would be a circle. We take the virtual center and radius of the inside of the circle and a circular band with little depth that includes the edge circle line. Then, all images outside the circular band are eliminated to extract the image of the outside edge line of the expanded hole.

In Figure 4, we explain in detail how to determine a virtual center and radius inside of the circle and the circular band with little depth. First, we take an arbitrary center position inside the circle and draw three vector lines in the $\pm 0°$ (horizontal), $\pm 90°$ (vertical), and $\pm 45°$ directions. These three lines must meet the outside edge line of the expanded hole twice to yield the six intersection points, as shown in Figure 4(a). To search for the intersection between the vector line in the $\pm 45°$ direction and the extracted image of the outside edge of the expanded hole, we use a stepwise searching scheme.

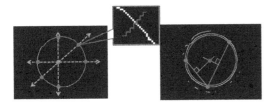

Figure 4.   Detection of edge outline and center position: (a) crossing direction of any six points (b) finding center using three data points not adjoining each other.

Using the three points of data which do not adjoin each other among these six crossing points, we can find the virtual center and mean radius R_m, as explained in Figure 4(b). Using this virtual center and radius, we draw a virtual circle line with a slightly smaller radius than the extracted image of the outside edge line of the expanded hole by using the three points apart from each other.

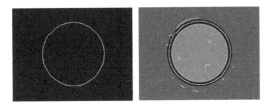

Figure 5.   Final image obtained from imaging process: (a) deleted section, (b) final image.

Then, the circular band is produced by taking an inner virtual circle with a radius of R_m-5Px and an outer virtual circle with a radius of R_m+5Px. Here, Px is the pixel size of a window. By taking the circular band, all noise and non-useful edge lines excluding the outside edge line of the expanded hole (i.e., the area of concern) are finally eliminated.

Figure 5 shows the final image obtained from the above image processing.

2.4   Pattern recognition

In the Japanese standard method, the sheared edge of the punched hole is placed upward against the punch during the hole expansion test. Thus, the fracture over the material ductility limit initiates at the outside edge of the expanded hole and propagates into the inside edge. When an edge crack occurs, the extracted image of the circular shape pattern deviates from the circular shape or shows a sharp vertex on the circular line, as indicated in Figure 6.

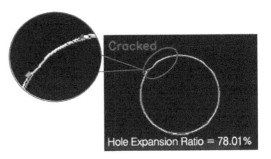

Figure 6.   Result of crack occurrence judgment.

In this study, we judge the occurrence of edge cracks by considering the maximum radius measured at the vertex of an edge line with a crack and the difference in radius Δr (= maximum radius at vertex − averaged radius of edge line with crack) at every position along the circular line. In this study, we choose $\Delta r = 2Px$ as the criterion for edge crack occurrence, so if the difference in radius at any position along the circular line is over $2Px$, an edge crack has occurred. Figure 6 presents one example for the judgment of edge crack occurrence. The HER is obtained by the ratio of the averaged radius of the edge line with a crack and the radius of the initial punched edge line.

3   MATERIALS AND EXPERIMENT

We conduct hole expansion tests according to the Japanese standard method and apply the pattern recognition method to evaluate the HER of the tested sheets.

We use four different kinds of steel and aluminum: automobile hot-rolled steel plate (SAPH440), FB540, FB590, DP590, and aluminum 6061. Table 1 lists the mechanical properties of the tested materials. Here, YS, TS, El, n, and R_m represent the yield strength, tensile strength, fracture elongation, work hardening coefficient, and anisotropic Rankford value, respectively.

Table 1. Mechanical properties of tested specimens.

Material	t mm	YS MPa	TS MPa	El. (%)	n	R_m
SAPH440	2.3	350	441	29	0.119	1.07
FB540	2.3	490	543	25	0.125	0.98
BF590	2.9	530	615	21	0.107	0.93
DP590	2.3	450	685	26	0.183	1.04
Al6061	2.0	214	311	26.3	0.238	0.56

The developed pattern recognition technique is runnig in Windows XP, and the program languages of Visual Studio 2005.net C++ and OpenCV are used. In order to capture the test images, a webcam with a resolution of 300,000 pixels is used as the CCD camera.

Figure 7 presents a screen capture of the main frame of the executed program before the developed hole expansion test. The main frame contains several dialog boxes of key functions for pattern recognition of the hole expansion test: received image, image preprocessing, image processing and control, result folder, reset, and start. In the image processing and control dialog box, we can modify and change the critical values used in the Gaussian filter and Canny edge detector and the Δr value denoting the criterion for edge crack occurrence.

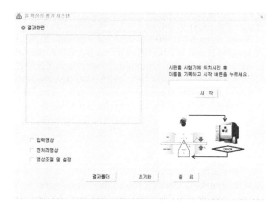

Figure 7. Screen view of main frame of executed program before developed hole expansion test.

For the pattern recognition system installed in the universal sheet metal forming simulator, we place a specimen between the die and blank holder, key in the specimen name, and push the start button in the dialog box. Then, the punch moves and expands the hole at a constant speed until edge failure is detected. At the initial stage of the test, the pattern recognition system automatically determines the radius of the initial punched edge line. The developed pattern recognition system is linked to the data acquisition system NI cDAQ-9172 (National Instruments).

During the test, the visual system shows the image of the specimen during hole expansion. When any edge (axial) crack occurs during the test, the simulator stops automatically, and the captured edge image and HER are saved in the hard disk of the computer.

Figures 8 and 9 show screen captures of the crack occurrence and HER values for the FB590L and FB540 specimens, respectively. The HER is 133.74% for the FB590L specimen and 119.17% for the FB540 specimen.

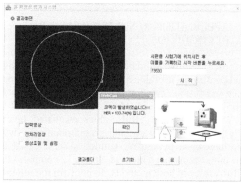

Figure 8. (a) Screen view of main frame of executed program showing crack occurrence in developed hole expansion test program and (b) cracked specimen after test for FB590L specimen.

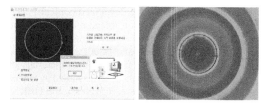

Figure 9. (a) Screen view of developed program after test, (b) cracked specimen after test for FB540 specimen.

Figures 10 and 11 show screen captures of the crack occurrence and HER values for the DP590L and Al6061 specimens, respectively. The HER is 104.43% for the DP590L specimen and 56.46% for the Al6061 specimen.

Figure 10. (a) Screen view of developed program after test, (b) cracked specimen after test for DP590L specimen.

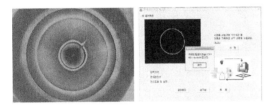

Figure 11. (a) Screen view of developed program after test, (b) cracked specimen after test for Al6061 specimen.

4 APPLICATION RESULTS OF PATTERN RECOGNITION TECHNIQUE

In order to further validate the proposed methodology and confirm its usefulness, we perform conical-punch hole expansion tests for the five sheet materials of FB590L, DP590L, SAPH440L, FB540, and Al6061.

We compare the HER values evaluated manually with the human eye and programmatically with the developed pattern recognition system based on the image processing method. Figure 12 compares the HER values of both methods. To confirm the reliability of the developed test method, the five specimens are examined in a preliminary test, and the results are averaged. The maximum variations of the tested materials dose not exceed 3.4%; thus, we conclude that the developed test method is reliable.

The difference between the two methods is within 10%; thus, we consider the developed conical-punch hole expansion test based on image processing to be satisfactory.

The three steel sheets of FB590L, DP590L, and SAPH440L are laser cut, and the sheet materials of FB540 and Al6061 are punched. As discussed by Hance et al. (2013) and Watanabe et al. (2006), the HER is strongly affected by material damage related to changes in the microstructure around the hole edges and die clearance from the previous punching operation.

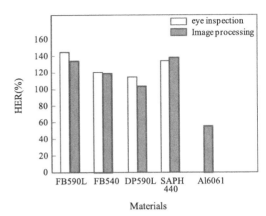

Figure 12. Results of HER values determined from human eye inspection and automatic evaluation based on proposed image processing method.

Much research into various edge finishing processes has revealed that the HER is sensitive to sheared edge finishing. Moreover, a punched edge experiences severe plastic shear forming and shows a great deal of damage with regard to the mechanical properties on and near the edge. Thus the HER of a punched edge is much lower than that of a fine edge for which the punched edge is removed by milling. Laser cutting also enhances the HER because only a narrow area of the hole edge is heat-affected, and most of the rest of the area is not damaged at all.

Even though FB540 has superior mechanical properties compared to FB590L with regard to the fracture elongation and work hardening coefficient, FB590L shows a relatively high hole expansion performance because its hole is prepared by laser cutting. The HER for FB590L is 12% higher than that for FB540.

5 CONCLUSIONS

In this study, we develop a hole expansion test system that uses a pattern recognition technique with image processing to improve the reliability of the conical-punch hole expansion test results. We can automatically identify the punched edge line of the specimen and the occurrence of edge cracks with the developed method. Using this technique, the HER is automatically calculated regardless of variations in the manual method. A comparison between human eye inspection and the developed method shows that the latter reliably and satisfactorily evaluates the HER of sheet materials in the conical-punch hole expansion test.

At present, however, improvements in the developed method depend upon the image resolution. This is the most important input for determining the accuracy of the HER value and depends on the optical surface properties of the specimen, such as the surrounding brightness, surface color of the specimen, camera focus, and vibration of the equipment. This will be examined in future research.

ACKNOWLEDGMENTS

This work was supported through a National Research Foundation of Korea (NRF) grant funded by the Korean government (2014R1A2A2A01005903). The authors gratefully acknowledge this support.

REFERENCES

[1] Bradski, C., Kaebler, A., 2008, Learning OpenCV, O'Reilly Media, California.

[2] Canny, J., 1986, A computational approach to edge detection, IEEE Trans. Pattern Anal. Mach. Intel. 8(6), p679–698.

[3] Chung, K.S., Ma, N., Park, T.J., Kim, D.G., Yoo, D.H., Kim, C.M., 2011, A modified damage model for advanced high strength steel sheets, Int. J. Plast. 27 p1485–1511.

[4] Dunckelmeyer, M., Karelova, A., Krempaszky, C., Werner, E., 2009, Instrumented hole expansion test, Proc. Int. Doctoral Seminar, p411–419.

[5] Fang, X., Fan, Z., Ralph, B., Evans, P., Underhill, R., 2003, The relationships between tensile properties and hole property of C-Mn steels, J. Mat. Research 38, p3877–3882.

[6] Hance, B.M., Comstock, R.J., Scherrer, D.K., 2013, The influence of edge preparation method on the hole expansion performance of automotive sheet steels, SAE paper No. 2013-01-1167.

[7] Hatanaka, N., Yamaguchi, K., Takakura, N., Iizuka, T., 2003, Simulation of sheared edge formation process in blanking of sheet metals, J. Mater. Proc. Technol. 140(1-3), p628–634.

[8] ISO 16630:2009. Metallic materials – sheet and strip – hole expanding test, 2009.

[9] Kacem, A., Krichen, A., Manach, P.Y., Thuillier, S., Yoon, J.W., 2013, Failure prediction in the hole-flanging process of aluminium Alloys, Eng. Fract. Mech. 99, p251–265.

[10] Kadarno, P., Mori, K.I., Abe, Y., Abe, T., 2014, Punching process including thickening of hole edge for improvement of fatigue strength of ultra-high strength steel sheet, Manuf. Rev. 1(4), p1–12.

[11] Konieczny, A. A, Henderson T., 2007 Product design considerations for AHSS displaying lower formablility limits in stamping-with sheared edge stretching, U.S. Steel Corporation, Great Designs in Steel Seminar.

[12] Kumar, S., Deva, A., Mukhopadhyay, S., Kumar, B., 2011, Assessment of forability of hot-rolled steel through determination of hole-expansion ratio, Mater. Manuf. Proc., 26, p37–42.

[13] Mole, N. and Stok, B. 2009, Finite element simulation of sheet fine blanking process, Int. J. Mater. Form. 2(1) p551–554.

[14] Lee, S.B., Cho, Y.R., Chin, K.G., 2007, Analysis of stretch-flangeability using a ductile fracture model, POSCO Tech. Report 10(1), p104–115.

[15] Lee, W.S., Kwon, T.W., Cho, Y.R., Kim, D.U., 2004, Development of automotive chassis parts by application of hot rolled high strength steel, Proc. Korean Soc. Auto. Engng. Conf., p1476–1481.

[16] Kim, K.H., 1994, Coating and mechanical properties of TiN on the steel tools by plasma assisted chemical vapor deposition, KOSEF 931-0800-018-1.

[17] Park, J.K., Park, B.C., Kim, Y.S., 2010, A study of automobile product design using hole expansion testing of high strength steel, Proc. Korean Soc. Tech. Plast. Conf., p282–287.

[18] Picart, P., Lemiale, P.V., Touache, A., Chambert, J., 2005, Numerical simulation of the sheet metal blanking process, VIII Int. Conf. Comp. Plast. CIMNE, Barcelona, p1–4.

[19] Takuda, H., Mori, K., Fujimoto, H., Hatta, N., 1999, Prediction of forming limit in bore-expanding of sheet metals using ductile fracture criterion. J. Mater. Proc. Technol. 92(93), p433–438.

[20] Schwich, V., Hirschmanner, F., Jaroni, U., 2005, Innovative steel sheets and products for applications in vehicles, Int. Conf. Steels Cars Trucks, p1–8.

[21] Shimizu, T., Funakawa, S., Kaneko S., 2004, High strength steel sheets for automobile suspension and chassis use–high strength hot rolled steel sheets with excellent press formability and durability for critical safety parts, JFE Tech. Report, No. 4, p 25–31.

[22] Soderberg, M., 2006, Finite element simulation of punching, Master Thesis of Luleå University of Technology, Sweden.

[23] Watanabe, K., Tachibana, M., 2006, Simple prediction method for the edge fracture of steel sheets during vehicle collision (1st report), LS-DYNA Anwenderforum, Ulm, p B-1-9–14.

Application of CR images for a CAD of pneumoconiosis for images scanned by a CCD scanner

R. Miyazaki & K. Abe
Interdisciplinary Graduate School of Science and Engineering, Kinki University, Osaka, Japan

M. Minami
Kanazawa Gakuin University, Ishikawa, Japan

H. Tian
Graduate School of Engineering, Kobe University, Hyogo, Japan

ABSTRACT: This paper presents an application of the computer-aided diagnosis of pneumoconiosis for chest X-ray images obtained from a CCD scanner for chest X-ray CR images. When we reported the system before, we showed the performance of the system for chest X-ray images obtained by a CCD scanner. However, since density distribution of the CR images is quite different from the images obtained by a CCD scanner, it is necessary to equip some pre-processing into the system for applying the system to the CR images. In this paper, as the first trial for the application, by drawing rib edges and additional lines on the CR images by a tablet PC, we examine whether the system can be interactively applied to the CR images. Besides, proposing a method for extracting ribs from the CR images should consider full automation of the system, so we have compared the performance of the two methods.

1 INTRODUCTION

Pneumoconiosis is a kind of interstitial lung disease caused by inhalation of fine particles (e.g., coal pneumoconiosis). In recent years, in addition to coal workers, dental technicians also suffer from the disease. However, since it is difficult for even experts on pneumoconiosis to diagnose pneumoconiosis, disagreements between diagnosticians often exist. Besides, the experts teach techniques for the diagnosis to immature diagnosticians based on just their experience. For the reasons, Computer-Aided Diagnosis (CAD) systems for pneumoconiosis have been required as a second opinion for diagnosticians. CAD systems for pneumoconiosis have been reported since 1970s (e.g., R.P. Kruger et al. 1974; H. Kobatake & K. Ohnishi 1987; A.M. Savol et al. 1980; T. Kouda & H. Kondo 2001). Their measurements of abnormalities for pneumoconiosis broadly are two ways: one measures the abnormalities of extracting features by texture analysis (R.P. Kruger et al. 1974; H. Kobatake & K. Ohnishi 1987), the other extracts small round opacities and measures their size and number as well as the real diagnosis by diagnosticians (A.M. Savol et al. 1980; T. Kouda & H. Kondo 2001). All the systems are proposed to images obtained by a custom-made scanner (e.g., a drum scanner or a film scanner). Therefore, if the systems are applied in general clinics, the clinics will have to deploy the special scanner, or order scanning chest radiographs to a printing company in spite of high costs.

In order to enhance cost-performance of CAD for pneumoconiosis, a CAD system for pneumoconiosis using images obtained from a common CCD scanner is reported (M. Nakamura et al. 2009; K. Abe et al. 2013). This system is composed of a CCD scanner and a tablet PC and discriminates pneumoconiosis by measuring abnormalities in rib areas. Since the images obtained from X-ray pictures from a CCD scanner are extremely unclear and it is hard to extract rib areas automatically, in this system, the user draws rib edges of the images manually using the tablet PC. The reason why only X-ray pictures are used in the diagnosis of pneumoconiosis is there is a criterion that diagnosticians have to diagnose pneumoconiosis by comparing X-ray pictures with the standard X-ray pictures of pneumoconiosis provided by the International Labor Organization (ILO). Similarly, in Japan, the Ministry of Health, Labor and Welfare provided another set of the standard X-ray pictures. And, both of the ILO and the ministry opened new standard images in 2011 (ILO 2011). Both of the new standard sets are digital images taken by computed radiography (CR). Since a picture taken by CR is saved as a digital file, CR enables diagnosticians to read X-ray pictures on a monitor. CR has been already regular radiography in Japan.

With the revision of the standard pictures of X-ray pictures to CR images, this paper examines whether the prior CAD system for pneumoconiosis using the X-ray images obtained from a CCD scanner can be applied to chest CR images. Basically, X-ray images scanned by a CCD scanner are unclear. On the other hand, since the shadows in CR images appear clearly, the density distribution of the CR image is quite different from the X-ray image obtained from a CCD scanner. Hence, if we consider applying the CAD to the CR images, we need to customize the CAD equipping some pre-processing. Therefore, this paper presents two methods for applying the CAD to the CR images. The one method is a manual pre-processing of drawing rib edges of the CR images before the diagnosis by the CAD, and another method is a proposal of extracting rib areas from the CR images automatically considering a complete automated CAD system. If the system can be applied to either of chest CR images and chest X-ray images obtained from CCD scanner, the system is useful in the diagnosis for either of them at a low cost.

2 PNEUMOCONIOSIS CLASSIFICATION

The level of pneumoconiosis is indicated by a profusion of small round opacities, where categories 0-3 are established. Figure 1 shows samples of the standard images provided by the Japanese ministry. Normal radiographs belong to category 0, where the opacities are not observed visually. And, abnormal ones belong to category 1, 2, or 3, where the most serious level is category 3 and the opacities are observed most in category 3. Although diagnosticians diagnose pneumoconiosis comparing chest X-ray pictures with the standard pneumoconiosis images prepared in every category, their own experience much depends on the diagnosis. For example, gray levels at the opacities are very similar to intersections of vascular shadows.

In Japan, the criteria of the classification are defined in the Japanese Pneumoconiosis Law in accordance with criteria of pneumoconiosis in the ILO.

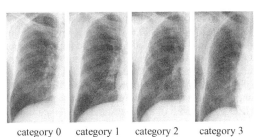

category 0   category 1   category 2   category 3

Figure 1. Standard images of pneumoconiosis (provided by the Ministry of Health, Labour, and Welfare in Japan).

3 CAD FOR PNEUMOCONIOSIS X-RAY IMAGES SCANNED BY A CCD SCANNER

3.1 Overview

In the CAD system for pneumoconiosis X-ray images scanned by a CCD scanner (K. Abe et al. 2013), abnormalities of pneumoconiosis are measured by extracting characteristics of the density distribution in rib areas. The rib areas are manually designated using a tablet PC. First, the right lung area in chest X-ray pictures is digitized by a CCD scanner. Next, the chest image is displayed on the tablet PC and the user draws curves along the edges for the shadows of ribs on the image using the tablet PC. And then, the rib areas are designated by the drawn curves. Finally, abnormalities of pneumoconiosis are extracted from the rib areas. The extracted abnormalities are used as valuables for discrimination of chest X-ray images into normal or abnormal cases in pneumoconiosis.

3.2 Preprocessing

The right lung area in chest X-ray pictures (35 cm × 35 cm) is digitized by a CCD scanner. The digitalized chest image is configured with 300 dpi and 256 gray levels. Next, the image is resized into 1000 pixels in height without changing the aspect ratio. And then, the range of the gray value in the image is standardized by the linear histogram stretching.

3.3 Drawing of rib edges on the X-ray image with tablet PC and extraction of rib areas

The image is displayed on the tablet PC and the user draws 8 curves with white color along the edges of 4 ribs on the image using the tablet PC. The manual for drawing the curves is designed as below.

[Manual for drawing the rib edges]

1 Open the right lung image in a paint tool.
2 Select the round brush and set the thickness of the circle as 4 pixels.
3 Select white color to the line color.
4 Look at the curve shown in Figure 2(b) as the guide of start points for drawing rib edges.
5 Image the curve on the image by yourself, decide the start point for the drawing and start draw the curve along an edge of a rib.
6 Stop drawing the curve at a location where the edge cannot be seen.
7 Until you finish drawing all couples of the edges for 4 ribs, repeat 5 and 6.

As shown in 4 and 5, Figure 2(b) is shown to the user as the guide for setting the location of the start point of the drawing. Figure 2(c) shows an example of the curves drawn to Figure 2(a). Then, as shown in Figure 2(d), the rib areas R_1–R_4 are extracted by the drawn edges (Refer to (K. Abe et al. 2013)).

As shown in Figure 1 and Figure 2, we can see that the images scanned by a CCD scanner (Figure 2) are extremely unclear than CR images.

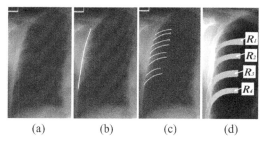

(a)　　　　(b)　　　　(c)　　　　(d)

Figure 2.　Drawing of rib edges according to the manual and rib areas R_1–R_4 extracted from the edges.

3.4 Extraction of abnormalities and discrimination of pneumoconiosis

Figure 3 shows a rib area R_m extracted above, where the first line is regarded as k-th scanning line ($k = 1$) and the value k is added by 1 whenever the scanning line is shifted to the right by 1 pixel. Besides, the uppermost pixel on k-th scanning line is regarded as $j = 1$ and the value j is added by 1 whenever the pixel goes down by 1 pixel. The linear histogram stretching is applied to every scanning line. In the scanning on the k-th scanning line, all the pixels on the line are divided into the upper set and the lower set. The boundary of them is obtained by the discriminant analysis for the pixels. In regarding $t_m[k]$-th pixel counted from the uppermost edge as the boundary $t_m[k]$, the upper set is composed of pixels of $j = 1 \sim t_m[k] - 1$ and the lower set is $j = t_m[k] + 1 \sim heiR_m[k]$.

As the scanning is conducted from the upper edge to the lower edge along the scanning line regarding $v[j]$ as the gray value of the j-th scanning spot, by using the parameters shown in Figure 3, the abnormality $Abn(R_m)$ of k-th scanning line in R_m is defined as

$$Abn(R_m[k]) =$$

$$\left[\frac{1}{heiR_m[k]-1} \left(\sum_{j=1}^{heiR_m[k]} |v[j]-avg[k,j]|^2 \right) \right]^{\frac{1}{2}} \quad (1)$$

where

$$avg[k,j] =$$

$$\begin{cases} if\ 1 \leq j \leq t_m[k],\ \left(\sum_{i=1}^{t_m[k]-1} v[j] \right) / (t_m[k]-1) \\ if\ j = t_m[k],\ v[j] \\ if\ t_m[k] < j \leq heiR_m[k],\ \left(\sum_{i=t_m[k]+1}^{heiR_m[k]} v[j] \right) / (heiR_m[k]-t_m[k]) \end{cases} \quad (2)$$

Next, shifting the scanning line in R_m shifts one by one to the scanning line at the most right side horizontally, the abnormality in R_m is defined as

$$Abn(R_m) = \frac{1}{widR_m} \left(\sum_{k=1}^{widR_m} Abn\left(R_m[k] \right) \right) \quad (3)$$

where $widR_m$ is the number of the horizontal shifts. Then, an abnormality in the whole rib areas $AbnR$ is defined as

$$AbnR = \frac{1}{L_R} \left(\sum_{m=1}^{4} \sum_{k=1}^{widR_m} Abn\left(R_m[k] \right) \right) \quad (4)$$

where

$$L_R = \sum_{m=1}^{4} widR_m \quad (5)$$

Besides, the maximum value $AbnRMAX$ in $Abn(R_m)$ ($m = 1 \sim 4$) is represented as another abnormality.

$$AbnRMAX =$$
$$\max\left\{ Abn(R_1), Abn(R_2), Abn(R_3), Abn(R_4) \right\} \quad (6)$$

Thus, the abnormality $AbnR$ represents an overall abnormality of pneumoconiosis in rib areas and $AbnRMAX$ represents local abnormalities of pneumoconiosis in all the rib areas.

After the extraction of the abnormalities, regarding the abnormalities as variants for discriminant machines, the diagnosis of pneumoconiosis is conducted by discriminating between normal cases and abnormal cases.

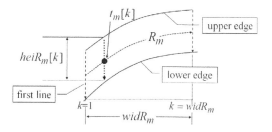

Figure 3.　A rib R_m and parameters used in the equations of the abnormalities.

4 PROPOSED METHODS

4.1 Method 1: Customization of the manual for CR images

Figure 4(b) shows a result that a user draws the rib edges to Figure 4(a) according to the manual shown in Sect. 3. Thus, differing from the case of the unclear images scanned by CCD scanner, the user can confidently draw the edges of the whole ribs. However, due to the drawing, the extracted rib areas would contain the clavicular shadow or the lung marking (a mass of linear shadows appeared in the lung area of chest X-ray images) and they could become noises in extracting the abnormalities from the rib areas. Therefore, to draw the edges on the CR images without containing them, the manual is customized for the CR images as follow.

[*Manual for drawing the rib edges (for CR Images)*]

1 Open the right lung image in a paint tool.
2 Select the round brush and set the thickness of the circle as 7 pixels.
3 Draw the lower edge of the clavicular shadow with a color except white (as Ⓐ in Figure 4(c)).
4 Draw a curve with a color except white at the left side of the lung marking and separate the lung area into the lung marking and outside of the lung area (as Ⓑ in Figure 4(c)).
5 Draw 4 couples of rib edges with white from the boundary between the lung area and its outside (i.e., left side of the right lung) until reaching either of the curves drawn in 3 and 4 as shown in Figure 4(c).

Figure 4(d) shows the rib areas extracted from Figure 4(c).

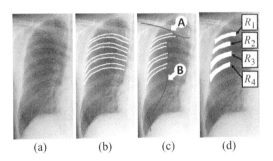

(a)　　　(b)　　　(c)　　　(d)

Figure 4. Drawing of the rib edges ((a): original image (b): drawn according to the old manual, (c): drawn according to the customized manual), and (d) is the rib areas R_1-R_4 extracted from image (c).

4.2 Method 2: Automated extraction of rib areas by extracting rib edges

There are reports on the automated extraction of rib areas from chest X-ray images (K.S. Fu et al. 1975; M. Loog & B. Ginneken 2006). However, there is high possibility that both of the methods contain obstacles for extracting the abnormalities such as shadows of blood vessels and other organs into the rib areas. Hence, this paper proposes a novel method for

extracting the rib areas which are suitable for extracting the abnormalities.

4.2.1 Pre-processing
First, the lung area is extracted from the discriminant analysis of the original image $G(x,y)$, and then, the left half of the lung area $L(x,y)$ is extracted.

Second, to remove the upper part of the clavicular shadow from $L(x,y)$, the vector convergence index filter (Y. Yoshinaga & H. Kobatake 2000) is applied to $G(x,y)$. The filter outputs convergence degree $con(x,y)$ of line segments and the value $ang(x,y)$ of line angle at every pixel in $G(x,y)$, where $-1 \leq con(x,y) \leq 1$ and $0 \leq ang(x,y) < 180$. The binarized image $S(x,y)$ is created by the discriminant analysis for $con(x,y)$. Scanning $S(x,y)$, pixels which satisfy the condition of $0 \leq ang(x,y) < 120$ are transformed into black pixel, and the image $S_l(x,y)$ is created. And then, the pixel $H_0(x_0,y_0)$ which is a white pixel and has the minimum of $x + y$ is obtained from $S_l(x,y)$. Next, $S_l(x,y)$ is scanned from the column whose x-coordinate is $x_0 + 1$ to lower pixels and by shifting the scanning line to the right. In the scanning, the pixel $H_n(x_n,y_n)$ ($1 \leq$ n $\leq$ the maximum value of x-coordinates) whose y-coordinate satisfies the following condition is searched in each scanning line and the upper pixels of $H_n(x_n,y_n)$ are removed from $L(x,y)$.

$$y_n = \begin{cases} y_n & (y_{n-1} + 20 > y_n \,\& \, S_1(x_n,y_n) = 255) \\ y_{n-1} & (otherwise) \end{cases} \tag{7}$$

Third, scanning $S(x,y)$, pixels which satisfy the condition $60 < ang(x,y) < 180$ are transformed into black pixel, and the reference image $Ref(x,y)$ is created.

Figure 5 shows the images used in the preprocessing, where (a) is an original image $G(x,y)$, (b) is the left half of the lung area $L(x,y)$, (c) is the image of $con(x,y)$ where the range of array values has been linearly transformed from $-1 \sim 1$ to $0 \sim 255$, (d) is $S(x,y)$ obtained from $con(x,y)$, (e) is $S_l(x,y)$ obtained from $S(x,y)$, and (f) is the reference image $Ref(x,y)$ of $G(x,y)$.

4.2.2 Extraction of rib areas for the diagnosis
The tracking of rib edges starts from pixels whose left pixel is black pixel in $L(x,y)$ and which is a white pixel in $Ref(x,y)$. Figure 6(a) shows the start points extracted from Figure 5(b) and Figure 5(f). Next, regarding each start point as the left bottom of the range of 5 × 5 pixels, the range is scanned to every of the start points. Among the 25 pixels in the range, pixels which are white pixel in $Ref(x,y)$

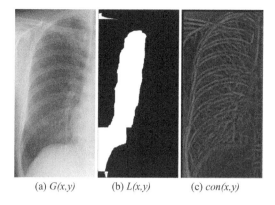

| (a) $G(x,y)$ | (b) $L(x,y)$ | (c) $con(x,y)$ |

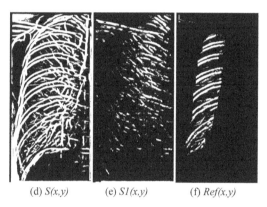

| (d) $S(x,y)$ | (e) $S1(x,y)$ | (f) $Ref(x,y)$ |

Figure 5.   The image arrays in the preprocessing.

are extracted. And then, to every of the extracted pixels, the same scanning is conducted and a candidate image for the lower edges of ribs is created as $R(x,y)$. Figure 6(b) shows $R(x,y)$ for Figure 5(a). Among all the white pixels in $R(x,y)$, the pixels which satisfy either of the following conditions (Eq.8–Eq.10) are removed as a noise from $R(x,y)$, and the transformed image is defined as $R_1(x,y)$.

$$\sum_{y=y-20}^{y-1} G_2(x,y) < \sum_{y=y+1}^{y+20} G_2(x,y) \tag{8}$$

$$\sum_{x=x-5}^{x-1} G_2(x,y) < \sum_{x=x+1}^{x+5} G_2(x,y) \tag{9}$$

$$\sum_{y=y-20}^{y-1}\sum_{x=x-25}^{x+25} G_2(x,y) < \sum_{y=y+1}^{y+20}\sum_{x=x-25}^{x+25} G_2(x,y) \tag{10}$$

where $G_2(x,y)$ is the image created by transforming 256 gray levels of $G(x,y)$ into 16 gray levels.

Figure 6(c) is $G_2(x,y)$ for Figure 5(a) and Figure 6(d) shows the image $R_1(x,y)$ for $R(x,y)$ of Figure 6(b).

After that, the largest 4 white regions are extracted from $R_1(x,y)$ as candidates for lower edges of rib areas. The transformed image is defined as $R_2(x,y)$. Figure 6(e) shows the image $R_2(x,y)$ obtained from $R_1(x,y)$ of Figure 6(d) (In Figure 6(e), $R_2(x,y)$ is overlapped on the original image of Figure 5(a).). And, the region which is at the highest location in the regions is removed from $R_2(x,y)$ as a noise. The reason why the region is removed is because contrast in the upper area of the lungs is not always high and there is possibility the region does not represent the rib edge as a noise. On the other hands, the thinning is conducted in the 3 regions. Since the lines created by the thinning would have some branches in addition to rib edges, the shortest curve is extracted between the pixel at the right end and the pixel at the left end in each region. Besides, thickness of the curves is increased by conducting the expansion twice. After the procedure shown above is conducted to $R_2(x,y)$, the image of the lower edges for 3 ribs is created. Figure 6(f) shows the image of the lower edges for Figure 5(a) and Figure 6(e).

Shifting each lower edge in the upper direction vertically at a height of between 20 and 50 pixels, the number of white pixels which are covered by the edge in $Ref(x,y)$ is counted at every shifting. And, finding the location of the shifting where the number is the maximum, the edge is copied as its upper edge. Figure 6(g) shows 3 couples of the rib edges obtained from Figure 5(f) and Figure6(f). Finally, as well as 3.3, the rib areas between the couples of the edges are extracted by the method proposed in (K. Abe et al. 2013). Figure 6(h) shows the rib areas for the diagnosis obtained from Figure 6(g).

5   EXRERIMENTAL RESULTS

The abnormalities are extracted from 51 CR chest X-ray images (resolution: 300 dpi, gray level: 8 bits, format: bitmap image). The number of images which belong to category 0 (normal case) is 36; the number in category 1 is 6; the number in category 2 is 6 and the number in category 3 is 3. To obtain the rib areas by Method 1, three non-experts (represented as userA, userB, and userC) in pneumoconiosis diagnosis draw the rib edges on all the images with a tablet PC (ThinkPad 200Tablet 4184-F5J). The total time used for drawing all the images was about 60 minutes in every user. Table 1 shows the mean and standard deviation (SD) of $AbnR$ and $AbnRMAX$ extracted from each of the normal and abnormal cases in each of Method 1 and Method 2,

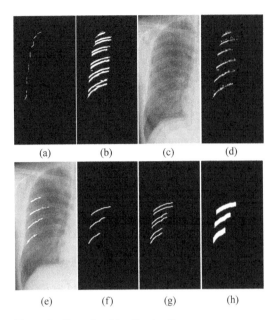

(a)   (b)   (c)   (d)

(e)   (f)   (g)   (h)

Figure 6.   Example of the rib extraction.

from Table 1, we can see both of two abnormalities in abnormal cases are higher than normal cases, and the abnormalities are appropriately extracted for the discrimination of pneumoconiosis in both of the proposed methods.

Next, we examine the performance of the proposed methods by the discrimination of the images into normal or abnormal (i.e. pneumoconiosis) cases by regarding the abnormalities as variants. As the way of the discrimination, Random Trees (RT), a Neural Network (NN), a linear Support Vector Machine (SVM) were applied. Training set and test set were chosen by the cross validation (F. Mosteller 1948) in all the discriminations. The method of cross validation is often applied to increase the number of trials when the number of samples is small for evaluations. In fact, every discrimination is conducted as the following procedure:

1   Choose one image from all the images as test data, and use the other images as training set.
2   Discriminate the test data between normal and abnormal cases.
3   Repeat the procedure from 1 to 2 to every combination changing the test data.

Table 2 shows the recall (*Rec.*) and the precision (*Pre.*) in the discriminations. *Rec.* and *Pre.* are defined as

$$Rec. = \frac{X_{h \cap c}}{X_h} \times 100 \qquad (11)$$

$$Pre. = \frac{X_{h \cap c}}{X_c} \times 100 \qquad (12)$$

where X_h is the set of the correct answers; X_c is a set of images discriminated by the proposed method and X is the number of images of a set X. Table 2 shows the ratios which have been more than 80% in most cases in any of Method 1 and normal cases in Method 2, and the ratios for abnormal cases in Method 2 has been worse.

6   DISCUSSIONS

6.1   *Performance of method 1*

Figure 7 (a), (b), (c) shows the distribution maps of all data by the abnormalities extracted by Method 1 (the number of users is 3 of A, B, and C), where ○ is normal case and × is abnormal case. From the figures, we can see the two cases are roughly separated without depending on users, and this suggests that Method 1 could be sufficiently applied to the CAD (K. Abe et al. 2013).

Figure 8 shows the result of the drawing for an image by each user. From the results, we can see that there are differences of the drawing between the users. In addition, although there are no significant differences in the discrimination ratios shown in Table 2 between the users, it is suggested that the abnormalities could depend on the user if the drawing depends on users. Therefore, it is necessary to reduce the difference between users by improving the manual more strictly.

Table 1.   The mean value and the standard deviation of the abnormalities in the proposed methods.

case	Method Type	AbnR		AbnRMAX	
		mean	SD	mean	SD
normal	Method 1 (userA)	2.3	0.2	2.5	0.3
	Method 1 (userB)	2.2	0.2	2.5	0.3
	Method 1 (userC)	2.1	0.2	2.4	0.3
	Method 2	2.4	0.4	2.7	0.3
abnormal	Method 1 (userA)	3.1	0.5	3.5	0.6
	Method 1 (userB)	3.0	0.4	3.5	0.5
	Method 1 (userC)	3.0	0.5	3.3	0.5
	Method 2	3.3	0.6	3.6	0.8

6.2 Extraction of abnormalities and discrimination of pneumoconiosis

As well as Method 1, Figure 7(d) shows the distribution map of all the data by the abnormalities extracted by Method 2. Besides, Table 3 shows inter-class variance and intra-class variance between normal and abnormal cases in the space distance of each figure shown in Figure 7. From Figure 7 and Table 3, we can confirm that an inter-class variance in Method 2 is smaller than the ones in Method 1 and an intra-class variance in Method 2 is larger than the ones of Method 1. These phenomena could be roughly brought by three patterns. The first one is a normal case where the lung marking exists in the rib areas. The abnormalities of the normal case are $AbnR = 4.1$ and $AbnRMAX = 4.1$, and we can see they are close to abnormal cases from the mean values and standard deviations shown in Table 1.

The two patterns are the case where the symptom is appeared at a part except the extracted rib areas and the case where the size of the rib areas is not enough. Regarding a case of the former pattern, the abnormalities are $AbnR = 2.4$ and $AbnRMAX = 2.5$, and we can see it is close to normal cases from the mean values and standard deviations shown in Table 1. Since the proposed method extracts only three rib areas, the whole extracted area might be still small for diagnosing any abnormal cases. To cope with this problem, we need to consider increasing the number of ribs extracted by the proposed method in addition to the size of rib areas. Regarding a case of the latter pattern, the abnormalities are $AbnR = 2.8$ and $AbnRMAX = 2.9$, and we can see it is close to normal cases from the mean values and standard deviations shown in Table 1 as well. The reason of error is because the size of rib areas extracted by Method 2 has been much smaller than Method 1 as Method 2 has been much smaller than Method 1 as shown in Figure 9 and the areas could not include symptom parts. In the future, it is necessary to propose to expand the edges by interpolations.

Table 2. Experimental results of the discrimination of pneumoconiosis by the proposed methods.

Method Type	method	Normal		Abnormal	
		Rec.	Pre.	Rec.	Pre.
Method 1 (userA)	RT	97.2%	92.1%	80.0%	92.3%
	NN	97.2%	89.7%	73.3%	91.6%
	SVM	91.7%	94.3%	86.7%	81.3%
Method 1 (userB)	RT	91.6%	94.2%	86.6%	81.2%
	NN	94.4%	97.1%	93.3%	87.5%
	SVM	88/8%	100%	100%	78.9%
Method 1 (userC)	RT	91.7%	97.0%	93.2%	82.4%
	NN	94.4%	97.2%	93.2%	87.5%
	SVM	94.4%	97.2%	93.2%	87.5%
Method 2	RT	80.6%	85.3%	66.7%	58.8%
	NN	91.7%	89.2%	73.3%	78.6%
	SVM	88.9%	91.4%	80.0%	75.0%

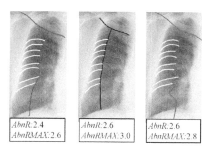

| AbnR:2.4 AbnRMAX:2.6 | AbnR:2.6 AbnRMAX:3.0 | AbnR:2.6 AbnRMAX:2.8 |

Figure 8. Results of the drawing for an image by the 3 users according to Method 1 (left: userA, middle: userB, right: userC).

Table 3. Inter-class variance and intra-class variance in Method 1 and 2.

Method Type	Intra-class variance	Inter-class variance
Method 1 (userA)	0.127	0.369
Method 1 (userB)	0.107	0.360
Method 1 (userC)	0.081	0.359
Method 2	0.170	0.337

In the extraction of the rib areas, two types (three images) are found as incorrect results. The first type is the case where the upper edges are not correctly extracted due to low contrast at the upper edges as shown in Figure 10. In the case when the contrast is low around the edges, the edges do not appear in the reference image. To cope with the case, we need to

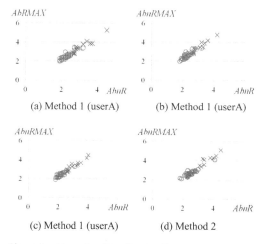

(a) Method 1 (userA)   (b) Method 1 (userA)

(c) Method 1 (userA)   (d) Method 2

Figure 7. Example of the rib extraction.

consider making the reference image where the upper edges appear without depending on contrast degree. And, the second one is the case where a part of the lung marking is tracked as a rib edge as shown in Figure 11 because the density distribution of the lung marking is similar to rib edges. Regarding the second problem, we need to consider recognizing the lung marking.

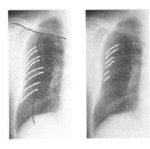

Figure 9.   Result of the extraction for an image (left: Method 1(userA), right: Method 2).

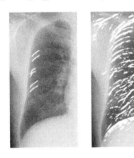

Figure 10.   Example 1 of incorrect result (left: result by Method 2, right: the reference image of the left image).

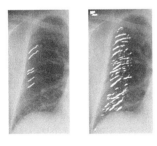

Figure 11.   Example 2 of incorrect result (left: results by Method 2, right: the reference image of the left image).

7   CONCLUSIONS

For the sake of application of the CAD system for pneumoconiosis for images scanned with CCD scanner to CR chest X-ray images, this paper presents two methods for customizing the system. As Method 1, this paper proposes a customization of the manual for drawing rib edges used in the system. And then, as Method 2, this paper proposes a method for extracting rib areas from CR chest X-ray images considering an automated CAD for pneumoconiosis. Experimental results of examining the proposed methods using 51 CR chest images shows that Method 1 is effective enough for the diagnosis and performance of Method 2 is worse than Method 1 though the discrimination ratios for normal cases has been more than 80%.

As future works, it is necessary to improve the manual shown in Method 1 more strictly to reduce differences between users, and then, to enhance performance of Method 2 by recognizing area of the lung marking and expanding rib areas which are not enough for extracting the abnormalities.

REFERENCES

[1] R.P. Kruger, W.B. Thompson & A.F. Turner 1974. Computer diagnosis of pneumoconiosis. Trans. on Systems, Man. and Cybernetics SMC-4(1): 40–49.

[2] H. Kobatake & K. Ohishi 1987. Automatic diagnosis of pneumoconiosis by texture analysis of chest x-ray images. Proc. IEEE ICASSP: 633–636.

[3] A.M. Savol, C.C. Li & R.J. Hoy 1980. Computer aided recognition of small rounded pneumoconiosis opacities in chest X-rays. IEEE Transactions on PAMI, 2(5): 479–482.

[4] T. Kouda & H. Kondo 2001. Computer-aided diagnosis for pneumoconiosis using neural network. Biomed. Soft Comput. Hum. Sci. 7(1): 13–18.

[5] M. Nakamura, K. Abe & M. Minami 2009. Quantitative Evaluation of Pneumoconiosis in Chest Radiographs Obtained with a CCD Scanner. Proc. of the 2nd International Conference on the Applications of Digital Information and Web Technologies (ICADIWT 2009): 673–678.

[6] K. Abe, T. Tahori, M. Minami, M. Nakamura & H. Tian 2013. Computer-aided diagnosis of pneumoconiosis X-ray images scanned with a common CCD scanner. Automation Control and Intelligent Systems 1(2): 24–33.

[7] International Labour Organization (ILO) 2011. Guidelines for the use of ILO international classification of radiographs of pneumoconiosis (Revised edition 2011). Geneva: International Labour Office.

[8] K.S. Fu, Y.P. Chien, & E. Person 1975. Computer Systems for the Analysis of Chest X-rays. Proc. EASCON '75 record: IEEE Electronics and Aerospace Systems Convention: 72A–72P.

[9] M. Loog & B. Ginneken 2006. Segmentation of the posterior ribs in chest radiographs using iterated contextual ppixel classification. IEEE Trans. on Medical Imaging 25(5): 602–611.

[10] Y. Yoshinaga & H. Kobatake 2000. The line detection method with robustness against contrast and width variation applied in gradient vector field. Systems and Computers in Japan 31(3): 49–58.

[11] F. Mosteller 1948. A k-sample slippage test for an extreme population. The Annals of Mathematical Statistics 19(1): 58–65.

Testing and Measurement: Techniques and Applications – Chan (Ed.)
© 2015 Taylor & Francis Group, London, ISBN: 978-1-138-02812-8

A study of the frequency converter control method for rotatory screen

H. Xu, J. Zheng, M.D. Li & X.S. Che

School of Information Science and Engineering, Shenyang University of Technology, Shenyang, Liaoning, P. R. China

ABSTRACT: The conventional rotatory screen usually adopts constant speed uniform rotation or two-speed uniform rotation. Long-term work will bring some problems, such as large power consumption, removing the dirt attached to the filter ineffectively, and the overloading of the revolving mechanism of the rotatory screen. The motor speed is adjusted in this design based on the water level before and after the rotatory screen in the method of closed-loop. The rotational speed of the rotatory screen changes according to the sine rule. In this way the dirt on the rotatory screen could loose under the work of inertia, and then the rinse can have a better cleaning of the dirt. A control system is composed by PLC, frequency converter, level sensors and conventional low-voltage electrical components and the expected functions are achieved.

1   INTRODUCTION

Rotatory screen system is an important component device on the industrial cooling water system. It can effectively intercept the debris as well as garbage from rivers and the sea with the use of upstream trash rack. It can also improve the water purity, so the rotatory screen is an important barrier in water treatment process. When the rotatory motor starts with power frequency, since the rotor is stationary in a stationary state, the stator rotating magnetic field cuts the rotor conductor with the fastest relative speed. A large rotor electromotive force and rotor current voltage drop are induced in the rotor winding which will affect the other loads connected to the same power.

This paper proposed a method which applies AC variable frequency to control the rotatory screen motor to start with variable frequency, and carries out the variable speed operation with the sine rule. The life of the rotatory screen motor is extended; the system efficiency is improved; and the energy is saved with the variable speed operation mode.

2   FREQUENCY ADJUSTMENT METHOD

In all types of three-phase asynchronous motor speed control systems, the variable voltage and variable frequency speed control system is the most efficient and the best performing system. The variable voltage variable frequency speed regulation system of the three-phase asynchronous motor is generally referred to frequency converter. The universal frequency converter can be divided into the step speed regulating and stepless speed regulating, according to the way of speed regulation control; it can also be divided into over base frequency and under base frequency according to the speed regulation range.

2.1   *The choice of speed control range*

For the rotatory screen filtration system, the dirt will always adhere to the rotatory screen, and the loads of the three-phase asynchronous motor are gradually increased or constant. By the formula:

$$P \times 9550 = T \times n \qquad (1)$$

where P is electrical power; T is torque, and n is speed. According to the electric machinery knowledge, adjusting speed above the base frequency is a speed regulation with constant power. In this speed regulation, torque T reduces when speed n increases. This will make the motor lack of the capacity of loading, and this can result in overloading of the motor. As a result, it can not achieve the purpose of saving energy. Therefore, the speed will be regulated under the base frequency in this system.

2.2   *The choice of speed control mode*

The step speed regulating mode is the frequency of the three-phase asynchronous motor to discretely change by control of the frequency converter, and the connection between each frequency value could be linear mode or S type mode to realize acceleration and deceleration. In general, the frequency converter can output frequency values of 16 segments. While the stepless speed regulating mode can output the continuously varying frequency, it means that motor speed changes continuously. According to formula (1), when speed regulation is under base frequency, as the flux of electromagnetic is constant, the torque can be considered as constant fundamentally. So the electrical power is proportional to the motor speed. To the three-phase asynchronous motor, the speed formula is,

$$n = 60(1-s) \times f/p \qquad (2)$$

where f is frequency; p is number of poles; s is slip. The electrical power is proportional to the motor's frequency by formula (1) and (2).

At step speed regulating mode, the speed is set as 2 segments, and the frequency's output timing is showed in the Figure 1a.

The initial frequency is set as 25Hz, that is to say when the motor starts operation, the rotational frequency is 25Hz. 0 to T is its speed regulation process, $T1$ to $T3$ is the process for acceleration, $T4$ to $T2$ is the process for deceleration, and 0 to $T1$, $T3$ to $T4$ and $T2$ to T is the time for holding. Acceleration process and deceleration process should not be too fast or too slow in order to prevent the motor from appearing the fault of stalling and over-current. This system selects the linear mode in which the slope of the rise process (acceleration) and the decrease process (reduction) is 1. As mentioned above, the integration of the frequency over the time (the area of the frequency over the time) is proportional to the power, and the rotatory screen that works in the unit time is also named proportional to the power consumption. Its area is:

$$S1 = 25T + \frac{(T2-T1)+(T2-T1-50)}{2} \times 25 \qquad (3)$$

As its rise (acceleration) and fall (deceleration) processes are set in a linear equation in which the slope is 1, so $T3-T1 > 25$, $T2-T4 > 25$, and $T4-T3>0$. $T2-T1 > 50$, thus $S1>25T+625$. It can be easily proved that, when speed is set more than two segments, the power it did in unit time is greater than the speed of the work done during two segments.

At stepless speed regulating mode, a sinusoidal variation of speed regulation mode is used in the system, and the frequency of the output timing diagram is shown in Figure 1b.

The initial frequency is also set as 25Hz, and the time of 0 to T is set as its speed regulation process. Its analytical expression is,

$$f(t) = 25(\sin t + 1) \qquad (4)$$

where t is time. $f(t)$ is output frequency. By mathematical knowledge, its area is,

$$S2 = \int_0^T 25(1+\sin t)\,dt = 25T + 25(1 - \cos T) \qquad (5)$$

Because $(1-\cos t) \leq 2$, $S2 \leq 25T+50$. By comparing the two equations, it can be got $S1>S2$. It means that it is more conducive to save energy by using stepless speed regulating during the unit time. In the process of stepless speed regulating, its rotation rate changes from time to time, and its acceleration also changes constantly. This change will make inertia force to be generated in the course of the rotation of the filter acceleration, and then make the stubbon dirt that is attached to the filter loose under the work of inertia. As a result the dirt can be washed off easily.

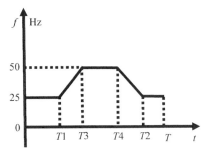

a) The step speed regulation timing diagram

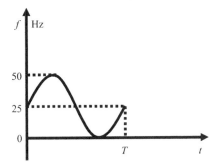

b) The stepless speed regulation timing diagram

Figure 1.   The frequency of the output timing diagram.

3   SYSTEM COMPONENTS AND WORKING PRINCIPLE

The control system consists of PLC, frequency converter, motor, level sensors and other peripheral devices. System block diagram is shown in the Figure 2. The sea level and water inlet level are measured by the water level sensors over the water surface in real-time, and the level values are fed into the PLC. When the water level difference between the sea level and the water inlet level is greater than the setting value, the PLC drives the rotatory screen to work. The water level difference will fall back to a range under the set value after the rotatory screen runs a certain time, and the PLC controls the rotatory screen to an idling rotation. Namely, motor is running in the lowest speed.

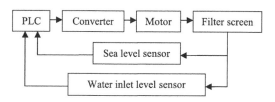

Figure 2. System block diagram.

4 MOVEMENT REGULATION OF THE REVOLVING FILTER SCREEN

When the system starts running, frequency converter output frequency is controlled by PLC from 0Hz up to 25Hz with the sine rule. In this case the rotatory screen motor is idled and maintains at this frequency. The traditional rotatory screen level control mode is mostly single value comparison mode, and it's easy to damage equipment. The system uses the difference between the way back to the level values to control the motor running, and it's a hysteresis comparison. The way of control is shown in Figure 3. When the level difference Δp between sea level and water inlet level exceeds the set value D2, PLC gives the frequency converter sinusoidal drive signal to control the output frequency in accordance with sinusoidal motion. When the water level difference is detected less than $\Delta 2$ and greater than $\Delta 1$, the speed of rotatory screen is still changed by sinusoidal. When the water level difference Dp is detected back to the set value $\Delta 1$, the idle drive signal is given to control the frequency converter idle running by PLC. So the rotatory screen motor runs at the lowest speed, and the purpose of saving energy can be achieved.

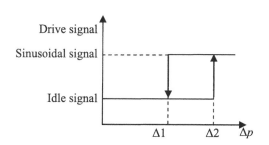

Figure 3. Hysteresis comparison control mode.

5 SYSTEM HARDWARE CONNECTION

The system consists of the Siemens smart series PLC, RS485 signal board, analog input module (EM231), circuit breakers and ABB's ACS510 series frequency converter. The water level difference data which is handled by PLC is compared with the set value, and then determines the working status of the rotatory screen at this time and controls the speed of motor through the drive signal. The system is divided into automatic and manual modes. When the switch is put on the manual position, the rotatory screen is controlled by the button, and the output frequency is adjusted by a potentiometer. It also means that the output frequency of the frequency converter is adjusted by the analog voltage. When the switch is put on the automatic position, the rotatory screen can be adjusted according to the output of the frequency converter real-time water level difference signal.

The system is connected with frequency converter though a RS485 signal board to communicate via Modbus protocol. Comparing to external analog frequency, RS485 interface has a stronger anti-interference ability, higher transfer rate, available auxiliary parts and other advantages. The PLC is set as the main station, frequency converter is set as slave, and sinusoidal frequency value is written from the main station to slave station via PLC programming software. Frequency converter operates according to frequency values written in the register. The principle is shown in Figure 4.

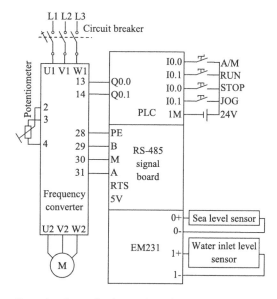

Figure 4. System hardware schematics.

6 SYSTEM SOFTWARE DESIGN

In the system software, the program mainly includes the process of analog signal, the hysteresis comparison of the water level difference signal, the generation of sine wave and the communication with the frequency converter via Modbus protocol. The generation of sine wave numerical and Modbus communication are the key to the system.

6.1 The generation of sine wave

The period of sine wave generator is set at 20s, and its frequency is 0.05Hz. According to the sampling theorem, the sampling frequency should be greater than 0.1Hz; the sampling period is less than 10s. Two 100ms timer T205 and T204 are used in the system to combine the pulse generator whose cycle is 2s. Namely, the sampling period is 2s. In the cycle of 20s of the sine wave, our sampling points are 10, namely, in a period of this sine wave of 0 to 2π, each sampling point spacing is $2\pi/10$, approximately 0.628. By using the counter C2 to count the 2s pulse generator, counter will reset when it reaches 11. For floating-point sine operation instruction comes from Siemens STEP7 software, we can calculate the corresponding sine value in the 10 sampling points. Finally, through standardization, rounding and other operations obtain the sine wave formula $f(t)=25(\sin t+1)$.

6.2 Modbus communication

When the PLC communicates with frequency converter via Modbus protocol, firstly, communication baud rate and parity need to be set. The PLC master station is set in the program and the frequency converter slave station is set on the operation panel. After the basic setup, the frequency converter internal control word register and the transmission parameters need to be set. In particular, when the system is powered on, depending on the difference of the operation mode (automatic/manual), different command parameters and control word are sent from PLC master station to frequency converter slave station. The control word registers are set as shown in Table 1.

In manual mode, the frequency converter output frequency is adjusted by potentiometer. In automatic mode, the PLC will store a cycle of sine wave values, and then the values are written into the frequency converter slave station through Modbus master station command. Frequency converter slave station write requesting enables bit pulse interval to be consistent with the sampling interval of sine wave or greater than it to avoid the communication errors.

When the system needs to control multiple rotatory screens, the PLC shall read and write converter slave station in step-by-step and time-sharing,

Table 1. Control word register setting.

| | Operating mode | |
Control word registers	Automatic mode	Manual mode
Start, stop and direction control parameters are 41001	8	0
External control to choose parameters are 41102	8	1
Given value choosing parameters are 41103	10	2
Profile control parameters are 40001	1142	1151

it means reading and writing with only one frequency converter slave station at the same time, and we should use the completion bit (Done) of above instruction to activate the enable bit (EN) of the next instruction for multiple read and write instructions, to ensure that there will be no communication error in all read and write instruction process cycle in the program.

7 CONCLUSION

A high efficiency, energy saving control of rotatory screen is completed which is based on frequency converter technology, and the stepless speed regulating mode are proposed according to the sine rule of output frequency. Theoretically, it is proved that the stepless speed regulating mode according to the sine rule of change relative to the step speed control mode has an advantage of saving electricity. Modbus communication protocol is used to control frequency converter to output the sine wave signal that is realized. According to the factory feedback, this system has a good effect in energy saving, continuous operation ability and other aspects.

REFERENCES

[1] Zheng, J. Gao D Y Ning Y D & Sun X L. 2014. The control method of rotatory screen based on frequency conversion technology, *The Eleventh Shenyang Science Symposium. shenyang,26 June,2014.*
[2] Sun, H. X., Dong, Y. & Zheng, Y. 2011. *Electric Drive and Realize Frequency Conversion Technology,* Beijing: Chemical Industry Press: 208–212.
[3] Wang, Y. M. 2010. *The Application of the Frequency Changer and PLC in the Multiple-speed Control of the Conveyer Belt,* Master Degree Thesis. Suzhou: Soochow University.
[4] Liao, C. C. 2012. *The PLC programming and applications,* Beijing: Mechanical Industry Press: 163–167.

Testing and Measurement: Techniques and Applications – Chan (Ed.)
© 2015 Taylor & Francis Group, London, ISBN: 978-1-138-02812-8

Vision displacement sensing system using digital image correlation for X, Y, and theta micro-positioning stage

D. H. Lee
Department of Mechanical Design and Mechatronics, Graduate School of Hanyang University, Haengdang 1-dong, Seongdong-gu, Seoul, Korea

M.G. Kim
Department of Mechanical Design and Production Engineering, Graduate School of Hanyang University, Haengdang 1-dong, Seongdong-gu, Seoul, Korea

S.W. Baek & N.G. Cho
Department of Mechanical Engineering, Hanyang Universiy, Sa 3-dong, Sangrok-gu, Ansan-si, Gyeonggi-do, Korea

ABSTRACT: In this paper, a vision based measurement system for precise feedback control of a three degree of freedom stage was proposed. A method of measuring 3 degree of freedom displacements with two vision cameras and two reference images was also presented. The vision system was applied to a flexure based precise three degree of freedom stage and evaluation test was implemented. It was confirmed that the proposed measurement system is valid for evaluation of a three degree of freedom stage.

1 INTRODUCTION

The importance of precision positioning stage has been increasing in precision measurement instruments, semiconductor manufacturing instruments, and high precision manufacturing machines. The flexure hinge consists of not many parts but one part itself and has a merit of smooth and continuous movement by elastic deformation. Piezoelectric actuators have been widely used as fine positioning devices because of their advantages such as high positioning resolution, high generative force, high speed of response and easy miniaturization [1]. Therefore, many research results about designing and manufacturing high precision stages using flexure hinges and piezoelectric actuators have been presented.

The end-effector of the stage with flexure hinge and the piezoelectric actuators has a several times longer working distance by the lever ratio than that of the piezoelectric actuator. By amplifying the working range, the stage can be used for wafer alignment, ultra-precision machining, and others. However, the manufactured stage has driving error due to machining tolerance, numerical analysis limit, and others. To calibrate the stage for the error, the displacement needs to be measured by a measurement system. Capacitance sensors are generally used for the calibration [1]. Capacitance sensors have fine resolution but very narrow measuring range. Besides, capacitance sensors can only measure displacement of steal surfaces. Therefore, they are difficult to apply to measure large range stages.

In this paper, a vision based measurement system to feedback a three degree of freedom stage is presented. The system consists of two vision cameras and two reference images.

2 VISION DISPLACEMENT SENSING SYSTEM USING DIGITAL IMAGE CORRELATION

The measurement system consists of two vision cameras, two telescopic lens, and two reference images. By measuring the two degree of freedom displacements, using digital image correlation (DIC) [2, 3], of each reference image which is attached on the moving part of the stage, the three degree of freedom displacements of the stage can be calculated. The measurement system schematic is shown in Fig. 1(a) and the principle to calculate three degree of freedom displacement of the stage by applying DIC with the two reference images is shown in Fig. (b). White and black random patters were used in the images.

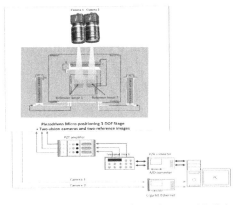

(a) Schematic drawing for the experimental setup with digital image correlation technique.

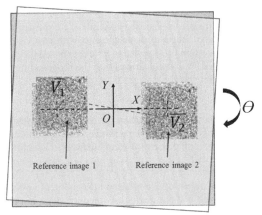

(b) Theta measurement using digital image correlation technique.

Figure 1. A schematic of the experimental setup for full traveling stage with two-vision cameras and two-reference images.

3 EXPERIMENTS

To verify the proposed method, an experimental system was set up as shown in Fig. 2. The used cameras (ALLIED Vision Technologies, Mako G-223 PoE) have 2048 x 1088 resolution and its pixel size is 5.5 μm. The working distance is 63.3 mm and magnification is x1.0 lens of the lens (OPTO engineering, TC23009, tele-centric lens).

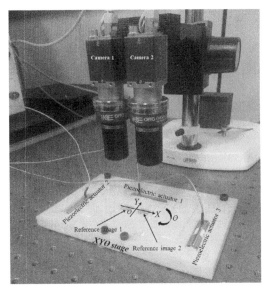

Figure 2. Experimental setup with two-vision cameras and two-reference images.

For the verification of the measurement system, a three degree of freedom stage was used, which had been developed using a 3D printer by the authors of this paper. The working distance is about 230 μm × 230 μm and ± 1000 arcsec. The X and Y axis resolution is about 50 nm. An experiment to calibrate the lever ratio over the overall working range was implemented.

Ten levels of input voltages over 0 to 100 V were input to the piezoelectric actuators. The displacements of end-effector were measured and are shown in Table 2 and 3. The average lever ratio for full-scale operating range is about 3.8 and its standard deviation is ±0.04. Comparing the measured lever ratio with result of FEM analysis, the real measured lever ratio was a bit smaller but the similar with the simulation result. A steady lever ratio for the full operating range was checked.

Another experiment for theta rotation over full-scale operating range was also implemented by driving the actuator 2 and 3. The results are in Table 4 and 5. Average lever ratio is about 16.4 μm/arcsec and its standard deviation is ± 0.17 μm/arcsec. By the experiment, it was checked that the stage operating ranges are 236 μm × 246 μm and ± 1030 arcsec. A higher resolution camera is required for more precise experiments of hysteresis and resolution.

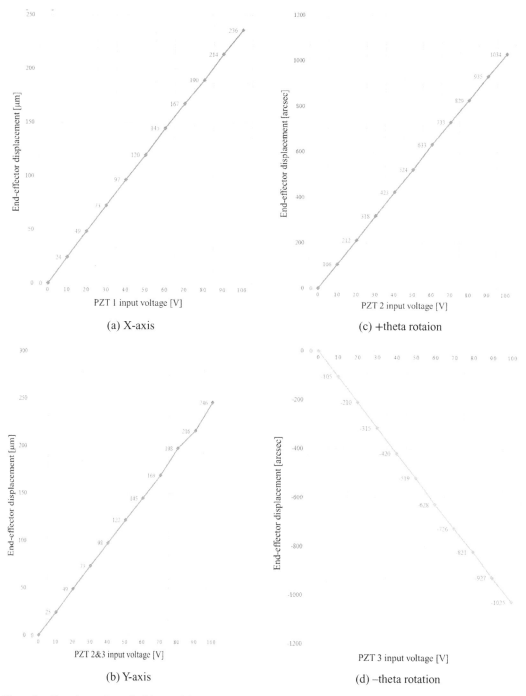

Figure 3. Experimental resutl with two-vision cameras and two-reference images.

4 CONCLUSION

A vision based measurement system was developed for precise feedback control of a three degree of freedom stage. A method of measuring 3 degree of freedom displacements with two vision cameras and two reference images was presented. The vision system was applied to a flexure based precise three degree of freedom stage and evaluation test was implemented. As a result, it was confirmed that the proposed measurement system is valid to evaluate a three degree of freedom stage.

ACKNOWLEDGEMENTS

This work was supported by the Basic Science Research Program through the National Research Foundation of Korea (NRF) funded by the Ministry of Science, ICT & Future Planning (NRF-2013R1A1A2013139)

REFERENCES

[1] Kwon, Kihwan, Nahmgyoo Cho, and Woojin Jang. "The design and characterization of a piezo-driven inchworm linear motor with a reduction-lever mechanism." JSME International Journal Series C 47.3 (2004): 803–811.

[2] Hild, Francois, and Stéphane Roux. Digital image correlation. Wiley-VCH, Weinheim, 2012.

[3] Pan, Bing, et al. "Two-dimensional digital image correlation for in-plane displacement and strain measurement: a review." Measurement science and technology 20.6 (2009): 062001.

Testing and Measurement: Techniques and Applications – Chan (Ed.)
© *2015 Taylor & Francis Group, London, ISBN: 978-1-138-02812-8*

On statistical adaptation of the order filters for periodic signals processing

V.I. Znak

The Institute of Computational Mathematics and Mathematical Geophysics of SB RAS, Russia

ABSTRACT: In this paper, restoration of the quality of noise periodic and frequency-modulated signals under condition of preservation of the waveform of an original signal is considered. For this purpose it is offered to use the Weighted Order Statistics (WOS) filters. However, the analytical estimation of the behavior of the WOS filters is complicated enough because of their nonlinearity, what allows us to consider the WOS filter response as a casual event. For this reason, the statistical trials method seems to be promising in selecting more qualitative WOS filters projects. For solving the corresponding task, a set of the WOS filters (filters bank) is attracted under condition of work frequencies variation of the CoPh WOS filters. The facility used is a specialized interactive graphical computer system interface. We intend to propose some experience in gained field.

1 INTRODUCTION

We will consider some issues of adaptation of the weighted order statistics (WOS) filters for processing the periodic signals. The periodic and the frequency-modulated (FM) signals are widely used. Let a periodic signal be a one-dimensional time series $Y=\{y_1,\ldots, y_N\}$ recorded at discrete instants of time $t_1,\ldots, t_N, (t_{i-1}-t_i=\Delta t=const, i=2,\ldots, N)$.

Specific features of the order filters were considered by Justusson (1981). They include the following aspects: 1) a remarkable ability of the impulse noise removal, 2) noise robustness, 3) preservation of steps for a signal in the form of a telegraphic sequence, 4) the response of the filter tends to zero while the filter length (n) approaches an integer number of signal periods, 5) nonlinearity.

Taking into account the restriction on the length of the WOS filters in the processing of periodic signals, the development of the co-phased WOS (CoPh WOS) filters was given a special attention by Znak (2005). An informal definition of such filters is illustrated in Figure 1. It is the following: the equality to zero of the weights of all the terms on the filter input except the weight of the central element (CE) and the weights of such terms, which are apart from one another, by the length of the period harmonic function (co-phased) provided the filter length is equal to m periods (where m is even ($m=2,4$, etc.).

Because of the complication of the analytical estimation of the behavior of the WOS filters, it can be supposed that their response is a casual event in the general case. The numerical method of solving mathematical tasks with attracting modeling the casual events, or statistical trials, for selecting the most effective project of the WOS filter is of interest.

The use of different filters projects and variation of the corresponding filters parameters allows us to

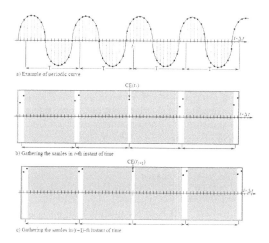

Figure 1. Sampling of periodic signal values closest to the CE phase at different instants of time.

change the quality of the results of signal processing. Thus, the task of interest is the monitoring of the quality of a restored signal depending on using different projects of filters and their parameters. For this purpose, we propose to use a specialized computer system (Znak, 2013) and to attract a certain set of the WOS filters pooled in the corresponding filters bank.

2 THE BASIC DEFINITIONS OF THE WOS FILTERS

Before the considering of the research into projects of the WOS filters, it is worthwhile to the give basic definitions of such objects.

Let a sequence $Y = \{y_j: j=1, \ldots, n\}$ present a signal that includes n quantities of numerical data or samples. Here, the rth–order statistics $y_{(r,n)}$ is defined as the rth quantity in size. Let a ratio $\alpha=(r-1)/(n-1)$ be a measure of the size of a corresponding sample on the sequence Y. Let this ratio be called the percentile of the WOS filter. Then the idea of the order statistics is transformed to the percentile form. A sample of the sequence Y is called α-order statistics, denoted as $y_{(\alpha,n)}$. If there exists a number $n_\alpha=(n-1)\times\alpha$ of $y_{i(\alpha)}$ values, and a number $n_\beta=(n-1)\times(1-\alpha)=(n-1)\times \beta$ of $y_{i(\beta)}$ values are provided:

$$\begin{cases} y_{i(\alpha)} \leq y_{(\alpha,n)}, \\ y_{i(\beta)} \geq y_{(\alpha,n)}, \\ y_{(\alpha,n)}\overset{n_\alpha}{\underset{1}{\bigcup}}y_{i(\alpha)}\overset{n_\beta}{\underset{1}{\bigcup}}y_{i(\beta)} = Y, \end{cases} \quad (1)$$

and $\alpha+\beta=1$

Now, a formal definition of the order filtration procedure can be presented as a sequence of the following operations:

1 $Y=\{y_i, i=0,\pm1, \ldots, \pm v,\}$
   is the sampling of $n=2v+1$ signal values, where $Y\subset X$ (n is odd),
2 $\tilde{Y} = \tilde{y}_{c-v} \leq \ldots \leq \tilde{y}_c \leq \ldots \leq \tilde{y}_{c+v}$,
   is construction of a variational row, where the term $\tilde{y}_i$ is statistics of a corresponding order, $i = \overline{1,n}$.
3 $\text{RANK}(y_1,\ldots, y_{(n-1)/2},\ldots, y_n)= \tilde{y}_r$
   is the operation of replacement of the central term (CE) $y_i\in Y$ by the statistics $\tilde{y}_r \in \tilde{Y}$ (v, c, $i\in Z$).

In a special case, if $r=(n1)/2$ (i.e., $\alpha=0.5$), a filter is a median one $MED_n(y_1,\ldots, y_{(n-1)/2},\ldots, y_n)= \tilde{y}_c$.

Let us have a set W of the quantities w_i ($i=1,\ldots, n$), each quantity w_i being associated with the sample $y_i\in Y$. This $w_i\in W$ is called a weight and can be considered to be a number of copies of the corresponding sample $y_i\in Y$. Weights are introduced for emphasizing some elements of the sequence. The extended sequence Y thereby gains a new quality as a set with a number of elements $N = \sum_{i=1}^{n} w_i$. At the same time, N is also odd.

According to the concept of weights, the corresponding set W of the CoPh WOS filter includes such a subset $W^*\subset W$ of weights w^*, where $\forall w^*\in W^*$: $w^*= 0$.

According to Znak (2005), the algorithm for calculating the weights of the CoPh WOS filter is as follows:

1 Computation of the length of the filter as $[\bar{n}]$, where $n=2RT/\Delta t+1$;
2 Realization of assignments: $w_0 =1$ for CE and $w_{\pm i} =0$ for other components of the sequence X: $i=1,\ldots, (n-1)/2$.
3 Computation in the loop $j=1,\ldots,R$ of the indices of components with nonzero weights, and

computation of the corresponding weights according to the rules:

i. $Vj = jT / \Delta t$;
ii. if $V_j = [\hat{V}_j]=[\check{V}_j]$, then $w_{\pm V_j} = 1$; else:
iii. $w_{\pm[\check{V}_j]} = V_j - [\check{V}_j]$;
iv. $w_{\pm[\hat{V}_j]} = [\hat{V}_j]-V_j$.

Here T is the period of a signal, the symbol $[\hat{x}]$ denotes the smallest integer (that is greater or equal to x or the nearest from above) and $[\check{x}]$ is the largest integer, which is less or equal to x (the nearest from below). The corresponding frequency, which defines values of the filter weights, will be called the work frequency.

3 THE BASIC PROCEDURES OF THE STATISTICAL TRIALS METHOD OF THE WOS FILTERS PROJECTS

It is possible to note the following parameters of the WOS and the CoPh WOS filters that influence the quality of signal processing:

values of the weights $W=\{w_1, w_2, \ldots\}$ of the WOS filters as well as the distribution of zero weights which are the functions of frequencies in the case of the CoPh WOS filters;
the size of the aperture or the window analysis;
a sequence of operations in the multistage process of filtration;
the type of operations;
percentiles of the WOS and the CoPh WOS filters.

As was established by Znak (2005), the CoPh WOS filter of the two-period length of a FM signal will save in its band the width $\Delta f(\text{CoPh WOS})\in 1.0\div1.5$ Hz. That is, if the frequency band $\Delta f(\text{FM}) > \Delta f(\text{CoPh WOS})$, it is needed to attract a special technique for restoring a source signal in the whole frequency band. Such a case was studied by Znak (2012), but now we will not discuss it.

The algorithm of the corresponding signals processing, which provides a variation of the above data and parameters in the course of signal processing, is presented in Figure 2.

Here, points of the basic notations are the following:

− $\{filter_i: i=1,\ldots, I\}$ is a filters set,
− filter(beg)/filter(end) is the 1st/last filters used;
− $filter_i=\{node_{k(i)}: k(i)=0,\ldots, K(i)-1\}$ is a sequence of data of consequent steps of the i-th filter of the corresponding bank;
− $\{f_l: l=1,\ldots, L\}$ is a frequencies set;
− $f(\text{beg})/f(\text{end})$ is the 1st/last work frequency;
− Δf is the step size of frequency updating;
− $s_{\Delta f}$ is the sequence steps of frequency updating, $\{s=0,\ldots, S_{\Delta f}-1\}$;

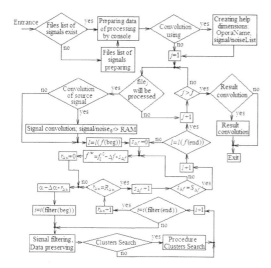

Figure 2. The algorithm of monitoring the filters projects.

- f^w is the current work frequency;
- $\Delta\alpha$ is the step size of procentile updating;
- $r_{\Delta\alpha}$ is the sequence steps of procentile updating, $r = 0, \ldots, R_{\Delta\alpha} - 1$;

Other notations are used in the course of the cluster analysis and will not be discussed here.

In the general case, a filter includes a sequence of processing nodes. Here, the list of available filter operations is as follows:

0 Transfer: data$\in$N1=>N3;
1 Data composition: (data$\in$N1+data$\in$N2)=>N3;
2 CoPh WOS filter;
3 Coupled CoPh WOS filter;
4 Co-phased average;
5 Coupled co-phased average;
6 Standard WOS filter;
7 Coupled standard WOS filter;
8 Standard average;
9 Coupled standard average.

The filter node as such is a sequence of the following data: $N_{operation}$, L1, $W1$, $\alpha1$, L2, $W2$, $\alpha2$, N1, N2, N3. Here, $N_{operation}$ is order number of the current operation; L1, L2 are lengths of the filters; $\alpha1$, $\alpha2$ are procentiles of the filters; $W1$, $W2$ are distributions of weights of the filters; N1, N2, N3 are numbers of work files, where N1, N2, N3 $\in \{0, 1, \ldots, 101\}$.

Thus, the user can employ both nonlinear and linear operations. The efficiency of such an approach was demonstrated by Neuvo (1994).

In the course of data processing, each step of action, which is connected with a corresponding node, can be presented by the scheme: processing data of the work file N1 with a consequent transfer of the results obtained the work file N3. The new work

files are created in the course of running the filtering procedure. Some nodes of a filter can include a coupled action of the same type, but with different data. Then, data L2, $W2$, $\alpha2$ are attracted, and the results of both actions are composed.

The work file of the source node of each filter is associated with number "0". At the same time, the work file of the last filter node is associated with number "101".

The processing operations sequence and corresponding nodes data are defined by using a computer console in the interactive mode. A flow-chart of creating the procedure of filters bank is presented in Fig. 3.

The basic features of graphical interface are presented by Znak (2013) and will not be considered here.

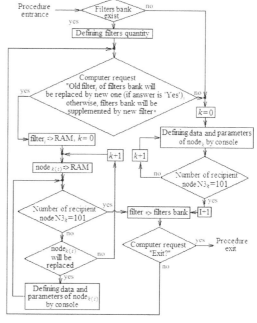

Figure 3. A flow-chart of the procedure of creating the filters bank.

4 SOME RESULTS OF SIGNAL PROCESSING

The methodology in question was investigated on the models of linear frequency-modulated signals using numerical modeling. Parameters and characteristics of the corresponding signals are as follows: 1) the sound signal (opora): start time is 0 s; 2) the noise signal: the time of signal start is 4 s. At the same time, the period of signal existence is 1200 s, a bandwidth is 7.2 Hz ÷ 8.2 Hz, the digitization frequency $\Delta t = 0.08$ s. in both cases. The white noise with zero average Gaussian distribution was used for obtaining a noise signal. At the same time, the filters bank, which contained 12 filters with operations from the above list, was used.

The results of the experiments conducted are presented in Table 1, where $š/ξ$ is the signal-to-noise ratio obtained as a result of convolution of the noise signal with the opora. At the same time, $š/ξ_0$ is the above ratio and s/n is the estimation of the ratio of the mean square deviation of the opora and the noise signal data before the processing.

Table 1. Results of convolution of filtered noise signal with opora.

$š/ξ_0=100.1 -$ s/n=0.2	$š/ξ_0=87.6 -$ s/n=0.1	$š/ξ_0=10.1 -$ s/n=0.01
$f\,\mathrm{Hz} - s/ξ$	$f\,\mathrm{Hz} - s/ξ$	$f\,\mathrm{Hz} - s/ξ$
7.605 – 120.9	7.7 – 93.0	7.65 – 14.4
7.6075 – 111.3	7.71 – 96.8	7.67 – 15.9
7.6125 – 114.2	7.72 – 94.1	7.68 – 10.2
7.6175 – 121.1	7.73 – 00.4	7.69 – 10.1
7.62 – 130.6	7.74 – 91.0	

5 CONCLUSION

In this paper, the approach to the statistical adaptation of the WOS filters for processing the frequency-modulated signals is proposed. The basic points of this approach are: attracting a filters bank and a variation of filters data and parameters of the processing.

The corresponding algorithm of filtering the FM signals in the conditions of statistical trials of the WOS filters projects was developed, and the procedure of creating the filters bank is presented. The results of the experiments conducted demonstrate the dynamics of dependence of the signals quality on the value of work frequency of a filter and its structure.

REFERENCES

[1] Justusson, B.I. 1981 Two-dimensional digital signal processing II: transforms and median filters. *Median filtering: statistical properties*, 161–196, Springer-Verlag: Berlin,.

[2] Znak V. I. 2005. Co-Phased Median Filters, Some Peculiarities of Sweep Signal Processing. *Mathematical Geology*, 37(2): 207–221.

[3] V. Znak. 2013. On statistical adaptation of the order filters for processing periodic signals using graphical interface. *Pattern Recognition and Image Analysis: New Information Technologies; Conference Proc. of the 11th International Conference (PRIA-11-2013), Samara, The Russian Federation, 23–28 September 2013*. Samara, 483–486.

[4] Vladimir I. Znak. 2012. Something about processing, analysis and restoration of periodic signals, In M. Petrou (ed), *Proc. of the Ninth IASTED Intern. Conference on Signal Processing, Pattern Recognition and Applications (SPPRA 2012), Crete, Greece, 18-29 June 2012*. ACTA Press, 154–161.

[5] Yin and Y. Neuvo. 1994. Fast adaptation and performance characteristics of fir-wos hybrid filters. *IEEE Transactions on signal processing*, 42(7), 1610–1628.

A study on the automated checking system for labels of medicine bottle

X.S. Che & Y. Li
School of Information Science and Engineering, Shenyang University of Technology, Shenyang, Liaoning, P. R. China

D.J. Liu
Shenyang Women and Children's Hospital, Shenyang, Liaoning, P. R. China

ABSTRACT: The manual checking method in the injection pharmacy admixture preparation process often causes large workload and error prone. In this article, we proposed an automated checking system in order to solve this problem. The proposed system consists of two parts. The hardware part uses ARM as the core control device including an image acquisition circuit, an LCD display circuit and an Ethernet interface circuit. The software algorithm part includes image preprocessing part and label image recognition part based on the method of feature extraction and image matching. The checking results for the bottle labels display on the user interface. Experimental results show that this system can effectively check the bottle label in the intravenous medicine preparation process in order to reduce the workload of manual checking and improve the efficiency of the preparation.

1   INTRODUCTION

Intravenous administration is an important clinical medical treatment method. Before intravenous infusion solution, powder therapy drugs need to be added to the media such as saline and glucose   which is called pharmacy admixture. However, this preparation process is not yet automated. Checking the bottle label by the pharmacists manually leads to the increase of the workload for pharmacists greatly and there exists a possible safety hazard. Therefore, this paper studied a system for automatically checking the medicine bottle label in the preparation process.

There is a variety of methods to get the information on the medicine bottle labels. One method is to extract and recognize the characters and numbers on the labels. However, the bottle is a curved surface and there are characters compressed and missed on both sides of the label image, which causes difficulty for image processing. Another method is a matching algorithm based on the histogram of gray label images. It is difficult to select an appropriate threshold for these related algorithms based on pixel gray values because the statistical characteristics of pixel gray value is not obvious and the accuracy of the results is not high. After studying these methods, we proposed a novel algorithm based on the local statistical characteristics of label image pixels used to match the label image with the known template. We also designed a hardware system to acquire the label images.

2   HARDWARE SYSTEM DESIGN

2.1   *Hardware system circuit structure*

The hardware circuit part of the automated checking system for labels of medicine bottle is mainly used to acquire label images. The system consists of the ARM processor and peripheral circuits, image sensor and image cache circuits, LCD display circuit as well as an Ethernet interface circuit. The core control device in this system is the chip named STM32F407, which has low power consumption, high performance, large storage space, and fast data transmission speed. The image sensor used in this system is OV9655, which supports a variety of image resolution and image output formats such as RGB565, RGB555, YUV422 and YCbCr422. In addition, OV9655 also contains a number of image processing functions and can control the quality of the image well. The screen used in this system is a TFT touch screen with a resolution of 320×240. The screen GYTF024TM81TP-37B has many features including fast response speed, good brightness, high contrast and bright color. The main device in the Ethernet interface circuit is DP83848 integrated Ethernet controller chip. This chip is a 10/100Mbit/s single port physical layer device with low power consumption, several intelligent power down states and supports MII and RMI interface mode. The block diagram of the hardware system is shown in Figure 1.

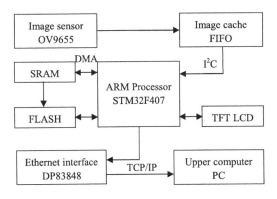

Figure 1. The block diagram of the hardware system.

2.2 *Hardware system program design*

At the beginning of the program, various program modules need to be initialized, including the STM32F407 chip, the image sensor and the LCD TFT screen. We set the size of the image acquired by the camera to 320×240 by setting the camera registers and the output image format is also set to RGB565. The reading and writing operations to the camera register is achieved by using I²C bus communication. In the initialization configuration process of LCD TFT screen, we need to set the display direction of the image so that the image acquired by the camera can be displayed on the LCD screen fully. After initialization, we can turn on the camera. In this way, the system can run the preview function. This function is achieved by reading the image data from OV9655 register directly to the LCD buffer in order to display real-time images collected by the camera. The next stage, the system waits for the command to take image. If we click the taking image area on the touch screen, the program will acquire a frame of the image and save the image data. Due to the slow speed of flash operation, the image data need to be stored in SRAM first by DMA mode. It can be calculated that the size of one frame of image is 320×240×16bit, that means 150K. We know that the address of SRAM of STM32F407 is discontinuous. So when a frame of image is obtained, it is divided into 16 parts and the first 12 parts is saved in the SRAM and other 4 parts in CCM data RAM. Part of the program in language C is as follows:

```
DMA_Cmd ( DMA2_Stream1, DISABLE);
DCMI_Cmd ( DISABLE);
DMA_ClearFlag(DMA2_Stream1,0x2F7D0F7D);
DMA_ClearFlag(DMA2_Stream1,0x0F7D0F7D);
DMA_ClearITPendingBit(DMA2_Stream1,DMA_
IT_TCIF1);
    datablock = 0;
```

```
DMA2_Stream1->M0AR = 0x20000000 + 9600 *
(datablock % 12);
    DMA_SetCurrDataCounter(DMA2_Stream1,
(320 * 240 * 2 / 4) / 16);
    DCMI_Cmd(ENABLE);
    DMA_Cmd(DMA_Stream1, ENABLE);
```

The saved image data needs to be packaged by TCP/IP protocol, which can be transmitted to the upper computer through the Ethernet interface for further processing. The flow chart of program is shown in Figure 2.

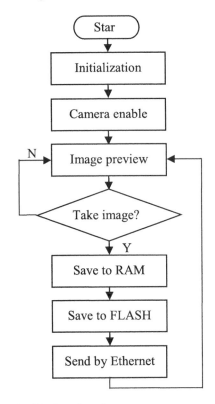

Figure 2. The flow chart of program.

3 SOFTWARE ALGORITHM DESIGN

The image data acquired by hardware system need to be processed on the upper computer in order to complete the check-in process for label image and this process can be achieved in the environment of Matlab. The first step is the image generating process which changes the pixel data of RGB565 format acquired by the image acquisition hardware system to an image of BMP format. The second step is an image preprocessing process, which makes the label image easy for processing in the next stage. The third step is the image recognizing process, which can be achieved by the method of template matching. The last step is getting the checking results and displaying the results in the user interface.

3.1 Label image generating process

In the hardware system, the output format of the label image acquired by camera sensor is set to RGB565. That means, each of the pixels of the image is stored as a transmitted in hexadecimal format, in which the red, green and blue components are respectively stored in high five bits, middle six bits and low five bits. This image format cannot be displayed directly on the upper computer and label image generating process extracts these color components and generates a new 24 bits BMP image. Part of the software algorithm program achieved by Matlab is shown as follows:

```
r8 = bitand (bitshift (rgb_565, -11), 31);
r8 = bitor(bitshift (r8, 3), bitshift ( r8, -2) );
g8 = bitand ( bitshift ( rgb_565, -5), 63);
g8 = bitor ( bitshift (g8, 2), bitshift (g8, -3) );
b8 = bitand (rgb_565, 31);
b8 = bitor (bitshift ( b8, 3), bitshift (b8, -2) );
```

Among them, the rgb_565 represents a pixel in the format of hexadecimal and the r8, g8, b8 represent the color components respectively.

3.2 Label image preprocessing process

The whole image acquired by the hardware system includes label and background. In this process, the label part needs to be extracted for further image process. The clustering algorithm is used to extract the label image according to the color differences between label image and background image. Next the label image needs to be corrected because the image acquired by the camera is usually accompanied by barrel distortion. Studies show that the distortion of the label image acquired is related to the r which represents the distance each pixel in the image is in the center of the image. The location relationship of pixels before and after correction is shown in Equation 1:

$$\begin{cases} u = x(1+k_1 r^2) \\ v = y(1+k_1 r^2) \end{cases} \tag{1}$$

where k_1 = correction coefficient and $r^2 = x^2 + y^2$.

In the next step, the label image needs to be converted to binary image. In the binary process, an appropriate threshold is selected by the method of global threshold and used to transform the gray image into a binary image aiming at reducing the amount of calculation. The last step is adjusting the size of the label image because the size of the image is greatly different which may cause great difficulty in the further image recognition process. So the image needs to be adjusted to the same size for the next label image recognition process.

3.3 Label image recognition process

This process uses the method of template matching to achieve the recognition of label image. Each batch of medicine bottle must be registered before using as well as the label image. In this stage the label image of various drugs must be processed and corrected to the standard template image for the next label image matching process. All the standard template images need to be numbered, denoting as M_1, M_2, ..., M_n and each of the template image has a unique label information corresponding to the number. We call the label image acquired by hardware system as the target image marked as T. In order to check the label image of a medicine bottle, the target image needs to match with the template image in the template image database. In the image matching algorithm, the feature of both the target image and template image needs to be extracted and the feature matrix of label images is obtained. By matching the feature matrix of target image acquired in real time with the feature matrix of the template we already know, we can finally obtain the corresponding information of the label image which needs to be checked.

3.4 Checking results getting process

At this stage, the checking results are obtained and the results are shown. In the process, the template image named M_1, M_2, $\cdots$, M_n are read sequentially and then the similarity between the feature matrix of the target image and each of the feature matrix of template image are calculated separately. The results of similarity are sequentially stored in the matrix A, the nth element in the matrix A corresponds to the nth similarity results with the template image. The formula to calculate the similarity of feature matrix is given in Equation 2 below:

$$corr2 = \frac{\sum_i \sum_j (T_{ij} - \overline{T})(M_{ij} - \overline{M})}{\sqrt{(\sum_i \sum_j (T_{ij} - \overline{T})^2)(\sum_i \sum_j (M_{ij} - \overline{M})^2)}} \tag{2}$$

Among them, T represents the feature matrix of target image while M represents feature matrix of template image. The ij represents the location of the pixel. Next, we check each of the similarity results stored in matrix A and find out the max value of A as well as the coordinate of max value. For example, the coordinate of max value is n, then the matching results for the target is template M_n, the label information corresponds to M_n including the name of the label, and the production batch number and production data can be acquired easily.

4 EXPERIMENTS AND RESULTS

In this section, the label image of Clindamycin Phosphate for injection is used as the target image, for example, to verity the proposed automated checking system for labels of medicine bottles. In this experiment, four kinds of label images are used as a sample space which including Clindamycin Phosphate for Injection, Rocuronium Bromide Injection, Cefuroxime Sodium for Injection and Cefazolin Sodium Pentahydrate for Injection. First of all, we process the label image acquired by the automated checking system using the method proposed in the label image preprocessing stage. The target image and the template corresponding to it can be seen in the Figure 3.

a) The target label image b) The template label image

Figure 3.   The target and template label images.

In the label image recognition stage, the size of the target and the template image are adjusted to 200×180 which is close to the actual size of different kinds of label images. At the same time, the size of unit area is 10×6 pixels in order to extract the feature matrix. The similarity between the target image and each of the template images is calculated out and then the similarity is drawn into a curve. In this experiment, 10 groups of target images of the same kind of label but are acquired from different medicine bottles by the hardware system are used. The similarity curve is given in Figure 4.

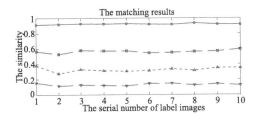

Figure 4.   The similarity curve for the matching results.

As shown in the Fig. 4, the similarity results represented by the solid line show the result of the template images corresponding to the target image and the other three dotted line represent the matching results with another three kinds of label images. The similarity results of correct template are always greater than another three kinds of template images. Theoretically, the results with the corresponding template images can be 1 while exact matching. But in fact, there exists a number of factors which would influence the matching results such as gray values changing under the influence of light, the distortion due to the camera sensor as well as the character on the label (clear/blur). Therefore, we believe that when the similarity result is higher than 0.9, the template image corresponding to the target image is the right checking results.

5 CONCLUSION

In this article, a new automated checking system is proposed for the labels on the medicine bottle. The label image is acquired by the hardware system and is transmitted to the upper computer for further process. In the upper computer, the software algorithm is designed to recognize the label image in the experimental environment of Matlab. The feature matrix is calculated by the statistical characteristics of the pixel values in each unit area, and this method is suitable to certain scale variation and small angle deviation in photographing. The template matching method proposed based on feature matrix is simple in calculation and has a small amount of computation. Experimental results show that this method can quickly and efficiently check the information on the bottle label and can be used in clinical intravenous drug preparation process instead of manually checking the information on the bottle label. In other words, it can reduce the workload of manual checking and dispensing error rate and has wide applicability.

REFERENCES

[1] Allwood, M.C. 1994. Practical guides.I: Central intravenous additive services, *Journal of Clinical Pharmacy and Therapeutics* 19 (3): 137–145.
[2] Xu, B. & Huang, K.Z. 2014. Graphical lasso quadratic discriminant function and its application to character recognition, *Neurocomputing* 129 (4): 33–40.
[3] Zhang, H.M., Gao, W. & Chen, X.L. 2006. Object detection using spatial histogram features, *Image and Vision Computing* 24 (4): 327–341.
[4] Shen, D.G. 2006. Image registration by local histogram matching, *Pattern Recognition* 40 (4): 1161–1172.
[5] Wang, Y.G., Yang, J. & Zhou,Y. 2006. Region partition and feature matching based color recognition of tongue image, *Pattern Recognition Letters*, 28 (1): 11–19.

Visualization makes array easy

R. Z. Ramli, A.Y. Kapi & N. Osman
Universiti Teknologi MARA, Kuala Pilah, Negeri Sembilan, Malaysia

ABSTRACT: In recent years, programming has become the most influential programming paradigm either in the industry or in education but the teaching of programming remains difficult. (Kölling, 1999). Students beginning their first programming class often find it hard to grasp the abstract nature of the coding and concepts involved in objects. In this project, we discussed the difficulties students may encounter in learning array objects and its operation and proposed visualization and interaction technique as an approach to teaching introductory programming course. Students are unable to visualize the concept of array behaviors and its operations. The difficulties in learning array are affecting student's interest towards the subject. Algorithm visualization system as mentioned by Boisvert (2009) are used in teaching as they can show the steps of execution in an animation to allow a viewer to construct a mental model of how the process leads to the required result. Therefore, we develop "Virtualization makes Array Easy" (VAE) that offers good array visualization, constructed analogy and direct interaction with its operation. VAE is able to attract the student's interest through graphical representation of the array and provide clear illustration. Hence, the students can obtain a new learning experience with a better understanding to the concepts of array's operation and thus perform better in their assessments.

1 INTRODUCTION

Nowadays, technology has become vital in today's world. It involves a rapid production of various smart phones, massively huge databases, high specification of computers, system and application software. Hence, programming skill is becoming inevitable since there are many demands for it. Programming skill takes a lot of practice and high enthusiasm from the students' part. Students need additional supports to attract and help them in the learning process. In this study, we proposed an e-content application, "Virtualization makes Array Easy" (VAE) that can attract students' interest to learn programming method through visualization, constructed analogy and interactive quiz. VAE also helps the students to understand the concept of programming generally and array objects specifically.

The following section will discuss techniques in motivating students' interest that exists in previous literature. The methodology in developing and testing the VAE is discussed in Section 3 while the results are presented in Section 4. The conclusion is in Section 5.

2 LITERATURE REVIEW

In teaching and learning process, Enzai et al. (2011) stated that students that learn computer programming find it hard to understand and visualize the nature of programming. In other research, Yang et al. (2014) claimed that the delivery of teaching and learning programming faced many challenges from both teacher and students' part. Watson et al. (2011) tackled the regular challenges and problems faced by the students. Law et al. (2010) investigate factors that motivate students in e-Learning that tested in system called **P**rogramming **A**ssignment a**S**sessment **S**ystem (PASS). The results showed that three motivating factors; 'individual attitude and expectation', 'challenging goals', and 'social pressure and competition' have a good relationship with effectiveness.

In order to address regular challenges in learning programming, many researches have existed in the previous literature. Enzai et al. (2011) suggested that creativity aspects such as analogy, games and mnemonic can be applied in the learning process. Yang et al. (2014) developed an interactive test system to enhance the students' learning outcomes.

Many tools have been developed to ease the learning and teaching process of computer programming. One of the popular tools is Alice that promotes interactive visual and animation (Wanda et al., 2011). Hsiao et al. (2012) proposed social comparative visualization that lets the students compare their achievements with their peers in the same classroom. The outcomes from these studies have shown positive results from the students. In addition, Liu et al. (2011) proposed a simulation games that results in successful technique to help students in learning computation problem solving skills.

However, our VAE is developed to help students have a better understanding in learning array concept.

Instead of using the textbooks and a PowerPoint presentation, students can learn in a different way through the visualization and interaction technique. VAE as an e-content application can vary the teaching style and teachers can also save time creating new teaching materials for students. It can also incite the interest of novice students to learn and love programming course.

Since the purpose of e-content is to support learning, the challenge is to make sure that it is usable to the students. One way to assess the e-content is by asking the students to evaluate the prototype. It can be evaluated on five factors which are learnability, memorability, satisfaction, efficiency and free errors (Nielsen, 1994). Nielsen's attribute usability (NAU) questionnaire has been used widely by many researches in e-learning (Ardito et al., 2006; Granic & Cukusic, 2011).

3 METHODOLOGY

VAE is developed based on ADDIE (Analyze, Design, Develop, Implement, and Evaluate) model which is an instructional design model. In analysis stage, technique of visualization is identified as the main factor that can lead the student to easily understand a programming concept. Conventional note in presentation slides is referred to in designing VAE. The process of design takes about one week and a storyboard is created. Figure 1 shows the storyboard of VAE. The storyboard is used as a guide to develop the VAE. In development process, two main software are used to develop VAE which are Video Scribe and MS Powerpoint with i-Spring feature. Figure 2 and 3 show some of the screens in VAE.

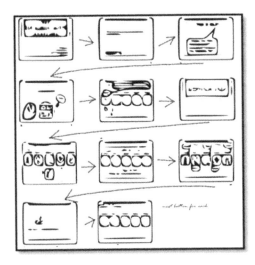

Figure 1.   VAE storyboard.

Figure 2.   VAE screen 1.

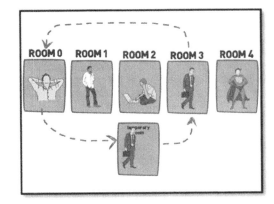

Figure 3.   VAE screen 2.

Once VAE is ready, the e-content application is uploaded to the university's online portal in order to make it accessible to the students. The final version of VAE consists of multiple graphics and 70% animation. It is identified that animation is able to attract students attention (Ramli, Sahari, Zin, Othman, & Osman, 2013). The students need to answer three sets of quizzes in the application. Results of the quizzes are then sent automatically to their lecturer's email.

Students are given 20-30 minutes to run the VAE. Then, they need to answer NAU questionnaire at **http://tinyurl.com/surveyVAE** to evaluate VAE's effectiveness.

4 RESULT

The questionnaire was constructed based on the five factors identified in the NAU questionnaire as learnability, memorability, satisfaction, efficiency and free errors on the scale 1 – 7 to denote the status evaluated from bad to good as shown in Table 1.

Table 1. Summary of respondents responses.

Likert Scale	1	2	3	4	5	6	7
Learnability				15%	25%	30%	29%
Efficiency				12%	25%	32%	25%
Memorability			5%	11%	27%	25%	30%
Accuracy			4%	12%	21%	36%	27%
Satisfaction			4%	12%	22%	27%	34%

Using a high score of 6 and 7 to denote the level of good evaluation, 59% of the students think that VAE and its content can be easily learned while 61% is satisfied with the performance and its e-content. The e-content has minimum error as 63% of the students evaluated it as accurate with minimum error. 57% of the students judged VAE as efficient and able to meet its objectives.

Based on Figure 4, only 5% and less evaluated VAE on a lower scale of 3 on memorability, accuracy and satisfaction. None of the respondents give bad scores on any of the five factors using 1 or 2 scale.

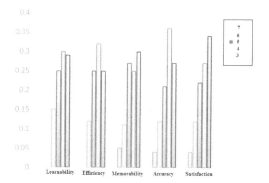

Figure 4. Graphs of respondents responses.

The average evaluation using a score of 4 and 5 shows the responses from 33% to 40% of respondents on all five factors. This result can be summarized as respondents strongly found that VAE has learnability, efficiency, memorability, accuracy and satisfactions features in their evaluation of the prototype as a learning aid in their programming class.

5 CONCLUSION AND FUTURE WORK

VAE brings benefit to the students and lecturers in learning and teaching introductory programming course. This study indicated that overall the respondents found VAE e-content application to be effective in helping them to learn programming concept through visualization, simulation and interactive quiz. Features such as learnability, efficiency, memorability, accuracy and satisfactions are used as contributing factors in determining the effectiveness of the prototype VAE.

For the students, with the aid of VAE, they can have good object visualization, which helps them in understanding better the concept of programming generally and array objects specifically. Thus, it will help them in getting a better assessment results.

Students' results have shown significant improvement this semester after being shown VAE prior to the lecture and also again after the lecture on the array. Hence it can be deduced that VAE has helped them in getting a better understanding to lead to a better result.

In future works, VAE can be further expanded to cover the entire topics in the introductory programming course beside array as it proves to be beneficial as a teaching and learning aid. More quizzes can be added to test the students understanding at every stage of the topic with varied, interesting animations and scenarios as an analogy to make it meaningful to the students and therefore memorable.

REFERENCES

[1] Ardito, C., Costabile, M. F., De Marsico, M., Lanzilotti, R., Levialdi, S., Roselli, T., & Rossano, V. (2006). An approach to usability evaluation of e-learning applications. Universal access in the information society, 4(3), 270–283.

[2] Granic, A., & Cukusic, M. (2011). Usability Testing and Expert Inspections Complemented by Educational Evaluation: A Case Study of an e-Learning Platform. Educational Technology & Society, 14(2), 107–123.

[3] Nielsen, J. (1994). Usability inspection methods. Paper presented at the Conference companion on Human factors in computing systems.

[4] Ramli, R. Z., Sahari, N., Zin, N. A. M., Othman, N., & Osman, S. (2013). Development and Validation of Game Interface with Culture Questionnaire: Graphic and Animation. Procedia Technology, 11, 840–845.

[5] Enzai, N.I.M.; Che Kar, S.A.; Husni, H.; Ahmed, N., "Harnessing creativity in teaching and learning introductory programming," Research and Development (SCOReD), 2011 IEEE Student Conference on , vol., no., pp.431,435, 19-20 Dec. 2011 Hsiao, I. H., Guerra, J., Parra, D., Bakalov, F.,

[6] König-Ries, B., & Brusilovsky, P. (2012, May). Comparative social visualization for personalized e-learning. In Proceedings of the International Working Conference on Advanced Visual Interfaces (pp. 303–307). ACM.

[7] Law, K. M., Lee, V., & Yu, Y. T. (2010). Learning motivation in e-learning facilitated computer programming courses. Computers & Education, 55(1), 218–228.

[8] Liu, C. C., Cheng, Y. B., & Huang, C. W. (2011). The effect of simulation games on the learning of computational problem solving. Computers & Education, 57(3), 1907–1918.

[9] Wanda P. Dann, Stephen Cooper, & Randy Pausch. 2011. Learning to Program with Alice (W/ CD Rom) (3rd ed.). Prentice Hall Press, Upper Saddle River, NJ, USA.

[10] Watson, C., Li, F., & Lau, R. (2011). Learning Programming Languages through Corrective Feedback and Concept Visualisation. Advances in Web-Based Learning-ICWL 2011, 11–20.

[11] Yang, T. C., Yang, S. J., & Hwang, G. J. (2014, July). Development of an Interactive Test System for Students' Improving Learning Outcomes in a Computer Programming Course. In Advanced Learning Technologies (ICALT), 2014 IEEE 14th International Conference on (pp. 637–639). IEEE.

Testing and Measurement: Techniques and Applications – Chan (Ed.)
© 2015 Taylor & Francis Group, London, ISBN: 978-1-138-02812-8

A single-end mechanism to measure available bandwidth under busty cross traffic environments

Y.Q. Zhou & B. Chen
Nanjing University of Aeronautics and Astronautics, Nanjing, Jiangsu, China

ABSTRACT: Available bandwidth is an important metric that indicates the network's dynamic characteristic. Most existing methods are based on the fluid cross traffic model; the accuracy is limited in the busty cross traffic network environments. In this paper, a probing-position packet train is constructed to track the changeable position of bottleneck link periodically and improve the probing rate. Besides, a novel mechanism is proposed based on analyzing the challenges brought by busty cross traffic, high accuracy of available bandwidth is obtained from the output dispersion of probing-position packet trains on the pre-bottleneck and post-bottleneck link. Moreover, we utilize" classify and double average" method to filter the measurement data and take into packet loss. Finally, the performance of our mechanism is verified by Network Simulator (NS-2) under different network environments. Results demonstrate that our mechanism measure available bandwidth more accurate than other well-known available bandwidth measurement tools such as Spurce and Pathchrip.

1 INTRODUCTION

As one of the key parameters of network performance, available bandwidth provided quantity of Service(QoS) for private networks. It has attracted attention, and several techniques and tools have been developed.

Most of the existing tools are classified into two categories: the Probe Gap Model (PGM) [1] and Probe Rate Model (PRM) [2]. Regrettably, both of them assume that the cross traffic follows the fluid cross traffic model [3], which is not applied in power networks where cross traffic is dynamic. Thus, the accuracy of available bandwidth must be improved.

Hu et al. [4] proposed a tool named self-loading decreasing rate train (SLDRT), and the input rate, when the one-way delay (OWDs) remains stable, was regarded as available bandwidth. SLDRT is quite effective, but it injects more probing packets. A novel algorithm [5] employed not only the gaps of any two consecutive probing packets but also nonadjacent probing packets. Besides, two stages, filtering and moving averages were used to reduce the measurement error, thus, it can effectively filter out probing noise and achieve the good accuracy of available bandwidth. Regrettably, procedures must be installed on the remote node.

In this paper, we propose a novel mechanism to obtain high accuracy of available bandwidth without deploying any measurement procedures on the receiver. The contributions of this paper are summarized as follows:

1 A probing-position packet train is constructed to obtain its output dispersion on preceding or later bottleneck link and track the changeable bottleneck link periodically.
2 A novel mechanism is proposed to achieve good accuracy of available bandwidth without deploying any procedures on the remote node.
3 "Classify and double average" method is used to filter the measurement data before computing the available bandwidth.
4 Packet loss is considered in our mechanism.

The rest of this paper is organized as below. Section II analyzes the problem and constructs a probing packet. Section III develops our mechanism. Section IV evaluates our mechanism by doing some experiments. Finally, Section V concludes the paper and considers the future work.

2 PROBLEM STATEMENT

In this section, we enumerate problem statements about the challenges of measuring available bandwidth under bursty cross traffic environments. We assume links before and after the bottleneck link are respectively called the pre-bottleneck and post-bottleneck link.

Firstly, as shown in Figure 1, the output interval on pre/post-bottleneck link (g_{i-1}, g_i) would be expanded when bursty cross traffic is inserted into M_1 and M_2 before the arrival of M_1, and be compressed when bursty cross traffic had queued before the arrival of

M_1 and M_2. In these cases, results [5] may encounter measurement error.

Secondly, available bandwidth was insensitive to the bursty cross traffic [6]. To solve these problems, we construct a probing-position packet train based on packet-pair, and we respectively capture the output dispersion of probing packets on pre/post-bottleneck link. We add position packets to track the position and capacity of bottleneck link periodically. If there exist changes, Time to Live (TTL) of probing packets is reinitialized and the capacity of the bottleneck link is re-obtained.

Finally, most measurement tools assumed that the buffer is infinite and underestimated without considering the packet loss. Actually, with the probing rate increased, the router buffer queue would overflow and packet loss is produced. So, we add the packet loss rate into our mechanism.

Figure 1.   The fluid cross traffic model.

3   MEASUREMENT MECHANISM

Before proposing our mechanism, some assumptions are introduced. Firstly, routers would respond to ICMP timeout packets when TTL is exhausted. Moreover, ICMP penetrates the firewall. Secondly, the position (O) and capacity (C) of bottleneck link are obtained by iPathneck [7]. Finally, the route from source node to destination node does not change during one probing period.

3.1   Probing-position packet train

In Figure 2, the probing-position packet train is constructed by some ICMP, including two probing units, with each of units consists of load packets (L), measurement packets (M) and position packets (P), and they are distinguished by identifier in IP. The size of them are respectively 1500B, 40B, 40B and denoted as S_L, S_M, S_P. M is added to compose four-packets structure and decrease the probing traffic, and L and M are sent in back-to-back way to reduce the probability of link in idle state. P is added to track the changeable position of bottleneck link timely, and the number of them is decided by the hops of the network. The interval and initial sending rate of M or probing unit are respectively denoted as g_m and R_m, g_u and R_u. In our simulation, at the beginning of measurement, the probing packets are sent in back to back way, and then the initial interval is set to the minimum value:

$$g_m = \frac{S_L + S_M}{0.02C}, g_u = \frac{2S_L + 2S_M + hS_P}{0.01(4+h)C}.$$

TTL of L in the first and second probing unit are respectively set to the hops of the pre/post-bottleneck link (i-1, i). If TTL of them are exhausted, L are discarded and ICMP timeout packets will be sent to the sender, and M is still sent to the remote node, and then g_{i-1} and g_i will be obtained from the dispersion of these ICMP timeout packets because the smaller length makes average interval of M keep unchanged. Secondly, when P passes through the bottleneck link, the delay from time sequence of ICMP timeout packets extremely increases, and the turning point (L_p) is the position of bottleneck link. If L_p changes, iPathneck will be used to re-probe.

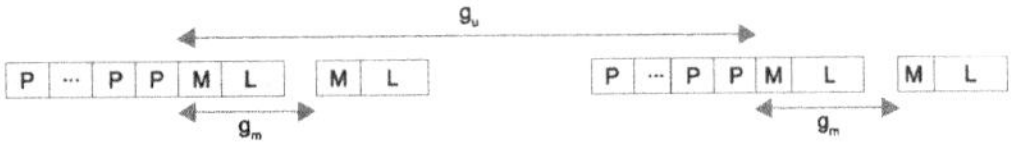

Figure 2.   The structure of probing-position packet.

3.2   Mathematical model

Let Rc be the rate of cross traffic, g_b be the transmission delay of probing packets on bottleneck link, N_{reply}, N, $N-N_{reply}$ respectively be the number of reply ICMP packets, probing packets and the packet loss, and ABW be the available bandwidth. As illustrated in Figure 1, when $Ru < ABW$, if the transmission of cross traffic has completed before the arrival of $M2$ within g_{i-1}, and g_i equals to g_m, then R_m should be improved. When $Ru \geq ABW$, whether g_{i-1} and g_i are expanded or compressed, we always have the capacity of bursty cross traffic $R_c g_{i-1}$ as long as the transmission delay of cross traffic on bottleneck link $g_c = g_i - g_b$ exceeds the moment of M2 arrival or M1 and M2 arrival. It is expressed as:

$$R_c g_{i-1} = g_c C = (g_i - g_b)C. \tag{1}$$

Where,

$$g_b = \frac{S_L + S_M}{C}$$

The interval between probing-position packet train and post-bottleneck link g_i is inferred as:

$$g_i = \begin{cases} g_{i-1} & R_u \leq ABW \\ g_b + \dfrac{R_c g_{i-1}}{C} & R_u > ABW \end{cases} \tag{2}$$

386

When $R_u \geq ABW$, then available bandwidth [4] is expressed as (without considering the packet loss):

$$ABW = C - R_C. \tag{3}$$

Combine (2) with (3)

$$ABW = C - \frac{g_i - g_b}{g_{i-1}} C \tag{4}$$

When $N_{reply} < N$, we replace g_{i-1} and g_i of lost M with the neighbour's. Besides, packets loss rate N_{reply}/N is added into the available bandwidth.

$$ABW = \left(C - \frac{g_i - g_b}{g_{i-1}} C \right) e \frac{N_{reply}}{N}, \frac{N_{reply}}{N} \in [0,1]. \tag{5}$$

3.3 Mechanism

Measurement data from existing methods showed irregular trend [8] which reduce the accuracy, so we come up with way to deal with this problem in our mechanism.

Firstly, we design "classify and double average" to filter measurement data. g_{i-1}, g_i are split into multiple categories according to the degree of distortion of interval: compressed interval sequence Q_c, normal interval sequence Q_n, and expanded interval sequence. Then, the average value of each category $\overline{Q_c}$, $\overline{Q_n}$, $\overline{Q_e}$ are calculated. Moreover, in the set of average points, we find the minimum (min) and maximum value (max). Additionally, we screen measurement data in every category to judge whether they are in [min, max], and then we make average the remaining measurement data in the range. Finally, equation (5) is used to calculate the available bandwidth.

Secondly, we timely adjust R_u according to the current ABW. If $R_u < 0.2ABW$, R_u is increased to $0.2ABW$; otherwise, R_u remains unchanged because [9] refers that the interaction between probing units is ignored when $R_u < 0.2ABW$. The algorithm is described as follow:

1 Construct the probing-position packet
2 Use iPathneck and traceroute to obtain O, C, h
3 Send probing-position packet trains to obtain g_{i-1}, g_i, N_{reply}, L_p
4 Use Classify and average filter to deal with g_{i-1}, g_i
5 Compute the packet loss
6 Compute available bandwidth
7 Adjust the probing rate

4 PERFORMANCE EVALUATIONS

In this section, we use NS-2 to evaluate our mechanism under different scenarios. The topology [10] is shown in Figure 3, and related parameters are denoted in Table I to verify the response speed to the various cross traffic and the accuracy of available bandwidth when the bottleneck link changes. Our mechanism is compared with Spruce [11] and Pathchirp.

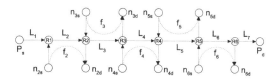

Figure 3.   The topology of simulations.

Table 1.   Related parameters in simulations.

Link (Capacity)	Time (Sec)	L2	L3	L4	L5	L6	P
Cross traffic (Mbps)	0-200	20	15	5	15	20	L4
	200-400	25	20	10	20	25	L3
	400-600	35	30	15	30	35	L5

4.1 Accuracy and response speed with bursty cross traffic

Measurement error (ε) is used to analyze the accuracy by computing the gap between measurement value and actual value, so the smaller measurement error is, the higher accuracy of available bandwidth becomes. It is denoted as:

$$\varepsilon = \frac{|ABW_e - ABW|}{ABW}. \tag{6}$$

Where ABW_e is the actual value of ABW.

In Figure 4, ABW, measured by Spruce and Pathchirp, has a lager jitter than the actual value, and their measurement error (16.4%, 23.7%, 69.7%) (13.6% 24.5% 34.8%) are higher than our mechanism (8.62% 10.47% 12.3%). The main reason is that they are based on the fluid model and the bursty cross traffic is not taken into account. In our mechanism, the probing-position packet train is constructed, and TTL of L is set to obtain output intervals on pre/post-bottleneck link. Besides, packet loss and "classify and double average" filters are considered. Consequently, results in our mechanism are close to the actual value and has a quickly respond to bursty cross traffic.

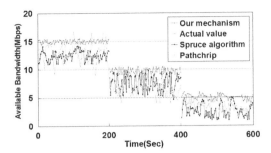

Figure 4. Comparison of accuracy between our mechanism and existing techniques with bursty cross traffic.

4.2 Accuracy and response speed with changeable bottleneck link

As shown in Figure 5, when the position of bottleneck link changes between 200s and 400s, deviation increases obviously by Spruce and Pathchirp, and their measurement error (22.7%, 46.3%, 50.6%), (23.5%, 24.3%, 22.7%) are higher than our mechanism (6.73%, 6.64%, 8.09%) because in our mechanism, the position of bottleneck link is tracked periodically, which reduces the effect of changeable bottleneck link on the accuracy of available bandwidth. Then, available bandwidth shows a smooth trend and less deviation, and the measurement error is beyond *10%*.

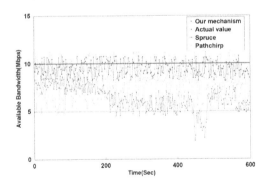

Figure 5. Comparison of accuracy between our mechanism and existing techniques with changeable bottleneck link.

4.3 Measurement duration

As shown in Figure 6, the measurement duration of Pathchirp has a large fluctuation, since the convergence algorithm greatly reduces the efficiency when the link is congested, while our mechanism is more smooth than Pathchirp due to the fact that L is discarded before the arrival of the receiver. Compared with Spruce, the measurement duration of our

mechanism is much smaller because eLare is inserted into measurement packets, and the size of them equals to 1540B when they are sent in back-to-back way. In this way, R_m has been increased in premise of unchangeable g_m.

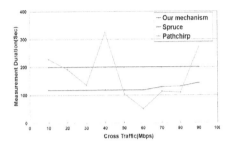

Figure 6. Comparison of measurement duration between our mechanism and existing techniques under bursty cross traffic.

In a word, our mechanism achieves higher accuracy of ABW and smaller measurement error, but also reduces the measurement duration.

5 CONCLUSIONS

This paper constructed the probing-position packet train to track the changeable position of bottleneck link and improve the probing rate without reducing gaps, and proposed a mechanism to achieve good accuracy of available bandwidth under bursty cross traffic environments. Moreover, "classify and double average" is utilized to filter the measurement data, and packet loss is considered. Finally, evaluations show that our mechanism improves the accuracy of available bandwidth and tracks the changeable bottleneck link, but also reduces the measurement duration. In the future, our mechanism is expected to perform well without the support of the intermediate routers. Besides, how to make ICMP penetrate the firewall is one of the key issues.

REFERENCES

[1] D. Croce, M. Mellia, and E. Leonardi, "The quest for bandwidth estimation techniques for large-scale distributed systems," ACM SIGMETRICS Performance Evaluation Review, vol. 37, no. 3, pp. 20–25, 2010.

[2] J. C. Kim and Y. Lee, "An end-to-end measurement and monitoring technique for the bottleneck link capacity and its available bandwidth," Computer Networks, vol. 58, no. 14, pp. 158–179, 2014.

[3] S. Y. Nam, S. Kim, J. Kim, "Probing-based estimation of end-to-end available bandwidth," Communications Letters. IEEE, vol. 8, no. 6, pp. 400–402, 2004.

[4] Z. Hu, D. Zhang, A. Zhu, Z. Chen, and H. Zhou, "SLDRT,: A measurement technique for available bandwidth on multi-hop path with bursty cross traffic," Computer Networks, vol. 56, no. 14, pp. 3247–3260, 2012.

[5] M. Li, Y.-L. Wu, and C.-R. Chang, "Available bandwidth estimation for the network paths with multiple tight links and bursty traffic," Journal of Network and Computer Applications, vol. 36, no. 1, pp. 353–367, 2013.

[6] C. D. Guerrero and M. A. Labrador, "Traceband: A fast, low overhead and accurate tool for available bandwidth estimation and monitoring," Computer Networks, vol. 54, no. 6, pp. 977–990, 2010.

[7] L. W.-W. Zhu Li, "Combined tool for measuring of available bandwidth and locating internet bottleneck-ipathneck," Computer Systems And Applications, vol. 21, no. 11, pp. 174–176, 2012.

[8] U. C. Nguyen, D. T. Tran, and G. V. Nguyen, "A taxonomy of applying filter techniques to improve the available bandwidth estimations," in 8th the International Conference on Ubiquitous Information Management and Communication Proceedings. ACM, 2014, p. 18.

[9] K. Ravindran and D. Loguinov, "What signals do packet-pair dispersions carry," in 24th annual joint conference of the IEEE computer and communications societies Proceedings. IEEE INFOCOM, 2005, pp. 281–292.

[10] H. Li and Y.S. Zheng, "Methodology for measuring available bandwidth on arbitrary links," Journal of Software, vol. 20, no. 4, pp. 998–1013, 2009.

[11] E. Goldoni and M. Schivi, "End-to-end available bandwidth estimation tools, an experimental comparison," in Traffic Monitoring and Analysis. Springer, 2010, pp. 171–182.

Design of production information management system for automotive enterprise

F.Q. Pei, Z.H. Tang, Y.X. Wang* & Y.F. Tong
Nanjing University of Science and Technology, School of Mechanical Engineering, Nanjing, People's Republic of China

ABSTRACT: After analyzing the performance of the production information management system for automotive enterprise, the overall functions of the designed system were proposed and discussed in detail. The information flow involved in the production information management system was analyzed. Finally, the database of the production information management system was designed. The above researches can lay a good foundation for the development of production information management system for automotive enterprise and hence improve the management and informatization level.

1 INTRODUCTION

Since China launched the policy of reform and opening, her economy has sustained a rapid development and the total GDP has been growing by about 10 percent in recent years (Clem 2009). China is known as the "world factory" with manufacturing thriving vigorous development (Zhang 2006). China's manufacturing industry has achieved such amazing achievements in recent thirty years. However, some serious problems still remain serious to be solved. Chinese people attach importance to local conditions and thus the role of the management system is weakened. Rapid development with highly dependent on the environment will not lead to production cycle (Blackburn 2012).

Chinese enterprises should pay more attention to the above problems, especially information technology. The modern industries must implement new information technologies as an efficient operation mode, which can enhance the core competitiveness of Chinese enterprises, and reduce business waste (Zhao 2010).

2 SYETEM REQUIREMENT ANALYSIS

System requirements analysis is an important process in the software development. A comprehensive and accurate requirement analysis is fundamental in developing a practical system. An information system can only be full of vitality when it's in accordance with the actual needs of users. After investigations of current enterprise management, system requirements are deeply analyzed.

The performance requirement of the production information management system includes not only the basic performance requirement of common management system, but also the security, reliability, integration and expansibility which are detailed as follows (Tong et al. 2009):

1 The information management system should be able to optimize the information management process. Enterprises focus on the rapid development of the products as well as the operational strategy to obtain the competitive advantage and the enterprises' ability.
2 The information management system should be able to ensure the security and reliability of system data.
3 The information management system should be open, extensible and integrated.

3 FRAMEWORK OF PRODUCTION INFORMATION MANAGEMENT SYSTEM FOR AUTOMOTIVE ENTERPRISE

A framework of the production information management system for automotive enterprise is proposed, it mainly covers the following aspects:

1 MRP (Material requirement planning)

The subsystem of MRP focuses on the management of all materials based on BOM (Bills of materials), including the functions of BOP (Bills of projects), material management and inquiry, as shown in Fig.1. Here, BOP aims to manage (add, delete or update) of projects and products.

---

*Corresponding author.

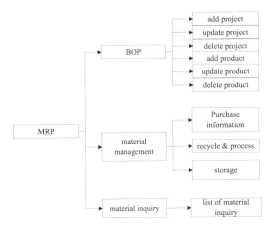

Figure 1.   Subsystem of MRP.

2   Production planning

The subsystem of production planning is mainly for the management of all production plans in an enterprise, including production information, production scheduling and warehousing, as shown in Fig.2. Production information contains all the production related information during the product life cycle, such as order information, delivery information, production management and production cycle table. Production scheduling module contains human resource scheduling, equipment scheduling, and capital allocation. Warehouse aims to manage the stock--in, stock--out and stock transfer information.

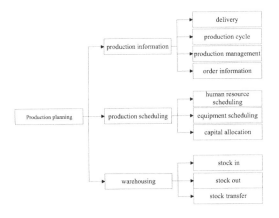

Figure 2.   Subsystem of production planning.

3   Quality inspection and management

The subsystem of Quality inspection & management is mainly for quality assurance as well as quality information (e.g. total quantities, qualified & unqualified information and so on that are available

and convenient to check and inquiry). Quality manual management is as shown in Fig.3, providing functions of quality control (QC), quantity inspection and quality manual compiling.

Figure 3.   Subsystem of quality inspection and management.

4   Process management

As shown in Fig. 4, subsystem of process management focuses on the business production process, including production process management, machining management, and manpower control management with the workflow, production work hour and various production consumptions incorporated.

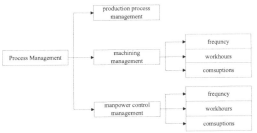

Figure 4.   Subsystem of process management.

5   Equipment management

Equipment management is mainly for malfunction & maintenance management and benefit analysis of equipments. Malfunction & maintenance can realize the management of containing malfunction/ maintenance frequency, malfunction/maintenance costs and warranty management. Benefit analysis includes general energy consumption ratio, high energy consumption ratio and negative earnings ratio.

Figure 5.   Subsystem of equipment management.

Production planning is the core of the entire production process, which is the link between the company's bond sales and production. It obtains customer orders from the sales department, and then sends the information to the production department for production scheduling. Then production scheduling is issued to the workshop to form a list of material requirements, equipment, inventory management, quality control lists, and the production process. These lists are generated and marks the completion of the entire production plan. Fig.6 shows the entire process of production:

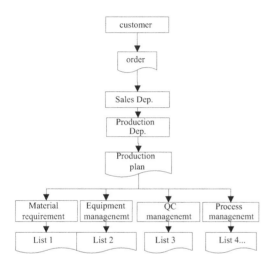

Figure 6. Subsystem of equipment management.

4 DATABASE DESIGN

In our research, a relational database called SQL Server 2000 is used for database management system (Harinath et al. 2011). The information involved in the proposed system is shown in the following tables:

Table 1. Information of workers.

No.	Item	Type	Width	Remark
1	ID	Char	5	id of worker
2	Name	Char	8	
3	Gender	Char	2	
4	Age	Nchar	6	
5	Job title	Char	10	
6	Seniority	Int	6	
7	Equipment ID	Char	20	
8	ID of the list	Char	20	

Table 2. Information of equipments.

No.	Item	Type	Width	Remark
1	Service life	Char	5	
2	Continous working period	Char	8	
3	ID	Char	2	ID of equip.
4	Name	Nchar	6	Name of equip.
5	Type	Char	10	Type of equip.
6	Functions	Int	6	
7	price	Char	20	
8	List ID	Char	20	
9	date of starting operation	Char	20	

Table 3. Information of wire harness.

No.	Item	Type	Width	Remark
1	Product ID	Char	5	
2	Name	Char	8	of the product
3	Product specification	Char	2	
4	Batch	Nchar	6	
5	price	Char	10	of the product
6	Level	Int	6	
7	Equipment ID	Char	20	
8	List ID	Char	20	

Table 4. Information of various lists.

No.	Item	Type	Width	Remark
1	List ID	Char	5	ID of the list
2	Name	Char	8	Name of the list
3	Conclusion	Char	2	
4	Batch	Nchar	6	
5	Demands	Char	10	
6	Type	Int	6	type of report

Table 5. Information of warehouse.

No.	Item	Type	Width	Remark
1	ID	Char	5	of the warehouse
2	Name	Char	8	of the warehouse
3	Product ID	Char	2	
4	List ID	Nchar	6	

Table 6. Information of the order

No.	Item	Type	Width	Remark
1	ID	Char	5	of the order
2	Product name	Char	8	
3	Product ID	Char	2	
4	List ID	Nchar	6	

5 CONCLUSION

The presented research aims to improve the management level of automotive enterprise by implementing an information system. It requires a comprehensive implement of various modern industrial engineering (IE) methodologies and technologies (e.g. 5S management, 5W1H, process analysis, business process reengineering, etc.) to improve the operation efficiency of enterprises. Only by this can it finally improve national management level of industries.

ACKNOWLEDGEMENTS

This work was financially supported by "excellence plans-Zijin star" Foundation of Nanjing University of Science. The supports are gratefully acknowledged. Also the authors would like to owe thanks to Mr. Xu Haiping from the automotive industry.

REFERENCES

[1] Blackburn, W. R. (2012). The Sustainability Handbook:" The Complete Management Guide to Achieving Social, Economic and Environmental Responsibility". Routledge.

[2] Clem Tisdell (2009). Economic Reform and Openness in China: China's Development Policies in the Last 30 Years Development. ECONOMIC ANALYSIS & POLICY, VOL. 39 NO. 2:271–294.

[3] Harinath, S., Zare, R., Meenakshisundaram, S., Carroll, M., & Lee, D. G. Y. (2011). Professional Microsoft SQL server analysis services 2008 with MDX. John Wiley & Sons.

[4] Tong, Y., He, Y., Li, D., & Shen, Y. (2009). Product Lifecycle Oriented Knowledge Management System for Product Rapid Development. In Management and Service Science, 2009. MASS'09. International Conference on (pp.1–4). IEEE.

[5] Zhang, K. H. (Ed.). (2006). China as the world factory. Routledge.

[6] Zhao, J. (2010). Key research areas of ecological and environmental science & technology. In Ecological and Environmental Science & Technology in China: A Roadmap to 2050 (pp. 85–141). Springer Berlin Heidelberg.

Testing and Measurement: Techniques and Applications – Chan (Ed.)
© 2015 Taylor & Francis Group, London, ISBN: 978-1-138-02812-8

Application of flow visualization and image processing techniques to the study of coherent structure in vegetated canopy open-channel flows

K. Dai, J. Yan, Y. Chen & M. Zhang
State Key Laboratory of Water resources and Hydraulic Engineering, Hohai University, Nanjing School of Water Conservancy and Hydropower Engineering, Hohai University, Nan Jing, China

ABSTRACT: Most fluids are transparent, thus their flow patterns are invisible. Flow visualization is the process of making the physics of fluid flows visible with some special method and help to understand the specific flow structure. In natural rivers, a lot of aquatic plants are observed and they have changed coherent structure both in the vegetated and non-vegetated zones. In the present study, we investigated turbulent coherent structures in vegetated canopy open-channel flows, on the basis of flow visualization technology and digital image processing theory. The experiment results show that the whole flow region was divided into three zones and the characteristics of the coherent structure are remarkably different. Flow visualization is an effective method to investigate the turbulent structure in vegetated flows.

1 INTRODUCTION

Visualization technology is a method to make the flow pattern, which cannot be seen, can be displayed and visualized directly. Through the way like this, the flow states and the development of flow process can be indicated; even more, some visualization methods can get the quantitative results of flow parameters. Flow visualization has a history more than 100 years. It is developed with the development of flow mechanics. As we know, the definition of Reynolds number, the proposal of Prandtl's boundary-layer, the study of Karman vortex streets and the discovering of coherent structure of flows, all these great findings in flow mechanics are based on the visualization technology.

With the rapid development of technology, the visualization methods become more precise and diversified. The application of visualization techniques keeps growing. Measuring range of velocity developed from natural low velocity to supersonic. It can not only apply on constant flow, but also the non-constant flow. However, visualization technology has its weakness. Although its theory is simple, but it is not easy to operate, for example, if the exposure is too long, coherent structures cannot be seen. And the quantitative outcome is not enough.

Vegetation element is an important role in most rivers, its biological characteristics improved habitats for microorganism, aquatic plants and animals. It also aggravated fluctuation inflows, which are responsible for the removal of sediment and pollutant (Tang et al. 2007). Therefore, to improve the water management and eco-system, it is important to investigate

turbulent coherent structures and their changing process in flows.

The study of coherent structures enriches the understanding of turbulence, verify the turbulence is a collection of repeat and order structure (Fan 2011). Aquatic vegetation canopies totally changed turbulent structures in the river. Most studies (Nezu, 2008, Nepf, 2000, 2008, Poggi, 2004, Yan, 2008) use the modern flow measurement technology to measure the mean velocity, turbulence intensity and other turbulence characteristics.

In the present study, we investigate turbulence coherent structures in vegetated canopy open channel flows, on the basis of flow visualization, the timespace characteristics of tracer concentration field and digital image processing technology.

2 EXPERIMENT SETUP

2.1 *Experimental flume and vegetation model*

The experiment was conducted in a 12 m long, 42 cm wide and 0.7 m deep tilting flume in the hydraulic laboratory of the school of Hohai University, Nanjing. The water level was controlled by a tailgate at the downstream end of the flume. The discharge was measured using ultrasonic flow meter on the supply pipe and the water depths were measured using a point gauge. A propeller flowmeter was placed at the release point of tracer to measure the outlet velocity.

The flume bed was placed 5 blocks of gray plastic plate, each of them is 2 m long, 42 cm wide and 1 cm deep. The board drilled 6 mm diameter hole, which was used to insert the vegetation model. The elements of vegetation model were composed of rigid cylindrical aluminum rods. The size of vegetation element was $h=6$ cm height, $d=6$ mm. S_x and S_z are the stream wise and span wise spacings between the neighboring vegetation elements respectively (Fig.1). The vegetation density a, was the dimensionless factor in Eq. (1).

$$a=nhd/s \qquad (1)$$

Where d = diameter of vegetation rods; s = total area of the vegetation zones; n = number of vegetation; h = height of vegetation elements. In this study, $S_x =5$ cm; $S_z=2$ cm.

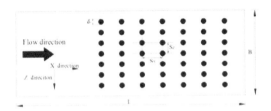

Figure 1.   The allocation of vegetation.

2.2   Flow visualization system

The experiments adopt dye injection technology to visualize the changing of coherent structures. The system placed three dye injection points at different height in flows. The period and transfer occurred in coherent structures, described by the time series of tracer dye trajectory, are identified and characterized. Though flow condition is turbulence, which will make the tracer dye diffused rapidly and oscillated acutely, we could captured coherent structures clearly by means of enhanced viscosity of dye and increased lighting. Further, use the digital image processing technology to analyze the frame-by-frame images to quantify the results.

The dye release system is composed of a square bottom tank to hold the tracer dye, a L-type draft tube, whose interior diameter is 5mm, and a butterfly valve to control the outlet of tracer dye. The axis of tracer dye outlet is placed upon the middle of the board and parallel to the longitudinal axis of flume. The dye solution head can be reckoned as constant and the discharge of dye keep steady during the experiment process. To ensure that the dye discharging velocity at the tube exit is equal to the local velocity at that point, we placed tank at the height of 15cm upon the surface flow, which was on the basis of hydraulic energy balance equation. Besides, we control the butterfly valve and propeller flowmeter to make it happen.

Fluorescein sodium liquid is selected as dye because it's good visualization, non-toxic effect

and the density is equal to the density of water (Tang et al. 2008). Under the light intensity of 36w, Fluorescein sodium can be motivated and visualized as yellow-green. The whole experiment process was captured by SONY HDR-CX350. The system is showed in Fig.2.

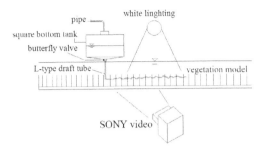

Figure 2.   The flow visualization system.

2.3   Digital image processing system

The experiment video is captured in the image frame by frame by Adobe Premiere Pro CS4 software. The size of a single image is 1440*1080 pixels. We quantify the image by Matlab.

The whole name of Matlab is Matrix Laboratory, which was a software to turn the image into digital signals. Matlab has its advantage in matrix operations (Yang et al. 2013). Digital image indicated as two-dimensional arrays of computer and the unit is pixels. Usually corresponds to each image pixel in the two-dimensional space is a particular location, and by one or more of the associated sample values and the composition of that point, so it will be the most intuitive results which were described using the digital image matrix is.

The image acquired by the SONY digital video is the true RGB color image; we need to convert the image information into the data information. Firstly, we reversed the image, which is more recognized. Then, we enhance the contrast of the image and transfer the image into gray level. Finally, we color the coherent structures and make it distinct with other irregular flow structures.

2.4   Calibration

The size of coherent structures in an image can be expressed as pixels. In order to transfer the image dimension in pixel into cm, calibration must be performed before every experiment. Firstly, placed a rule in front of the video section, and then adjust the camera to make the scale on rule as clearly as possible. Due to the linear relation between pixel/mm, we can get the following equation: 1 pixel=0.32 mm.

2.5 Release points of tracer dye and experiment conditions

In experiments, we placed three typical release points along the depth direction. The first located at half the height of vegetation (inner region), the second layout at the top of vegetation and the third placed at the half of the distance between the top of the vegetation and the water surface (outer region).

Table 1. Experiment conditions.

Case	H	Um	Q	I	a	R	Re	Fr
	cm	cm/s	l/s	1/1000		m	*10⁴	
A1	12	5	2.52	0.65	0.34	0.08	0.29	0.05
A2	12	10	5.04	0.67	0.34	0.08	0.58	0.09
A3	12	15	7.56	1.05	0.34	0.08	0.88	0.14

In Table 1, the temperature is $10.4°C$. And the mean velocity $Um=Q/(BH)$, $Re=UmR/v$, $Fr=Um/(Hg)^{0.5}$.

3 RESULTS AND DISCUSSION

3.1 Coherent structures in different region

In inner zone, the ejections are more dominated in this area (Fig.3). The coherent ejection motion rises to the edge of canopies, and then it was carried by the sharply shear layer motion and high speed vortex which occurred near the top of the canopy, finally it combined to be bigger vortex and developed in downstream till burst up.

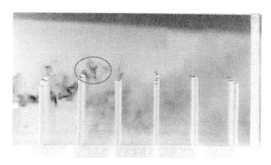

Figure 3. The ejection motion in inner zone (section A3).

Unlike open-channel flow, it is dominated by single ejections. The ejection angle is nearly 2 times compared with open-channel flows, and the maximum of ejection height is 3 times to the open-channel conditions. Lateral section coherent structures occurred in inner zone has three different states, include normal state, deflected state and the combined state.

Normally, flow transition from laminar flow to turbulence flow because of laminar flow disturbed and continued to enlarge. Near the vegetation edge, however, the bypass transition (Pan, 2011) dominated this zone, and it is suggested strongly that the reason of bypass transition is the vegetation edge generated coherent eddies, which govern the vertical transport of momentum. The disturbance growth and broken time of vortex in bypass transition is faster than the normal transition. The disturbance always occurred due to the vegetation edge generated absolute instability.

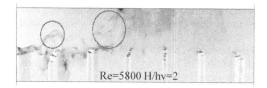

Figure 4. Group burst phenomenon near the top of the vegetation (case A2).

Near the top of the vegetation, it reflects the ejection instability process of low speed stripes completely. The process has four stages, include the formation of stripes, the vortex uplifting, the oscillation and the breaking. Fig.4 reveals the group ejections, which also happened near the vegetation edge.

Not only the burst phenomenon (boundary-layer), but also the vortex pairing (mixing-layer) occurred in the top portion of the canopy. Firstly, the vortex discretized. Then the two discrete vortex interactions with each other and rolled up in a bigger one. As time passed, the bigger vortex keeps enlarging in downstream. Finally, it was broken. It is the traditional characteristics in free shear mixing-layer (Lin, 1995), which also occurs repeatedly near the top of the vegetation. When the flow velocity is larger, pairing and merging occur frequently and appeared at different position.

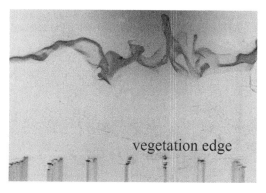

Figure 5. Coherent structures in outer zone (section A1).

In the outer zone, coherent structures are analogous to one in boundary layer and open-channel flows with rough beds. As Fig.5 shows, the mainly characteristics in outer zone is the vortices stretches in streamwise and the intermittent of rolled up vortices, which is also effected by vegetation edge.

3.2 *Partition of coherent structures*

As the results shows, turbulent coherent structures have different characteristics along the depth. So it can be divided into three zones: inner zone, canopy edge zone and the outer zone. Coherent structures in inner zone are affected by diameter and the allocation of vegetation. At the top of the canopy, vortices change quickly and it has a positive correlation to the burst frequency, velocity, submergence and the height of ejection. The outer zone vortices are analogous to the classical boundary-layer region. Each zone has its traits and likely co-exists in space.

4 CONCLUSIONS

On the basis of flow visualization and image processing techniques, we investigated turbulent coherent structures in the vegetated canopy open-channel flows. Experimental consideration revealed that the coherent structure was complicated and had its partition along the depth direction. The whole flow region can be divided into three zones: the inner zone, in which the wake-burst phenomenon and simil-Karman vortex streets are controlled; at the top of the canopy, the coherent structure is complicated, it has burst phenomenon and large eddy pairing phenomenon which generally occurred in freely mixing layer; and the outer zone is analogous to the free shear turbulence flow.

ACKNOWLEDGEMENTS

The work is supported by School of Water Conservancy and Hydropower Engineering, Hohai University, Nan Jing, the National Natural Science Foundation of China (Grant No: 51109066, 51479070, 51125034).

REFERENCES

[1] B. Chen. 2011. Investigation of near-wake flow characteristics of two circular cylinders close to a wall based on PIV technology. *Doctoral Dissertation of Huazhong University.* Wuhan.

[2] B.C. Fan , G. Dong and et al. 2011. *Principles of turbulence control.* Beijing: National Defense Industry Press.

[3] Heidi NEPF and Marco GHISALBERTI. 2008. Flow and transport in channels with submerged vegetation. *Acta Geophysica* 56(3):753–777.

[4] J.Z. Lin. 1995. *Turbulence coherent structures.* Beijing: China Machine Press.

[5] Nepf HM and Vivoni ER. 2000. Flow structure in depth-limited, vegetated flow. *Journal of Geophysical Research* 105(C12):28547–28557.

[6] Nezu I and Sanjou M. 2008. Turburence structure and coherent motion in vegetated canopy open channel flows. *Journal of Hydro-environment Research* 2(2):62–90.

[7] C. Pan, J.J. Wang. 2011. Progress in bypass transition induced by free-stream disturbance. *Advances in Mechanicals* 41(6): 668–685.

[8] Poggi D, Porpotato A and RidolfiL. 2004. The effect of vegetation density on canopy sub-layer turbulence. *Boundary-Layer Meteorology* 111:565–587.

[9] H.W. TANG, J. YAN, S.Q. LV. 2007. Advances in research on flows with vegetation in river management. *Advances in Water Science* 18(5):785–792.

[10] C. WANG, C. WANG. 2010. Turbulent characteristics in open-channel flow with emergent and submerged macrophytes. *Advances in Water Science* 21(6):816–822.

[11] F.S. WU. 2010. *Dynamic Characteristics of Open Channel Flow with Vegetation.* Nanjing: Southeast University Press.

[12] J. YAN. 2008. Experimental Study on Flow Resistance and Turbulence Characteristics of Open Channel Flows with Vegetation. *Doctoral Dissertation of Hohai University.* Nanjing.

[13] D. Yang, H.B. Zhao and et al. 2013. *Matlab image processing and its application.* Beijing: Qinghua University Press.

Testing and Measurement: Techniques and Applications – Chan (Ed.)
© *2015 Taylor & Francis Group, London, ISBN: 978-1-138-02812-8*

Urban transition in Yangtze River Delta based on TM image

G.F. Chen, Q.X. Zhang, X. Lu & T. Qin

China institute of water resources and hydropower research, Beijing, China

ABSTRACT: According to the economic and social development, the quick urban sprawl has changed the land use pattern, especially occupied cultivated land. In order to realize a healthy and sustainable development, a planned use of land should be built and the basic land use pattern should be supplied. Remote sensing images are ideally used to monitor current land cover changes thanks to their rapid up-date capability. Taking the Yangtze River Delta as an example, the article interpreted the urban transition process in the year 2000s and 2010s based on TM image. The results showed that the urban area has been increased from 5185km² to 14901km². The urban region ration has reached 10.38% in the study area. A scientific and sustainable land planning should be taken in this region.

1 INTRODUCTION

Based on TM image from the website of the United States Geological Survey[1], the article interpreted the urban transition process in the year 2000s and 2010s using ARCGIS and ERDAS. It provided the urban area and spatial distribution which can provide basic information for scientific and sustainable land planning in the Yangtze River Delta.

Remote sensing images are ideally used to monitor current land cover changes thanks to their rapid up-date capability. The urban land cover interpretation can show the urban transition quickly and provide useful information for urban research[2].

The Urban land cover interpretation using remote sensing data has many related studies[3-4]. Taking the Yangtze River Delta as an example, the article interpreted the urban transition process in the year 2000s, and 2010s based on TM image, which provides useful information for scientific and sustainable land planning.

2 STUDY AREA AND METHOD

2.1 Study area

The Yangtze River Delta is located on the eastern coast, and belongs to the most developed region in China. It is one of the national city group and Shanghai is the core city which is one of the biggest cities in the world.

The study area belongs to the core area of the Yangtze River Delta, with an area of 13131 km² and it contains the eight main cities in this region. See fig. 1.

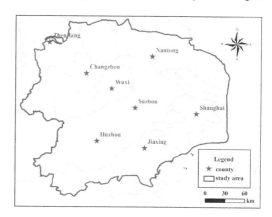

Figure 1. Study area of the Yangtze River Delta.

2.2 Method

The land use was classified by supervised classification method and the results were amended by visual calibration. The details are as follows: the pre-interpretation was taken for each image by mosaic, mask and subset. Five images for each period (the year 2000 and the year 2010) was taken and the basic image was prepared for classification. See figure2 and figure 3.

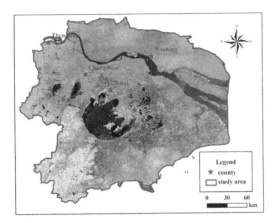

Figure 2. TM image of 2000.

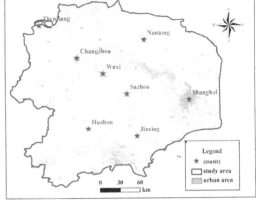

Figure 4. Urban distribution of 2000.

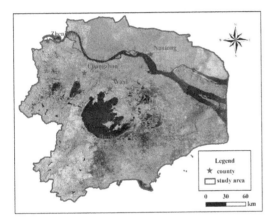

Figure 3. TM image of 2010.

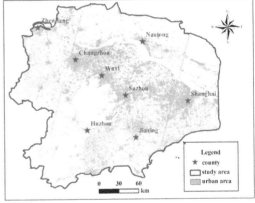

Figure 5. Urban distribution of 2010.

Classification models were built for each period and the supervised classification was taken. Considering the characters of study areas, four types of land model were taken as water, urban area, farmland and forest.

The classification results were taken from raster to shape and the small cells affection was deleted.

Visual calibration. According to google earth image with high resolution, the classification results were amended and the urban areas were taken. See figure 4 and figure 5.

3 RESULTS AND DISCUSSION

3.1 Urban area increasing

The urban area was 5185 km²in the year 2000, which takes a ratio of 10.38% of the study area, while it reached 14901km²which takes a ratio of 29.85% of the study area. From 2000 to 2010, the urban area has increased dramatically.

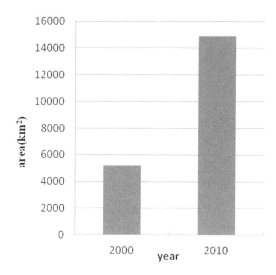

Figure 6.   Urban area of the year 2000 and the year 2010.

3.2 *Urban cover transition*

According to the transfer matrix of land use in the study area, in 2010, 8485 km² were transferred from farmland of 2000 and 649 km² were transferred from water. The urban is sprawled mainly by occupying farmland.

Table 1.   The transfer matrix of land use from 2000 to 2010 units: km².

2010 / 2000	farmland	water	urban	forest	total
farmland	11831	301	8486	506	21124
water	842	5731	650	16	7239
urban	1607	63	2494	182	4346
forest	980	57	311	3616	4964
total	15260	6152	11941	4320	37673

3.3 *Affection by image resolution*

In this study, the article used TM image with 30m resolution as source data to analyze the land use and urban transition, the results were affected by image resolution and high resolution data can increase the precision of the results. Besides, four types of land cover were taken from image. If possible, according to character of the study area, it can build more detailed model to class the land use cover.

4 CONCLUSIONS

The article interpreted the land use cover of the Yangtze River Delta and the urban transition was analyzed with the results data. According to the results, urban area sprawled dramatically by economic and social development. In order to keep sustainable development of the study area, a land use planning should be built to guide the land user.

ACKNOWLEDGMENT

This research was financially supported by the National Science Foundation (51309244).

REFERENCES

[1] http://glovis.usgs.gov/
[2] Peng Guangxiong, Xu Bing, Shen Wei, Li Jing. Extracting Urban Land-use on the TM Imagery. REMOTE SENSING TECHNOLOGY AND APPLICATION.2006, 21(1): 31–36.
[3] Yang cunjian, Zhou chenghu. Extracting residential areas on the TM imagery. Journal of remote sensing. 2000, 4(2): 146–150.
[4] Wen gongjian, li deren, Ye Fen. Automatic extraction of urban area from satellite panchromatic remote sensing images. Geomatics and information science of Wuhan University. 2003, 28(2): 212–217.
[5] Sheng Hongya, Zhang Yongbin, Pang Qinggang. Analysis of land use pattern in Tangshan City based on remote sensing and GIS. Mine Surveying. 2011 (6).

TL-diversity: Type of *L*-diversity for privacy protection of the clients within the cloaking area

D.H. Song, J.W. Sim, B.S. Kim & K.J. Park
Department of Information Communication Engineering, Wonkwang University, Iksan-shi, Republic of Korea

J.M. Kang, D.H. Sin & I.J. Lee
Department of Electrical Engineering, Wonkwang University, Iksan-shi, Republic of Korea

M.B. Song
Samsung Electronics, Suwon, Republic of Korea

ABSTRACT: This paper proposes a cloaking algorithm called *TL*-diversity that produces a cloaking area by considering both the number and type of buildings within the cloaking area. In location-based services (LBS), users send queries to the LBS servers along with their locations, but this location information can be maliciously used by adversaries. The proposed *TL*-diversity strengthens the privacy protection of the querying user by additionally maintaining the diversity with regard to the types of buildings in a cloaking area. That is, it reduces the probability of inferring the user's current activity or whereabouts based on the types of the facilities in the cloaking area (e.g., hospital).

1 INTRODUCTION

Location-Based Services (LBS) provide information on the geographical position of a user to respond to the user's spatial queries. Typical LBS applications include friend finder, nearest point of interest (POI) query, road navigation, emergency call location, etc. The use of these services, however, poses privacy issues as the user locations and queries are exposed to untrusted LBS entities. Some of them could be malicious and the exposure of the location information might reveal users' identity or other sensitive information [1].

Since it is a key challenge to efficiently preserve user's privacy while accessing LBS, several techniques have been proposed to protect the location privacy of a user. Most user privacy techniques are based on cloaking, which achieves location *k*-anonymity [2-10]. However, the size of the cloaking area created in a setting where the clients are densely populated is generally small so an adversary can easily infer the client locations. The solution proposed in [2] provides more stringent privacy guarantees for densely populated POIs via A_{min} that indicates the required minimum area of the cloaked spatial region. In addition, the *L*-diversity technique that ensures at least *L* buildings within a cloaking area was proposed in [3]. The *TL*-diversity proposed in this paper takes into account the types of buildings in a cloaking area in addition to the number of buildings that is addressed in the previous *k*-anonymity work in [3]. For example,

suppose that a cloaking area containing *L* buildings is created for the querying user who is in a food court. Although location *L*-diversity is satisfied, it is very likely that the types of the buildings in the cloaking area are mostly 'restaurant'. This information helps adversaries infer what the user is doing even if they are not aware of the exact location of the user. To discourage this kind of location privacy intrusion, the proposed *TL*-diversity prevents the buildings of a same type from being grouped into the same cloaking area.

The rest of the paper is organized as follows. Section 2 introduces the related work in this section. Section 3 describes the proposed *TL*-diversity algorithm. Section 4 presents the experiment results of the *TL*-diversity. Finally, conclusions are given in Section 5.

2 RELATED WORK

LBS profiles may include sensitive information of clients and they can help identify a person, which makes them susceptible to many attacks. The profile based anonymization model suggested in [4] ensures anonymity by generalizing both location and profiles to the extent specified by the user. The drawback of this model is that some loss of data about a client is inevitable, thus decreasing the effectiveness and value of LBS.

There exist two cloaking approaches that use location *k*-anonymity and location *L*-diversity as two quantitative metrics to model the location privacy requirements of a user [3, 5]. First, the PrivacyGrid framework provides dynamic bottom-up and top-down grid cloaking algorithms for *k*-anonymity and *L*-diversity in a mobile environment [5]. In this approach, the entire space is divided into equal-sized grid cells and the number of users in each cell is stored and maintained. This approach seeks to satisfy *L*-diversity after satisfying *k*-anonymity, so explored regions can become very big when search directions are not appropriate.

The cloaking algorithm in [3] gives priority to *L*-diversity. This algorithm creates the minimum cloaking region by finding *L* number of buildings (*L*-diversity) and then finds *k* number of users (*k*-anonymity). For example, given that *L*=3 and the querying user *q* is outside the buildings, the three buildings *L1*, *L2*, and *L3* that are adjacent to *q* are searched. In each building, a single user who is inside the building and the most adjacent to *q* is found. A minimum bounding rectangle that includes the discovered adjacent users becomes a temporary cloaking area. If the *k*-anonymity requested by the user is satisfied, the temporary cloaking area is confirmed as the cloaking area. Otherwise, additional adjacent cells are explored in every directions (up, down, left, and right) until the requested *k*-anonymity is satisfied. But they neglect building types, such as hospital, church, restaurant, etc.

3 *TL*-DIVERSITY

This section describes the creation of a cloaking area with regard to the querying user *q* and the buildings around *q*. In Figure 1, the solid bounding rectangle is a cloaking area produced according to the existing *L*-diversity technique, and the dotted rectangle is a cloaking area produced according to the *TL*-diversity proposed in this paper.

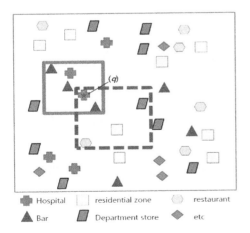

Figure 1. Cloaking area generation according to the *L*-diversity and *TL*-diversity.

Assume that user *q* requests for a query inside a hospital and *L*=5. In the solid rectangle, the number of buildings included in the region is 5, satisfying *L*-diversity. However, this group exhibits homogeneity in the building type, so the buildings in the cloaking area are distilled into two types, hospitals and bars. The proposed *TL*-diversity creates a cloaking area by considering both the number and type of buildings. Hence, its cloaked region (the dotted rectangle in Figure 1) has five buildings, each of which is of a different building type. With *TL*-diversity, the chance that the type of building is disclosed is reduced to 1/5. The algorithm of the proposed *TL*-diversity technique is as follows.

Algorithm 1. Generation of a cloaking area satisfying *TL*-diversity.

Input: The location of the user, the locations and types of the buildings

Output: The minimum cloaking area that satisfies the *L* and *TL* parameter values given by the querying user

01: Check the locations and types of the buildings;
02: Check the location of the user;
03: Check the *L*-diversity and *TL*-diversity values specified by the user;
04: **for** (measure the distances of the buildings from the user; sort the buildings in ascending order by distance)
05: **if** (the number of buildings satisfies *L* && the number of different types of buildings satisfies *TL*)
06: Generate a cloaking area by grouping the buildings satisfying the *L* and *TL* condition with a minimum bounding rectangular
07: result= *TL* ∪ *L*;
08: **return** result;

---

4 EXPERIMENTAL EVALUATION

The performance of the proposed *TL*-diversity was examined in the experiments. The experiments were performed on the computer with a 2.9 GHz processor and 4 GByte memory. C++ was used to implement the experiments. The users requesting queries were stationary, and the distance of a single grid was set to 10*m*. The buildings were randomly distributed in the grid cells. Table 1 shows the experimental parameter settings.

Table 1. Experimental setup.

Parameter	Values
Number of grids	50 × 50
Number of buildings	200
TL-diversity	1, 3
L-diversity	3, 4, 5, 6

The experiment results of the two different *TL* values (*TL*-diversity=1 and *TL*-diversity=3) are compared. These results are the average of the values obtained by executing the experiments 10,000 times.

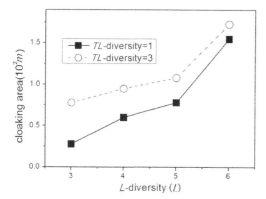

Figure 2. Cloaking area sizes (*TL*-diversity=3, *TL*-diversity=1).

Figure 2 presents the cloaking area sizes of the two different *TL*-diversity settings while *L*-diversity increases from 3 up to 6. The cloaking areas generated when *TL*-diversity=1 are smaller than those generated when *TL*-diversity=3. The gap between the *TL*-diversity graphs is bigger when *L* is small. Even if *L*-diversity is satisfied, the *L* value is increased further until *TL*-diversity is satisfied. That is, more buildings need to be included in the cloaking area until the requested *TL*-diversity is achieved, thereby increasing the overall size of the cloaking area.

5 CONCLUSION

This paper proposes the *TL*-diversity algorithm that considers both the number and type of the buildings in the cloaking area. The *TL*-diversity initially finds a cloaking area based on *L*, the number of buildings specified by the querying user, and then enlarges the cloaking area by including more buildings until *TL*, the number of different types of buildings specified by the user, is satisfied. Compared to the existing method considering only *L*-diversity, the proposed *TL*-diversity produces a larger cloaking area but it improves the protection of user privacy in LBS by adding

another layer of diversity with regard to the building type in a cloaked region. In the future, an efficient index structure that exploits the locations of buildings will be studied.

ACKNOWLEDGEMENT

This research was supported by Basic Science Research Program through the National Research Foundation of Korea(NRF) funded by the Ministry of Education, Science and Technology (2014R1A1A2054315)

REFERENCES

[1] Muntz, W. R. Barclay, T. Dozier, J. Faloutsos, Maceachren, A. Martin, J. Pancake, C. and Satyanarayanan, M.2003. IT Roadmap to a Geospatial Future, The National Academics Press.

[2] Sweeney, L. 2002. k-anonimity: A model for protecting privacy", International Journal of Uncertainty, Fuzziness and Knowledge Based Systems 5(10): 557–570.

[3] Kim, J. H. Chang, J. W. Um, J. H. 2009. An Algorithm for generating Cloaking Region Using Grids for Privacy Protection in Location-Based Services, Journal of Korea Spatial Information Society, 2(11): 151–161.

[4] Shin, H. Atluri, V. Vaidya, J. 2008. Profile Anonymization Model for Privacy in a Personalized Location Based Service Environment, Proc. intern. Conf., Mobile Data Management, Beijing, 27–30 April 2008.

[5] Mokbel, M. F. Chow, C.-Y. and Aref, W. G. 2006. The New Casper: Query Processing for Location Services without Compromising Privacy, Proc. intern. Conf., Very Large Data Bases, Seoul, 12-15 Sept. 2006.

[6] Kim, J. Lee, A. r. Y. Kim, K. J. Um, H. Chang, J. W. 2008. Cloaking Method supporting K-anonymity and L-diversity for Privacy Protection in Location-Based Services, Journal of Korea Spatial Information Society, 4(10): 1–10.

[7] Kim, J. Y. Jeong, E. H. Lee, B. -K. 2011. A Design of Cloaking Region using Dummy for Privacy Information Protection on Location-Based Services, Journal of Communications and Networks 8(36): 929–938.

[8] Park, S. Bai, Park, J. S. 2011. An Anonymization Technique of Continuous Query and Query Log for Privacy in Location-Based Services, Journal of Computing Science and Engineering 2(38): 65–71.

[9] Lee, J. 2013. Hierarchical Clustering-Based Cloaking Algorithm for Location-Based Services, Journal of Korea Information and Communications Society 8(8): 1155–1160.

[10] Bamba, B. and Liu, L. 2008. Supporting Anonymous Location Queries in Mobile Environments with PrivacyGrid, Proc. intern. Conf., International World Wide Web Conference Committee, Beijing, 21–25 April 2007.

Testing and Measurement: Techniques and Applications – Chan (Ed.)
© 2015 Taylor & Francis Group, London, ISBN: 978-1-138-02812-8

The test and measurement technologies in the study of vital capacity of the students of Tajik and Zang nationality

W.T. Hao, Y.L. Zhang & C. Yue

Physical Culture Institute of normal university of Hainan, Haikou, China

ABSTRACT: In this paper, we have done some study on the vital capacity of 1440 children and adolescents (aged from 7 to 18) from Zang nationality in Lhasa of Tibet and Tajik nationality in the Taxkorgan autonomous county of Xinjiang Uygur Autonomous Region using test and measurement technologies. After the test, statistical analysis and comparative research, we can provide some practical suggestions for the physical education of Zang nationality and Tajik nationality.

KEYWORDS: The test and measurement technologies, Zang nationality, Tajik nationality, vital capacity.

1 INTRODUCTION

Study Goal: In this paper, we have used many different assessment methods, including subjective test and objective text. We have effectively observed and measured some defined contexts and indexes. In other words, we have used numbers and symbols to illustrate some features to the study objects and made the social phenomenon digitized and stylized. We have done some study on the vital capacity of 1440 children and adolescents (aged from 7 to 18) from Zang nationality in Lhasa of Tibet and Tajik nationality in the Taxkorgan autonomous county of Xinjiang Uygur Autonomous Region using test and measurement technologies.

Study Methods: In this paper, we used the study methods of field survey, logical analysis, mathematical analysis, and documentation method. We used SPSS17.0 software to analyze the data and obtained T values and P values. Then we analyzed the values and come to the conclusion.

2 STATISTICAL ANALYSIS ON THE VITAL CAPACITY OF ZANG AND TAJIK STUDENTS

From Table 1 and Figure 1, we can see that the vital capacities of Zang and Tajik boy students generally increase with the growth of their age. The basic trends for boy students from these two nationalities are: a) from age 7 to 12, the vital capacity of Tajik boy students is better than that of Zang students; b) from age 13 to 18, the vital capacity of Zang boy students is better than that of Tajik students. Although there are some differences in the average values of the boy students of two nationalities, but by using the T test, we can see that the difference in the values in age 16, 17 and 18 is very obvious (P<0.01), the difference of the values in age 15 is obvious (P<0.05), and the difference in the values in the other age group is not obvious (P>0.05).

Table 1. In 2010, the Tibetan nationality boys lung capacity comparison of tower.

Age	N	Tajik $\overline{X}$	S	N	Zang $\overline{X}$	S	T	P
07	30	942.67	151.88	30	869.00	224.27	1.490	0.142
08	30	1067.17	170.41	30	974.13	337.29	1.348	0.183
09	30	1132.90	156.51	30	1052.13	316.15	1.254	0.215
10	30	1223.67	235.55	30	1137.27	372.47	1.074	0.287
11	30	1439.50	322.15	30	1350.67	258.44	1.178	0.244
12	30	1594.40	422.61	30	1491.43	371.48	1.002	0.320
13	30	1606.13	298.85	30	1773.27	530.31	−1.504	0.138
14	30	1828.33	347.82	30	2015.57	644.24	−1.401	0.167
15	30	1985.00	415.47	30	2263.70	538.57	−2.244	0.029
16	30	2070.33	270.05	30	2447.77	670.83	−2.859	0.006
17	30	2231.67	416.83	30	2780.30	587.07	−5.283	0.000
18	30	2342.67	677.47	30	2868.23	603.43	−3.173	0.002

From Table 2 and Figure 2, we can discover that the vital capacity of Zang and Tajik girl students generally increase with the growth of their age. The basic trends for girl students of the two nationalities are: from age 7 to 14, the vital capacity of Tajik girl students is better than that of Zang students; but from age 15 to 17, the vital capacity of Zang girl students is better than that of Tajik students. Although there are some differences between the average values of the boy students of two nationalities, but by using the

T test, we can find that the difference in the values in age 10 is very obvious (P<0.01), the difference of the values in age 8, and 12 is obvious (P<0.05), and the difference in the values in other age groups is not obvious (P>0.05).

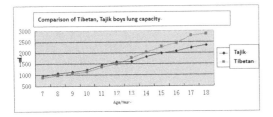

Figure 1. In 2010 the tower family, Tibetan boys lung capacity trend graph.

3 COMPARATIVE ANALYSIS

After analyzing the vital capacity of the children and adolescents of Zang and Tajik nationalities aging 7~18, we can find the following reasons.

3.1 The influence of altitude

The vital capacity samples for Zang boy students are from students in Lhasa. Lhasa is the provincial capital city of Tibet by the People's Republic of China. It is also the largest city in Tibet, the famous historical and cultural city in China, and Tibet's political, economic, cultural and religious center. It is an ancient city with a history of over 1300 years. It is the highest city in the world with an average altitude of 4500~5000 meters. High altitude makes Tibet such an environment with thin air, low air pressure and less oxygen. The data from the tables and figures tells us that the vital capacity of boy students of Zang nationality is far smaller than that of boy students of Tajik nationality. Therefore, we can know that the high altitude of Tibet is the major cause for the differences.

The vital capacity samples for Tajik boy students are from the Taxkorgan autonomous county. Although Lhasa and Taxkorgen are both in the area with high altitude and thin air, the altitude of Taxkorgen is still lower than that of Lhasa. So the oxygen in Taxkorgan is more than in Lhasa. This explains why the vital capacity of Tajik students is bigger than Zang students in age 7~12.

The vital capacity of Zang adolescents who live in high altitude area also increases with the growth of their age. Therefore, the vital capacity of Zang adolescents is higher than that of Tajik adolescents.

3.2 Climate influence

Tibet has its unique plateau climate. The average temperature among years changes very little, but the temperature between day and night changes greatly. Besides, the altitude is very high and the air is very thin. Therefore, in Tibet, there is a saying "four seasons in one mountain, different weathers in adjacent places".

Firstly, the climate of Taxkorgen County is generally cold drought or half drought. The winter is drought with little rain and much sunshine and it is very cold; the spring and autumn are short and windy with little rain; the weather is summer is very difficult to tell. In the rough, we can divide a year into two seasons –hot season and cold season. Secondly, it has high altitude and low air pressure. Thirdly, the oxygen content is 170—180g/cubic meter. Lastly, the average year humidity is 30—35%.

Table 2. In 2010, the Tibetan nationality girls lung capacity comparison of tower.

	Tajik nationality			Tibetan nationality				
Age	N	$\overline{X}$	S	N	$\overline{X}$	S	T	P
07	30	864.60	180.15	30	808.17	213.95	1.105	0.274
08	30	936.57	184.79	30	825.77	212.13	2.157	0.035
09	30	1045.77	227.68	30	934.17	292.88	1.648	0.105
10	30	1292.27	286.27	30	1054.07	271.89	3.305	0.002
11	30	1339.77	245.91	30	1204.77	283.36	1.971	0.054
12	30	1486.83	338.11	30	1263.73	316.08	2.640	0.011
13	30	1579.57	384.83	30	1487.70	362.27	0.952	0.345
14	30	1598.33	326.96	30	1614.57	478.29	−0.153	0.879
15	30	1619.53	325.10	30	1711.37	313.02	−1.115	0.270
16	30	1710.33	427.12	30	1816.80	427.07	−0.965	0.338
17	30	1859.33	391.49	30	2041.43	457.63	−1.656	0.103
18	30	1878.33	583.81	30	1876.53	355.58	0.014	0.989

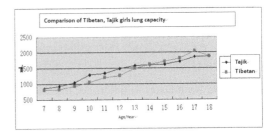

Figure 2. In 2010 the tower family, Tibetan boys lung capacity trend graph.

3.3 The influence of the social politic and economic development

On May 23rd 1951, the Central People Government and Tibet local government signed an agreement– *The agreement of how to liberate Tibet peacefully by Central people government and Tibet local government,* which is generally called *the 17 agreements.* Then Tibet was liberated peacefully. On Februaly 9th 1955, in the 7th Chinese State Council's plenary meeting, the *State Council's decision about founding the preparatory committee of Tibet Autonomous Region* was passed. On April 22nd 1956, the preparatory committee of Tibet Autonomous Region held its inaugural meeting. Darai Lama Tenzing Gia Zo served as chairman and Panchen Lama Gyaltsen served as first vice chairman, and Guohua Zhang as second vice chairman. While Ngapoi Awang Jigme served as secretary general.

Taxkorgan autonomous county of the Xinjiang Uygur Autonomous Region was founded earlier than Tibet. The education system is relatively more developed than that of Tibet's, and the physical education developed faster, and its executive power is stronger. Those are good conditions for the adolescents in the Taxkorgan Autonomous County. Therefore, the vital capacity of Tajik boy students in age 7~12 is higher than that of Zang students in the same age.

4 CONCLUSION AND SUGGESTIONS

Based on the comparison of the physical qualities of students of Zang and Tajik nationalities and the analysis of the study materials with test and measurement technology, we can draw the following conclusions and give some suggestions.

The students' vital capacity of Zang and Tajik nationality from 7~12 years old generally increases with the growth of their age. But taking into account the factor of gender and age, the increase rates, ranges and sensitive ages are different. As Zang nationality lives in high altitude area, the education authorities in Tibet should pay more attention to its school physical education (P.E.) in middle schools and primary schools. An overall improvement in their physical playground and equipment conditions is needed and more P.E. classes are also required.

The boy students and girl students of Zang and Tajik nationalities are greatly different. In the adolescence puberty, girl students' vital capacity is smaller than boy ones. Especially after age 13, as the influence of adolescence puberty to the girls is much bigger than that to the boys, the physical qualities between the boys and girls differ greatly. Therefore, we should enhance the physical education in the adolescence puberty for the students', especially for the girls.

ACKNOWLEDGEMENT

Funding Project: Research on later philosophy and social science funding project of Ministry of Education (13JHQ061)

Writer Introduction: Wenting Hao, Doctor of Beijing Physical Education University,master supervisor,research fields on nationality physical quality and school physical education

Corresponding Writer Introduction: Yaling Zhang,Female,Professor,master supervisor in Hainan Normal University

REFERENCES

[1] Fang Yuan. Sociological research methods course[J]. 2004. The sixth chapter. page165.
[2] Wenting Hao. Chinese Nationality Students Physical and Health research [M]. Shenzhen: Chinese Education and Culture Press,2004.
[3] Yaling Zhang, Wenting Hao. The Test and Analysis of the Physical Quality of Tajik Nationality High School Students [J]. Beijing Physical Education University Journal,2005/4.
[4] Yaling Zhang. The Survey Report of the Physical Quality of Tajik Nationality High School Students [J]. Beijing Physical Education University Journal,2004/3.

Feedback systems with pseudo local loops

V. Zhmud
Novosibirsk State Technical University, Novosibirsk, Russia

V. Semibalamut
Siberian Branch of Geophysical Service of SB RAS, Novosibirsk, Russia

A. Vostrikov
Novosibirsk State Technical University, Novosibirsk, Russia

ABSTRACT: The paper proposes the method for designing regulators for unstable problematic objects. The method basis is introducing compensative feedback directly into the model of object. As a result, the structure should be transformed to get single-loop regulators. The method has been investigated with modeling in the program *VisSim*, and its effectiveness has been demonstrated. The effectiveness of regulators for some kinds of objects, for example, such as objects with deep non-linear positive feedback, depends on the accuracy of the realization of compensating elements.

1 INTRODUCTION

The design of regulators for objects containing problematic circuits is a difficult task, especially in the presence of non-linear elements in the case of positive feedback on the individual elements of the structure of an object. Positive feedbacks are the most dangerous when the loop gain is greater than the unit is.

To solve this problem, the modeling method is proposed and investigated. It consists of two stages.

In the first stage, compensating feedback loops are introduced into the model of the object. These loops allow to convert the non-linear element into the linear one, or to make its model close to a linear one. Also, these loops can convert Integrator into a periodic link. The resulting modified object is easy to control. So the design of the regulator for it is carried out simply on the basis of automatic control theory.

In the second stage, the feedbacks loops introduced to the structure must be recalculated into corresponding elements of the successive regulator. In this case, the loops inside the regulator itself are not problems; they can be preserved in the model of the regulator. The problem is only introduced loops in the structure of an object, because they can't be realized directly. So these loops should be recalculated into the equivalent fragments of the successive regulator structure.

Despite the fact that this method is intuitive and clear, with simulated desired positive effect is not always achieved. Simulation with the use of program *VisSim* has the advantage that the calculation of derivatives and integrals is then carried out on the same

or similar algorithms, which should take place in the digital regulator. This allows to identify possible problems of implementation, and to find an effective solution to overcome them.

2 THE CONTROL OF OBJECTS WITH TWO INTEGRATORS

Theoretically, the control of an object, which is two series-connected integrators, is not of great complexity [Zhmud V.A. 2009; Zhmud V.A.., Frantsuzova G.A., Vostrikov A.S. 2014]. But this example can demonstrate the essence of the method and its efficiency.

Let the model of the object to be given by the following transfer function:

$$W(s) = \frac{X(s)}{U(s)} = \frac{1}{s^2} \tag{1}$$

This object can be represented as two serially connected first-order integrators:

$$W(s) = \frac{1}{s} \cdot \frac{1}{s} \tag{2}$$

If we introduce a negative feedback around one of the integrators, for example, a proportional link with the gain equal to two, it will be converted into an aperiodic link. Remaining Integrator can play the role of an integral regulator for a static control, i.e. it will reduce the static error to zero.

Thus, to control the object in question it would be sufficient to use the specified local negative feedback, serial regulator with the unit gain and a single global negative feedback, as shown in *Fig.* 1. In this figure, the traditional notation is used for input, output, and control signals. Any element, which does not belong to the object, is a part of regulators.

In the implementation of this structure a problem appears: the signal $z(t)$, required for switching of local feedback is unavailable for measurements. A simulation of an object, namely – the regulator, provides a model of the signal as the signal $z'(t)$, as shown in *Fig.* 2.

Fig. 3 shows the structure for simulation of the system accordingly with the structure shown at *Fig.* 2, with the use of the program *VisSim*. This *Fig.* 3 also shows the resulting graph of the transition process in response to a single job step jump $v(t)$.

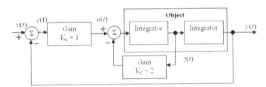

Figure 1. Block diagram for the control of the object with the help of the local loop.

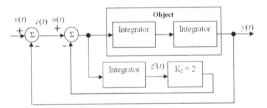

Figure 2. Block diagram for object control using the pseudo local loop.

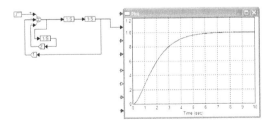

Figure 3. Stabilization of object, consisting of two series-connected integrators: a first Integrator is stabilized with the helps of the pseudo local loop, a second integrator of the object is not stabilized, and it performs the function of an integral regulator; transient shows the absence of overshoot, a static error is zero because of the presence integrator circuit.

In the example, robust control is achieved, because the object is linear. The exact values of the coefficients of regulator do not matter much. The proposed method is efficient, intuitive and easy to implement in the case of a linear object with small amount of a series-connected integrators without feedback (in our example object is the two integrators).

3 THE CONTROL OF OBJECTS WITH TWO INTEGRATORS AND NON-LINEAR POSITIVE FEEDBACK

Let complicate the problem by introducing a positive feedback, which contains a non-linear element in the form of an element that is raising the input signal into the third power. The structure of such an object is shown in *Fig.* 4. The effective regulator is required to be found. Introduced positive feedback destroys the stability of the object to a great extent. If the output signal is greater than unit, then in the local loop of the object the positive feedback is greater than unit arises. The cubing increases this problem. Methods of numerical optimization of PID-regulator [Zhmud, V, Yadrishnikov, O., Poloshchuk, A., Zavorin, A. 2012; Zhmud, V., Liapidevskiy, A., Prokhorenko, E. 2010; Zhmud, V., Zavorin, A. Yadrishnikov, O., 2013a. Non-analytic method; Zhmud, V, Zavorin, A. Yadrishnikov, O, 2013b. Fractiona] can not be used to calculate regulator for the control of such an object.

If we virtually introduce the negative feedback into the object, repeating the existing non-linear relationship, but with an opposite sign, then these two loops should compensate to each other. As a result, we obtain an object that was discussed in the previous section. The controlling of this derived object can be accomplished by the previously demonstrated way. *Fig.* 5 shows the corresponding block diagram.

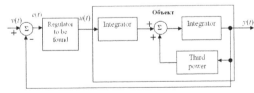

Figure 4. The task of controlling of the object with an internal nonlinear unstable loop.

After that, the local non-linear loop must also be converted into a pseudo-local loop to provide a regulatory structure, which does not use the signals from the inside of the object and does not use non-existent inputs of the object. Block diagram obtained by means of the necessary equivalent transformations is shown in *Fig.* 6 Simulation fully confirms the effectiveness of the proposed method to the object in question. Simulation scheme and the results for different values of the task $v(t)$ are shown in *Fig.* 7.

The resulting structure of the regulator can also be simplified with the aid of equivalent transformations. When using the digital controller it is not required because the according program can provide this structure.

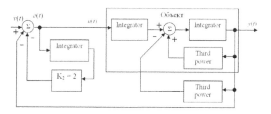

Figure 5. The block diagram for controlling the object using one local and one pseudo local loop.

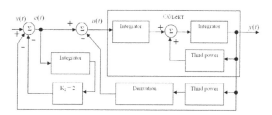

Figure 6. Block diagram for controlling the object using the two pseudo-local loops.

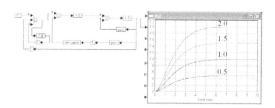

Figure 7. The simulation results of the object: the object is fully stabilized by the pseudo-local feedback; transients are shown for different values of the task (from 2 to 0.5).

The resulting system is fairly robust, that is, small changes in the coefficients do not lead to loss of stability. Also, the stability of the system is not affected by changing the method of integration (i.e. the method of computation of integrals and derivatives of the signals used).

4 THE CONTROL OF OBJECTS WITH THREE INTEGRATORS AND NON-LINEAR POSITIVE FEEDBACK

Let us complicate the problem by introducing into the object of another series-connected integrator.

Two variants can be offered for the use of the proposed method.

The first variant assumes deeper negative feedback than inside the object. This makes the problem loop enveloped by the aggregate negative feedback instead of the positive one, as shown in Fig. 8. In the simulation of the system stability when input value jump to about unity value has been provided. The nonlinear system may have different levels of quality of transient processes according to the input signal value, for example, it may be stable at low input signals and unstable at high input signals.

Simulation according to the structure of Fig. 8 reveals a number of problems. Particularly, we recommend choosing a simple Euler method among the possible methods of integrating. This method uses the calculation of the integral of a function with the helps of the sum of the values of this function on all the interval of integration, taken with the regular period, multiplied by the duration of such period. Specified above interval is the value of the integration step. Other methods of integration do not give the desired effect.

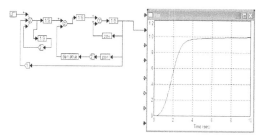

Figure 8. The providing of the stable control by means of the introduction of local loops after the conversion of the primary loop: a first integrator is stabilized by pseudo-local loop, the second block, which consists of two integrators, the latter of which is enveloped by a non-linear positive feedback, is stabilized by a local loop.

Fig. 9 shows the results of the simulation of the converted system, wherein the regulator structure is transformed into a structure with a single main loop (the local control loops inside the regulator are not forbidden). In this structure, naturally, the coefficient k in the regulator after involution into the third power can be equal to unity or can be less. In the case of a unit value of k almost complete compensation of nonlinearity occurs, but the loop has two integrators, making it unstable. Therefore, we use different coefficients k greater than unity. With the gain $k = 2$, the transient process in response to a single step jump has the best form, as can be seen from the graphs in *Fig.* 9. If the value of this jump is reduced, there is an overshoot in the system. Particularly, when a jump is 0.6, the overshoot amount is about 30%. In the case of an increase in this value, in the system arises reverse process, which can be called conditionally "undershoot" as an antonym to "overshoot". In particular, when the input signal is equal to the output 1.2 the signal first rather rapidly approaches the value of 1 and then moves slowly to the desired value of 1.2. *Fig.* 10 shows the transient processes depending on the coefficient on the negative feedback k.

As we can see, if the input jump is equal to unity, then for $k = 1.5$ overshoot exceeds 20%, for $k = 2$, no overshoot, when $k = 2.5$ undershoot about 20% has place in the system, with a further increase of this gain, undershoot slowly rises.

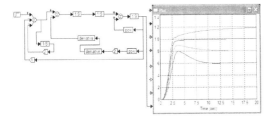

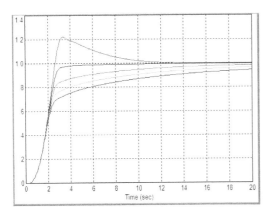

Figure 9. The result of the conversion of the previous structure into the structure with one main loop and graphics of the transient processes when at the input step is equal to 1.2; 1.0; 0.8 and 0.6, correspondingly.

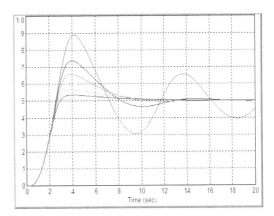

Figure 11. The same values with the input shock equal to 0.5; the gain k is changing from the top to the bottom of the graph, respectively: 1.5; 2; 2.5; 3; 3.5 and 4.

With the reducing of the size of the input signal to a value of 0.5, this dependence varies. Corresponding transients are shown in Fig. 11.

When k = 1.5 overshoot exceeds 80%, for k = 2, overshoot is 50%, with k = 2.5, it is 30% and with further increasing of gain k the overshoot is reduced.

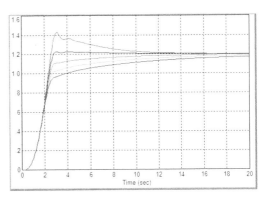

Figure 12. The same values with a jump at input equal to 1.2; the gain k is, from the top to the bottom of the graph, respectively: 1.4; 1.6; 1.8; 2 and 2.2.

With the increase of the value of the input signal to a value of 1.2, this dependence is also changed. Corresponding transient processes are shown in Fig. 12. When k = 1.4 the overshoot is close to 20%, with k = 1.6 overshoot is negligible, when k = 1.8 or more, undershoot occurs. If k < 1.2, then the system is unstable.

Figure 10. Transient processes in the system with the different gains after the block of cubing: upper graph is with k = 1.5, further - k = 2; k = 2.5; k = 3 and k = 3.5.

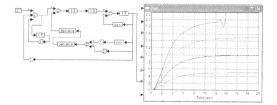

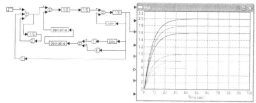

Figure 13. The result of the simulation of systems with almost full compensation of nonlinearities: transient processes for small values of the input signal are corresponding to jump in a linear system with high quality control; problems arise only when the input jump is 1.8 or more.

The second variant of controlling uses the attempt of the total compensation of the nonlinearity. In this case, the ratio discussed above should be equal to unity: k = 1.

But in this case, the two series-connected integrators become not enveloped by stabilizing feedbacks. Consequently, additional proportional feedback enveloping one of the integrators is necessary.

Fig. 13 shows the structure and the simulation result in program VisSim accordingly with this variant of the proposed method. If the jump of the input signal does not exceed 1.8 units, the system is close to linear. If this value will increase, the system becomes unstable. Namely, when the output signal value is 1.9, then rapidly growing amplitude oscillations occur (see. upper graph in Fig. 13).

Fig. 14 shows the results of attempts to find better values of the coefficient of proportional feedback (parallel to the compensatory nonlinear loop). Both increase and decrease of this ratio do not increase the stability of the system.

Fig. 15 shows the result of reducing the integration step from 0.1 s to 0.01 s. It can be seen that the stability of the system has been raised. Now the system has become stable with the jump of 2.0 units, as well as with any smaller value of this jump. The reason of it this restoration of stability is better fidelity of the model of compensating loop to the model of the initial compensated loop. The compensating loop comprises two additional integrator and two additional differentiating units. Each integrator introduces a delay whose value is the integration step.

Figure 15. Illustration that the success occurs at lower values of the integration step from 0.1 s to 0.01 s, and the linearity of the system is achieved when the input jump is 2 units or less.

Thus, in the compensating circuit implicitly contained delay link by an amount equal to twice of the integration step.

This rule is confirmed by further modeling. Indeed, with further increase in size of the jump up to 4 units, the stability of the system becomes broken again, as Fig. 16 demonstrates with the transient processes in it. But it is enough to reduce -integration step again from 0.01 s to 0.001 s, and the stability of the system for these values of the input signals is again restored, as shown by the graph in Fig. 17.

Therefore, in the case of the implementation of the control system based on digital technology, ADC and DAC to ensure the best stability, it requires the use of the devices of the highest possible speed.

However, even this does not solve the problem completely for the following reasons:

1 These integrators are included into the object model, and increasing of the regulator speed does not change the speed mismatch problem of the problematic loop to compensate its influence fully.

2 The model of an object can be known with insufficient accuracy.

3 The object model coefficients may vary with time during its operation.

4 The implementation of coefficients can also be carried out only within a given error, even with quite small one.

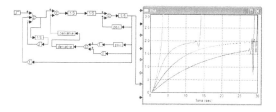

Figure 14. Illustration that any attempts to find other feedback coefficients do not lead to success.

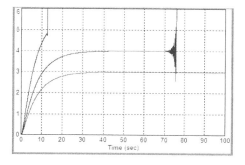

Figure 16. Violation of stability with increasing input jump to a value of 4 units.

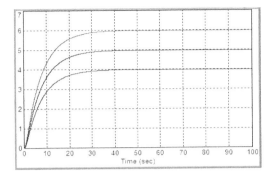

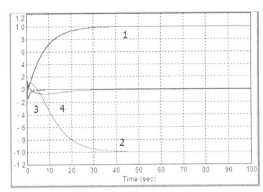

Figure 17. Restoring of the stability when changing the integration step from 0.01 s to 0.001 s.

Figure 18. Graphs of transient processes, including internal values of the object: 1 is output value, 2 is the output of the second Integrator, 3 is controlling signal, 4 is the output of the first Integrator.

Therefore, for all kinds of calculations it is useful to endeavor to provide robust control, i.e. the calculation of such regulators, in which the system remains stable even in the case of a small deviation of the true parameters of the object from the values of these parameters, is used in the calculation of the regulator.

For what considered in the last section problematic case, the design of robust regulator can be fundamentally impossible.

The reason of this problem can be understood by considering the static state and the system.

To the situation when the output signal of the object would be in steady state, i.e. not changed (and in this case it would be equal to the input specified value), the third (last) integrator signal must be equal to zero.

However, signal equal to the cube of the output value of the object is applied through the adder and the nonlinear element to the input of the integrator signal. Therefore, to compensate for this quantity, the output signal of the second (penultimate) integrator must also be equal to the cube output magnitude but with a minus sign. And with the aim of keeping by this integrator of its output value in the static mode, the output signal of the first integrator must also remain zero at the end of the transition process. Also at the end of the transient process zero control signal of the regulator output must take place. This is a necessary but not sufficient condition. Furthermore, these signals are related by relations, according to which some of these signals are the integrals of the other or the differences of the other signals in agreement with a mathematical model of the object. The respective signals are shown in *Fig.* 18.

In this case, all of these signals are generated only from the shape of the control signal $u(t)$. From this it can be seen how complicated are the requirements for this signal, and thus for the regulator, which generates the signal from the prescribed value $v(t)$ and the output value $y(t)$.

In order to investigate the robustness of the resulting the regulator gain in the compensating tract has been changed. These changes did not cause significant changes in the transient processes with the setting value equal to 12 units or less, the system remain stable. If the value is more than 13 units, it is not stable.

A method for the design of regulator for problem objects has been proposed and investigated. The object model can include many integrators and unstable internal loop. The method has been investigated on the example of non-linear object with a positive feedback, which includes a non-linear element elevating the input signal into a cube.

The efficiency of the systems designed by this method has been demonstrated with the simulation. In the case of using the method based on the full compensation of the nonlinearity, the error of the coefficient about 10% does not cause the loss of stability or a significant change in the quality of the transient process. This is true within the limited magnitude of the input signals (up to 12 reference units, and corresponding output signal 1728 at the output cubed link).

5 CONCLUSIONS

A method for the design of regulator for problem objects has been proposed and investigated. The object model can include many integrators and unstable internal loop. The method has been investigated on the example of non-linear object with a positive feedback, which includes a non-linear element elevating the input signal into a cube.

The efficiency of the systems designed by this method has been demonstrated with the simulation. In the case of using the method based on the full compensation of the nonlinearity, the error of the coefficient about 10% does not cause the loss of stability or a significant change in the quality of the transient process. This is true within the limited magnitude of the input signals (up to 12 reference units, and corresponding output signal 1728 at the output cubed link).

This work was financially supported by the Ministry of Education and Science of the Russian Federation on the state task №2014/138, theme: "New structures, models and algorithms for the management of breakthrough technical systems based on high technology intellectual property".

REFERENCES

[1] Zhmud V.A. 2009. Simulating and optimization of system for the control of laser light in program VisSim. School book. Novosibirsk. Russia. Published by NSTU. (*In Russian:*В.)

[2] Zhmud V.A, Frantsuzova G.A., Vostrikov A.S. 2014. Dynamics of machatronic systems. School book. Novosibirsk. Russia. Published by NSTU. (In Russian: НГТУ, 2014. – 176 c. ISBN 978-5-7782-2415-5).

[3] Zhmud, V., Yadrishnikov, O., Poloshchuk, A., Zavorin, A. 2012. Modern key technologies in automatics: Structures and numerical optimization of regulators Proceedings - 2012 7th International Forum on Strategic Technology, IFOST 2012.

[4] Zhmud, V, Liapidevskiy, A, Prokhorenko, E. 2010. The design of the feedback systems by means of the modeling and optimization in the program vissim 5.0/6 Proceedings of the IASTED International Conference on Modelling, Identification and Control.

[5] Zhmud, V, Zavorin, A. Yadrishnikov, O., 2013. Non-analytic methods for calculating the PID controllers. School book. Novosibirsk. Russia. Published by NSTU. (In Russian: В.А.)

[6] Zhmud, V, Zavorin, A. Yadrishnikov, O., 2013. Fractional Power PID-regulators. School book. Novosibirsk. Russia. Published by NSTU. (In Russian: В.А.)

Design of robust energy-saving regulators by means of optimization software

V. Zhmud
Novosibirsk State Technical University, Novosibirsk, Russia

V. Semibalamut
Siberian Branch of Geophysical Service of SB RAS, Novosibirsk, Russia

A. Vostrikov
Novosibirsk State Technical University, Novosibirsk, Russia

ABSTRACT: The paper discusses the method for the design of regulators for an object, containing one or several consecutive integrators in addition to other elements. The method provides stable astatic control with small overshooting and minimal power consumption, which is specified as square of controlling signal. The method has been tested upon several examples.

1 INTRODUCTION

Negative feedback with serial controller allows successful control of objects, providing the desired values of the output values even in the presence of disturbances. Such control loops are widely used in science and technology. The success of such controlling way can be achieved only with the correct regulator design based on the object model and partly on the conditions of its validity. Due to the complexity of the analytical calculation of the regulator, a method based on numerical optimization can be often applied for these purposes more successfully [Zhmud V.A. 2009]. Often regulator model is known only approximately. This is why the properties of the real system are very different from the results of preliminary modeling or simulation. We have proposed various methods for more effective resolving of the problem of the regulator optimizing taking into account this understanding, one of which is to bring into the object model of additional delay link. This allows us to calculate regulator which will ensure the stability of the system even if the object model is worse than the used one in the numerical optimization.

2 TASK STATEMENT

Often saving of the controlling resource in the system is necessary, which can be, for example, described by the square of the control action [2–3]. To solve this problem, we propose a modified method of numerical optimization based on the programs for modeling using specially introduced optimality criteria and

artificial deterioration of high-frequency part of the object model. We will consider only the single-loop system with negative feedback loop, a simplified block diagram of which is shown in Fig. 1 [Zhmud V.A. 2009]. This loop includes object, regulator and element of comparison. Here, the output value Y(t), which should be as close as possible to the prescribed value V(t). Control error E(t) is the difference of these values; the regulator transforms the error E(t) into a control signal U(t), which acts on the object input to do its output value equal to that prescribed value. As a result, the regulator with the controlling loop reduces the error E(t) to zero.

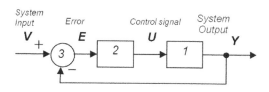

Figure 1. Block diagram of a basic system with negative feedback: 1 – object, 2 – regulator, 3 – element of comparison.

Numerical optimization methods allow simulating of the whole system by changing the parameters of the regulator (for given its structure). As a result of optimization such parameters are found to give the minimum value of a positively definite cost function, which is a criterion of optimality. The cost function is used as arguments the signals in the system, e.g., it is equal to the integral of the modulus of the error. Complications of the cost function allow to correct the optimization process, for example, to eliminate overshoot or periodic fluctuations in output around the steady state about the prescribed value. Some changes of the cost function allow success executing of the optimization of regulator in those cases where optimization procedure cannot be successfully completed without such changes, for example, due to an unlimited increase of one of the signals in the system.

Single-loop system in this case, we call such a system in which no links from any point in the object model or to any point of this model. Thus the regulator can be implemented with any structures with any loop in the regulator itself, in which case the system ceases to be a single-circuit, since it has a progressive connection of regulator and the object and element of comparison in a single closed loop.

The most effective energy savings take place in the case where the object contains an integration link in its model. Although the most efficient regulators are generally considered to be PID-regulators (i.e., regulators having proportional, integrating and derivative links), in the presence of the integrator in the object, the integrator in the regulator is not necessary. Therefore, we consider the task of designing PD-regulators (containing only the proportional and derivative links).

3   THE PROPOSED PRINCIPLES FOR THE TASK RESOLVING

The following principles of design of energy-saving systems are proposed.

1   Although earlier for the majority of problems in our papers we recommend as a cost function the integral of the modulus of the error is multiplied by the factor of which is the time since the beginning of the transient process to its end, for the design of energy-saving regulators we offer to eliminate this specified factor t, leaving only the integral of the modulus of the error. The reasons for this are two considerations. First, this coefficient was necessary to reduce overshoot and fluctuations in the transient process, but it does not require to be introduced into the cost function, because the energy efficiency is the most important. Second, we can see that other features of the cost function accomplish this aim, and the final process is rather good.

2   Since in most problems of energy saving controls the initial form of the transient process is not critical, it is advisable to cut this initial stage, starting with the integration of a previously specified time, after some time interval from the beginning.

3   The term equal to the integral of the square of the control signal applied to the object with weight gain is proposed to be introduced into the cost function, which means the power spending [Zhmud V.A, Frantsuzova G.A., Vostrikov A.S. 2014].

4   We propose the introduction into the structure of the object model used in the numerical optimization some delay link. The value of this delay should be chosen such that a significant phase shift would rise in the frequency range in which the amplitude-frequency characteristic of the object is defined in the identification with no sufficient accuracy. This should ensure a small effect on the delay to the phase response of a model of the object in the range where the model of the object has been determined with sufficient accuracy. This artificial deterioration of the performance of the object model can effectively achieve the goal, even if the real object has what some features at these frequencies, where the object model is not accurate enough. The usefulness of this approach has been confirmed by simulation.

5   As for the simulation software, a program VisSim is chosen, which works very well for solving of the tasks of this class.

6   As the test signal in the case of a linear object unit step can be used. Non-linear objects in this paper are not considered.

4   THREE TERMS OF THE COST FUNCTION

The cost function is proposed in the form of the sum of the three terms:

$$\Psi = \psi_1 + \psi_2 + \psi_3, \tag{1}$$

The first term is the integral of the error module during the time from t_1 to T:

$$\psi_1 = \int_{t_1}^{T} |e(t)| \, dt, \tag{2}$$

Here T – is the time before the achievement of the finish of the controlling task; t_1 – is given time, before which the form of the transient process is not critical. In the simulation, it is more convenient to use the following form of this term:

$$\psi_1 = \int_{0}^{T} |e(t)| \sigma(t - t_1) \, dt \tag{3}$$

Here $\sigma(t - t_1)$ - is the step function from zero to one at the time moment t_1.

The second term is the energy specified as the integral of the square of the control action with a weighting factor:

$$\psi_2 = \int_0^T k_2 u^2(t)\,dt, \qquad (4)$$

Here, the coefficient k_2 is necessary to bring all the cost items in the same scale. This factor is required for all terms except one, since multiplication of cost function as a whole by a constant factor does not change its actions, so for simplicity, a first factor in (3) we can put the unit: $k_1 = 1$.

The third term – is modulus of the derivative of the error of control during a given time, or the integral of this module at a given time interval:

$$\psi_3 = \left| \frac{d}{dt} e(t_1) \right|, \qquad (5)$$

$$\psi_{3'} = \int_{t_1}^T \left| \frac{d}{dt} e(t) \right| dt \qquad (6)$$

5 THE EXAMPLE OF THE USE OF THE PROPOSED METHOD

Let consider an object having a transfer function of the form:

$$W(s) = \frac{1}{s^2(4s^2+4s+1)} \cdot W_x(s) \qquad (7)$$

Here $Wx(s)$ – is an unidentified component of the transfer function of the object.

For example, it may be delay link, by an amount $\tau = 2$:

$$W_x(s) = \exp\{-2s\} \qquad (8)$$

It can also be high-pass filter:

$$W_x(s) = \frac{1}{(0.3s+1)(0.3s+1)} \qquad (9)$$

Let suppose, that according to the technology of application of object it is required saving of control resource with the condition of the demand to achieve the objective of the control at time $t_1 = 50$ s.

We propose to carry out the optimization of PD-regulator with the cost function of the form (1) under the terms of (3), (4) and (5). Then the equation of the regulator has the form:

$$W_R(s) = k_P + k_D s \qquad (10)$$

Fig. 2 shows a diagram of the program VisSim for optimizing of regulator (10).

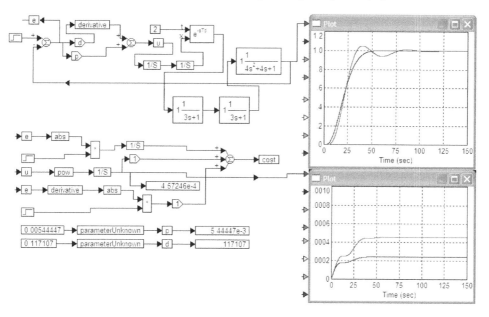

Figure 2. Scheme in the program VisSim to optimize the regulator (10) and to investigate the results of optimization: during the optimization, the delay link and two filters are excluded.

Since the function $W_x(s)$ is unknown, optimization can alternatively be performed with the model which does not contain this feature. In this case, the proposed procedure gives the following coefficients: $k_P = 0.005444$; $k_D = 0.1171$. Energy costs at the same time are $P = 0.00024$ *units*. The transient process has no overshoot. If the time constant will increase to the value of $\tau = 2.2\,s$, then the system will loss the stability. Further, we can carry out simulation of the system with the object, which contains this transfer function as a product of functions (8) and (9). In this case, we obtain the other curves, which are shown in *Fig.* 3. In this case, the system generates an overshoot of about 2%, energy consumption increases to a value $P = 0.00045$ *units*.

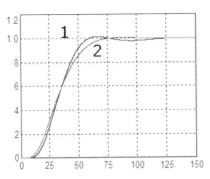

Figure 3. Result of optimization of regulator (10): during the optimization, the delay link has delay value $\tau = 2$ s, and two filter circuits are excluded from the loop.

The modified procedure is the optimization with the introduction into the model of delay link with the time constant $\tau = 2\,s$ (filters are excluded). This procedure gives the following coefficients: $k_P = 0.0021455$; $k_D = 0.0718$. Energy costs at the same time are $P = 8.3 \cdot 10^{-5}$ *units*. Transient process has negligible overshoot of less than 1%, its appearance is shown in *Fig.* 3 (line 1). If together with the delay link we would apply filter, the transient process would not change significantly (line 2), and the cost would even be reduced to a value of $P = 6.5 \cdot 10^{-5}$ *units*, as is shown by the red line in *Fig.* 3.

6 CONTROL OF OBJECT WITH THREE INTEGRATORS

Let consider an object having a transfer function of the form:

$$W(s) = \frac{1}{s^3} \cdot W_x(s) \qquad (11)$$

The task of control of such object can not be effectively resolved by simple PD-regulator. Therefore, we

introduce the structure additional derivative channel connected in the output of the first derivative link. In the result we obtain control with dual derivative, together with a single one. Such regulators in various literatures are named as PDD-regulators or PD²-regulators. The transfer function of such regulator has the form.

$$W_R(s) = k_P + k_D s + k_{DD} s^2 \qquad (12)$$

The result of optimization and the structure are shown at *Fig.* 4.

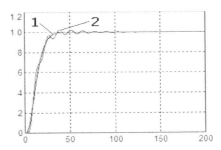

Figure 4. Scheme and the result of optimizing of the regulator for the object (11) and for investigating of the results of optimization: for the case when in the optimization of the delay link was excluded, the result is blue graph; the result for the optimization with the delay link is red graph.

Fig. 5 shows the optimization results when delay link with the time constant 2 s is introduced into the object model. When optimizing object without delay we obtain the following regulator coefficients: $k_P = 0.006025$; $k_D = 0.0851$; $k_{DD} = 0.45068$. The transition process has no overshoot, energy consumption is $P = 8 \cdot 10^{-4}$ units. When optimizing with a delay link $\tau = 2$ we obtain the transient process with small damped oscillations (red line in Fig. 4), energy consumption increases by 10 times. If the time constant of the increase to the value of $\tau = 2.2$ s or more, the system losses stability.

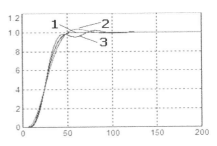

Figure 5. The result of the optimization of regulator for object (11): optimization with the delay link (line 1); the result without delay link (line 2); the result with the delay link which delay is increased to a value 5 s (line 3).

The resulting coefficients of the regulator are: k_P = 0.000432; k_D = 0.0118; k_{DD} = 0.144. Energy consumption with $\tau = 2\,s$ is $P = 3 \cdot 10^{-6}\,units$. If we remove the delay link, energy consumption is reduced by half. But even if the time constant of delay link increases to $\tau = 5\,s$, the stability of the system is preserved, energy costs are rising up to $P = 5.2 \cdot 10^{-6}\,units$.

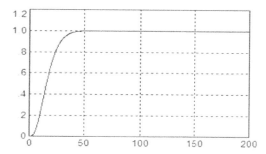

Figure 6. The result of optimizing the regulator for the object (13): the optimization has been done with the delay element excluded.

7   CONTROL OF OBJECT WITH FOUR INTEGRATORS

Consider an object having transfer function of the form:

$$W(s) = \frac{1}{s^4} \cdot W_x(s) \qquad (13)$$

For such object double derivation in the controller is not enough, control of the form (12) can not ensure stable control. We propose to use virtual negative feedback loop around one of the integrators, as suggested in [5]. In this case, the transfer function of the regulator takes the following form:

$$W_R(s) = (k_P + k_D s + k_{DD} s^2) \frac{1}{1 + 2/s} \qquad (14)$$

This transfer function can be transformed to the following form:

$$W_R(s) = (k_P + k_D s + k_{DD} s^2) \frac{0,5s}{0,5s + 1} \qquad (15)$$

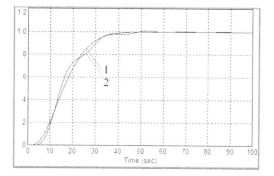

Figure 7. The result of the regulator optimization for the object (11): the optimizing result with the delay link excluded is curve 1, the result with the use of delay link is curve 2.

Fig. 6 shows the results of optimizing of regulator of the form (14) or (15) for the object (13) using a model without additional delay element. Values of the regulator coefficients are: k_P = 0.000572; k_D = 0.09866; k_{DD} = 0.660. The transient process has no overshoot. Energy consumption is $P = 1.25 \cdot 10^{-5}$ units. With the introduction of the delay link into object model it also preserves its stability only for small values of its time constant. Starting with a value of about $\tau = 2.2$, the stability of the system is broken. When τ = 2, the transient process has fluctuations, as shown in Fig. 7 (line 2). At the same time energy consumption increases twice.

Fig. 8 shows the results of optimizing of regulator (14) or (15) for the object (13) using the object with an additional delay link with a time constant τ = 2. The values of the regulator coefficients are: k_P = 0.000884; k_D = 0.0235; k_{DD} = 0.2936. The transient process has no overshoot. Energy consumption is P = 3.4·10⁻⁷ units.

If we remove delay link, energy consumption become reduced to $P = 2.9 \cdot 10^{-7}\,units$, and in the transient process it has small overshoot of about 2%. With increase of the time constant of delay link to $\tau = 5\,s$, the stability of the system will remain, and several fluctuations in the transition process of amount about 2% arises.

Fig. 8 shows the results of optimizing of regulator (14) or (15) for the object (13) using the object with an additional delay link with a time constant τ = 2. The values of the regulator coefficients are: k_P = 0.000884; k_D = 0.0235; k_{DD} = 0.2936. The transient process has no overshoot (line 1). Energy consumption is P = 3.4·10⁻⁷ units. If we remove delay link, energy consumption become reduced to $P = 2.9 \cdot 10^{-7}\,units$, and in the transient process it has small overshoot of about 2% (line 2). With increase of the time constant of delay link to $\tau = 5\,s$, the stability of the system will remain, and several fluctuations in the transition process of amount about 2% arises (line 3).

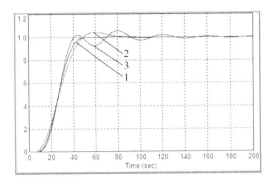

Figure 8. The result of the optimization of regulator for the object (13) when the optimization has been made with the delay link excluded is curve 1, the same result when system is simulated also without delay link is curve 2, the result with the link delay is 2.5 times greater is curve 3.

Energy consumption increases to $P = 4.25 \cdot 10^{-7}$ *units.* Corresponding transient processes are shown in *Fig.* 8.

8 CONCLUSIONS

The proposed method for designing of regulators has been studied by simulation on a set of examples. This method saves energy when the indicator of energy consumption at the same time is considered to be the integral of the control action. The effectiveness of the previously proposed modification of the object in the optimization has been also tested. This modification provides a more reliable obtaining of the useful results, because with this consistently the model of the object delay link is included. The use of the proposed method allows significant reduction of power consumption to provide more sure achievement of the controlling aim.

This work was financially supported by the Ministry of Education and Science of the Russian Federation on the state task №2014/138, theme: "New structures, models and algorithms for the management of breakthrough technical systems based on high technology intellectual property".

REFERENCES

[1] Zhmud V.A. 2009. Simulating and optimization of system for the control of laser light in program VisSim. School book. Novosibirsk. Russia. Published by NSTU. (*In Russian: НГУ, 2009. – 116 c.*).
[2] Zhmud V.A, Frantsuzova G.A., Vostrikov A.S. 2014. Dynamics of machatronic systems. School book. Novosibirsk. Russia. Published by NSTU. (In Russian: НГТУ, 2014. – 176 c. ISBN 978-5-7782-2415-5).

Testing and Measurement: Techniques and Applications – Chan (Ed.)
© *2015 Taylor & Francis Group, London, ISBN: 978-1-138-02812-8*

Numerical and experimental analyses of the modal characteristics of framed structures subjected to earthquakes

F.C. Ponzo, R. Ditommaso, G. Auletta, C. Iacovino, A. Mossucca, A. Nigro & D. Nigro
School of Engineering, University of Basilicata, Potenza, Italy

ABSTRACT: Damage detection approach based on dynamic monitoring of structural properties over time has received a considerable attention in recent scientific literature. In the earthquake engineering field, the recourse to experimental research is necessary to understand the mechanical behavior of the various structural and non-structural components. A new methodology to detect and localize a possible damage occurred in a framed structure after an earthquake is discussed in this paper which resumes the main outcomes retrieved from many numerical nonlinear dynamic models of reinforced concrete framed structures characterized by 3, 5 and 8 floors with different geometric configurations and designed for gravity loads only. In addition, the main results of experimental shaking table tests carried out on a steel framed model are also shown to confirm the effectiveness of the proposed procedure.

1 INTRODUCTION

Structural Health Monitoring, especially for structures located in seismic prone areas, has assumed a meaning of great importance last year, for the possibility to make a more objective and rapid estimation of the damage occurred in buildings after a seismic event. Last year, many researchers worked to set-up new methodologies for Non-destructive Damage Evaluation based on the variation of the dynamic behavior of structures under seismic loads (Dinh et al. 2012; Omrani et al. 2011a,b; Ponzo et al. 2010). The NDE methods for damage detection and evaluation can be classified into four levels, according to the specific criteria provided by the Rytter (1993). Each level of identification is correlated with specific information related to monitored structure: when increasing the level, it is possible to obtain more information about the state of the health of the structures, to know if damage occurred on the structures, and to quantify and localize the damage and to evaluate its impact on the monitored structure. Pandey et al. (1991) discussed the possibility to use the mode shape curvature to localize damage on structural elements. Sampaio et al. (1999) extended the idea of Pandey et al. (1991) by applying the curvature-based method to frequency response function instead of mode shape and demonstrated the potential of this approach by considering real data. The techniques for damage identification based on vibration and, in particular, those based on changes in modal parameters have been widely applied to the assessment of the health status of the existing structures, Doebling et al. (1996);

Limongelli (2014). Roy & Ray-Chaundhuri (2013) provided a mathematical basis to show the correlation between a structural damage and a change in the fundamental mode shape and its derivatives. In order to increase the performance level of damage detection and localization on monitored structures, it is necessary to support the theoretical criteria with numerical and experimental tests on both real and scaled structures, used in laboratory and in situ tests. Last year, in order to localize and quantify the damage occurred on both single structural elements and structures, several authors proposed to use the mode curvature variation over time (Pandey et al. 1991; Sampaio et al. 1999; Ditommaso et al. 2012; Roy and Ray 2013; Cao et al. 2014; Limongelli 2014). Practically, compared to the geometric mode shape curvature exhibited by the elements, and/or by the structure, over time it is possible to localize the damage position (Ditommaso et al., 2014a).

In this paper a new procedure for damage detection in framed structures based on changes in mode curvature is discussed. The proposed approach is based on the use of Stockwell Transform, a special kind of integral transformation that became a powerful tool for nonlinear signal analysis and then to analyze the nonlinear behavior of a general structure (Stockwell et al. 1996). The aim of this paper is to show, through practical examples of framed structures, how it is possible to identify and to localize damage on a structure, comparing mode shapes and the related curvature variations over time, before, during and after an earthquake. Furthermore, the main scientific results retrieved from an experimental campaign of shaking table tests performed on a 1:15 scaled structure and conducted at the Seismic Laboratory of the University of Basilicata (SISLAB) are described.

2 METHODOLOGY

In this paper, using fewer sensors installed inside a structure (one three directional accelerometer for each floor) it defines a new methodology that is able to assess the presence of any damage on the structure, and provide information about the related position and severity of the damage. It is based on a band-variable filter that we are able to extract the nonlinear response of each mode of vibration.

The Band-Variable Filter, (Ditommaso et al. 2012), is used to extract the dynamic characteristics of systems that evolve over time by acting simultaneously in both time and frequency domain. The filter is built using the properties of convolution, linearity and invertibility of the S-Transform. It gives the possibility to extract from a nonstationary and/or nonlinear signal just the energy content of interest preserving both amplitude and phase in the region of interest as discussed by Ditommaso et al., (2012).

The S-Transform of a function $h(t)$ is defined as:

$$S(\tau, f) = \frac{|f|}{\sqrt{2\pi}} \int_{-\infty}^{+\infty} h(t) \cdot e^{-\frac{(\tau-f)^2 \cdot f^2}{2}} \cdot e^{-i \cdot 2 \cdot \pi \cdot f \cdot t} \, dt \qquad (1)$$

where t = time; f = frequency, and τ = a parameter that controls the position of the Gaussian window along the t axis. The seismic structural behavior is analyzed using a band-variable filter (Ditommaso et al. 2012) based on the S-Transform.

So the complete process can be written as:

$$h_f(t) = \int_{-\infty}^{+\infty} \left(\int_{-\infty}^{+\infty} [S(\tau, f) \cdot G(\tau, f)] d\tau \right) \cdot e^{-i \cdot 2 \cdot \pi \cdot f \cdot t} df \qquad (2)$$

In this section, we will discuss about the opportunity to use the band variable filter to evaluate the mode shapes during the non-stationary phase. As mentioned before, the basic idea is to isolate, by means the band-variable filter, the fundamental mode shape over time and evaluate its changes in terms of shape and related curvature. Here we show how, using the proposed band variable filter, it is possible to extract the mode shapes of a system during the phase of maximum nonlinearity.

The proposed procedure has been applied on reinforced concrete framed structure to detect and localize the damage occurred after an earthquake.

The algorithm involves the following steps:

- Evaluation of the structural response acceleration at the last floor;

- Defining the filtering matrix on the vibration mode considered (calibrated signal recorded on the last floor);
- The convolution of the filtering matrix with the Stockwell transform of the signals recorded at each level and in the same direction;
- Evaluation of the mode shape over time and its curvature.

When we consider the fundamental mode shape of a framed structure as a beam displacement, following Cao et al. (2014), it is possible to localize the damage analyzing the singularity on the curvature of the fundamental mode shape. In this paper, following the evolution over time of the singularity on the curvature related to the fundamental mode shape the damage is detected and quantified.

Figure 1 shows the frequency evolution of the fundamental mode extracted by means of the band variable filter and the time-point from which the mode shapes are evaluated. To apply the proposed procedure, it is necessary to focus the attention on three most important instants for a structure subjected to an earthquake: (A) one instant before the earthquake, (B) the time-instant where the damaging structure exhibits the minimum fundamental frequency and (C) one instant after the earthquake. Comparing the mode shape characteristics evaluated in the instant A, B and C, it is possible to understand how damage occurs after the earthquake and localizes it on the structure. Instant A is the reference instant and it is necessary to compare the difference in terms of mode curvature between B – A.

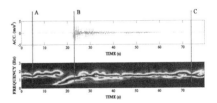

Figure 1. Normalized S-Transform and selection of the instants A, B and C.

It is worth noting that using the standard approach, it would have been possible to evaluate only the starting and final mode shapes, on the contrary, using the band variable filter, it is possible to evaluate also the mode shape related to the minimum frequency recorded during the maximum excursion in the plastic field. Therefore, being able to evaluate the mode

Curvature during the maximum excursion in nonlinear field and isolating it from superimposed signals, we can achieve a better understanding of the mechanisms of damage as well as a more precise location of the damage on the structure.

3 NUMERICAL MODELS

The main outcomes retrieved from many numerical nonlinear dynamic models of reinforced concrete framed structures characterized by 3, 5 and 8 floors with different geometric configurations and designed for gravity loads only are resumed in Fig.2 (more details in Ditommaso et al. 2014b).

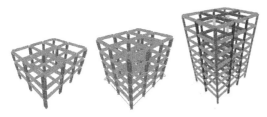

Figure 2. Numerical models regular in plan.

The numerical campaign is carried out using both natural and artificial accelerograms compatible with the Italian code for a soil type B and a soft soil type D (NTC2008). The natural accelerograms are extracted from the European database (European strong motion database). The artificial accelerograms are generated with the program SeismoArtif starting from the response spectrum.

In this section the main outcomes retrieved from structure regular in plan subject to natural accelerograms compatible with the Italian code (NTC2008) are presented.

The following figures show the curvature differences among floors evaluated over time in the time-instant where the damaging structure exhibits the minimum fundamental frequency (B).

Figures 3-4-5 show that drift agrees with the curvature difference indeed the maximum inter-story drifts are in correspondence to the second and the third floors. These parameters allow to achieve a better understanding of the mechanisms of damage as well as a more precise location of the mainly damaged floor.

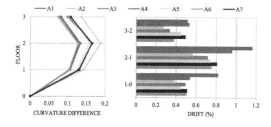

Figure 3. Curvature differences among floors and maximum inter-story drift in the time instant (B) of minimum fundamental frequency for the structure with 3 floors.

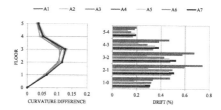

Figure 4. Curvature differences among floors and maximum inter-story drift in the time instant (B) of minimum fundamental frequency for the structure with 5 floors.

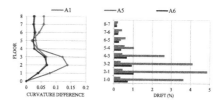

Figure 5. Curvature differences among floors and maximum inter-story drift in the time instant (B) of minimum fundamental frequency for the structure with 8 floors.

Another important parameter is the difference in terms of mode curvature between the time-instant where the damaging structure exhibits the minimum fundamental frequency (B) and the instant before the earthquake (A). A correlation between maximum inter-story drift and the maximum curvature difference has been defined. Figure 6 shows that the outcomes for the structure regular in plan with 5 floors subjected to natural accelerograms (A1, A5, A6) refer to a soil type B and an artificial accelerogram refer to a soft soil type D to grow seismic intensity.

As shown in Figure 6, the curvature difference is strongly related to the maximum inter-story drift, a useful indicator for structural and non-structural damage occurred on a structure.

It is interesting to observe a first range of maximum inter-story drift where the structure exhibits a quasi-linear behavior.

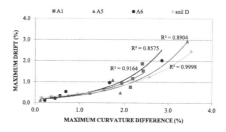

Figure 6. Correlation among maximum inter-story drift and maximum curvature difference for the structure with 5 floors.

427

After that it is possible to note a slight nonlinear behavior that increases with the importance of nonlinear effects occurred in the damaging structure. It is very important to highlight the limited dispersion of the data. This kind of trend has been found for all analysed structures: it is probably due to the fact that the used procedure is based on the evaluation of the curvature from the fundamental mode of vibration while the drift is evaluated considering the overall dynamic response, defined by all the modes of vibration. Therefore, the two parameters are directly proportional as long as the structure remains linear elastic. As the growth of the seismic intensity and increasing nonlinearity on structural elements, the curvature differences grow no longer proportionally to the drift because the higher modes of vibration become more important.

4 EXPERIMENTAL MODELS

In order to test and verify the algorithm for damage localization, a five-story 1:15 scaled model has been realized. The model consists of two spans and two frames (Mossucca 2008) in the x direction and by one span and three frames in the Y direction. It is regular in plan and in elevation (Ponzo et al., 2014). The model has been designed using elements that allow to easily change the mass, stiffness and geometric configuration. It is made by means of modular elements in steel and aluminium bars, differently tapered, replaceable and resistance and calibrated stiffness. The designed and framed structure can be assembled following several kinds of configuration in order to reproduce the seismic behavior of several kinds of reinforced concrete framed structures: a) designed using different codes; b) several number of floors; c) to simulate different collapse mechanism; d) changes the regularity characteristics both in plan and in elevation.

The experimental model is tested under dynamic conditions on the shaking table available in the Seismic Laboratory of the University of Basilicata.

In order to acquire the dynamic behavior of the model during the shaking table tests several kinds of accelerometric sensors (using cable and wireless network) have been distributed both on the structure and on the basement of the table. In addition, potentiometric transducers have been installed in order to acquire the displacement at all levels of the tested structure. The experimental campaign is conducted using natural accelerograms selected from the ITACA 1.1 version, a website collecting accelerometer Italians data. In the preliminary phase seven earthquake characterized by response spectra compatible with the target spectrum provided by the Italian seismic code (NTC 2008) related to Potenza City and soil type B. After the selection of the seismic database (7 earthquake), in order to reduce the number of the

shaking table test, following the criterion based on the target spectrum described before, only 3 earthquake have been selected and as shown in Figure 7, there is a very good agreement between the average of the selected earthquake and the target spectrum. With the aim to take into account the scale factor used for the framed structure, the entire selected earthquake database has been scaled in the time domain through a constant equal to the square root of the scale model factor. Figure 8 shows the possibility to use the modal curvature variation and also to detect anomalies in the stiffness distribution of the monitored structure using weak motion recorded signals. In fact, the floors up the third level of the scaled structure are characterized by a reduced section of the pillars (compared with those of the first and second level).

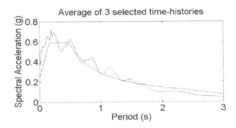

Figure 7. Average of the elastic acceleration response spectra selected for shaking table tests.

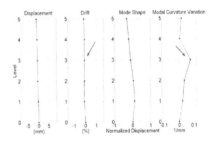

Figure 8. Displacement, drift, mode shape and curvature variation for low intensity input.

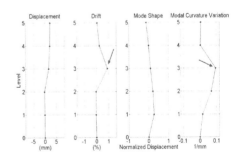

Figure 9. Displacement, drift, mode shape and curvature variation for high intensity input.

428

During the strong motion phase (when the structure is damaging) the maximum inter-story drift occurs on the third floor, according with the maximum curvature variation evaluated before and during the earthquake (Fig. 9).

In order to test the possibility to use the maximum curvature variation as damage index, Figure 10 shows the correlation between the maximum curvature difference between floors and the maximum inter-story drift evaluated using the accelerogram ter6 with different intensities (from 25% to 175%) of the original time-history.

It is worth noting from Figure 10 the good correlation existing between the mentioned parameters (as discussed also for the numerical analyses). For the considered steel framed structure a slight non-linear behavior starts after 0.5% (in the plane maximum curvature variation-maximum inter-story drift). Further analyses are necessary to confirm this kind of results for different input motions and different kind of representations.

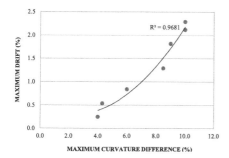

Figure 10. Correlation among maximum inter-story drift and maximum curvature difference for the model.

Figure 11 shows the damaged scaled model after the strong motion shaking table tests: the maximum damage occurred at the third floor.

Figure 11. Mechanism pillars plasticized on the third floor.

5 CONCLUSIONS

In this paper, the results of the methodology show the possibility to localize the damage occurred in framed structures subjected to strong motion earthquakes through analysis of parameters such as the mode curvature variation over time related to the first mode of vibration of a monitored structure. The proposed method is based on the evaluation of fundamental mode shape evaluated directly from the acceleration time-histories (and not from the displacements), so it is possible to avoid problems of divergence in the operation of double integration.

The aim of the work is to validate using several shaking table tests performed on a five-story 1:15 scaled structure a fast procedure for damage localization on framed building subjected to earthquakes. Starting from the results obtained from the numerical and the experimental campaign, it is worth noting the possibility to use the maximum curvature variation (over time) to detect and localize the damage occurred in a monitored structure just after a strong motion earthquake. Further analyses are necessary to confirm the preliminary results and to better calibrate the proposed procedure also considering the typological characteristics.

ACKNOWLEDGMENTS

This study was partially funded by the projects DPC-RELUIS 2010-2013 and Special Project "Monitoraggio" 2014 financed by the Italian Department of Civil Protection.

REFERENCES

[1] Cao, M. Xu, W. Ostachowicz, W. & Su, Z. 2014. Damage identification for beams in noisy conditions based on Teager energy operator-wavelet transform modal curvature, Journal of Sound and Vibration. 333, pp. 1543–1553.

[2] Dinh, H.M. Nagayamaz, T. & Fujinoy, Y. 2012. Structural parameter identification by use of additional known masses and its experimental application, Struct. Control Health Monit. 19, 436–450. DOI: 10.1002/stc.444.

[3] Ditommaso, R., Mucciarelli, M. & Ponzo, F.C. 2012. Analysis of non-stationary structural systems by using a band-variable filter, Bulletin of Earthquake Engineering, DOI: 10.1007/s10518-012-9338-y.

[4] Ditommaso, R.. Ponzo, F.C. Auletta, G. 2014a. Damage detection on framed structures: modal curvature evaluation using Stockwell transform under seismic excitation. Accepted for publication on Earthquake Engineering and Engineering Vibration.

[5] Ditommaso, R. , Ponzo, F.C. Auletta, G. & Iacovino, C. 2014b. Testing a new procedure for damage detection

on framed structures subjected to strong motion earthquakes. Second European Conference on Earthquake Engineering and Seismology. Istanbul, August 24–29, 2014.

[6] Doebling, S.W. Farrar, C.H. & al., 1996. Damage identification and health monitoring of structural and mechanical systems from changes in their vibration characteristics: a literature review, Los Alamos National Laboratory Report, New Mexico.

[7] Gabor, D. 1946. Theory of communications, J Inst Elect Eng 93:429–457.

[8] Limongelli, M. P. 2014. Seismic health monitoring of an instrumented multistory building using the interpolation method, Earthquake Engng Struct. Dyn. 2014.

[9] Mossucca, A. 2008. Modellazione in Scala di Strutture Intelaiate: Analisi delle Distorsioni ed Applicazioni Sperimentali. Tesi di Dottorato di Ricerca in Rischio Sismico XXI Ciclo, Università degli Studi della Basilicata – Potenza.

[10] Omrani, R. Hudson, R.E. & Taciroglu, E. 2011a. Story-by-story estimation of the stiffness parameters of laterally-torsionally coupled buildings using forced or ambient vibration data: I. Formulation and verification, Earthquake Engng Struct. Dyn, DOI: 10.1002/eqe.1192.

[11] Omrani, R. Hudson, R.E. & Taciroglu, E. 2011b. Story-by-story estimation of the stiffness parameters of laterally-torsionally coupled buildings using forced or ambient vibration data: II. Formulation and verification, Earthquake Engng Struct. Dyn, DOI: 10.1002/eqe.1193.

[12] Pandey, A.K. Biswas, M. & Samman, M.M. 1991. Damage detection from changes in curvature mode shapes, Journal of Sound and Vibration, Vol. 145: Issue 2, pp. 321–332.

[13] Ponzo, F.C. Ditommaso, R. Nigro, A. Mossuca, A. & Nigro, D. (2014). Validating of a new procedure for damage localization using shaking table tests on a 1:15 scaled structure. Second European Conference on Earthquake Engineering and Seismology. Istanbul, August 24–29, 2014.

[14] Ponzo, F.C, Ditommaso, R. Auletta, G. & Mossucca, A. 2010. A Fast Method for Structural Health Monitoring of Italian Strategic Reinforced Concrete Buildings, Bulletin of Earthquake Engineering, Volume 8, Number 6, pp. 1421–1434. DOI: 10.1007/s10518-010-9194-6.

[15] Rytter, A. 1993. Vibrational based inspection of Civil Engineering Structures, Ph.D. Thesis, University of Aalborg, Denmark.

[16] Roy, K. & Ray-Chaudhuri, S. 2013. Fundamental mode shape and its derivatives in structuraldamage localization, Journal of Sound and Vibration, 332 (2013), 5584–5593.

[17] Sampaio, R.P.C. Maia, N.M.M. & Silva, J.M.M. 1999. Damage detection using the frequency-response-function curvature method, Journal of Sound and Vibration, 226 (5), pp. 1029–1042.

[18] Stockwell, R.G. Mansinha, L. & Lowe, R.P. 1996. Localization of the complex spectrum: the S transform, IEEE Trans Signal Process 44:998–1001.

Testing and Measurement: Techniques and Applications – Chan (Ed.)
© 2015 Taylor & Francis Group, London, ISBN: 978-1-138-02812-8

Research on method of shipboard anti-sway electronic weighing

G.F. Zhang, Y. Chen & J. Luan
Dalian Ocean University, Dalian, China

X.W. Zhao & J. Liang
Zhangzidao Group CO. LTD, Dalian, China

ABSTRACT: In this paper, based on dynamic weighing mass method, a shipboard weighing method (acceleration-calibrating method) was proposed to resolve the problem that shipboard weighing is inaccurate due to ship sway. In the method, a weight was employed to calibrate the acceleration of ship sway and filter it during weighing to remove the effect of the ship sway so as to increase weighing measurement precision. Finally, the weight data were dealt with weighted moving average filtering method and calculated the average value to obtain the measured weight. The test result shows that the weight measured error can be controlled within 0.5% for the objects weighing 0.5~25kg, and the method is simple, convenient, and easy to implement.

1 INTRODUCTION

The measurement on weight is divided into static and dynamic weighing. Static weighing is performed when the measured object and measurement environment is in a static state, so it has a high precision and the relative measurement error can reach the very few (such as 0.001%) [1]. Dynamic weighing method will be employed to resolve the precision problem when measured object or weighing environment is in non-static state, and it has a low precision with a larger error [21]. Weighing on ship belongs the dynamic weighing. It is very difficult to do weighing on ship for ship sway, and it is more difficult to guarantee the accuracy of weighing. As a result, the shipboard trading is unfair, and the study and management of aquaculture and fishing farming is difficult to implement.

Studies on dynamic weighing were performed and some helpful methods were proposed, in which typically include: M Haiimie [2-4] used linear Gauss method, Calman is filtering and fuzzy logic method to process data and improve the speed and precision of weighing a block object. W. Balachandran [5,6] researched dynamic weight sorting machine with the theory and method of fuzzy controllers. S. Almodarresi [7] introduced feature extraction and two layer artificial neural network to evaluate the weight under the influence of vibration noise. Daniel [8] researched how to improve weighing speed and precision with artificial neural network in linear and nonlinear dynamic weighing system. Danaci [9] researched the application of nonlinear regression method in dynamic weighing. Yoshihiro [10] proposed a high precision weighing method for weight measurement in movement state. Yang Qing [11] introduced an adaptive estimation weight method in the powdery material proportioning system and carried out and verified with 8031 SCM. Dong Wei [12] developed intelligent weighing electronic scale and expanded the application of electronic scale. Shi Yongkun [13] proposed a method to improve speed and efficiency of the constant weighing system with fuzzy controller. Zhang Haiqing [14] established mathematic model and parameter identification model for quantitative feeding problem. Yuan Mingxin [15] established nonlinear model using radical basic function neural network to process dynamic weighing data. Song Aijuan [16] studied the structure and implement methods of dynamic weighing with DSP chip control. Gao ZH [17] proposed a parameter identification method based on the least square method and the prediction method to resolve the speed and precision problem in dynamic weighing. Bai Ruilin [18-20] researched dynamic weighing control method based on neural network technology.

In general, according to application environment and condition, the commonly used methods of dynamic weighing [21] are impulse method, volume method, radiation absorption method, gravity method and mass method. To resolve the inaccuracy problem in shipboard weighing due to ship sway, this paper studied a practical method named as acceleration-calibrating method. In the method, a weight was employed to calibrate the acceleration of ship sway and filter it during weighing to remove the effect of ship sway so as to increase weighing measurement precision, and finally the weight data was dealt with weighted moving average filtering method and calculated the average value to obtain the measured weight. The test result shows that the weight measured error can be controlled within 0.5% for the objects weighing 0.5~25kg.

2 PRINCIPLE

The method came out from mass dynamic weighing method in which Newton's second law $F = ma$ was used to get weight though measuring the force F and mass a. In this method, two electronic scales are located in the same position on the ship, one of which is used to measure the stress force F of object on an electronic scale, and the other is used to calibrate the acceleration a of the ship on the position. We set the actual weight of the measured object as W, electronic scale readings as W^*, the actual weight of calibration weight as W_0 (known in advance), and measured weight as W_0^*, the vertical acceleration of the position on the ship as a, according to Newton's second law, there will be:

$$W^* - W = \frac{W}{g}a, \; W_0^* - W_0 = \frac{W_0}{g}a \qquad (1)$$

Here, g is the acceleration of gravity. And we merge the two equations to remove the acceleration a and obtain the actual weight of the measured object as $W = \frac{W_0}{W_0^*}W^*$, afterwards the weight data is dealt with weighted moving average filtering method to improve measurement precision.

3 WEIGHING SYSTEM CONFIGURATION AND IMPLEMENTATION STEPS

As fig.1 shown, the weighing system mainly consists of four parts including a main electronic scale, a calibration electronic scale, a calibration weight and a pocket computer for data processing.

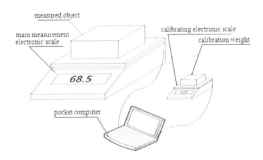

Figure 1. Weighing system configuration.

Firstly, we set the sampling frequency (such as 10HZ) of the pocket computer, take a 100g weight on the calibration electronic scale as calibration weight, and place the measured object on the main scale. When ship sway is slow down, the computer starts to weigh (reads weight data) and stop after a period of time (such as 10 second) in which a series of weight data of main electronic scale (W_i^*) and calibration electronic scale (W_{0i}^*) will be received and processed, accordingly a series of object weight can be calculated as $W_i = \frac{W_0}{W_{0i}^*}W_i^*$, and then we process data with the following method and steps to get the measured weight:

Step 1: calculate the average value of all weight data $\overline{W} = \frac{1}{n}\sum_{i=1}^{n}W_i$;

Step 2: calculate the error of sample data W_i and the average value $\overline{W}$, $\varepsilon_i = \left|W_i - \overline{W}\right| / \overline{W}$;

Step 3: set an error ε and filter the data with a big error ($\varepsilon_i > \varepsilon$);

Step 4: do smoothing processing with 5-points three polynomial symmetric smoothing method (one of weighted moving average filtering methods [2]) as following eq (2);

$$w_i = \frac{1}{35}(-3w_{-2i} + 12w_{-1i} + 17w_{0i} + 12w_{1i} - 3w_{2i}) \qquad (2)$$

Here, w_i is weight data after smoothing processing, $w_{-2i}, w_{-1i}, w_{0i}, w_{1i}, w_{2i}$ are the five weight values locating at before and after the ith data and sorted by their value.

Step 5: calculate the average of the data series w_i to get the final measured weight of the object.

4 INSTANCE AND ANALYSIS

Three kinds of weight 0.50kg, 4.92kg and 25.00kg were selected to test and verify the method above mentioned. To minimize the influence of error from the weighing device, all the real weight data was measured with the same device (weight scale) in the stable environment on land. The test result is analyzed as follows.

Figs. 2 and 3 is respectively the time history curve (abbreviated as THC) of measured weight of object and calibration weight. From the figures, it can be seen that the measured weight data fluctuates around the real weight value and the fluctuating rule of the weight data of object and calibration weight are mostly consistent. It is because that ship sway influences the weight measurement though acceleration, and the influence is uniform at the same position. Fig.4 shows the THC of an object's weight after calibrating, which is not big fluctuation and most influence from ship sway is filtered away. Fig. 5 shows the THC of error of measured object weight data and real weight which is controlled within 0.5%.

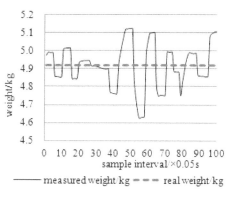

Figure 2.   THC (time history curve) of object's measured weight.

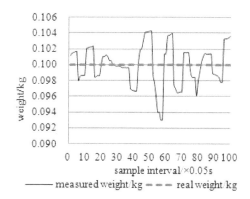

Figure 3.   THC of calibrating weight's measured weight.

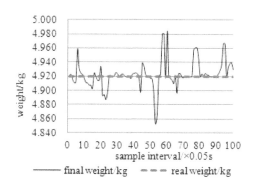

Figure 4.   THC of object's weight after calibrating.

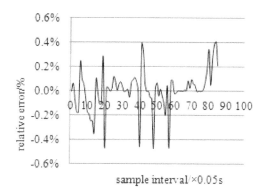

Figure 5.   THC of measurement error.

The measurement error (the relative error of final measured value and real value) and maximum error (the relative error of sample data after calibrating and real data) of various scales of weight are listed in the Tab.1 in which it can be seen the final measurement error is within 0.5% and the error will increase with weight increasing. The reason may be a bigger weight will magnify the influence of ship sway on weighing measurement.

Table 1.   Measurement error of various weight scales.

Real weight/kg	Measured weight/kg	Relative error/%	Maximum error/%
0.500	0.501	0.200%	0.391%
4.900	4.922	0.449%	1.367%
25.000	25.121	0.484%	1.790%

5   CONCLUSIONS

To resolve the inaccuracy problem in shipboard weighing due to ship sway, a calibration weight can be used to calibrate the acceleration caused by ship sway and filter the influence on weighing measurement. The 5-points, three polynomial symmetric smoothing methods (one of weighted moving average filtering methods) can be used to do the smoothing processing of the weight data to improve measurement precision. The test result shows that the weighing measurement error can be controlled within 0.5% for the objects weighing 0.50~25.00kg and the acceleration-calibration method mentioned in the paper resolved perfectly the problem of shipboard weighing. The defect of the method is that a weight scale and a computer should be added to process the data. It can be improved through writing the program of data processing into the processor chip in main electronic scale. The method is simple, convenient, and easy to implement, especially, fits for application in research and management in aquaculture field.

ACKNOWLEDGEMENTS

The authors acknowledge the supports of the National Science and Technology Planning Project of China (2012BAD18B0), and Horizontal Subject ("Optimal design of scallop aquaculture facility") from Dalian Zhangzidao Group CO. LTD, and express profound thanks for their financial support.

REFERENCES

[1] Yu Qikai. Research on dynamic weighing system[D]. Tianjin University of Science and Technology, 2003.

[2] M Halimic, W Balachandran. Kalman filter for dynamic weighing system[C].IEEE International Symposium on Industral Electronics, 1995:786–791.

[3] M Halimic, W Balachandran, Y Enab. Fuzzy logic estimator for dynamic weighing system[C]. IEEE Iniernational Conference on Fuzzy Systems, 1996:2123–2129.

[4] M Halimic, W Balachandran. Performance ImProvement of Dynamic Weighing Systems using Linear Quadratic Gaussian Controller[C]. IMTC2003-Instrumentation and Measurement Technology Conference 2003:1537–1540.

[5] W Balachandran, YM Enab, M Halimic, M Tariq. Intelligent robot-based dynamic weighing system[C]. Proceedings of SPIE, Intelligent Robots and Computer vision VIII: Algorithms and Computer vision, 1994, 2353:398–409.

[6] W Balachandran. Optimal digital control and filtering for dynamic weighing systems[C]. IEEE Instrumentation and Measurement Technology Conferences, 1995:293–298.

[7] S Almedarresi, N White. Application of artificial neural net works to intelligent weighing Systems[C]. Proceedings of the IEEE, 1999, 146:265–269.

[8] Daniel Massicotie, Bruno MBA Megner. Neural-network-based Method of Correction in a Nonlinear Dynamic Measuring system[C]. proceedings of the IEEE, 1999:1641–1646.

[9] Danaci. Detection of the starting of the newly applied mass for successive weighing[C]. Mathematical and Computational Applieations.1999, 4(l):53–59.

[10] Yoshihiro Fujioka, Jianxin Sun, Toshiro ONO. High Accurate weighing system Used under the Vibration-Like Moving Conditions[C]. SICE, 2002:483–488.

[11] Yang Qing, Li Huaizhang. Study on mathematical model of intelligent batch system[J]. Transactions of the Chinese Society of Agricultural Engineering. 1996, 12(4):48–52.

[12] Dong Wei, Zhang Jianjiu. Design of intelligent fixed quantity measurement electronic scale[J]. Manufacturing Automation, 1995 (2):34–36.

[13] Shi Yongkun, Dang Huanli, Zhu Chunhua. Application of fuzzy control in fixed output[J]. Weighing Instrument, 1999, 28(2):27–28.

[14] Zhang Haiqing, Li Baoan, Luo Xianhe. Dynamic weighing solution for predetermined filling material[J]. Acta Metrologica Sinica, 1998, 19(3):221–224.

[15] Yuan Mingxin, Zhang Yong, Zhang Yu. Design of dynamic weighing system based on RBF network[J]. Computer and Communications, 2003, 21(2):60–63.

[16] Song Aijuan, Yun Dongmei. Design of dynamic weighing system based on DSP[J]. Chinese Journal of Scientific Instrument, 2003, 24(4):115–117.

[17] Gao ZH, Mao JD. Dynamic weighing technology combining Parameter identification[C]. Proceedings of the Third International symposium on Instrumentation Science and Technology, 2004, 1:367·371.

[18] Bai Ruilin, Liang Hong, Li Jun. The study of fuzzy neural PID controller for instrument [J]. Chinese Journal of Scientific Instrument, 1999, 20(6):603–605.

[19] Bai Ruilin, Li Jun, Bai Ruixiang, Yan Xinzhong, Xu Hui. The study on dynamic quantitative weighing control strategy based on neural network technology [J]. Process Automatic Instruments, 2000, 21(7):8–10.

[20] Bai Ruilin, Yan Xinzhong, Li Jun. A study of quantitative scale based on fuzzy neural network technique[J]. Acta Metrologica Sinica, 2000, 25(2):127–130.

[21] Shi Changyan. State of arts and trends for the dynamic weighing and force measuring technology [J]. Acta Metrologica Sinica, 2001; 22(3):201–205.

Temperature sensing from luminescence of Eu^{3+} - doped $YAlO_3$ ceramics

V. Lojpur, B. Milićević, M. Medić, S. Ćulubrk & M. D. Dramićanin
University of Belgrade, Vinca Institute of Nuclear Sciences, Belgrade, Serbia

ABSTRACT: Temperature dependent photoluminescence of Eu^{3+} doped $YAlO_3$ sample synthesized by solid state reaction was investigated for high-temperature phosphor thermometry. The photoluminescence spectra were collected under excitation of 399nm, elevating the temperature gradually from room temperature to 800K. The characteristic emission lines of Eu^{3+} were detected and the following transitions: $^5D_1 \rightarrow {}^7F_1$ and $^5D_0 \rightarrow {}^7F_2$ were chosen for the temperature-dependence study using the fluorescence intensity ratio method. Emission decay curves measured at the strongest emission peak centered and 614 nm were recorded in a same temperature range. Data analysis showed that thermometry by the fluorescence intensity ratio method can be used over the temperature region of 300 to 800K with the absolute sensitivity of 2.95 K^{-1} and the relative sensitivity of $7.76*10^{-4}$ %K^{-1}. Temporal dependence of emission (lifetime) provides temperature sensing from 600 to 800K with the absolute sensitivity of 0.013 ms K^{-1} and the relative sensitivity of 1.79 %K^{-1}.

1 INTRODUCTION

Yttrium Aluminium Perovskite (YAP) is a famous material greatly explored in the last couple of decades. This material is very important for a large number of applications: doped with rare earth ions it can be used for lasers (YAP:Nd, (Fibrisch et al. 2014), YAP:Tm, (Han et al. 2014), YAP:Ho, (Wang et al. 2014) and inorganic scintillators with short response time (YAP:Ce, (Baccaro et al. 1998), YAP:Yb, (Zeng et al. 2005)). When doped with transition metals, for example YAP:Mn (Zorenko et al. 2012, Baran et al. 2012), it has application for thermoluminescent dosimetry of ionizing radiation, YAP:Co (Talik et al. 2007) is known for its magnetic properties while YAP:Cr (Sugiyama et al. 2013) is also scintillator material. Recently, YAP's application in cancer treatment has been demonstrated (Perra, in press).

Temperature is one of the fundamental thermodynamic state variables and the most measured physical property. Temperature sensors are extensively used accounting for about 75–80% of the world's sensor market. Luminescence thermometry is an optical contactless technique for temperature measurements. Temperature can be determined remotely by measuring changes in the luminescent properties, like the intensity of the luminescence, the decay time or the rise time of the emission, the luminescence excitation spectra, emission band maxima, and band shapes (Khalid et al. 2008, Brubach et al. 2011). If some of these properties changes with temperature, material could find application in optical temperature sensing devices. A large number of organic and inorganic materials have been investigated and used for temperature measurements, and among them Thermographic Phosphors (TPs) showed the best properties (Brites et al. 2012). TPs are usually ceramic materials doped with rare earth or transition metal ions, and their luminescence characteristics are significantly changing with the temperature fluctuations. They are commonly used in two types of luminescence temperature sensing methods: fluorescence intensity ratio (FIR) method and emission lifetime method. FIR method is based on comparison of two emission lines in photoluminescence spectrum. This method offers many advantages over the other luminescence thermometry methods, such as a wide temperature range (10-2000 K, depending on the used TP), elimination of errors that are a consequence of excitation light fluctuations, and relatively high sensitivity of measurements (Lojpur et al. 2013). Lifetime method, on the other hand, can be employed on a single emission band, and it is not affected by light scattering. It is important to mention that both methods are self-referencing.

Due to the importance of YAP many procedures for its synthesis have been developed so far. It is not that easy to obtain a pure intermediate perovskite phase, since there are two more phases in Y_2O_3-Al_2O_3 pseudo-binary system: monoclinic (YAM) and garnet (YAG) phases (Lojpur et al. 2013). Successful preparation of YAP phase is achieved by solid-state reaction (Zhydachevskii et al. 2014) while soft chemical reactions, such as sol-gel (Queiroza et al. 2010), combustion or spray pyrolysis (Mancic et al. 2012) usually produce the other two perovskite phases in small quantities.

Although YAP luminescence was widely explored and used, there is only a single report on its use in the luminescence thermometry (Kissel et al. 2013). On the other hand, YAG – the other important material from Y_2O_3-Al_2O_3 pseudo-binary system – has demonstrated the excellent potential for luminescence thermometry. With the use of YAG:Dy and YAG:Tm the highest temperature of 2000 K was measured by luminescence thermometry (Cates et al 2003).

In this paper, YAP doped with 4 at% of Eu^{3+} was investigated for the purpose of application in thermometry at elevated temperatures, up to 800 K using both FIR and lifetime measurement methods.

2 EXPERIMENTAL

YAP doped with 4 at% of Eu^{3+} was synthesized by solid-state reaction, starting from Y_2O_3 and Al_2O_3. Oxide reagents were carefully weighed corresponding to the chemical formula $Y_{1-x}Eu_xAlO_3$ (x=0.04) and mixed in an agate mortar. Europium was added during grinding to the mixture of oxides as a solution of $Eu(NO_3)_2$ prepared by dissolution of Eu_2O_3 in diluted nitric acid. After drying, powders were pressed (up to 200 MPa) into pellets and heated at ~1620 K for 5–6 h in air. The photoluminescence measurements were performed under excitation light of a 450 W Xenon lamp on the Fluorolog-3 Model FL3-221 (Horiba Jobin-Yvon) spectrofluorometer system (λ_{ex}=399 nm for YAlO$_3$:Eu^{3+}), elevating the temperature gradually in the temperature range (300 – 800 K) with step of 50 K. The samples were placed in a custom-made, temperature controlled furnace, and emissions were collected via an optical fiber bundle. The temperature of the samples was controlled within the accuracy of ±0.5°C by a temperature control system utilizing proportional-integral-derivative feedback loop equipped with T-type thermocouple for temperature monitoring.

3 RESULTS AND DISCUSSION

Emission spectra for YAlO$_3$ doped with 4 at% Eu^{3+}, Figure 1, were recorded over the temperature range of 290 - 800 K upon excitation at 399 nm. The five emission bands originating from spin-forbidden inter-shell f-f electron transition of Eu^{3+} ion can be observed: $^5D_0 \rightarrow {}^7F_0$ (581 nm), $^5D_0 \rightarrow {}^7F_1$ (590 nm), $^5D_0 \rightarrow {}^7F_2$ (614 nm), $^5D_0 \rightarrow {}^7F_3$ (655 nm), and $^5D_0 \rightarrow {}^7F_4$ (696 nm). Temperature increase affects the shape of emission spectra, namely the reduction in intensity can be observed for all transitions, indicating in this way high sensitivity of emission on temperature.

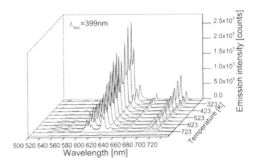

Figure 1. Emission spectra for YAlO$_3$: 4 at% Eu^{3+} under elevated temperatures.

3.1 Fluorescence intensity ratio method

The FIR method is based on the intensity ratio between two emission lines in the PL spectrum. This technique involves the use of emissions that originate from two closely spaced, "thermally coupled" excited energy levels. The relative population of these levels follows a Boltzmann type distribution and it is dependent on the temperature and the energy difference between levels (energy gap). Ideally, one of the emission lines should be independent of temperature (internal reference). Then, a calibration between the ratios of emissions is indicative of temperature. The main mechanism behind "thermally coupled" energy levels is thermalization: when two energy levels of the RE activator are closely separated by a difference of approx. 2000 cm^{-1} or less, the upper level will not fluoresce at low temperatures since electrons do not have enough energy to bridge the energy gap. As the temperature increases, the upper level becomes populated and hence the emission from this level gradually increases in intensity at the expense of the lower level population. The ratio of emission intensities from these levels, FIR, can be described by the Boltzmann - type equation:

$$FIR = C \exp\left(-\frac{\Delta E}{kT}\right), \tag{1}$$

where k = 0.69503476(63) cm^{-1}/K is the Boltzmann constant, ΔE is the energy gap between two excited levels, and C is the proportionality constant.

In the materials doped with Eu^{3+}, temperature can be determined using the FIR technique by observing emissions from $^5D_1$ and $^5D_0$ energy levels that are separated ~1700 cm^{-1} (depending on the host material). The important parameters for the applicability of the any sensing method are absolute and relative sensitivities of the measurements. The rate at which the fluorescence intensity ratio or lifetime changes in temperature is known as the absolute sensitivity S_a, which is given by following equation:

$$S_a = \left| \frac{dQ}{dT} \right| \qquad (2)$$

The relative sensor sensitivity, S_r can be found from:

$$S_r = 100\% \times \left| \frac{1}{Q} \frac{dQ}{dT} \right| \qquad (3)$$

where Q corresponds to sensor variable (FIR or τ). Dependence of emission intensities at 535 nm and 614 nm with temperature for $YAlO_3$: 4 at % Eu^{3+} are presented in Figure 2. The $^5D_1 \rightarrow \ ^7F_1$ transition is chosen as an internal reference since the peak at 535 nm is fairly dependent on the temperature. The intensity of emission centered at 614 nm gradually decreases with the temperature increase and shows the maximum value at 300 K.

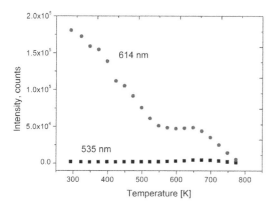

Figure 2. Temperature dependence variations of emissions centered at 535 nm and 614 nm for $YAlO_3$: 4 at % Eu^{3+}.

Figure 3 provides the experimentally derived FIR (symbols) of emission at 535 nm relative to emission 590 nm at different temperatures from 300 to 800 K. The solid line represents the FIR obtained by fitting experimental data to Eq. (1). The fitting procedure-provided the following values of the parameters: C= 4.59 and ΔE = 1716.7 cm^{-1}. The value of energy gap obtained by fitting closely matches the value obtained from emission spectra.

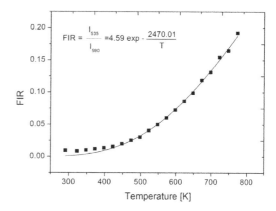

Figure 3. Intensity ratio of emission at 535 nm relative to emission at 590 nm. The experimental data are represented as symbols, whereas the theoretical curves were obtained using Eq.(1) and are represented with solid line.

Absolute and relative sensitivity, calculated according to equations (2) and (3) are presented in Figure 4. Absolute sensitivity is the highest at 773 K and its value is 2.95 K^{-1}, while at 300 K is 0.41 K^{-1}. Relative sensitivity is decreasing with the temperature, from 7.76 x 10^{-4} % K^{-1} to 2.78 x10^{-5} % K^{-1} at 773 K.

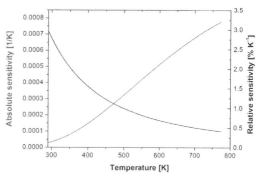

Figure 4. Temperature dependence of absolute and relative sensitivity of FIR thermometry.

3.2 Lifetime decay method

Lifetime from an excited-state level of rare earth ions depends on radiative (ω_r) and nonradiative (ω_{nr}) decay rates

$$\tau = \frac{1}{\omega_r + \omega_{nr}}. \qquad (4)$$

Radiative decay rate is not temperature dependent while non-radiative rate changes with the change of temperature and can be estimated using several models: Multiphonon Relaxation (MPR) model, the Arrhenius-type Mott equation and temperature quenching through a Charge Transfer State (CTS). In this research we have used Motts model (also known as the energy gap law model) that describes non-radiative relaxation rate in a following way:

$$\omega_{nr}^{Mott} = \omega_{nr}(0)\exp\left(-\frac{\Delta E}{kT}\right) \qquad (5)$$

Then, from Eqs. (4) and (5) temperature dependence of a lifetime can be described as

$$\tau = \left[\omega_r + \omega_{nr}(0)\times\exp\left(-\frac{\Delta E}{kT}\right)\right]^{-1} \qquad (6)$$

Absolute and relative sensitivity of lifetime the mometry is calculated and presented in Figure 5. Both sensitivities have the zero values until about 650 K when they start to increase until 800 K, indicating a constant value in a large part of the spectrum, which corresponds to a constant value of the lifetime in the same temperature region in figure 6. In the high-temperature region the maximal values of absolute sensitivity of 0.013 ms K^{-1} and relative sensitivities of 1.79 % K^{-1} are found. Fitting of experimental data according to equation (6) provided the following results: $\Delta E = 13210$ cm^{-1}, $\omega_r = 0.5584$ ms^{-1}, $\omega_{nr}(0) = 7.361\times10^9$ ms^{-1}. Again, the value of energy gap is in excellent agreement with the energy difference between the highest Stark level of the Eu^{3+} ground state ($^7F_6$) and the lowest Stark component of the first excited state ($^5D_0$).

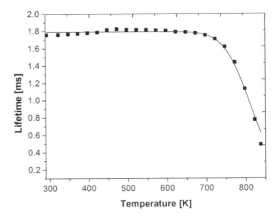

Figure 6. Lifetime vs temperature: experimental data (symbols) and fitting curve using Eq. (6) (line).

they are not affected by light scattering or reflection, by the intensity fluctuation of the excitation light field or inhomogeneous distribution of the phosphor. Also, for the lifetime thermometry measuring can be done at a single wave length. On the other hand, this method is sensitive only in a fairly limited temperature range in which lifetime rapidly decreases because of the thermal quenching effect.

4 CONCLUSION

YAlO$_3$ doped with 4 at% of Eu^{3+} was investigated for high temperature thermometry in the range from 300 - 800K using fluorescence intensity ratio and lifetime decay methods. For the FIR method emission lines at 535 nm and 614 nm are used and their ratio is fitted by Boltzmann-type equation. Maximum values of absolute and relative sensitivities were 2.95 K^{-1} (773 K) and 7.76x10^{-4} %K^{-1} (300 K), respectively. With lifetime method obtained absolute sensitivity is 0.013 ms K^{-1} and relative sensitivity is 1.79 % K^{-1}. Based on this results it can be concluded that YAP doped with Eu^{3+} ion can be used as material for luminescence temperature sensing for temperatures up to 800 K.

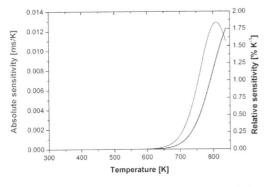

Figure 5. Temperature dependence of absolute and relative sensitivity of lifetime thermometry.

Lifetime-based methods have some advantages over the other luminescence thermometry methods:

ACKNOWLEDGEMENT

This work is supported by the Ministry of Education, Science and Technological Development of the Republic of Serbia (grant No. 45020) and the APV Provincial Secretetariat for Science and Technlogocal Development of the Republic of Serbia, through project no. 114-451-4787.

REFERENCES

[1] Baccaro, S. Cecilia, A. Montecchi, M Malatesta, T. de Notaristefani, F. Torrioli, S & Vittori, F. 1998. Refractive index and absorption length of YAP : Ce scintillation crystal and reßectance of the coating used in YAP : Ce single-crystal matrix. *Nuclear Instruments and Methods in Physics Research A.* 406:479–485.

[2] Baran, M. Zhydachevskii, Y. Suchocki, A. Reszka, A. Warchol S. Diduszko, R. & Pajaczkowska, A. 2012. Sol–gel synthesis and luminescent properties of nanocrystalline YAP:Mn. *Optical Materials.* 34:604–608.

[3] Brites,C.D.S Lima, P.P. Silva, N.J.O. Millan, A. Amaral, V.S. Palacio, F. & Carlos, D.L. 2012. Thermometry at the nanoscale. *Nanoscale.* 4:4799–4829.

[4] Brubach, J. Kissel, A. Frotsher, M. Euler, M. Albert, B. & Dreizler, A.. 2011. A survey of phosphors novel for thermography. *Journal of luminescence.* 131:559–564.

[5] Cates, M.R Allison, S.W. Jaiswal, S.L. & Beshears, D.L. 2003. YAG:Dy and YAG:Tm fluorescence to 1700 °C. *Proceedings of the International Instrumentation Symposium.* 49:389–400.

[6] Fibrich, M. Hambalek, T. Nemec, M. Šulc, J. & Jenlinkova, H. 2014. Multiline generation capabilities of diode-pumped Nd:YAP and Nd:YAG lasers, *Laser Physics* 24:035803–035807.

[7] Han, L Yao, B Duan, X. Li, S. Dai, T. Ju, Y. & Wang, Y. 2014. High power slab Tm: YAP laser dual-end-pumped by fiber coupled laser diodes. *Optical and Quantum Electronics.* in press.

[8] Khalid, A.H. & Kontis, K. 2008. Thermographic phosphors for high temperature measurements: principals, current state of the art and recent applications. *Sensors.* 8:5673–5744.

[9] Kissel, T. Brubach, J. Euler, M. Frotscher, M. Litterscheid, C. Albert, B. & Dreizler, A. 2013. Phosphor thermometry : On the synthesis and characterisation of $Y_3Al_5O_{12}$:Eu (YAG:Eu) and YAlO$_3$:Eu (YAP:Eu). *Materials Chemistry and Physics.* 140:435–440.

[10] Lojpur, V. Egelja, A. Pantic, J. Ðorđevic, V. Matović, B. & Dramićanin, M.D. 2014. $Y_3Al_5O_{12}$:Re^{3+} (Re=Ce, Eu, and Sm) Nanocrystalline Powders Prepared by Modified Glycine Combustion Method. *Science of Sintering.* 46:75–82.

[11] Lojpur, V. Nikolić, M. Mančić, L. Milosević, O. & Dramićanin, M.D. 2013. Y_2O_3:Yb,Tm and Y_2O_3:Yb,Ho powders for low-temperature thermometry based on up-conversion fluorescence. *Ceramics International.* 39:1129–1134.

[12] Mančić, L. Lojpur, V. Barroso, I. Rabanal, E.M. & Milosević, O. 2012. Synthesis of Cerium-Activated Yttrium Aluminate Based Fine Phosphors by an Aerosol Route. *European Journal of Inorganic Chemistry.* 16:2716–2724.

[13] Perra, A. Kowalik, M.A. Ghiso, E. Columbano, G.M.L. Tommaso, L. Angioni, M.M. Raschioni, C. Testore, E. Roncalli, M. Giordano, S. Columbano, A. in press. YAP activation is an early event and a potential therapeutic target in liver cancer development. *Journal of Hepatology.*

[14] Sugiyama, M. Yanagida, T. Totsuka, D. Yokota, Y. Futami, Y. Fujimoto, Y. & Yoshikawa, A. 2013. Crystal growth and luminescence properties of Cr-doped YAlO$_3$ single crystals. *Journal of Crystal Growth.* 362:157–161.

[15] Talik, E. Kruczek, M. Zarek, W. Kusz, J. Wójcik, K. Sakowska, H. & Szyrski, W. 2007. XPS characterization of YAlO$_3$:Co single crystals. *Crystal Research and Technology.* 42:1341–1347.

[16] Wang, Z. Ma, X. &Li, W. 2014. Efficient Ho:YAP laser dual-end-pumped by Tm fiber laser. *Optical Review.* 21: 150–152.

[17] Zeng, X. Zhaoa, G. Xua, X. Lia H. Xua, J. Zhaoa, Z. Hea, X. Panga, H. Jiea, M. & Yana, C. 2005. Comparison of spectroscopic parameters of 15 at%Yb:YAlO$_3$ and 15 at% Yb: $Y_3Al_5O_{12}$. *Journal of Crystal Growth.* 274:106–112.

[18] Zhydachevskii, Y. Suchocki, A. Berkowski, M. 2012. Thermoluminescent properties of Mn-doped YAP ceramics. *Source of the Document International Conference on Oxide Materials for Electronic Engineering, OMEE.*6464742:241–242.

[19] Zorenko, Y. Gorbenko, V. Savchyn, V. Kuklinski, B. Grinberg, M. Bilski, P. Gieszczyk, P. Twardak, A. Mandowski, A. Mandowska, E. & Fedorov, A. 2012. Luminescent properties of YAlO$_3$:Mn single crystalline films. *Optical Materials.* 34:1979–1983.

[20] Queiroza, T.B. Ferraria, C.R. Ulbrichb, D. Doyleb, R. Camargoa, A.S.S. 2010. Luminescence characteristics of YAP:Ce scintillator powders and composites. *Optical Materials.* 32:1480–1484.

Testing and Measurement: Techniques and Applications – Chan (Ed.)
© *2015 Taylor & Francis Group, London, ISBN: 978-1-138-02812-8*

Measuring network user psychological experience quality

X.Y. Wu & P. Wang

School of Electronic and Information Engineering, Xi'an Jiaotong University, China

ABSTRACT: In the post WWW era, Web systems focus on facilitating intelligent services for human. The quality of user experience determines whether Web service would be accepted by users, but now little has been known about evaluating Web systems from users' psychological experience perspective. In this paper, we propose a quantitative approach of measuring users' psychological experience in the context of intelligent e-learning. The properties (components), which affect e-learners' psychological experience, are analyzed and quantified. Then the quality model of user psychological experience is built through the analytic hierarchy process. A case study indicates that the approach could analyze e-learners' experience.

KEYWORDS: User psychological experience, e-learning, analytical hierarchy process.

1 INTRODUCTION

Nowadays, information and computer technology provides various services which facilitate human work and life, and experience economy is emerging. Whether users are satisfied with the services, have good experience and are willing to take the services perpetually is an important problem which draws more and more attention from researchers in diverse fields.

User experience, which is a holistic and subjective psychological perception, is built by a user when the Web system provides a service for the user. Considerable researches have been conducted in diverse areas and from different views respectively, such as computer science, psychometrics and social informatics fields, about the art of the quantitative and qualitative properties (components) which influence user experience. There is a range of properties (components) that are commonly considered when Web systems' user experience is assessed. Some are from computer system itself and others from users or the environment between systems and users. The researches mostly focus on computer system itself rather than user psychological experience. How to assess and improve user psychological experience through taking into account user psychological and cognitive properties (components) is a problem.

In the paper, a quantitative approach is proposed to evaluate user psychological experience in intelligent e-learning. There are some key scientific problems as follows. 1) Identify properties (components) that may affect user experience. There are a variety of properties (components) that may influence users' experience, such as content construction form, vision affection [1], system delay, ease of use, and so forth.

As different applications have different needs, we analyze which properties (components) are relevant for e-Learning application. 2) How to measure the properties (components) above. In the paper, task-oriented approaches are applied to measure users' psychological experience. 3) Construct users' psychological experience model. The effect of these properties on the users' experience is unclear and quantitative analysis is needed. The analytic hierarchy process approach is applied to construct user experience, quantitative model and weights of the properties are obtained.

In this paper, we review the process of evaluating the users' psychological experience in intelligent e-learning. The rest of the paper is structured as follows. First, we survey users' experience research in diverse fields and describe a large set of properties that can be relevant for users' experience. Then we propose a quantitative method to access users' psychological experience in the context of intelligent e-learning. Finally, we describe the experiments and suggest protocols for experimentation.

2 RELATED WORK

The following sections review how to analyze, to quantify properties (components) and construct a user experience model, respectively.

2.1 Identify properties (components) that may affect user experience

Users' experience is a holistic psychological perception. There are a vast amount of research in diverse fields and from different views respectively, such as computer science, psychometric and social informatics fields, which have studied the quantitative and qualitative properties (components) of user experience.

To improve the design and evaluation of computer systems, James Lewis develops standardized subjective usability satisfaction measures (IBM Computer Usability Satisfaction Questionnaires) which has the components of users' satisfaction, such as system usefulness, information quality, and interface quality, to assess users' satisfaction with the system usability [2-4].

Totally, the properties (components) that affect users' experience are as follows: (1) usability (accessibility, functionality), which means the system is easy to use. The property is composed of other concrete and detailed properties, such as response speed, which means that users can access to the service rapidly and easily. It is easy to learn, which means that users can learn how to use the service easily, and it is easy to navigate, which means that user can obtain the service through the shortest route, and simple manipulation etc. (2) information quality, which means that the service is valuable for the users. The property is composed of other concrete and detailed properties, such as satisfying users' needs, improving work efficiency, etc. (3) other properties, such as users' characteristic, emotional state etc.

2.2 *Measure properties (components) through quantitative approaches*

In human computer interaction and interface field, the approaches are used commonly, such as observation, interview, questionnaire and role action etc., to analysis users' needs and habits. Baiducompany investigated users' behavior with the questionnaire. The approach is qualitative, easy to conduct, but it needs more time and resource is liable to subjective views.

The properties (components), which influence users' experience, are derived from users, the current context of use, etc. And the user experience is an individual perception which cannot be simulated. But there are some common and similar psychological properties from the view of the whole user group. The user experience can be improved according to the common properties first, and then it can be optimized according to the context of use.

There are some quantitative approaches in the literature, such as GOMS CPM-GOMS, NGOMSL, which emphasize operate time. GOMS approach constructs Goals, Objects, Method and Selection rules model, and disassembles users' behavior into behavior unit. It can evaluate the time cost in the specific scenario that an experienced user conducts. The model is simplified and CPM-GOMS model improves it, which is suitable for evaluating the complicated, overlapping and time-depended interface. GOMS Model is suitable for describing and forecasting an experienced user's usual, skilled conduct and it is not suitable for a fresh user, but it also focuses on the time efficiency of users' mechanical conduct in specific scenario, which does not consider the influence of human endurance and emotion on the users' experience.

2.3 *Model a holistic user experience*

There are some approaches, such as gray theory and analytical hierarchy process, to model a holistic user experience from systemic and comprehensive view. Gray theory is based on comparative analysis methods, which is suitable for evaluating multi-version systems. Analytic hierarchy process is suitable for evaluating single system.

3 A QUANTITATIVE APPROACH OF EVALUATING USER PSYCHOLOGICAL EXPERIENCE IN INTELLIGENT E-LEARNING

The quantitative approach of evaluating the users' psychological experience in this paper first identifies properties (i.e. components, elements, and indicator) that may affect users' experience, and then build the model of user psychological experience, designs quantitative methods to measure properties, finally calculates the holistic users' psychological experience. The architecture of the approach is shown in Figure 1.

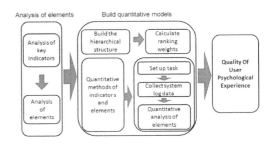

Figure 1. The architecture of quantitative approach of user psychological experience.

There are many elements that affect users' experience, usefulness and ease of use is most important to e-learning among them. The following, we will mainly study usefulness and ease of use.

First, usefulness refers to whether the services provided by the system to the user are useful and can promote the learning. We can use the following three elements to reflect the usefulness of the e-learning system.

1 Coverage of the resources: refers to the extent of resources provided by the e-learning system and covers the needs of users.
2 Recommend Hit rate: If the user clicks on the knowledge element recommended by the system for the user, and provide a good evaluation, we think that a recommend hit, the recommend hit rate is defined as the ratio of hit times and recommend times. Assume the times of the system recommended knowledge elements for the user is n, the times of hits is h, the RH equals the ratio of n and h, as shown in equation 1.

$$RH = \frac{h}{n} \qquad (1)$$

3 User loyalty: If the system is useful for the user, the user may use the system again, user loyalty is defined as how strong willingness the user have to re-use the system, it can be measured by calculating the frequency of users' visits from the system historical data records. It is defined as: If the user used the system n times in a week, according to experts' experience, defined the benchmark times is N, the user loyalty equals the ratio of use times and benchmark times, if n/N is more than 100%, then we take the user loyalty for 100%:

$$L = \begin{cases} 1, (n \geq N) \\ \frac{n}{N}, (n < N) \end{cases} \qquad (2)$$

Next, ease of use refers to that the user can get the services provided by the system quickly and easily, that the system is simple to operate, easy to learn, ease of navigation, and meets the user's habits, etc. In e-learning system, we use the following elements to reflect ease of use of the system:

1 Response speed: response speed can be calculated from the response time, response time refers to the time interval between users began to request service and the system rendered the service, it can be measured from the system log record.
2 Ease of navigation: define the definition of navigation as the ratio of the number of target knowledge elements and the total number when the user completes tasks in the e-learning system. If the user can directly find the target knowledge elements, the definition of navigation is 100%, if the user cannot find them, and the definition of navigation is 0%.
3 Ease of learn: define the efficiency of task as the ratio of the time spent on the target knowledge elements and the total time when users complete task in the e-learning system. The *Task* is a learning work requiring the user to complete in e-Learning system, for example it requires the user to learn one or more knowledge elements. The time user login T_{in} refers to the time when the user login the system, it can be got from the system log. The time task start T_s refers to the time when the system has finished loading, and begun to provide service, it can be got from the system log. Response time T_{re} refers to the time from the users' login to the system loaded, that is:

$$T_{re} = T_s - T_{in} \qquad (3)$$

Response speed RS: Define reference response time as TS. The response speed is calculated as follows:

$$RS = \begin{cases} 1, (\overline{T_{re}} < TS) \\ \frac{TS}{\overline{T_{re}}}, (\overline{T_{re}} > TS) \end{cases} \qquad (4)$$

Knowledge element e: The e-learning system provides knowledge elements as learning resources for users. The target knowledge element is that the user needs to learn in order to complete the task, and the set of target knowledge elements are defined as E, that is $E = \{e_1, e_2, \cdots\}$. The learning time T_{ek}: the time from that user clicked on the knowledge element e_k to that he clicked on the next, we define as the time that users use to learn knowledge element e_k. The minimal duration t_{ek}: the minimal duration users need to learn the target knowledge element e_k, t_{e1}, t_{e2} is the minimal duration corresponding to the target knowledge element e_1, e_2, t_{ek} can be defined according to the experts' experience, or calculated from the system log data. While the learning time T_{ek} that the user spent learning the element is longer than the minimal duration t_{ek}, we think the users learn the knowledge element e_k successfully. That is successfully learning knowledge element S_{ek}:

$$S_{ek} = \begin{cases} 1, (T_{ek} \geq t_{ek}) \\ 0, (T_{ek} \leq t_{ek}) \end{cases} \qquad (5)$$

Task success TS: If the user has learned N knowledge elements which belong to the set of target knowledge elements E, we think the user had completed the task successfully. N is the minimal number of knowledge elements defined according to the experts' experience, N will be different values according to different task. That is:

$$TS = \begin{cases} 1, (\sum S_{ek} \geq N, e_k \in E) \\ 0, (\sum S_{ek} < N, e_k \in E) \end{cases} \qquad (6)$$

The definition of navigation NC: Assume the user had learned M elements before he had completed the task, M is smaller, the navigation of the system is better. Define the definition of navigation as the ratio of N (the minimal number of knowledge elements) and M, that is:

$$NC = \frac{N}{M} \quad (7)$$

Time used to complete the task T_t: the duration from the system loaded to the task is completed, that is the sum of time that user spent to learn all elements before he completed the task:

$$T_t = \sum T_{en} \quad (8)$$

Efficiency of task ET: the efficiency of task is defined as the ratio of the time user spent on the target knowledge elements and the time spent on the task:

$$ET = \frac{\sum T_{ek}}{\sum T_{en}}, (e_k \in E) \quad (9)$$

4 EXPERIMENTS AND ANALYSIS

With the above method, this paper analyzes the users' quality of experience of junior undergraduate students who use intelligent e-learning system in a college of science and engineering. In accordance with section 3.3, we set up a task and invited 118 participants to experiment using the above method.

In this experiment, participants were asked to complete the task: "Using intelligent e-learning system to supplement learning circuit switching and packet switching."

Defining the set of target knowledge elements E: ["Packet Switching", "Message switching", "Packet ", "Line switching", "Data exchange technology", "Packet-switched", "Packet-switched", "Packet switching", "Message switching mode characteristics", "the advantages of Packet switching", "Message switching property", "the method of Packet switching", "Packet switching process", "Circuit switching, Message switching, Packet switching contrast", "Packet switching property", "Circuit switching", "Switching technology"].

According to the experts' experience, define task succeeds as the user who has learnt 4 knowledge elements that belong to E successfully. Define User loyalty reference number as 2. Define reference response time as 10 seconds.

Use the Analytical Hierarchy Process, build the model of the quality of user psychological experience based on usability and ease-of-use, the results of the above calculated is the total ranking weights of the six elements: "coverage of the resources", "the hit rate of recommended", "user loyalty", "response speed", "the definition of navigation", "efficiency of task", $W = (0.1937, 0.0785, 0.4778, 0.0262, 0.1593, 0.0646)$, as shown in figure 2. By combined consistency test, the comparison matrix of Elements is consistent.

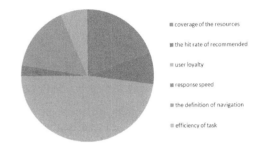

Figure 2. The total ranking weights of the elements.

Overall, the quality of users' psychological experience, which the e-learning system brings to user, the highest is 62%, the lowest is 27%, the average is 38%, and it means that quality of users' psychological experience of the system is not good.

5 CONCLUSION

The study analyzes the elements which affect the quality of e-learning users' psychological experience, design methods to quantify them, use the analytic hierarchy process to build a quantitative method to analyze quality of users' psychological experience. In the course of the study, we mainly analyzed two indicators, Ease of use and Usability (Usefulness). In future work, other factors in section 2.1 will be studied, (such as emotional, etc.), and the influence of quality of user experience will be researched, better method to quantitatively analyze the quality of user psychological experience will be built.

ACKNOWLEDGMENT

The research was financially supported in part by China National Key Technology R&D Program under Grant No. 2012BAI34B01; Ministry of Education of China Humanities and Social Sciences Project under Grant No. 12YJC880117.

REFERENCES

[1] Lindgaard, G., An Exploration of Relations Between Visual Appeal, Trustworthiness and Perceived Usability of Homepages. 2006.

[2] Lewis, J.R., IBM Computer Usability Satisfaction Questionnaires:Psychometric Evaluation and Instructions for Use. 1993, IBM Corporation.

[3] Gauch S, Speretta M, Aravind C, et al. User profiles for personalized information access[J]. The Adaptive Web Methods and Strategies of Web Personalization, 2007: 54–89.

[4] Zhicheng Dou, Ruihua Song, et al. Evaluating the Effectiveness of Personalized Web Search [A]. IEEE Transactions on Knowledge and Data Engineering, vol. 21, No. 8, August 2009.

Testing and Measurement: Techniques and Applications – Chan (Ed.)
© 2015 Taylor & Francis Group, London, ISBN: 978-1-138-02812-8

A proposal for teaching programming through the Four-Step Method

Y. Uchida & S. Matsuno
National Institute of Technology, Ube College, Ube, Japan

T. Ito
Hiroshima University, Higashi-Hiroshima, Japan

M. Sakamoto
University of Miyazaki, Miyazaki, Japan

ABSTRACT: We teach computer programming to students aged from 17 to 18 years old. In the course, a few students consider themselves to have insufficient understanding of programming or think that they are not good at programming. In response, we adopted and implemented a part of the Computer Science Unplugged (CS Unplugged) method, which is considered as an effective way of teaching information science. However, although CS Unplugged has generated considerable results in motivating students to learn and in initial learning, we feel that it is not sufficiently connected to full-fledged programming languages such as C and Java. Accordingly, we propose advancing from CS Unplugged to full-fledged programming through a new Four-Step Method. In this paper, we will describe the thinking and concepts behind this proposed method.

KEYWORDS: CS Unplugged, CS Plugged, programming, four-step method.

1 INTRODUCTION

The course we teach has about 40 students per class, aged from 17 to 18 years old. Since the students' major field of study is Management Information, they need to learn programming [1] techniques. However, the results of the survey described below show that not a few students consider themselves to have insufficient understanding of programming or think that they are not good at programming. We are examining ways to improve this situation. CS Unplugged [2] is a method of teaching information science without using computers, first proposed by Tim Bell of the University of Canterbury in New Zealand. While CS Unplugged is said to be effective in teaching information science [3], there have been concerns that its success or failure may be an effect of the skill and experience of the instructors. To address this topic, we implemented CS Unplugged for a group of students of different age groups, from fifth through ninth grades, and looked at their responses. The results showed that after first making sufficient preparations the method could generate results without necessarily depending on the skill or experience of the instructors. Other research underway in Japan includes a study on the use of teaching aids in learning about algorithms through CS Unplugged [4] and a study on learning the fundamentals of computer programming through a programming learning environment for beginners [5]. A study by Y. Feaster et al [6] concerns practice in teaching high-school students, and it has been reported to have some, albeit limited, success.

However, at present almost no research has been conducted on advancement from CS Unplugged to full-fledged programming languages. Accordingly, we proposed a new method of advancing from CS Unplugged to full-fledged programming. The proposed method begins with conducting a CS Unplugged activity, and then continues on writing a program on the same theme and further to its abstraction in Java. Accordingly, we propose advancing from CS Unplugged to full-fledged programming through a new Four-Step Method. The proposed method consists of the following steps: Step 1, A CS Unplugged activity; Step 2, A CS Plugged activity; Step 3, Preparing pseudocode; and Step 4, Writing Java source code.

2 BACKGROUND OF THIS STUDY

Every year we conduct a survey following our first-term course of two 90-minute sessions per week held over 30 weeks. Here we will look at the numbers of students who answered "yes" or "no" to this survey question, "Do you think you largely understand Java?"

As is seen in Fig. 1, the survey's results show that not a few students consider themselves to have insufficient understanding of programming or think that they are not good at programming. As such, there is a need to improve this situation.

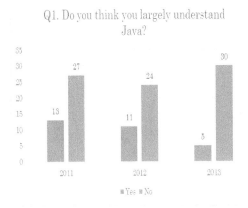

Figure 1.   Survey results on understanding of Java.

3   CS UNPLUGGED IN PRACTICE

For about 50 minutes on August 2, 2014 we used the CS Unplugged method for approximately 30 students from fifth through ninth grades as part of a summer-vacation junior science course. The activity we implemented was CS Unplugged's Image Representation activity. Fig. 2 shows a scene from this activity, while Fig. 3 shows examples of students' work.

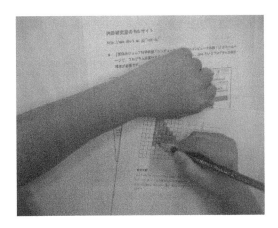

Figure 2.   Scene from course practice.

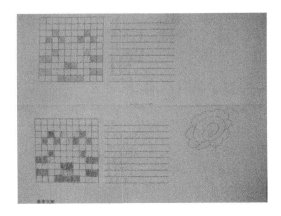

Figure 3.   Examples of students' work.

4   THE FOUR-STEP METHOD

In this paper, we use the CS Plugged to refer to the implementing of a CS Unplugged activity through a computer program. The goal is to advance to computer programming through using a computer program to conduct the work done by human beings in a CS Unplugged activity.

Accordingly, we propose advancing from CS Unplugged to full-fledged programming through a Four-Step Method. The proposed method consists of the following steps: Step 1, A CS Unplugged activity; Step 2, A CS Plugged activity; Step 3, Preparing pseudocode; and Step 4, Writing Java source code.

4.1   *Step 1: CS Unplugged*

First, students conduct the CS Unplugged activity as described in previous above.

4.2   *Step 2: CS Plugged*

Here we will look at the example of using a computer program to implement the CS Unplugged activity Image Representation. This is conducted through two activities. First, converting the image to code, which is represented in Fig. 4. When students click on squares in the grid on the left to draw a picture, the image is converted instantly to code as displayed on the right. The other activity is the reverse of this process, with students reproducing a picture from code (Fig. 5). They enter the code in the text boxes on the right in a format such as "7,1,8" and press the Enter key to display the resulting image instantly in the grid on the left.

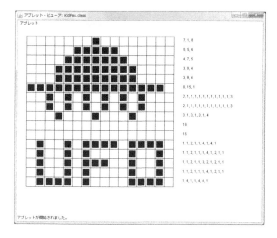

Figure 4. Applet for converting image to code.

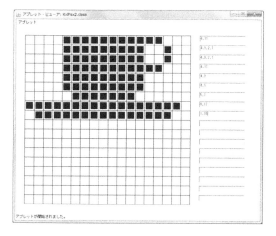

Figure 5. Applet for reproducing an image.

4.3 Step 3: Preparing pseudocode

In this step, the three elements of sequence, decision, and repetition through preparation of the trace table in the preceding step are extracted and represented in pseudo-language. In Japan, the national Information Technology Engineer Examinations employ pseudo-language [7]. A pseudo-language simulator is a type of software that makes such pseudo-language executable. In this paper, we used the freeware pseudo-language simulator SARA [8]. Fig. 6 shows the code written in this step and its execution.

```
 name: Image2Code
n: Display function (A)
  type: Subscript 1
  type: Subscript 2
  type: Count
er type: Array [2,4]
er type: Previous character

0,0] ← "□"
0,1] ← "■"
0,2] ← "■"
0,3] ← "■"
0,4] ← "□"
1,0] ← "□"
1,1] ← "□"
1,2] ← "□"
1,3] ← "□"
1,4] ← "■"

pt 1 ← 0
pt 1 ≦ 1
Previous character ← "□"
Count ← 0

Subscript 2 ← 0
Subscript 2 ≦ 4
    ● Display processing (Array [Subscript 1, Subscript 2]
    ● Subscript 2 ← Subscript 2 + 1

Subscript 2 ← 0
Subscript 2 ≦ 4
    ▲ Array [Subscript 1, Subscript 2] = Previous charact
    |     ● Count ← Count + 1
    +----------
    |     ● Display processing (Count)
```

Figure 6. Sample use of pseudo-language simulator.

4.4 Step 5: Writing Java source code

In this step students create a program by converting the pseudo-language from the previous step to Java source code. The subsequent debugging process is handled by going back and examining each previous step depending on the content of the error messages.

```
public class Image2Code {
  public static void main(String args[]) {
    char[][] image = {{'□', '■', '■', '■', '□'},
                      {'□', '■', '■', '■', '■'},
                      {'■', '■', '■', '■', '■'},
                      {'■', '■', '■', '■', '■'},
                      {'□', '■', '■', '■', '■'}
    };

    // Iterate only number of lines in an array
    for(int i = 0; i < image.length; i++) {
      char previous = '□';
      int count = 0;

      // Display the image in pixels
      for(int j = 0; j < image[0].length; j++) {
        System.out.print(image[i][j]);
      }
      System.out.print(" ");

      // Count number of adjoining pixels of same color and display in numerical
form (code)
      for(int j = 0; j < image[0].length; j++) {
        if (image[i][j] == previous) {
          count++;
        } else {
          previous = image[i][j];
          System.out.print(count + ", ");
          count = 1;
        }
      }
      System.out.println(count);
    }
  }
}
```

Figure 7. Java source code.

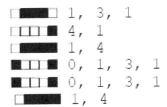

```
1, 3, 1
4, 1
1, 4
0, 1, 3, 1
0, 1, 3, 1
1, 4
```

Figure 9.   Results of running the above program.

5 SUMMARY

We have proposed a new method of advancing from CS Unplugged through the new process of CS Plugged to full-fledged computer programming languages, as a means of deepening understanding in computer programming education. We also have proposed a new Four-Step Method consists of the following steps: Step 1, A CS Unplugged activity; Step 2, A CS Plugged activity; Step 3, Preparing pseudocode; and Step 4, Writing Java source code. Topics for the future are putting together detailed procedures for abstraction of pseudocode from the trace table as well as continually implementing the proposed method and measuring its results.

REFERENCES

[1] D. Everitt, Introducing programming skills in education. Information on https://www.academia.edu/1478617/ Introducing_programming_skills_in_education.
[2] The Unplugged Community, Computer Science Unplugged. Information on http://csunplugged.org/.
[3] Y. Idosaka, Y. Kuno, S. Kanemune, Attempt and the Practice of Class Method Improvement Based on "Computer Science Unplugged", Journal of the Japan Society of Technology Education, vol.53, no.2, pp.115–123 (2011) (in Japanese).
[4] H. Manabe, S. Kanemune, M. Namiki, Effects of Teaching Tools in CSU Algorithm Education, IPSJ Journal, vol.54, no.1, pp.14–23 (2013) (in Japanese).
[5] T. Nishida, A. Harada, R. Nakamura, Y. Miyamoto, T. Matsuura, Implementation and Evaluation of PEN: The Programming Environment for Novices, IPSJ Journal, vol.48, no.8, pp.2736–2747 (2007) (in Japanese).
[6] Y. Feastery , L. Segarsz, K.S. Wahbay, O.J. Hallstrom, Teaching CS Unplugged in the High School (with Limited Success), Proceedings of the 16th annual joint conference on Innovation and technology in computer science education, pp.248–252 (2011).
[7] Information-technology Promotion Agency Japan, Kyotsu ni shiyo sareru gijigengo no kijutsukeishiki ("Symbolic conventions of commonly used pseudo-language"). Information on https://www.jitec. ipa.go.jp/1_13download/gijigengo_keisiki.pdf (in Japanese).
[8] S. Mimura, Gijigengo shimyureta SARA ("SARA pseudo-language simulator"). Information on http:// mimumimu.net/software/#sara (in Japanese).

Predator-prey coevolution applied in multi-objective optimization for micro-grid energy management

X.J. Yao
College of New Energy Engineering, Shen Yang University of Technology, Shen Yang, China

Y.C. Shao
Shenyang University, Shen Yang, China

H. Chen
Huachuang Wind Energy Company, Qing Dao, China

D.Z. Wang
Department of Electronic information and Electrical Engineering, Dalian University of Technology, Dalian City, Liaoning Province, China

L.W. Tian
Shenyang University, Shen Yang, China

ABSTRACT: The optimization of micro grid energy is a multivariable, nonlinear programming problem. We find a new multi-objective optimization algorithm based on Predator-prey Coevolution. In the algorithm, the constraint was transferred into a target function and a pair of relatively isolated predator and prey was selected as the active predator body and preybody, and with the cloning, recombination and mutation performed, we will obtain the convergence of ideal Pareto-front and a uniform distribution of the Pareto optimal solution. The simulation of the Micro grid Management was made utilizing the algorithm. The algorithm provides a new kind of efficient and stable constrained multi-objective optimization method for Micro grid.

1 INTRODUCTION

The smart grid by centralizing power combined with distributed generation, can achieve clean, efficient, safe and reliable operation power system. Its development and research have attracted the full attention of various countries. Micro grid, as an organic component of the smart grid, is based on the distributed generation and fused energy storage device, control device Integrated unit and protection device. Micro power access the Micro power through a certain interface (mainly the power electronic inverter interface). And energy management control system can control the optimization of micro power supply current, voltage and power [1-2].

From the viewpoint of technology and the economy, the optimization problem of the micro grid energy ultimately attributed to the typical multi variable, multi constrained mixed nonlinear planning problems. Processing method for this problem can be grouped into two categories. One is to convert the multi-objective problem into a single objective problem. This kind of method can only produce one solution each time. If we try to get a set of approximate Pareto optimal solutions by many runs, then each calculation will be independent of each other, so the results may be inconsistent. And in the process of optimization, considering the weight of the various factors in the total goal, the change of weight will affect the whole target of the optimization results, which is not very good for solving the multi-objective optimization problems. The other is to optimize the processing of the application of multi-objective optimization algorithms. This kind of method considers whether the optimization goals can converge to the Pareto front, and whether to obtain the uniformly distributed Pareto optimal solutions.

[1] This research is supported by the International S&T Cooperation Program of China (ISTCP) under Grant 2011DFA91810-5 and Program for New Century Excellent Talents in University of Ministry of Education of China under Grant NCET-12-1012 and is also supported by Program for Liaoning Excellent Talents in University

Mao guo Gong [3] simulate the phenomenon of immune response in the antibody diversity symbiosis, and choose a few relatively isolated, non dominated individuals as the antibody. Experiments show that this method can obtain Pareto ideal front convergence and uniformly distributed Pareto optimal solution [3-5]. In this paper, on the basis of the above algorithm, we added the treatment of constraints, and obtained the crowding distance nondominated neighbor constrained multi-objective optimization algorithm, and we applied the algorithm to the optimization of the micro grid energy management.

2 SYSTEM MODEL

The micro-grid structure using America model was proposed by CERTS. The optimization of energy management includes:

1 Static voltage stability;
2 The minimum network loss;
3 Operating power factor power maximum.

2.1 The voltage stability

For any electric power system in a line, there are:

$$
\begin{cases}
P_i = \dfrac{\left(P_i^2 + Q_i^2\right)R}{U_i^2} + P_j \\[3mm]
Q_i = \dfrac{\left(P_i^2 + Q_i^2\right)R}{U_i^2} + Q_j
\end{cases}
\tag{1}
$$

P_i and Q_i are the starting node of active and inactive i respectively; P_i and Q_i are the starting node of active and inactive j respectively; R and X are resistance and reactance of the line between the nodes i and j.

P_i, Q_i as variables, (1) the conditions of having a real solution:

$$
\frac{4\left(XP_j - RQ_j\right)^2 + \left(XQ_j + RP_j\right)U_i^2}{U_i^4} < 1
\tag{2}
$$

Scaling back voltage of 1, get:

$$
4\left(XP_j - RQ_j\right)^2 + XQ_j + RP_j < 1
\tag{3}
$$

Voltage stability index is defined as:

$$
L = 4\left(XP_j - RQ_j\right)^2 + XQ_j + RP_j
\tag{4}
$$

The smaller L is, the better the system voltage stability is; the higher L is, the worse the voltage stability of the system is. When the L is close to 1, the

system voltage will collapse. Voltage stability index of all systems get in the maximum voltage stability index of all branches, i.e.

$$
L = \max\left(L_1, L_2, ..., L_{N-1}\right)
\tag{5}
$$

N is the number of nodes. Voltage stability needs to be good that is the largest branch of the whole system voltage stability index minimum value.

$$
\min(L) = \min\left(\max\left(L_1, L_2, ..., L_{N-1}\right)\right)
\tag{6}
$$

2.2 The minimum network loss

By controlling the flow, we keep the active power loss of the system and reactive power loss at the minimum level. That is

$$
\min\begin{cases}
\min\sum_{i=1}^{n} P_i \\[3mm]
\min\sum_{i=1}^{n} Q_i
\end{cases}
\tag{7}
$$

N is the number of branch systems, P_i is the active power loss of i branch, Q_i is the inactive power loss of i branch.

2.3 The power of the highest efficiency

The objective function is to obtain the best economic benefits and promote the efficiency of the micro grid power based on the qualified voltage, and achieve the independent control of active and reactive power of micro source by changing the parameters of the control flow without considering the reactive power compensation case, In order to make the micro source power factor reached a maximum, we take the following objective function:

$$
\min(\text{PF}) = \frac{Q_{dgi}}{\sqrt{P_{dgi}^2 + Q_{dgi}^2}}
\tag{8}
$$

PF: power factor P_{dgi}^2: P article I a micro source active power Q_{dgi}^2: article i a micro source wattles power.

2.4 Constraint system

2.4.1 Variable constraint
Including the constraints of the output active power of micro sources and the compensation capacitor of the reactive power and the operational constraints of each nodes.

$$\begin{cases} P_k^{\min} \le P_k \le P_k^{\max} \\ Q_k^{\min} \le Q_k \le Q_k^{\max} \\ U_j^{\min} \le U_j \le U_j^{\max} \end{cases} \qquad (9)$$

$P_k^{\min}$, $P_k^{\max}$, $Q_k^{\min}$, $Q_k^{\max}$ are the minimum and maximum values of the output active power and the compensation capacitor $U_j^{\min}, U_j^{\max}$ represent the minimum and maximum operating voltage of j nodes respectively.

2.4.2 Flow constraints

$$\begin{cases} P_i = U_i \sum_{j \in i}^{n} U_j \left(G_{ij} \cos \theta_{ij} + B_{ij} \sin \theta_{ij} \right) \\ Q_i = U_i \sum_{j \in i}^{n} U_j \left(G_{ij} \sin \theta_{ij} - B_{ij} \cos \theta_{ij} \right) \end{cases} \qquad (10)$$

N is the number of nodes in the system; G_{ij}, B_{ij}, θ_{ij} is the susceptance and phase angle between the node i and node j. jI is the all of the nodes that connect node i.

3 PREDATOR-PREY COEVOLUTION MULTI-OBJECTIVE OPTIMIZATION ALGORITHM

For multi-objective optimization problems including an n decision variables, Objective function and a k m constraint are defined as follows:

$$\begin{cases} \min y = f(x) = (f_1(x), f_2(x), ..., f_k(x)) \\ s.t.: g(x) = (g_1(x), g_2(x), ..., g_l(x)) \le 0 \\ h(x) = (h_{l+1}(x), h_{l+2}(x), ..., h_m(x)) = 0 \\ x = (x_1, x_2, ..., x_n) \in X \\ X = \{(x_1, x_2, ..., x_n) \mid l_i \le x_i \le u_i\} \\ l = (l_1, l_2, ..., l_n), u = (u_1, u_2, ..., u_n) \\ y = (y_1, y_2, ..., y_n) \in Y \end{cases} \qquad (11)$$

where x is the decision variable, X is the decision space, l and u are lower and upper bounds; y is the objective function. Y as the target space; $g(x) = (g_1(x), g_2(x), ..., g_l(x)) \le 0$ the definition of the Line quality constraints: $h(x) = (h_{l+1}(x), h_{l+2}(x), ..., h_m(x)) = 0$ the definition of the M-L equality constraints. Obviously, the objective function and constraints are functions of the decision variable.

3.5 Constraint problem

The main idea of the constrained optimization problem is to treat the constraint condition ad one or more

target view and convert to multi-objective problems, they then use the method of multi-objective optimization to treat the problem. There are two ways. One is to convert the constrained optimization problem into a two objectives problem; the other is to treat the objective function and constraint conditions of constrained optimization problems as two different target views. In the first way, the goal function is the original function: F (x), second target individual constraint the degree: $G(x)$. For the second method, the optimization will have to p+1 target multi-objective problem is transformed, where p is the number of constraints of the original problem. This will be a new optimized vector $F(x) = ((f(x), f_1(x), ..., f_p(x)))$, where $f(x), f_1(x), ..., f_p(x)$ are the constraints of the original problem.

In this algorithm, the constraint function is:

$$G_j(x) = \begin{cases} \max\{0, g_j\}, & 1 < j < l \\ \max\{0, |h_j(x)| - \partial\}, & l+1 < j < m \end{cases} \qquad (12)$$

The δ is equality constraint conditions of tolerance values, positive and generally the smaller.

$$f_{k+1}(x) = G(x) = \sum_{j=1}^{m} G_{j(x)} \qquad (13)$$

After the above processing, the multi-objective optimization problem of a n decision variables, k objective function and an m constraint becomes a multi-objective optimization problem of non constraint n decision variables, objective function of k+1:

$$\begin{cases} \min y = f(x) = (f_1(x), f_2(x), ..., f_k(x), f_{k+1}(x)) \\ x = (x_1, x_2, ..., x_n) \in X \\ X = \{(x_1, x_2, ..., x_n) \mid l_i \le x_i \le u_i\} \\ l = (l_1, l_2, ..., l_n), u = (u_1, u_2, ..., u_n) \\ y = (y_1, y_2, ..., y_n) \in Y \end{cases} \qquad (14)$$

Optimization model corresponds to the micro grid energy management. Constraint conditions are as follows: (9) variable constraint type setting in the population variation range and variable constraint in the range of initial population; (10) the trend of constraints written

$$\min G_j = \begin{cases} \max\{0,\} \mid P_j - \partial, \\ \max\{0, |Q_j| - \partial\}, \end{cases} \qquad (15)$$

After the above processing, optimization model is transformed into a multi-objective optimization problem with 4 objectives:

$$\min F = \min(L, SP, F, G)^T \qquad (16)$$

453

3.6 Optimization algorithm

The Predator-prey Coevolution algorithm is applied to the micro grid energy management optimization system. The process is: to generate a number of the population vectors that the same as the number of micro power active, inactive variables (the population can be initialized population or the updated population), Populations assign a value to the micro power system model, and Micro grid begin the flow calculation when they obtain the input power supply for power and then get the objective function value and judge whether the the value is maximum, if the value is maximum then output or if not satisfied, then the application of Predator-prey Coevolution algorithm will update the populationvalues, and then enter the next cycle. Schematic diagram was shown in figure 1.

We can see from Figure 1, the key step in the optimization of population regeneration. The population was updated by Predator-prey Coevolution algorithm to achieve. The algorithm uses the Predator bodies as external population to achieve elitist strategy, and puts forward the concept of Predatorbody activity according to the crowding distance of none dominated. According to the proportion of clones the crowding distance activity of Predatorbody, recombination and mutation operation, in order to strengthen the front end when Pareto- is sparse region search.

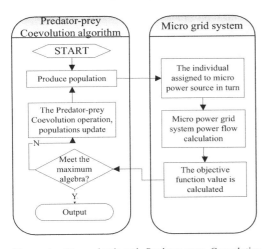

Figure 1. None dominated Predator-prey Coevolution algorithm to optimize the micro grid energy management diagram.

In the non-dominant Predator-prey Coevolution algorithm, there are several important concepts.

The no dominated antibodies: Predatorbody is divided into dominant and non-dominant Predatorbody.
The Predatorbody population B: an Predatorbody $b_i \in B(i=1,2,...,n)$ as the non-dominated antibodies condition is B. The Predatorbody population is

not another Predatorbody to satisfy the following conditions:

$$\left(\forall r = 1,2,...,k : f_r(e^{-1}(b_j)) \geq f_r(e^{-1}(b_i)))\right)$$
$$\wedge(\exists s = 1,2,...k : f_s(e^{-1}(b_j)) > f_s(e^{-1}(b_i)))$$

(17)

The crowding distance: a Predatorbody d crowding distance

$$I(d,D) = \sum_{i=1}^{k} \frac{I_i(d,D)}{f_i^{max} - f_i^{min}}$$

(18)

D for the Predatorbody group, namely the set of all Predatorbody; f_i^{max} and f_i^{min} is respectively in the current population of the i goal of maximum and minimum values.

Active Predatorbody: select few relatively isolated individuals as active Predatorbody.

The proportion of clones: proportion proportional cloning adaptive parameters according to (20) type selection ($I(a_i, A)$ A is the activity of Predatorbody crowding distance value, n_c clone population size expected value).

The population regeneration process is shown in Figure 2, in which B_i is the Predatorbody population,

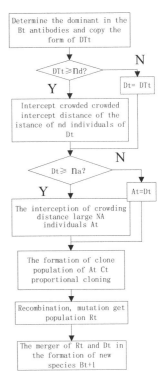

Figure 2. Population regeneration flow chart.

D_i, A_i are non-dominated Predatorbody population and activity of Predatorbody population on t moment; n_d, n_a are Predatorbody population size limit and activity of Predatorbody population limit.

4 CONCLUSIONS

This paper uses the Predator-prey Coevolution multi-objective optimization algorithm based on crowding distance to optimize the static voltage stability and the power factor of micro grid operation. The optimization results show that the micro grid has a good voltage stability, a small network loss and high power factor effect. The algorithm is a new optimization method of micro grid restraining multi-objective optimization.

REFERENCES

[1] Nikes Hatziargyriou, Hiroshi Asano, Reza Irvine, et al. Micro-grids [J]. IEEE Power and Energy Magazine (S1540–7977), 2007, 5(4): 78–94.
[2] Lassoer R H, Akhil A, Mar nay C, et al. Integration of Distributed Energy Resources: the CERTS Micro grid Concept [R]//Berkeley, CA,USA: USA Consortium for Electric Reliability Technology Solutions, 2002.
[3] Maoguo Gong. MultiobjectiveImmune Algorithm with Nondominated Neighbor-based Selection [J].IEEE Transactions on Evolutionary Computation, 2008, 16(2):225–255.
[4] Zixing Cai, Yong Wang. A Multiobjective Optimization-based Evolutionary Algorithm for Constrained Optimization [J]. IEEE transactions on evolutionary computation (S1089-778X), 2006,10(6): 658–675.
[5] Yong Wang, Zixing Cai. Combining Multiobjective Optimization with Differential Evolution to Solve Constrained Optimization Problems [J]. IEEE Transactions on Evolutionary Computation (S1089-778X), 2012, 16(1): 117–134.

Testing and Measurement: Techniques and Applications – Chan (Ed.)
© 2015 Taylor & Francis Group, London, ISBN: 978-1-138-02812-8

Impact of pre-construction planning and project execution on performance considering electrical project characteristics

D.Y. Kim

Department of Architectural Engineering, Dong-Eui University, Busanjin-gu, Busan, Korea

H.C. Lim

Department of Architectural Engineering, Changwon National University, Chagwon-si, Gyeongsangnam-do, Korea

ABSTRACT: The electrical market has been highly competitive since electrical construction requires more sophisticated and complex work with the scientific and technological advances. The primary goal of this paper is to identify the relationship between project characteristics, pre-construction planning, construction execution, and project performance. Several statistical analyses were conducted to model those relationships. Ultimately, this paper is able to help electrical contractors achieve better performance by understanding the relationship between project characteristics, pre-construction planning, project execution, and project performance.

KEYWORDS: Pre-construction planning, project execution, project performance, electrical construction, project characteristics.

1 INTRODUCTION

The electrical market has been highly competitive since electrical construction requires more sophisticated and complex work with the scientific and technological advances. Typically, electrical contractors are subcontracted by the general contractor with their specialized skills. By its very nature, electrical construction industry is uncertain and competitive. As a subcontractor, they have been facing unfavorable environments: constrained areas, schedule interference, and inefficient communication between other trades. Due to the adverse environment, electrical constructions are easily exposed to decreased productivity, inefficient work, and schedule delays (Guo 2002; Horman et al. 2006). It only takes one failed project to drive profit margins down significantly, when the projects are not managed efficiently. The current electrical construction requires more systematic and effective project management strategy to achieve desired goals. As a result, it is significant to investigate the relationship between project characteristics, pre-construction planning, project execution, and project performance.

The conceptual model developed through the previous research is shown in Fig. 1. This model acknowledges that there are many factors that can influence the performance of a project. Problems such as excessive changes, poor productivity, or numerous delays can have a negative impact on the outcome of a project. However, good planning that establishes systems that can be used to properly manage the problems during the execution stage can improve the chances of achieving a successful outcome. Therefore it is very important to understand the relationship between inherent project characteristics, planning, execution, and performance.

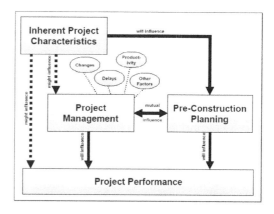

Figure 1. Conceptual model of the relationship between project characteristics, pre-construction planning, project management, and performance (Menches 2006).

1.1 *Methodology*

The primary goal of the research is to investigate and quantify the relationship between project characteristics, pre-construction planning, project execution, and performance. To achieve the goal, several statistical analyses were conducted. The research methodology includes five key steps: (1) Data Collection; (2) Bivariate Correlation Analysis; (3) the Kaiser-Meyer-Olkin(KMO) Analysis; (4) Principle Component Analysis (PCA); (5) Regression Analysis; and (6) Quantitative Model for the relationship between project characteristics, pre-construction planning, project execution, and performance. Fig.2 shows the detailed process for research methodology.

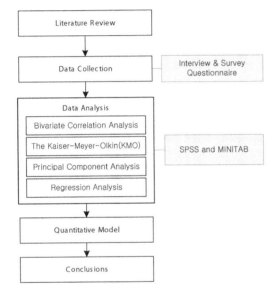

Figure 2. Summary of research methodology.

During the 25 comprehensive visits, 50 projects from across the U.S. were selected and asked to describe the characteristics of each project. Overall, 8 variables that characterize projects were identified, including:

1. Project type of construction
2. Project's level of complexity
3. Contract cost at award
4. Original estimated project duration for the electrical work
5. Original estimated total work-hours for the project
6. Estimated peak number of electricians for the project
7. Has the company worked for the owner in the past?
8. Has the company worked for the general contractor/CM in the past?

To improve understanding of inherent project characteristics, the data set was examined. Each variable of project characteristics was coded for further analysis. The results are summarized in Table 1.

Table 1. Variables for project characteristics.

Variable Number	Variable Name	Coding
1	Project Type	Commercial=1, Industrial=2, Institutional=3, Others=4
2	Project Complexity	Low=1, Average=2, High=3
3	Project Cost	Dollars ($)
4	Project Duration	Weeks
5	Estimated total work-hours	Hours (hrs)
6	Estimated peak number of electricians	# of Electrician
7	Experience for Owner	Yes=1, No=2
8	Experience for GC/CM	Yes=1, No=2

In addition, 10 pre-construction planning variables and 14-project execution variables were also used in the analysis. These variables are shown in Table 2 and 3.

Table 2. Variables for pre-construction planning.

Pre-Construction Planning	
1	Administrative setup
2	Budget preparation
3	Schedule development
4	Construction exeuction kickoff mtg
5	Material handling plan
6	Tracking and control
7	Buyout process
8	Layout and sequencing plan
9	Team selection and trurnover
10	Scope and contract review

Table 3. Variables for project execution.

Project Execution			
1	mobilization	8	Cost control and billing
2	Document management	9	Quality management
3	Material management	10	Safety management
4	Communication	11	Subcontract management
5	Coordination	12	Tool management
6	Scheduling	13	Scope change and control
7	Labor Management	14	Project closeout

2 SUMMARY

The primary goal of this paper is to identify the relationship between project characteristics, pre-construction planning, construction execution, and project performance. Several statistical analyses were conducted to model those relationships. Ultimately, this paper is able to help electrical contractors achieve better performance by understanding the relationship between project characteristics, pre-construction planning, project execution, and project performance.

ACKNOWLEDGEMENT

This research was supported by Basic Science Research Program through the National Research Foundation of Korea (NRF) funded by the Ministry of Education (NRF-2012R1A1A2044317)

REFERENCES

[1] S.-J. Guo, (2002). Identification and Resolution of Work Space Conflicts in Building Construction. Journal of Construction Engineering and Management, 128(2002), 287–295..

[2] M. J. Horman, M. P. Orosz, R. D. Riley. Sequence Planning for Electrical Construction. Journal of Construction Engineering and Management, 132(2006), 363–372.:

[3] C. L. Menches. Effect of Pre-Construction Planning on Project Performance," University of Wisconsin-Madison, Madison, 2006.

Author index

Printed and bound by CPI Group (UK) Ltd, Croydon, CR0 4YY

18/10/2024

01776219-0009